Plant Respiration

Advances in Photosynthesis and Respiration

VOLUME 18

The scope of our series, beginning with volume 11, reflects the concept that photosynthesis and respiration are intertwined with respect to both the protein complexes involved and to the entire bioenergetic machinery of all life. *Advances in Photosynthesis and Respiration* is a book series that provides a comprehensive and state-of-the-art account of research in photosynthesis and respiration. Photosynthesis is the process by which higher plants, algae, and certain species of bacteria transform and store solar energy in the form of energy-rich organic molecules. These compounds are in turn used as the energy source for all growth and reproduction in these and almost all other organisms. As such, virtually all life on the planet ultimately depends on photosynthetic energy conversion. Respiration, which occurs in mitochondrial and bacterial membranes, utilizes energy present in organic molecules to fuel a wide range of metabolic reactions critical for cell growth and development. In addition, many photosynthetic organisms engage in energetically wasteful photorespiration that begins in the chloroplast with an oxygenation reaction catalyzed by the same enzyme responsible for capturing carbon dioxide in photosynthesis. This series of books spans topics from physics to agronomy and medicine, from femtosecond processes to season long production, from the photophysics of reaction centers, through the electrochemistry of intermediate electron transfer, to the physiology of whole orgamisms, and from X-ray christallography of proteins to the morphology or organelles and intact organisms. The goal of the series is to offer beginning researchers, advanced undergraduate students, graduate students, and even research specialists, a comprehensive, up-to-date picture of the remarkable advances across the full scope of research on photosynthesis, respiration and related processes.

The titles published in this series are listed at the end of this volume and those of forthcoming volumes on the back cover.

Plant Respiration

From Cell to Ecosystem

Edited by

Hans Lambers
The University of Western Australia,
Crawley, WA, Australia

and

Miquel Ribas-Carbo
Universitat de les Illes Balears,
Palma de Mallorca, Spain

Springer

A C.I.P. Catalogue record for this book is available from the Library of Congress.

ISBN-13 978-90-481-6903-0 (PB)
ISBN-13 978-1-4020-3589-0 (e-book)

Published by Springer,
P.O. Box 17, 3300 AA Dordrecht, The Netherlands.

www.springeronline.com

Cover illustration painted by Josep M Barba (Lloret de Mar - Catalunya)

The camera ready text was prepared by
Lawrence A. Orr, Center for the Study of Early Events in Photosynthesis,
Arizona State University, Tempe, Arizona 85287-1604, U.S.A.

Printed on acid-free paper

Editorial

Advances in Photosynthesis and Respiration

Volume 18: Plant Respiration: From Cell to Ecosystem

I am delighted to announce the publication, in the *Advances and Photosynthesis and Respiration* (AIPH) Series, of a second book related to plant respiration (*Plant Respiration: From Cell to Ecosystem*), edited by Hans Lambers, and Miquel Ribas-Carbo. The first plant respiration book, *Plant Mitochondria: From Genome to Function,* was edited by David Day, Harvey Millar and James Whelan. Two earlier volumes, both edited by Davide Zannoni, deal with *Respiration in Bacteria and Archaea.* I list below all the AIPH volumes published during a decade:

Published Volumes (1994–2004):

- ***Molecular Biology of Cyanobacteria*** (Donald A. Bryant, editor, 1994);
- ***Anoxygenic Photosynthetic Bacteria*** (Robert E. Blankenship, Michael T. Madigan and Carl E. Bauer, editors, 1995);
- ***Biophysical Techniques in Photosynthesis*** (Jan Amesz and Arnold J. Hoff, editors, 1996);
- ***Oxygenic Photosynthesis: The Light Reactions*** (Donald R. Ort and Charles F. Yocum, editors, 1996);
- ***Photosynthesis and the Environment*** (Neil R. Baker, editor, 1996);
- ***Lipids in Photosynthesis: Structure, Function and Genetics*** (Paul-André Siegenthaler and Norio Murata, editors, 1998);
- ***The Molecular Biology of Chloroplasts and Mitochondria in Chlamydomonas*** (Jean David Rochaix, Michel Goldschmidt-Clermont and Sabeeha Merchant, editors, 1998);
- ***The Photochemistry of Carotenoids*** (Harry A. Frank, Andrew J. Young, George Britton and Richard J. Cogdell, editors, 1999);
- ***Photosynthesis: Physiology and Metabolism*** (Richard C. Leegood, Thomas D. Sharkey and Susanne von Caemmerer, editors, 2000);
- ***Photosynthesis: Photobiochemistry and Photobiophysics*** (Bacon Ke, author, 2001);
- ***Regulation of Photosynthesis*** (Eva-Mari Aro and Bertil Andersson, editors, 2001);
- ***Photosynthetic Nitrogen Assimilation and Associated Carbon and Respiratory Metabolism*** (Christine Foyer and Graham Noctor, editors, 2002);
- ***Light Harvesting Antennas*** (Beverley Green and William Parson, editors, 2003);
- ***Photosynthesis in Algae*** (Anthony Larkum, Susan Douglas and John Raven, editors, 2003);
- ***Respiration in Archaea and Bacteria: Diversity of Prokaryotic Electron Transport Carriers*** (Davide Zannoni, editor, 2004);
- ***Respiration in Archaea and Bacteria: Diversity of Prokaryotic Respiratory Systems*** (Davide Zannoni, editor, 2004);
- ***Plant Mitochondria: From Genome to Function*** (David A. Day, A. Harvey Millar and James Whelan, editors, 2004); and
- ***Chlorophyll a Fluorescence: A Signature of Photosynthesis*** (George C. Papageorgiou and Govindjee, editors, 2004).

The readers are requested to go to < http://www.springeronline.com > and search for the Book Series *Advances in Photosynthesis and Respiration* for further information and to order these books. Please note that the members of the International Society of Photosynthesis Research, ISPR (< http://photosynthesisresearch.org >) and authors receive special discounts.

Plant Respiration: From Cell to Ecosystem

The book is edited by two outstanding authorities on 'Plant Respiration': Hans Lambers (the University of Western Australia) and Miquel Ribas-Carbo (Universitat de Illes Balearic, Spain). It was initiated several years ago when I was visiting Joe Berry (at the Carnegie Institute of Washington, Stanford, CA), and Miquel was collaborating with Joe. The topic of the book, as provided by our distinguished editors, is: 'As in all living organisms, respiration is essential to provide metabolic energy and carbon skeletons for growth and maintenance of plants. As such, respiration is an essential component of a plant's carbon budget. Depending on species and environmental conditions, it consumes 25–75% of all the carbohydrates produced in photosynthesis — even more at extremely slow growth rates. Respiration in plants can also proceed in a manner that produces neither metabolic energy nor carbon skeletons, but heat. This type of respiration involves the cyanide-resistant, alternative oxidase; it is unique to plants, and resides in the mitochondria. The activity of this alternative pathway can be measured based on a difference in fractionation of oxygen isotopes between the cytochrome and the alternative oxidase. Heat production is important in some flowers to attract pollinators; however, the alternative oxidase also plays a major role in leaves and roots of most plants. A common thread throughout this volume is to link respiration, including alternative oxidase activity, to plant functioning in different environments.' This book is for the use of advanced undergraduates, graduates, postgraduates, and beginning researchers in the areas of plant and agricultural sciences, plant physiology, plant ecology, bioenergetics, cellular biology and integrative biology.

Plant Respiration: From Cells to Ecosystems has 13 authoritative Chapters, and is authored by 31 international authorities from 8 countries. The book begins with an introductory chapter on 'Regulation of Respiration in Vivo' by H. Lambers, S.A. Robinson and M. Ribas-Carbo (Australia and Spain). Chapter 2, by L.D. Hansen, R.S. Criddle and B.N. Smith , from USA, deals with 'Caloresspiratometry in Plant Biology.' M. Ribas-Carbo , S.A. Robinson and L. Giles (Spain, Australia and USA) provide, in Chapter 3, a summary of 'The Application of the Oxygen Technique to Assess Respiratory Pathway Partitioning.' In Chapter 4, V. Hurry, A.U. Igamberdiev, O. Keerberg, T.Parnik, O.K. Atkin, J. Zargoza-Castells and P. Gardestrom, from four countries (Sweden, Canada, Estonia and UK), present a discussion on 'Respiration in Photosynthetic Cells'; the chapter includes discussion on gas exchange, components, interactions with photorespiration and the operation of the mitochondria in light. K. Noguchi (of Japan) summarizes, in Chapter 5, 'Effects of Light Intensity and Carbohydrate Status on Leaf and Root Respiration.' In Chapter 6, J. Flexas, J. Galmés, M. Ribas-Carbo and H. Medrano (of Spain) discuss 'The Effects of Water Stress on Plant Respiration.' O. K. Atkin (of UK) and D. Bruhn (of Australia), in Chapter 7, discuss 'Responses of Plants to Changes in Temperature'; they include mechanisms and consequences of variations in Q_{10} values and the acclimation processes. Chapter 8 is by T. D.Colmer and H. Greenway (both of Australia); it focuses on 'O_2Transport, Respiration and Anaerobic Carbohydrate Catabolism in Roots in Flooded Soils.' R. Minocha and S.C. Minocha (both of USA) summarize, in Chapter 9, 'Effects of Soil pH and Aluminum on Plant Respiration.' In Chapter 10, T.J. Bouma (of the Netherlands) provides for the readers 'Understanding of Plant Respiration: Separation of Respiratory Components versus a Process-based Approach.' F. R. Minchin and J. F. Witty (of UK), discuss, in Chapter 11, 'Respiratory and Carbon Costs of Symbiotic N_2 Fixation in Legumes.' Chapter 12 , by D. R. Bryla and D. M. Eissenstat (both of USA), deals with 'Respiratory Costs of Mycorrhizal Associations.' Finally, M. A. Gonzalez and L. Taneva (of USA) discuss, in Chapter 13, 'Integated Effects of Atmospheric CO_2 Concentration on Plant and Ecosystem Respiration.'

A Bit of Early History

'It is a noble employment to rescue from oblivion those who deserve to be remembered' (Pliny the Younger, Letters V).

A 1927 Paper by Robert Emerson on Chlorella Respiration

Robert Emerson, who did his PhD work in the laboratory of Otto Warburg, and was later my advisor during 1956–1959, had published in 1927 the following historical paper: Emerson R (1927) The effect of certain respiratory inhibitors on the respiration of *Chlorella*. Journal of General Physiology, volume **x**, pp. 469–477). Here, Emerson compared the effects of respiratory inhibitors, including HCN and H_2S, on the green alga *Chlorella*. What Emerson showed in this 1927 paper is that 10^{-4} M HCN (and H_2S) stimulated, not inhibited, respiration in Chlorella. However, under1% glucose (heterotrophic condition), respiration was four times faster but inhibited by 50% by either H_2S or HCN. All these effects were fully reversible, and predated the later measurements on cyanide-insensitive respiration.

Early Plant Respiration Books

I mention here the books by F. F. Blackman; W. Stiles and W. Leach; and W. O. James. Although we may not remember the old, the new is built upon the old directly or indirectly, knowingly or unknowingly. On the personal side, my own training in Plant Respiration, during 1952–1956 at the University of Allahabad, was under Shri Ranjan, who had been a student of Felix Frost Blackman. Blackman studied both respiration and photosynthesis. G. E. Briggs (of Cambridge, UK) has published a nice book on the work of Blackman after his death (Late F. F. Blackman (1954) Analytic Studies in Plant Respiration. Cambridge, at the University Press.) A great deal of the experimental work in Blackman's laboratory was done by one of his students from India, P. Parija. Most of this work was done in the late 1920s at Cambridge. Regarding the importance of respiration, I quote Blackman and Parija (1928) '*Of all protoplasmic functions, the one which is, by tradition, most closely linked with our conception of vitality is the function for which the name of respiration has been accepted.*' In 1932, the year I was born, Walter Stiles and William Leach wrote their little (124 pages) book, *Respiration in Plants* (London, Methuen). The theories of two Nobel-laureates Otto H. Warburg and Heinrich O. Wieland on the oxidation-reduction and the enzymatic nature of respiration, both under aerobic and anaerobic conditions, were discussed in this book. It was the only book I had read during my student days. In 1953, one year after I obtained my B.Sc. degree, a new book was published (W. O. James (1952) *Plant Respiration.* Oxford, at the Clarendon Press). It is this book that I studied after I had obtained my M.Sc. degree. Again, on the personal side, I was thrilled to note that James discussed (see pp. 99 and 100) unpublished work of my Professor (Ranjan); this work was done while Ranjan was in Blackman's lab. It was a 282 page thorough and modern book. James also related respiration, although very briefly, to photosynthesis and commented '*Possible interactions with respiration have been the bugbear of photosynthetic measurements since their beginning.*' He did mention the related work of Bessel Kok, Robert Emerson, Jack Myers James Franck, Hans Gaffron, Melvin Calvin and Andy Benson, among others. A 19-page bibliography was very helpful to me in obtaining the necessary information.

The advancement made during the last 50 years is really remarkable and exciting. To understand the historical evolution of research in 'plant respiration,' I encourge the readers to consult the following three books: (1) Harry Beevers (1961) *Respiratory Metabolism in Plants* (Row, Peterson and Company, Evanston, Illinois); (2) Helgi Opik (1980) *The Respiration of Higher Plants* (London, E. Arnold); and (3) Roland Douce and David A. Day (Eds) (1985) *Higher Plant Cell Respiration. Encyclopedia of Plant Physiology, New Series*, Volume 18, Springer-Verlag, Berlin.

The Scope of the Series

Advances in Photosynthesis and Respiration is a book series that provides, at regular intervals, a comprehensive and state-of-the-art account of research in various areas of photosynthesis and respiration. Photosynthesis is the process by which higher plants, algae, and certain species of bacteria transform and store solar energy in the form of energy-rich organic molecules. These compounds are in turn used as the energy source for all growth and reproduction in these and almost all other organisms. As such, virtually all life on the planet ultimately depends on photosynthetic energy conversion. Respiration, which occurs in mitochondria and in bacterial membranes, utilizes energy present in organic molecules to fuel a wide range of metabolic reactions critical for cell growth and development. In addition, many photosynthetic organisms engage in energetically wasteful photorespiration that begins in the chloroplast with an oxygenation reaction catalyzed by the same enzyme responsible for capturing carbon dioxide in photosynthesis. This series of books spans topics from physics to agronomy and medicine, from femtosecond (10^{-15} s) processes to season-long production, from the photophysics of reaction centers, through the electrochemistry of intermediate electron transfer, to the physiology of whole organisms, and from X-ray crystallography of proteins to the morphology of organelles and intact organisms. The intent of the series is to offer beginning researchers, advanced undergraduate students, graduate students, and even research specialists, a comprehensive, up-to-date picture of the remarkable advances across the full scope of research on bioenergetics and carbon metabolism.

Future Books

The readers of the current series are encouraged to watch for the publication of the forthcoming books (not necessarily arranged in the order of future appearance):

- ***Discoveries in Photosynthesis Research*** (Editors: Govindjee, J. Thomas Beatty, Howard Gest, and John Allen);
- ***Photosystem II: The Water/Plastoquinone Oxido-reductase in Photosynthesis*** (Editors: Thomas J. Wydrzynski and Kimiyuki Satoh);
- ***Chlorophylls and Bacteriochlorophylls: Biochemistry, Biophysics and Biological Function*** (Editors: Bernhard Grimm, Robert J. Porra, Wolfhart Rüdiger and Hugo Scheer);
- ***Photosystem I: The NADP+/Ferredoxin Oxidoreductase in Oxygenic Photosynthesis*** (Editor: John Golbeck);
- ***Photoprotection, Photoinhibition, Gene Regulation and Environment*** (Editors: Barbara Demmig-Adams, William W. Adams III and Autar Mattoo);
- ***The Structure and Function of Plastids*** (Editors: Kenneth Hoober and Robert Wise); and
- ***Photosynthesis: A Comprehensive Treatise; Biochemistry, Biophysics Physiology and Molecular Biology. Part 1***(Editors: Julian Eaton-Rye and Baishnab Tripathy); and
- ***Photosynthesis: A Comprehensive Treatise; Biochemistry, Biophysics Physiology and Molecular Biology. Part 2*** (Editors: Baishnab Tripathy and Julian Eaton-Rye)

In addition to these contracted books, we are already in touch with prospective Editors for the following books:

- *Biophysical Techniques in Photosynthesis. part 2;*
- *Molecular Biology of Cyanobacteria. part 2;*
- *Protonation and ATP Synthases;*
- *Genomics and Proteomics;*
- *Protein Complexes of Respiration and Photosynthesis;* and
- *The Cytochromes*

Other books, under discussion, are: Molecular Biology of Stress in Plants; Artificial Photosynthesis; and Global Aspects of Photosynthesis and Respiration. Readers are requested to send their suggestions for these and future volumes (topics, names of future editors, and of future authors) to me by E-mail (gov@uiuc.edu) or fax (1-217-244-7246).

In view of the interdisciplinary character of research in photosynthesis and respiration, it is my earnest hope that this series of books will be used in educating students and researchers not only in Plant Sciences, Molecular and Cell Biology, Integrative Biology, Biotechnology, Agricultural Sciences, Microbiology, Biochemistry, and Biophysics, but also in Bioengineering, Chemistry, and Physics.

I take this opportunity to thank Hans Lambers and Miquel Ribas-Carbo for their excellent editorial work. I thank all the 31 authors of volume 18: without their authoritative chapters, there will be no book. I owe Jacco Flipsen and Noeline Gibson (both of Springer) special thanks for their friendly working relation with us that led to the production of this book. Thanks are also due to Jeff Haas (Director of Information Technology, Life Sciences, University of Illinois) and Evan DeLucia (Head, Department of Plant Biology, University of Illinois) for their support.

My special and particular thanks go to Larry Orr for his friendly and excellent work in typesetting and in producing this book, including its Index.

I wish all the readers a very fruitful New Year (2005).

January 5, 2005

Govindjee
Series Editor, Advances in Photosynthesis and Respiration
University of Illinois at Urbana-Champaign
Department of Plant Biology
Urbana, IL 61801-3707, USA
E-mail: gov@uiuc.edu;
URL http://www.life.uiuc.edu/govindjee

Series Editor Govindjee

Govindjee is Professor Emeritus of Biochemistry, Biophysics and Plant Biology at the University of Illinois at Urbana-Champaign (UIUC), Illinois, USA, since 1999. He obtained his B.Sc. (Chemistry and Biology) and M.Sc. (Botany, specializing in Plant Physiology) in 1952 and 1954, respectively, from the University of Allahabad, Allahabad, India. His advisor in India was Shri Ranjan, who was a former student of F. F. Blackman. Govindjee served as a Lecturer in Botany at the University of Allahabad from 1954–1956. From 1956–1959, he was a Fulbright Scholar and a graduate Fellow in Physico-Chemical Biology at the UIUC. During 1959-1960, he was a Research Assistant in Botany. He was a doctoral student, first of Robert Emerson (1956–February 4, 1959), and then of Eugene Rabinowitch (1959–1960). He received his Ph.D. in Biophysics from the UIUC in 1960, with a thesis on the *Action Spectra of the Emerson Enhancement Effect in Algae*. From 1960–1961, he served as a United States Public Health (USPH) Postdoctoral Fellow; from 1961–1965, as Assistant Professor of Botany; from 1965–1969 as Associate Professor of Biophysics and Botany; and from 1969–1999 as Professor of Biophysics and Plant Biology, all at the UIUC. Julian Eaton-Rye, Prasanna Mohanty, George Papageorgiou, Alan Stemler, Thomas Wydrzynski, Jin Xiong, Chunhe Xu and Barbara Zilinskas are among his more than 20 past PhD students. The late Jean-Marie Briantais, Christa Critchley, Adam Gilmore, Jack van Rensen and Wim Vermaas are among his more than 10 past Research Associates. His honors include: Fellow of the American Association of Advancement of Science (1976); Distinguished Lecturer of the School of Life Sciences, UIUC (1978); President of the American Society of Photobiology (1980–1981); Fulbright Senior Lecturer (1996–1997); and honorary President of the 2004 International Photosynthesis Congress (Montreal, Canada). Govindjee's research has focused on the function of 'Photosystem II' (water-plastoquinone oxido-reductase), particularly on the primary photochemistry; role of bicarbonate in the electron and proton transport; thermoluminescence, delayed and prompt fluorescence (particularly lifetimes), and their use in understanding electron transport and photoprotection against excess light. He has coauthored *Photosynthesis* (1969; John Wiley & Sons); and has edited (or co-edited) *Bioenergetics of Photosynthesis* (1975; Academic Press); *Photosynthesis* (in two volumes, 1982; Academic Press); *Light Emission by Plants and Bacteria* (1986; Academic Press); and *Chlorophyll a Fluorescence: A Signature of Photosynthesis* (2004; Springer), among several other books. He is a member of the American Society of Plant Biology, American Society for Photobiology, Biophysical Society of America, International Society of Photosynthesis Research (ISPR) and Sigma Xi. Govindjee's scientific interests, now, include Fluorescence Lifetime Imaging Microscopy (FLIM), and the regulation of excitation energy transfer in oscillating light. However, his real focus now is on the 'History of Photosynthesis Research,' and in 'Photosynthesis Education.' His personal background appears in Volume 13 (edited by B. Green and W. Parson); and contributions to photosynthesis and fluorescence in algae in Volume 14 (A. Larkum, J. Raven and S. Douglas, editors) of the *Advances in Photosynthesis and Respiration* (AIPH). He serves as the Series Editor of AIPH, and as the 'Historical Corner' Editor of *Photosynthesis Research*. For further information, see his web page at: http://www.life.uiuc.edu/govindjee. He can always be reached by e-mail (gov@life.uiuc.edu).

Contents

Preface

Respiration: From Cell to Ecosystem is the 18th volume in the series Advances in Photosynthesis and Respiration (Series Editor, Govindjee). It is one of four volumes dealing with respiration. The first two (volumes 15 and 16) were on Respiration in Bacteria and Archaea (edited by Davide Zannoni), and this volume complements another volume (17), edited by David A. Day, James Whelan and A. Harvey Millar, dealing with molecular and biochemical aspects of plant respiration.

Louis Pasteur (1822–1895) was probably the first scientist to study the link between respiration and growth (in yeast). His pioneering work was published in 1861 (Expériences et vues nouvelles sur la nature des fermentations. Comptes Rendus 52: 1260-1271). For a long time since, the significance of respiration for growth was far from clear. In his PhD thesis, entitled 'Energiemessungen bei Aspergillus niger mit Hilfe eines Mikro-Kompensations Calorimeter' (University of Groningen, the Netherlands), L. Algera (1932) concluded that 93–94% of all the energy generated during respiration was lost as heat. In fact, he considered that the 6–7% that was not lost as heat was insignificant, and probably due to experimental error. Therefore, he concluded that respiration was a wasteful process. Using data from a modern biochemical handbook, it can be calculated that at the very most only 8% of the chemical energy in glucose is transferred to the chemical energy in mycelium. If non-phosphorylating pathways, uncoupling proteins and respiratory paths required for transport are taken into account, that value of 8% is expected to be considerably less. Yet, W. Stiles and W. Leach (1932) in their book Respiration in Plants (Methuen & Co., Ltd., London, reprinted in 1936) stated that 'the supreme importance of respiration, being as it is one of the most universal and fundamental processes of living protoplasm, is recognized by all physiologists. In spite of this, students of botany frequently give respiration little more than a passing consideration.'

H.R. Barnell (1937) made the observation that of every three hexose molecules used by barley embryos two were used as carbon skeletons, and one in respiration (Analytical studies in plant respiration. VII—Aerobic respiration in barley embryos and its relation to growth and carbohydrate supply. Proceedings of Royal Society B. 123: 321-342). He suggested that respiration was probably needed to drive the growth process, but it wasn't until the details of energy-requiring metabolic and transport processes became known that the proof for his suggestion was delivered. Indeed, in his book entitled Plant Respiration (Clarendon Press, Oxford) W.O. James (1953) stated that there was no absolute proof yet that respiration was essential for life as we know it. To conclude that respiration is an absolute requirement since life without respiration has never been observed would be similar to the conclusions that 'pillar boxes could not perform their postal functions upon ceasing to be red'.

By the time Harry Beevers (1924–2004) published his comprehensive book, entitled Respiratory Metabolism in Plants (1961; Harper & Row, New York), the significance of respiration for plant growth and functioning had been soundly established. Many questions remained, for example, about the role of cyanide-resistant respiration, the efficiency of the respiratory process, and the quantitative links between respiration and growth. When the Encyclopedia of Plant Physiology, Volume 18, Higher Plant Cell Respiration (Series Editor, R. Douce and D.A. Day) appeared in 1985, enormous progress had been made, and a molecular approach to respiration research had been added to the more classical physiological, biochemical and biophysical way of doing experiments. In the almost 20 years since the publication of that book, vast progress has been made, in part made possible because of technological advances. The present volume presents the state of the art of plant respiration research, from cell to ecosystem. Since Barnell's research on barely embryos, our understanding about the quantitative significance of plant respiration has increased enormously. Plants use a large fraction of their daily produced photoassimilates for respiratory processes (0.25–0.70). As described in several chapters in this volume, this fraction strongly depends on environmental conditions, both abiotic and biotic. We have come a long way since Harry Beevers' book, when cyanide-resistant respiration was known, but its biochemical basis and physiological significance was entirely unknown. As in animal mitochondria, respiration proceeds via a cyanide-sensitive cytochrome pathway; however, plants also have a cyanide-resistant, alternative respiratory path, which may be responsible for a major part of their respiration. In-vivo techniques, based on oxygen-isotope fractionation, have been developed to assess the contribution of both the cytochrome and the alternative pathways. As reviewed in several chapters, the fraction of daily produced photoassimilates, that is used in respiration and the contribution of the alternative path, varies strongly as dependent on environmental conditions. The principal aim of both volumes 17 and 18 is to provide final-year undergraduate students, postgraduate and other researchers with an up-to date overview of plant respiration. Many people have contributed to the completion of this book, including Larry Orr (of Arizona State University), Govindjee, the Series Editor and several helpful professionals at Springer. However, we wish to single out two persons to thank profoundly for their support: Marion Cambridge and Pepi Martín Pina, without whose continuous support this book would not have been completed.

Hans Lambers
School of Plant Biology
The University of Western Australia
Crawley, WA 6009
Australia
http://hlambers@cyllene.uwa.edu.au

Miquel Ribas-Carbo
Departament de Biologia
Universitat de les Illes Balears
07122 Palma de Mallorca
Spain
http://mribas@uib.es

Hans Lambers

Miquel Ribas-Carbo

The volume editors, Hans Lambers and Miquel Ribas-Carbó, are Head of School of Plant Biology at the University of Western Australia, Perth, Australia, and Research Scientist at the Department of Biology at the University of the Balearic Islands, Palma de Mallorca, Spain, respectively.

Hans Lambers' main research is in the broad field of plant respiration and carbon metabolism, with a current focus on proteoid and dauciform roots, which are typically found in Proteaceae and Cyperaceae, respectively. Species belonging to these families typically occur on phosphate-impoverished soils. To access phosphorus in these soils, they exudate vast amounts of carboxylates which is associated with major changes in their carbon metabolism and (alternative) respiration. Hans Lambers received his undergraduate (1974, 1976) and doctorate (1979) degrees from the University of Groningen, the Netherlands, where he made pioneering studies on cyanide-resistant respiration and on respiration as linked with growth, maintenance and ion uptake. After his PhD, he had a number of postdoctoral positions in Australia and the Netherlands. Plant respiration was an important component of his research while holding these positions. He continued work on respiration when he accepted the Chair in Ecophysiology, when he focused on unraveling the physiological differences between inherently fast- and slow-growing herbaceous species, when a gain (alternative) respiration was a major component of his work. In 1998, he moved to the University of Western Australia, Perth, Australia. Currently, he is Professor of Plant Biology at this university.

Miquel Ribas-Carbo has dedicated his entire research life to study the regulation of the cyanide-resistant alternative pathway, its role and function. He received his undergraduate (1989) and doctorate (1993) from the University of Barcelona, Spain. He held a Fulbright Postdoctoral Fellowship at the Department of Botany at Duke University, Durham, USA, with Dr. James N. Siedow, and was a Research Associate at the Department of Plant Biology of the Carnegie Institution of Washington, USA, with Dr. Joseph A. Berry. Throughout all these years he has continuously focused on the development of the oxygen-isotope fractionation technique, currently, the only available technique to quantify the electron partitioning between the two mitochondrial respiratory pathways, and its application in different plant tissues and species. In 1997, he received the Spanish Society of Plant Physiology Award for Young Scientists. In 2002, he moved to the Universitat de les Illes Balears in Palma de Mallorca, Spain, where he joined a team headed by Prof. Hipólito Medrano interested in water-use efficiency and the effects of water stress on respiration. His goal is to develop a vibrant European group interested in the development and application of the oxygen-isotope technique to plant respiration studies.

Author Index

Chapter 1

Regulation of Respiration In Vivo

Hans Lambers*

School of Plant Biology, Faculty of Natural and Agricultural Sciences, The University of Western Australia, 35 Stirling Highway, Crawley WA 6009, Australia;

Sharon A. Robinson

Institute for Conservation Biology, Department of Biological Sciences, University of Wollongong, Wollongong, NSW, 2522, Australia

Miquel Ribas-Carbo

Departament de Biologia, Universitat de les Illes Balears, Ctra. Valldemossa Km. 7,5, 07122 Palma de Mallorca, Spain

*Author for correspondence, email: hlambers@cyllene.uwa.edu.au

H. Lambers and M. Ribas-Carbo (eds.), Plant Respiration, 1–15.

I. Introduction

A large fraction (0.25–0.70) of all the carbohydrates that a higher plant assimilates each day are expended in respiration in the same period (Van der Werf et al., 1992; Lambers et al., 1998a). This fraction increases with decreasing maximum relative growth rate (RGR) of a species (Poorter et al., 1991), because the specific costs of nitrogen acquisition are higher for plants with an inherently slow RGR, when plants are compared at near-optimum nutrient supply (Lambers et al., 1998a; Scheurwater et al., 1998; 1999). This fraction also strongly increases at a limiting supply of nutrients (N and P) for two reasons. Firstly, at a low supply of N or P, plants invest relatively more in roots, and less in photosynthetic organs (Lambers et al., 1998b). Secondly, the respiratory costs, presumably the cost that are associated with nutrient acquisition which account for 40–70% of root respiration at high nutrient supply (Poorter et al., 1991), increase substantially at low nutrient supply (Van der Werf et al., 1992). Up to 25% of all carbohydrates may also be required to support microsymbionts (Lambers et al., 1998b), in those higher plants that live symbiotically with dinitrogen-fixing microorganisms (Chapter 11, Minchin and Witty) or mycorrhizal fungi (Chapter 12, Bryla and Eisenstat).

Dark respiration produces energy (ATP), reducing equivalents (NAD(P)H) and carbon skeletons to sustain plant growth; however, in plants, a significant part of respiration may proceed via a nonphosphorylating pathway that is cyanide-resistant and generates less ATP than the cytochrome pathway, which occurs in both plants and animals. We have no satisfactory answer to date to the question why plants have a respiratory pathway that is not linked to ATP production, but several hypotheses are explored in this chapter and in following ones (Chapter 12, Bryla and Eisenstat; Chapter 6, Flexas et al.). If we seek to understand and model the carbon balance of a plant, and aim to understand plant performance and growth in different environments, then it is imperative first to try to obtain a clear understanding of respiration, including alternative path activity (Chapter 10, Bouma).

Abbreviations: AOX – alternative oxidase; CAM – crassulacean acid metabolism; DTT – dithiothreitol; KCN – potassium cyanide; NMR – nuclear magnetic resonance; PEP – phosphoenol pyruvate; Q – ubiquinone; Qr – reduced ubiquinone; RGR – relative growth rate; ROS – reactive oxygen species; RQ – respiratory quotient; SA – salicylic acid; SHAM – salicylhydroxamic acid; TCA – tricarboxylic acid cycle; TMV – tobacco mosaic virus

The types and rates of plant respiration are controlled by a combination of energy demand, substrate availability and oxygen supply. In the absence of oxygen (anoxia) or at low levels of oxygen (hypoxia), respiration cannot proceed by normal aerobic pathways, and fermentation starts to take place, with ethanol and lactate as major end-products (Chapter 8, Colmer and Greenway). The ATP yield of fermentation is considerably less than that of aerobic respiration. Temperature (Chapter 7, Atkin et al.), water stress (Chapter 6, Flexas et al.) and rhizosphere conditions such as pH and the presence of toxic levels of aluminum (Chapter 9, Minocha and Minocha) also affect respiration, depending on the time of exposure and species. In this chapter, we briefly discuss the control over respiratory processes, so as to provide the background for understanding later chapters dealing with species comparisons and effects of environmental conditions.

II. General Characteristics of the Respiratory System

A. The Respiratory Quotient

The respiratory pathways in plant tissues include glycolysis, which is located both in the cytosol and in the plastids, the oxidative pentose phosphate pathway, which is located in the plastids, the tricarboxylic acid (TCA) or Krebs cycle, in the matrix of mitochondria, and the electron-transport pathways, which are in the inner mitochondrial membrane.

The respiratory quotient (RQ, the ratio between the number of moles of CO_2 released and that of O_2 consumed) is an index of the types of substrates used in respiration, and the subsequent use of respiratory energy to support biosynthesis. In the absence of biosynthetic processes, the RQ in nonphotosynthetic tissues is expected to be 1.0, if sucrose is the only substrate for respiration and is fully oxidized to CO_2 and H_2O. For seeds of *Triticum aestivum* (wheat), in which carbohydrates are major storage compounds, RQ is close to unity, whereas for the fat-storing seeds of *Linum usitatissimum* (flax) RQ values as low as 0.4 are found (Stiles and Leach, 1936; Lambers et al., 1998b). The RQ can be greater than 1.0, if organic acids are an important substrate, because organic acids are more oxidized than sucrose, and, therefore, produce more CO_2 per unit O_2. Biosynthetic processes can also affect RQ. For example, if nitrate reduction proceeds in the roots, then the RQ is expected to be greater than one, because an additional two molecules of CO_2 are produced per molecule of nitrate reduced to ammonium; in contrast, a net production of organic acids will lower the RQ (Lambers et al., 1998b).

B. Glycolysis, the Pentose Phosphate Pathway, and the Tricarboxylic Acid (TCA) Cycle

The first step in the production of energy for respiration occurs when glucose (or other storage carbohydrates) is metabolized in glycolysis or in the oxidative pentose phosphate pathway. Glycolysis involves the conversion of glucose, via phospho*enol*pyruvate (PEP), into malate and pyruvate. In contrast to mammalian cells, where virtually all PEP is converted into pyruvate, in plant cells malate is the major end-product of glycolysis, and thus the major substrate for the mitochondria (Day and Hanson, 1977). Malate is formed from oxaloacetate, which is produced by PEP-carboxylase from PEP and HCO_3^-. Key enzymes in glycolysis are controlled by adenylates (AMP, ADP and ATP), in such a way as to speed up the rate of glycolysis when the demand for metabolic energy (ATP) increases.

Oxidation of one glucose molecule in glycolysis produces two malate molecules, without a net production of ATP. When pyruvate is the end-product, there is a net production of two ATP molecules in glycolysis. Despite the production of NADH in one step in glycolysis, there is no net production of NADH when malate is the end-product, due to the need for NADH in the reduction of oxaloacetate, catalysed by malate dehydrogenase.

Unlike glycolysis, which is predominantly involved in the breakdown of sugars and ultimately in the production of ATP, the oxidative pentose phosphate pathway plays a more important role in producing intermediates (e.g., amino acids, nucleotides) and NADPH. There is no evidence for a control of this pathway by the demand for energy.

The malate and pyruvate that are formed in glycolysis in the cytosol are exported to the mitochondria, where they are oxidized in the TCA cycle. Complete oxidation of one molecule of malate, yields five molecules of NADH and one molecule of $FADH_2$, as well as one molecule of ATP, NADH and $FADH_2$ subsequently donate their electrons to the electron-transport chain (Sec. II.C.1).

C. Mitochondrial Metabolism

The malate formed in glycolysis in the cytosol is imported into the mitochondria, and oxidized partly via malic enzyme, which produces pyruvate and CO_2, and partly via malate dehydrogenase, which produces oxaloacetate. Pyruvate is then oxidized in the TCA cycle, so that malate is regenerated. Whenever carbon skeletons, e.g., citrate or α-ketoglutarate, are drained from the TCA cycle, to be used in biosynthesis, the concerted action of PEP carboxylase and malate dehydrogenase replenishes the drained carbon, and sustains TCA-cycle activity.

Oxidation of malate and other NAD-linked substrates of the TCA cycle is associated with complex I (Sec. II.C.1). In mitochondria there are four major complexes associated with electron transfer, and one associated with oxidative phosphorylation, all located in the inner mitochondrial membrane. In addition, there are two small redox molecules, ubiquinone (Q) and cytochrome *c*, which play an important role in electron transfer. Finally, in plant mitochondria there is the cyanide-resistant alternative oxidase, also located in the inner membrane.

1. The Complexes of the Electron-Transport Chain

The role of the mitochondrial electron transport chain is to produce ATP. This ATP production is driven by the proton gradient across the inner mitochondrial membrane linked to the flow of electrons from reducing equivalents to O_2 (Fig. 1). The system responsible for ATP production is the ATPase (see Sec. II.E). Complex I is the main entry point of electrons from NADH produced in the TCA cycle or in photorespiration (glycine oxidation). Complex I is the first coupling site, or site 1, of proton extrusion, and this is linked to ATP production. Succinate is the only intermediate of the TCA cycle that is oxidized by a membrane-bound enzyme: succinate dehydrogenase. Electrons enter the respiratory chain via complex II, and are transferred to ubiquinone. NAD(P)H that is produced outside the mitochondria, also feeds its electrons into the chain at the level of ubiquinone. As with complex II, the external dehydrogenases are not connected with the translocation of H^+ across the inner mitochondrial membrane. Hence less ATP is produced per molecule of oxygen when succinate or NAD(P)H are oxidized in comparison with that of

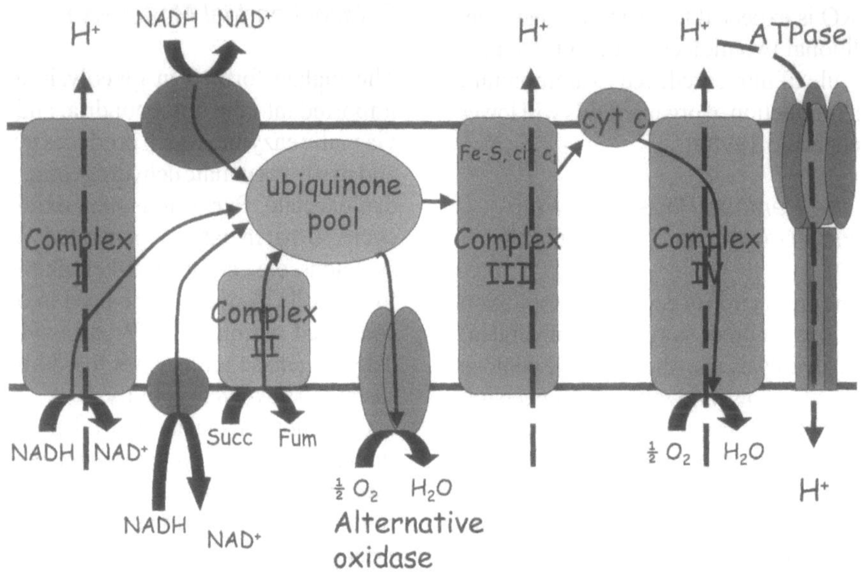

Fig. 1. The organization of the electron-transporting complexes of the respiratory chain in higher plant mitochondria. All components are located in the inner mitochondrial membrane. Some of the components are membrane-spanning, others face the mitochondrial matrix or the space between the inner and the outer. mitochondrial membrane. Ubiquinone is a mobile pool of quinone and quinol molecules; cyt = cytochrome oxidase; fum = fumrate; succ = succinate.

malate, citrate, or oxoglutarate. Complex III transfers electrons from ubiquinone to cytochrome *c*, coupled to the extrusion of four protons per electron pair to the intermembrane space and is therefore site 2 of proton extrusion. Complex IV is the terminal oxidase of the cytochrome pathway, accepting electrons from cytochrome *c*, and donating these to O_2. It also generates a proton-motive force which makes complex IV the third coupling site (Lambers et al., 1998b).

2. The Cyanide-Resistant Terminal Oxidase

Mitochondrial respiration of most plant tissues is not fully inhibited by inhibitors of the cytochrome path (e.g., KCN, antimycin). This is due to the presence of a cyanide-resistant, alternative electron-transport pathway, consisting of one enzyme, the alternative oxidase, firmly embedded in the inner mitochondrial membrane. Both the C terminus and the N terminus of the alternative oxidase face the matrix. The C terminus contains a binuclear iron centre; the N terminus contains, in most species, a conserved cysteine residue (Siedow and Umbach, 1995, 2000; Andersson and Nordlund, 1999; Berthold et al., 2000). The AOX protein is not membrane spanning, as was thought originally. The cysteine residue is involved in the dimerization of two alternative oxidase subunits, and in the binding of pyruvate and other keto-acids. The two ends are connected via an inter-membrane helix region, where the potential ubiquinone-binding site is located (Andersson and Nordlund, 1999).

The alternative oxidase is encoded by nuclear DNA, and involves a multigene family whose individual products can be separated on a protein gel (e.g., Whelan et al., 1996; Finnegan et al., 1997; Ito et al., 1997). The sequenced genes in the different species contain highly conserved regions (Vanlerberghe and McIntosh, 1997). There have been few studies aimed at unravelling the gene expression, biochemical properties or relative activities of the individual members of the AOX gene family (McCabe et al., 1998; Chapter 3, Ribas-Carbo et al.).

The branching point of the alternative path from the cytochrome path is at the level of ubiquinone, a component common to both pathways. Transfer of electrons from ubiquinone to oxygen via the alternative path is not coupled to the extrusion of protons from the matrix to the intermembrane space, and the energy is lost as heat. Hence, the transfer of electrons from NADH, produced inside the mitochondria, to O_2 via the alternative path yields only one third of the amount of ATP that is produced when the cytochrome path is used (Millenaar and Lambers, 2003).

3. Substrates, Inhibitors, and Uncouplers

Oxidation of glycine is of quantitative importance only in tissues exhibiting photorespiration (Chapter 4, Hurry et al.). Glycolysis may start with glucose, as discussed above, or with starch, sucrose, or any major transport carbohydrate or sugar alcohol imported via the phloem (Lambers et al., 1998b).

A range of respiratory inhibitors have helped to elucidate the organization of the respiratory pathways. To give just one example, cyanide effectively blocks complex IV and has been used to demonstrate the presence of the alternative path. Uncouplers make membranes, including the inner mitochondrial membrane, permeable to protons, and hence prevent oxidative phosphorylation. Concentrations of CO_2, in a range that is expected to occur within the next century, may inhibit leaf respiration, due to inhibition of, for example, cytochrome oxidase and succinate dehydrogenase (Chapter 13, Gonzàlez-Meler and Taneva). Do the high CO_2 concentrations that normally occur in soil also inhibit root respiration? There is a remarkable lack of information in the literature to answer this obvious question in a satisfactory manner (Chapter 8, Colmer and Greenway).

4. Respiratory Control of the Electron Transport Chain

To learn more about the manner in which plant respiration responds to the demand for metabolic energy in vivo, we refer to some classical experiments with isolated mitochondria (Lambers et al., 1998b). Freshly isolated intact mitochondria in an appropriate buffer, a condition referred to as 'state 1', do not consume an appreciable amount of oxygen; in vivo they rely on a continuous import of respiratory substrate from the cytosol. Upon addition of a respiratory substrate ('state 2') there is still not much oxygen uptake; for rapid rates of respiration to occur, addition (in vitro) or import (in vivo) of additional metabolites is required. As soon as ADP is present, a rapid consumption of oxygen can be measured. This 'state' of the mitochondria is called 'state 3'. In vivo, rapid supply of ADP will occur when a large amount of ATP is required to drive biosynthetic and transport processes. Upon conversion of all ADP into ATP ('state 4'), the respiration rate of the mitochondria declines again to the rate found before addition of ADP. Upon addition of more ADP, the mitochondria go into state 3 again, followed by state 4 upon depletion of ADP. This can be repeated until all oxygen in the cuvette is consumed. Thus, the respiratory activity of isolated mitochondria is effectively controlled by the availability of ADP: respiratory control is quantified in the 'respiratory control ratio' (the ratio of the rate at substrate saturation in the presence of ADP to that under the same conditions, but after ADP has been depleted). The same respiratory control occurs in intact tissues, and is one of the mechanisms ensuring that the rate of respiration is enhanced when the demand for ATP increases.

D. The Major Points of Control of Plant Respiration

We briefly discussed the control of glycolysis by 'energy demand' (Sec. II.B), and a similar control by 'energy demand' of mitochondrial electron transport, termed respiratory control (Sec. II.C.4). The effects of energy demand on dark respiration are a function of the metabolic energy that is required for growth, maintenance, and transport processes. When tissues grow fast, take up ions rapidly and/or have a fast turnover of proteins, they, therefore, generally have a fast rate of respiration. At low levels of respiratory substrate (carbohydrates, organic acids), however, the activity of respiratory pathways may be substrate-limited. When substrate levels increase, the respiratory capacity is enhanced and adjusted to the high substrate input, through the transcription of specific genes that encode respiratory enzymes (Chapter 5, Noguchi). Plant respiration is clearly quite flexible and responds rapidly to the demand for respiratory energy as well as to the supply of respiratory substrate. The production of ATP which is coupled to the oxidation of substrate may also vary widely, due to the presence of both nonphosphorylating and phosphorylating paths.

E. ATP Production in Isolated Mitochondria and in Vivo

The rate of oxygen consumption during the phosphorylation of ADP can be related to the total amount of ADP that must be added to consume this oxygen. This allows calculation of the ADP:O ratio in vitro. This ratio is around 3 for NAD-linked substrate and around 1.5 for succinate and external NAD(P)H (Lambers et al., 1998b). Nuclear Magnetic Resonance

(NMR) spectroscopy has been used to estimate ATP production and oxygen consumption in intact tissues (Roberts et al., 1984).

1. Oxidative Phosphorylation: The Chemiosmotic Model

During the transfer of electrons from various substrates to oxygen via the cytochrome path, protons are extruded into the space between the inner and outer mitochondrial membranes. This generates a proton-motive force across the inner mitochondrial membrane that drives the synthesis of ATP. The basic features of this chemiosmotic model are:

(1) protons are transported outwards, coupled to the transfer of electrons, thus giving rise to both a proton gradient (ΔpH) and a membrane potential ($\Delta\Psi$)

(2) the inner membrane is impermeable to protons and other ions, except by special transport systems;

(3) there is an ATP synthetase (also called ATPase), which transforms the energy of the electrochemical gradient, generated by the proton-extruding system, into ATP

The pH gradient, ΔpH, and the membrane potential $\Delta\Psi$, are interconvertible. It is the combination of the two which forms the proton-motive force (Δp), the driving force for ATP synthesis, catalysed by an ATPase:

$$\Delta p = \Delta\Psi - 2.3\ RT/F.\Delta pH \qquad (1)$$

where R is the gas constant (J mol^{-1} K^{-1}), T is the absolute temperature (K), and F is Faraday's number (Coulomb). Both components in the equation are expressed in mV. Approximately one ATP is produced per three protons transported.

2. ATP Production in Vivo

As mentioned above, ATP production in vivo can be measured using Nuclear Magnetic Resonance (NMR) spectroscopy (Roberts et al., 1984). This technique relies on the fact that certain nuclei, including ^{31}P, possess a permanent magnetic moment, owing to their nuclear spin. Such nuclei can be made 'visible' in a strong external magnetic field, in which they orient their nuclear spins in the same direction. NMR spectroscopy can be used to monitor the absorption of radiofrequency by the oriented spin population in the strong magnetic field. The location of the peaks in a NMR spectrum depends on the molecule in which the nucleus is present, and also on the 'environment' of the molecule (e.g., pH) (Roberts, 1984).

The resonance of specific P-containing compounds can be altered by irradiation with radiofrequency power. If this irradiation is sufficiently strong ('saturating'), then it disorientates the nuclear spins of that P-containing compound, so that its peak disappears from the spectrum. Upon hydrolysis of ATP, the γ-ATP phosphate atom becomes part of the cytoplasmic P_i pool. For a brief period, therefore, some of the P_i molecules also contain disoriented nuclear spins; specific radiation of the γ-ATP peak decreases the P_i peak. This phenomenon is called 'saturation transfer', which has been used to estimate the rate of ATP hydrolysis to ADP and P_i in vivo (Roberts et al., 1984).

If the rate of disappearance of the saturation effect in the absence of biochemical exchange of phosphate between γ-ATP and P_i is known, then the rate of ATP hydrolysis can be derived from the rate of loss of saturation. This has been performed for root tips for which the oxygen uptake was measured in parallel experiments. In this manner ADP:O ratios in *Zea mays* root tips exposed to a range of conditions have been determined (Roberts et al., 1984). The ADP:O ratios for *Z. mays* root tips supplied with 50 mM glucose are remarkably close to those expected when glycolysis plus TCA cycle are responsible for the complete oxidation of exogenous glucose, provided the alternative path does not contribute to the oxygen uptake. This is surprising, since root respiration proceeds to a large extent via the alternative path (Millar et al., 1998; Millenaar et al., 1998, 2000). Perhaps the alternative path is less active in root tips than in the rest of the root system, but this would need to be further investigated using the isotope-fractionation technique (Chapter 3, Ribas-Carbo et al.). So far, maize root tips are the only intact plant material used for the determination of ADP:O ratios in vivo. We cannot assume, therefore, that the ADP:O ratio in vivo is invariably 3. In fact, the ratio under most circumstances is probably far less than 3, due to alternative path activity (Sec. II.F.2).

F. Regulation of Electron Transport via the Cytochrome and the Alternative Paths

The existence of two respiratory pathways, both transporting electrons to oxygen, in higher plant mitochondria, raises the question if and how the partitioning of electrons between the two paths is regulated. This is important because the cytochrome path is coupled to proton extrusion and the production of ATP, whereas transport of electrons via the alternative path is not, at least not from the point where both pathways branch to oxygen (Vanlerberghe and McIntosh, 1997).

1. Competition or Overflow?

Initially, it was widely believed that the alternative path did not compete for electrons with the cytochrome path, and that it served as an overflow when the cytochrome path was (virtually) saturated with electrons (Bahr and Bonner, 1973). Much later, it was found that the activity of the cytochrome path increases linearly with the fraction of ubiquinone (the common substrate with the alternative path) that is in its reduced state (Q_r/Q_t) (Dry et al., 1989). By contrast, the alternative path showed no appreciable activity until a substantial (30–40%) fraction of the ubiquinone was in its reduced state, and then the activity increased exponentially. By that stage a sound biochemical explanation seemed available for the 'energy overflow' model; however, as more experimental results became available, the 'energy overflow' model was rejected.

2. The Intricate Regulation of the Alternative Oxidase

The alternative pathway changes its level of activity, so that it competes with the cytochrome pathway for electrons. When embedded in the inner mitochondrial membrane, the alternative oxidase exists as a dimer, with the two subunits linked by disulfide bridges. These disulfide bridges may be oxidized or reduced. If they are reduced, then the alternative oxidase is in its higher-activity state, as opposed to the lower-activity state when the disulfide bridges are oxidized. In vitro the change from the oxidized to the reduced state can be brought about by isocitrate and other organic acids. Note that this is not an effect of the organic acids being used as a respiratory substrate (Vanlerberghe and McIntosh, 1997). Moreover, citrate accumulation enhances expression of the gene(s) that encode(s) the alternative oxidase (Vanlerberghe and McIntosh, 1996). K. Noguchi and co-workers (personal communication), grew plants of the shade species *Alocasia macrorrhiza* under low-light conditions, and the shade leaves exhibited slow respiration rates. They then transferred these plants to stressful high-light conditions which was associated with a change in the alternative oxidase from the low-activity oxidized state to the high-activity state. This shows that changes in redox state *can* play a role in vivo. Is this the rule, or rather an exception?

Electron partitioning has been studied in mitochondria isolated from both green and etiolated cotyledons as well as from roots of soybean (*Glycine max*), using isocitrate and DTT (dithiothreitol) to modify the redox state of the disulfide bond, and pyruvate as activator of the alternative pathway. These showed the importance of activation by both α-keto acids and the redox status of the regulatory sulfhydryl-disulfide system in regulating the flux through the alternative pathway (Ribas-Carbo et al., 1997). In mitochondria from green cotyledons, the alternative oxidase is poised to compete with the cytochrome pathway, and either enhanced reduction of the disulfide bonds or increased activation by α–keto acids shift electrons to the alternative pathway. In mitochondria from etiolated cotyledons, the level of alternative oxidase protein was too low to allow effective competition with the cytochrome pathway, even under fully activated conditions. In contrast, root mitochondria contained high levels of alternative oxidase protein, but it was predominantly in the oxidized form. It was conclude that the AOX protein must be both reduced (by the addition of DTT) and activated (by the addition of pyruvate) to compete with the cytochrome pathway in these mitochondria (Ribas-Carbo et al., 1997).

In roots of seedlings of *Glycine max* the activity of the alternative pathway was very low, and part of the enzyme was oxidized (Millar et al., 1998). Within a few days, the growth rate and the cytochrome oxidase activity declined about four-fold. That is, when the contribution of the alternative path to root respiration was 35 to 55%. At that stage, AOX was in its fully reduced state, suggesting that the change from partly oxidized to fully reduced was responsible for a small part of the increased alternative oxidase activity (Millar et al., 1998). In intact roots of *Poa annua* and several other grasses, however, the alternative oxidase was invariably in its reduced, higher-activity state (Millenaar et al., 1998, 2000). At present, there is no clear evidence that changes in redox state of the alternative oxidase generally play an important regulatory role in vivo (Hoefnagel and Wiskich, 1998). However, this may reflect the way plant scientists, in general, carry out their experiments. That is, we tend to grow our experimental plants under favorable control conditions, which might require fast alternative respiration rates and full activation of the alternative oxidase.

The alternative oxidase's capacity to oxidize its substrate (Q_r) also increases in the presence of pyruvate and other α-keto acids (Millar et al., 1996). The effect of pyruvate is mainly on V_{max}, rather than on the K_m for ubiquinol (Hoefnagel et al., 1997). As a result, in the presence of the potent activator pyruvate the alternative path shows significant activity, even when less than 30% of ubiquinone is in its reduced state. Note, again, that this is not because pyruvate is used as a respiratory substrate. In other words, the alternative pathway becomes active even when the cytochrome pathway is not fully saturated. In intact tissues pyruvate levels appear to be sufficiently high to fully activate the alternative oxidase. This suggests that changes in the level of keto-acids may not play a major regulatory role in vivo (Hoefnagel and Wiskich, 1998; Millenaar et al., 1998), but see Gastón et al. (2003).

Whenever the alternative oxidase is in its higher-activity state and active at low levels of Q_r, there is most probably competition for electrons between the two pathways, both in vitro (Hoefnagel et al., 1995; Ribas-Carbo et al., 1995) and in vivo (Atkin et al., 1995). Competition for electrons between the two pathways appears to be the rule, rather than an exceptional situation, as it was once believed.

3. Can We Really Measure the Activity of the Alternative Path?

The application of specific inhibitors of the alternative path shows that the alternative path does contribute to the respiration of roots and leaves of at least some species. The decline in respiration, however, upon addition of an inhibitor of the alternative path (e.g., SHAM) frequently underestimates the actual activity of the alternative path. Since the two pathways compete for electrons, the inhibition by SHAM is less than the activity of the alternative path. Thus, any observed inhibition of respiration following the addition of an alternative pathway inhibitor indicates that some alternative pathway activity was present prior to inhibition, but provides no quantitative estimate of its activity (Day et al., 1996).

Stable isotopes can be used to estimate alternative path activity without the complications caused by the use of inhibitors. The alternative oxidase and cytochrome oxidase discriminate to a different extent against the heavy isotope of oxygen (^{18}O) when reducing O_2 to produce water (Guy et al., 1989, 1992; Robinson et al., 1992, 1995; Ribas-Carbo et al., 1995, 1997; Chapter 3, Ribas-Carbo et al.). This allows calculation of the partitioning of electron flow between the two pathways in the absence of added inhibitors, in both isolated mitochondria and intact tissues.

The development of a simplified aqueous-phase system for on-line measurements of oxygen-isotope fractionation (Ribas-Carbo et al., 1995; Chapter 3, Ribas-Carbo et al.) allowed direct experimental verification that the alternative oxidase can compete for electrons with the cytochrome oxidase, overturning the longstanding paradigm that the alternative pathway only becomes engaged when the cytochrome pathway is saturated (Bahr and Bonner, 1973; Lambers, 1980; Moore and Siedow, 1991). In mitochondria isolated from green soybean (*Glycine max*) cotyledons, electrons were partitioned to the alternative pathway under state 3 conditions in the presence of pyruvate, i.e. where the ubiquinone pool was relatively oxidized and the cytochrome pathway not saturated. Furthermore, in state 4, the alternative pathway was carrying 45% of the electron transport; but addition of SHAM did not inhibit the respiration rate, indicating that those electrons were redirected to the unsaturated cytochrome pathway (Ribas-Carbo et al., 1995).

III. The Ecophysiological Function of the Alternative Pathway

Why should plants produce and maintain a pathway that supports nonphosphorylating electron transport in mitochondria? Are there situations where respiration in the absence of ATP production could serve important physiological functions? In this section we briefly discuss the merits of hypotheses to explain the presence of the alternative path in higher plants. Testing of these hypotheses will require the use of mutants lacking alternative path activity, some of which have been produced with molecular techniques.

It will also be necessary to assess alternative path activity in plants exposed to a range of environmental conditions, to enhance our understanding of situations that turn the alternative oxidase on or off. It will also require investigation of the activities of the different members of the AOX gene family. Since we now know that inhibitors do not provide reliable tools to assess alternative path activity, the isotope-fractionation technique (Chapter 3, Ribas-Carbo et al.) will be essential for such research. Many experiments carried out in the past two or three decades are worth revisiting using this new technique.

A. Heat Production

An important consequence of the lack of coupling to ATP production in the alternative pathway is that the energy produced by oxidation is released as heat. More than 200 years have passed since Lamarck described heat production in *Arum* and more than 60 years since thermogenesis was linked to cyanide-resistant respiration (Laties, 1997). This heat production is ecologically important in some flowers (Knutson, 1974). These inflorescences may expand in early spring when air temperatures are low and can 'melt' their way through late-lying snow. Preceding the upsurge in respiration, generally referred to as the 'respiratory crisis', salicylic acid accumulates which triggers the increase in respiration in some parts of the flower (Raskin et al., 1987). During the 'respiratory crisis' the respiration rate of the *Arum* spadix increases to very high levels, and its temperature rises to approximately 10 °C above ambient, so that odoriferous amines are volatilised, pollinators are attracted and rates of ovule and pollen development increase (Meeuse, 1975). During heat production the respiration of the spadix is largely cyanide-resistant. This contributes to heat production, as the lack of proton extrusion coupled to electron flow allows a large fraction of the energy in the substrate to be released as heat.

Heat production also occurs in the flowers of several South American *Annona* species, *Victoria amazonica* (Amazon water lily) and *Nelumbo nucifera* (sacred lotus), presumably also linked to activity of the alternative path. These flowers regulate their temperature with remarkable precision (Seymour et al., 1998). While the air temperature varies between 10 and 30 °C, the flowers remain between 30 and 35 °C The stable temperature is a consequence of increasing respiration rates in proportion to decreasing temperatures. This phenomenon of thermoregulation in plants is known for only two other species: *Philodendron selloum* and *Symplocarpus foetidus* (skunk cabbage: Knutson, 1974). It has been suggested that heat production in the sacred lotus is an energetic reward for pollinating beetles. These are trapped overnight, when they feed and copulate, and then carry the pollen away when the flower opens the following day (Seymour and Schultze-Motel, 1996).

Can the alternative oxidase also play a significant role in increasing the temperature of leaves, for example during exposure to low temperature? There is indeed some evidence for increased heat production (7–22% increase) in low-temperature resistant plants (Moynihan et al., 1995). However, such an increase in heat production *cannot* lead to a significant temperature rise in leaves (less that 0.1 °C; Breidenbach et al., 1997; Lambers et al., 1998b), and hence is unlikely to play a role in any cold-resistance mechanism. To explain the contribution of the alternative path in respiration of nonthermogenic organs other ecophysiological roles must be invoked.

Does the alternative path also play a role in the respiration of 'ordinary' tissues, such as roots and leaves? The application of specific inhibitors of the alternative path shows that the alternative path does contribute to the respiration of roots and leaves of at least some species. The discrimination technique has also shown that the alternative pathway may account for up to 50% of all respiration. If the role of the alternative path in nonthermogenic plants is not that of heat production, what might its role be?

B. The Alternative Pathway as an Energy Overflow

As pointed out above, the quantitative significance of the alternative path might increase when the production of organic acids is not matched by their oxidation, such that they accumulate. This observation led to the 'energy overflow hypothesis' (Lambers, 1980). It states that respiration via the alternative path only proceeds in the presence of high concentrations of respiratory substrate. It considers the alternative path as a 'coarse control' of carbohydrate metabolism, but not as an alternative to the finer control by adenylates (Secs. II.A and II.B).

The continuous employment of the alternative oxidase under normal 'nonstress' conditions might ensure a rate of carbon delivery to the root that enables the plant to cope with 'stress'. However, it has not been successfully demonstrated that 'stress' requires greater carbon demand. In fact if growth slows down under stress, the carbon requirements might actually decrease (Chapter 6, Flexas et al.). However, the demand for carbon or metabolic energy of a tissue may also change suddenly when a stress is imposed. For example, a decrease in rhizosphere water potential increases the roots' carbon demand for synthesis of compatible solutes for osmotic adjustment (Lambers et al., 1981). Similarly, attack by parasites and pathogens may suddenly increase carbon demands for tissue repair and the mobilization of plant defenses (Simons and Lambers, 1999; Simons et al., 1999). The alternative oxidase activity may also prevent the production of superoxide and/or hydrogen peroxide. Superoxide is produced when electron transport through the cytochrome path is impaired (e.g., due to low temperature or desiccation injury), and this is partly due to a reaction of ubisemiquinone with molecular oxygen (Purvis and Shewfelt, 1993). Superoxide, like other reactive oxygen species (ROS), can cause severe metabolic disturbances. So far, the various interpretations of the physiological function of an 'energy overflow' remain speculative.

C. Using the Alternative Pathway in Emergency Cases

The activity of the cytochrome path may be elevated upon increased availability of ADP, at the expense of the activity of the alternative path. Addition of nitrate to two-week-old *Pisum sativum* (pea) roots, grown without nitrate appeared to show this effect, but this interpretation should be checked, as it was based on inhibitor studies (De Visser et al., 1986). A sudden increase in energy demand for nitrate uptake should increase the concentration of ADP in the cell, thus increasing glycolytic activity. This then might lead to a greater input of electrons into the respiratory chain than could be accommodated by the cytochrome path, if this was already operating at its maximum capacity. As a result, the alternative path might become engaged. Because more electrons are fed into the mitochondrial electron-transport chain than can be accepted by the cytochrome path, this model was termed the 'energy overcharge model'. It could apply to nonsteady state conditions, as in the above experiments, and/or to conditions when the activity of the cytochrome path is controlled more by substrate supply than by adenylates.

D. NADH-Oxidation in the Presence of a High Energy Charge

If cells require a large amount of carbon skeletons (e.g., citrate or oxoglutarate) but do not have a high demand for ATP, then the operation of the alternative path could prove useful. However, can we envisage such a situation in vivo? Whenever the rate of carbon skeleton production is high, there tends to be a great need for ATP to further metabolize and incorporate these skeletons. When plants are infected by pathogenic microorganisms, however, they may produce phytoalexins which might well require engagement of the alternative path (Simons and Lambers, 1999; Simons et al., 1999). Also, in roots that exude vast amounts of carboxylates at a time when their ATP demand is low, enhanced alternative path activity might be required. This might account for increased expression of AOX in the cluster roots of *Lupinus albus* (Kania et al., 2003) and *Hakea prostrata* (Shane et al., 2004). Other examples include the infected cells in root nodules of legumes; however, these have less alternative path capacity than any other cells from the same nodules or other tissues of the same plants (Millar et al., 1997). Leaf guard cells also have the capacity to rapidly synthesize malate, which plays a role in stomatal opening. Guard cells have a very high respiratory capacity when compared to adjacent mesophyll cells. However, the cyanide-resistant component of respiration only constitutes about 38% of total respiration in the guard cells, as opposed to 66% in mesophyll cells (Vani and Raghavendra, 1994). Hence, rapid synthesis of organic acids in guard cells is probably not associated with greater alternative path capacity. However, since these experiments were performed with inhibitors, they may underestimate the contribution of the alternative pathway.

There may be a need for a nonphosphorylating path to allow rapid oxidation of malate in CAM plants during the day (Lance et al., 1985). Unfortunately, there are no techniques available to assess alternative path activity in the light. If measurements are made in the dark by briefly darkening tissues for up to one hour during the normal light period, then malate decarboxylation in CAM plants is indeed associated with increased engagement of the alternative path (Robinson et al., 1992). Malate decarboxylation, however, naturally occurs in the light. It therefore remains to be confirmed that the alternative path plays a vital role in Crassulacean acid metabolism.

E. Continuation of Respiration when the Activity of the Cytochrome Path Is Restricted

Naturally occurring inhibitors of the cytochrome path (e.g., cyanide, sulfide, carbon dioxide and nitric oxide) may reach such high concentrations in the tissue that respiration via the cytochrome path is partially or fully inhibited (Palet et al., 1991, 1992; Millar and Day, 1997).

Dry seeds, including those of species such as *Cucumis sativus*, *Hordeum vulgare*, *Oryza sativa* and *Xanthium pennsylvanicum* contain cyanogenic compounds, such as cyanohydrin, cyanogenic glycosides and cyanogenic lipids. Such compounds liberate free HCN by hydrolysis during imbibition. Upon imbibition and triggered by ethylene, seeds producing these cyanogenic compounds produce a mitochondrial β-cyanoalanine synthase that detoxifies HCN (Hagesawa et al., 1995). Despite this detoxifying mechanism, toxic levels of HCN might be present in the mitochondria of some tissue (Millenaar and Lambers, 2003).

Some plants produce sulfide (e.g., species belonging to the Cucurbitaceae; Rennenberg and Filner, 1983). Sulfide is also produced by anaerobic sulfate-reducing microorganisms. It may occur in high concentrations in the phyllosphere of aquatic plants or the rhizosphere of flooded plants. In such flooded soils, carbon dioxide levels also increase (Chapter 8, Colmer and Greenway).

Measurements with the oxygen-isotope-fractionation technique have also shown that some growth-inhibiting allelochemicals increase electron partitioning to the alternative pathway (Peñuelas et al., 1996). When green soybean (*Glycine max*) cotyledons were treated with cinnamic acid, the partitioning of electrons to the alternative pathway increased from 39 to 62%; there was also a slightly increased partitioning with α-pinene, but not with quercetin or juglone. However, this increase in partitioning to the alternative pathway was due to a reduction in the rate of cytochrome pathway respiration, rather than an absolute increase in alternative pathway activity.

Leaf respiration in the dark might be inhibited at elevated concentrations of CO_2 in the atmosphere, such as predicted to occur in the next century. Although some of these reports should be dismissed as artifacts, the phenomenon does exist (Drake et al., 1999; Chapter 13, Gonzàlez-Meler and Taneva). Such inhibition may be partly due to inhibition of the cytochrome path because in vitro cytochrome oxidase is inhibited by CO_2 concentrations in the range known to inhibit respiration of intact leaves (Gonzàlez-Meler et al., 1996; Chapter 13, Gonzàlez-Meler and Taneva). The presence of an alternative path, which is unaffected by inhibitors that block the cytochrome path, may allow continued respiration and ATP production, albeit with low efficiency, under such conditions.

In addition, when the activity of the cytochrome path is restricted by low temperature, the alternative path might increase in activity. In fact, exposure to low temperature enhances the amount of alternative oxidase in mitochondria of *Zea mays* (Stewart et al., 1990) and *Nicotiana tabacum* (Vanlerberghe and McIntosh, 1992). Such an induction is also achieved when the activity of the cytochrome path is restricted in other ways [e.g., by application of inhibitors of mitochondrial protein synthesis (Day et al., 1996), or of inhibitors of the cytochrome path (Wagner et al., 1992)]. Interestingly, only those inhibitors of the cytochrome path that enhance superoxide production lead to induction of the alternative oxidase.

In maize (*Zea mays*), total leaf respiration, measured during recovery at 25 °C, was not affected by a chilling treatment at 5 °C in either a chilling-sensitive or a chilling-resistant cultivar, but electron partitioning to the alternative pathway was significantly higher in the more stressed, chilling-sensitive cultivar (Ribas-Carbo et al., 2000). This suggests that the alternative pathway activity is related to the level of stress that the plant is subjected to, rather than to the cold treatment itself. The cytochrome pathway appears more sensitive to chilling stress than the alternative oxidase, and greater activity of the alternative pathway may be required to prevent over-reduction of the ubiquinone pool, and formation of reactive oxygen species.

Moreover, superoxide itself can also induce expression of the alternative oxidase. This has led to the suggestion that reactive oxygen species, including H_2O_2, are part of the signal(s) communicating cytochrome path restriction in the mitochondria to the nucleus, thus inducing alternative oxidase synthesis (Wagner and Krab, 1995). The key question is, of course, if enhanced expression of the alternative oxidase leads to greater activity of the alternative path. In *Vigna radiata* this appears to be the case, but such a response is not found in *Glycine max* (Gonzàlez-Meler et al., 1999) or in *Nicotiana tabacum* (Lennon et al., 1997).

The role of the alternative oxidase in protecting plants from over-reduction of the mitochondrial electron transport chain has been examined in tobacco plants treated with salicylic acid (Lennon et al., 1997). Salicylic acid (SA) is a signal molecule in systemic-acquired resistance, and induces the alternative oxidase protein in tobacco (*Nicotiana tabacum*). Treatment of tobacco leaves with SA had no effect on the overall respiratory rate, or on the relative contribution of either the cytochrome or alternative pathways to this respiration, despite a nine-fold increase in the level of AOX protein. The rate of cyanide-resistant oxygen uptake (V_{KCN}) increased (two-fold) with this increase in AOX protein, but the relationship was not linear. However V_{KCN} was equal to the total respiration rate, suggesting saturation of the respiratory capacity in these leaves. AOX protein also increased in tobacco leaves in response to infection by tobacco mosaic virus (TMV), in both inoculated and systemic leaves, and in leaves of *Arabidopsis thaliana*, infected by *Pseudomonas syringae* (Simons et al. 1999). AOX protein levels were four-fold higher in direct lesion areas and 1.5-fold higher in areas adjacent to lesions when compared to uninfected control leaves. Consequently, in this study, the increase in AOX protein did not affect either total respiration or partitioning between the two pathways in these infected leaves. The increase in AOX protein may be required during the oxidative burst that follows infection and for the hypersensitive response; in this case fractionation measurements would have to be made immediately following infection.

The effect of herbicides that inhibit branched-chain amino acid biosynthesis on alternative and cytochrome pathway respiration was recently examined in soybean roots (Gastón et al., 2003). Herbicide treatment resulted in a decrease in cytochrome pathway activity, whilst alternative pathway activity was maintained or increased. Although AOX protein increased in the herbicide-treated roots, the protein level did not correlate with alternative pathway activity. The pyruvate concentration was significantly higher in the herbicide-treated roots, and pyruvate may have acted as a substrate and/or as an allosteric activator of the enzyme. The alternative oxidase may thus be important in maintaining the carbon balance in stressed roots, and preventing pyruvate fermentation.

In the absence of an alternative oxidase, inhibition or restriction of the activity of the cytochrome path would inexorably lead to the accumulation of fermentation products, as found in transgenic plants lacking the alternative oxidase (Vanlerberghe et al., 1995). In addition, it might cause the ubiquinone pool to become highly reduced, which might well lead to the formation of reactive oxygen species and concomitant damage to the cell (Purvis and Shewfelt, 1993). Further work with genotypes lacking the alternative path is an essential avenue of future research on the ecophysiological role of the alternative path in plant function.

IV. Concluding Remarks

The regulation of the alternative pathway is complex, probably species dependent, and may relate to a number of factors including protein concentration, pyruvate concentration (cellular carbohydrates, adenylate control, ubiquinone reduction, etc.). However, evidence is accumulating to support a role of the alternative oxidase in the prevention of over-reduction of the ubiquinone pool and the formation of reactive oxygen species.

Acknowledgments

This work was partially supported by the Spanish Ministry of Science and Technology (Grant BFI2002-00772).

References

Andersson ME and Nordlund P (1999) A revised model of the active site of alternative oxidase. FEBS Lett 449: 17–22

Atkin OK, Villar R and Lambers H (1995) Partitioning of electrons between the cytochrome and the alternative pathways in intact roots. Plant Physiol 108: 1179–1183

Bahr JT and Bonner WD (1973) Cyanide-insensitive respiration. II. Control of the alternate pathway. J Biol Chem 248: 3446–3450

Berthold DA, Andersson ME and Nordlund P (2000) New insight into the structure and function of the alternative oxidase. Biochim Biophys Acta 1460: 241–254

Breidenbach RW, Saxton MJ, Hanson LD and Criddle RS (1997) Heat generation and dissipation in plants: can the alternative oxidative phosphorylation pathway serve a thermoregulatory role in plant tissues other than specialized organs? Plant Physiology 114: 1137–1140

Day DA and Hanson JB (1977) Pyruvate and malate transport and oxidation in corn mitochondria. Plant Physiol 59: 630–635

Day DA, Krab K, Lambers H, Moore AL, Siedow JN, Wagner AM and Wiskich JT (1996) The cyanide-resistant oxidase: To inhibit or not to inhibit, that is the question. Plant Physiol 110: 1–2

De Visser R, Spreen Brouwer K and Posthumus F (1986) Alternative path mediated ATP synthesis in roots of *Pisum sativum* upon nitrogen supply. Plant Physiol 80: 295–300

Dry IB, Moore AL, Day DA and Wiskich JT (1989) Regulation of alternative pathway activity in plant mitochondria. Non-linear relationship between electron flux and the redox poise of the quinone pool. Arch Biochem Biophys 273: 148–157

Drake BG, Azcon-Bieto J, Berry, J, Bunce J, Dijkstra P, Farrar J, Gifford RM, Gonzàlez -Meler MA, Koch G, Lambers H, Siedow J and Wullschleger S (1999) Does elevated atmospheric CO_2 concentration inhibit mitochondrial respiration in green plants? Plant Cell Environ 22: 649–657

Finnegan PM, Whelan J, Millar AH, Zhang Q, Smith MK, Wiskich JT and Day DA (1997) Differential expression of the multigene family encoding the Soybean mitochondrial alternative oxidase. Plant Physiol 114: 455–466

Gaston S, Ribas-Carbo M, Busquets S, Berry JA, Zabalza A and Royuela, M (2003). Changes in mitochondrial electron partioning in response to herbicides inhibiting branched-chain amino acid biosynthesis in soybean. Plant Physiol 133: 1351–1359

Gonzàlez-Meler MA, Ribas-Carbo M, Siedow JN and Drake BG (1996) Direct inhibition of plant mitochondrial respiration by elevated CO_2. Plant Physiol 112:1349–1355

Gonzàlez-Meler MA, Ribas-Carbo M, Giles L and Siedow JN (1999) The effect of growth and measurement temperature on the activity of the alternative respiratory pathway. Plant Physiol 120: 765–772

Guy RD, Berry JA, Fogel ML and Hoering TC (1989) Differential fractionation of oxygen isotopes by cyanide-resistant and cyanide-sensitive respiration in plants. Planta 177: 483–491

Guy RD, Berry JA, Fogel ML, Turpin DH, and Weger HG (1992) Fractionation of the stable isotopes of oxygen during respiration by plants—the basis of a new technique to estimate partitioning to the alternative path. In: Lambers H and Van der Plas LHW (eds) Molecular, Biochemical and Physiological Aspects of Plant Respiration, pp 443–453. SPB Academic Publishing, The Hague

Hagesawa R, Muruyama A, Nakaya M and Esashi Y (1995) The presence of two types of β-cyanoalanine synthase in germinating seeds and their response to ethylene. Physiol Plant 93: 713–718

Hoefnagel MHN and Wiskich JT (1998) Activation of the plant alternative oxidase by high reduction levels of the Q-pool and pyruvate. Arch Biochem Biophys 355: 262–270

Hoefnagel MHN, Millar AH, Wiskich, JT, and Day DA (1995) Cytochrome and alternative respiratory pathways compete for electrons in the presence of pyruvate in soybean mitochondria. Arch Biochem Biophys 318: 394–400

Hoefnagel MHN, Rich PR, Zhang Q and Wiskich JT (1997) Substrate kinetics of the plant mitochondrial alternative oxidase and the effects of pyruvate. Plant Physiol 115: 11145–1153.

Ito Y, Saisho D, Nakazone M, Tsutsumi N and Hirai A (1997) Transcript levels of tandem-arranged alternative oxidase genes in rice are increased by low temperature. Gene 203: 121–129

Kania A, Langlade N, Martinoia E and Neumann G (2003) Phosphorus deficiency-induced modifications in citrate catabolism and in cytosolic pH as related to citrate exudation in cluster roots of white lupin. Plant Soil 248: 117–127

Knutson R M (1974) Heat production and temperature regulation in eastern skunk cabbage. Science 186: 746–747

Lambers H (1980) The physiological significance of cyanide-resistant respiration in higher plants. Plant Cell Environ 3: 293–302

Lambers H, Blacquière T and Stuiver CEE (1981) Interactions between osmoregulation and the alternative respiratory pathway in *Plantago coronopus* as affected by salinity. Physiol Plant 51: 63–68

Lambers H, Scheurwater I, Mata C and Nagel OW (1998a) Root respiration of fast- and slow-growing plants, as dependent on genotype and nitrogen supply: a major clue to the functioning of slow-growing plants In: H Lambers, H Poorter and MMI Van Vuuren (eds) Inherent Variation in Plant Growth. Physiological Mechanisms and Ecological Consequences, pp 139–157. Backhuys, Leiden

Lambers H, Chapin FS III and Pons TL (1998b) Plant Physiological Ecology. Springer-Verlag, New York

Lance C, Chauveau M and Dizengremel P (1985) The cyanide-resistant path of plant mitochondria. In: R Douce and DA Day (eds) Encyclopedia of Plant Physiology, pp 202–247. Springer-Verlag, Berlin

Laties GG (1997) The discovery of the cyanide-resistant alternative path and its aftermath. In: Yang S-Y and Kung S-D (eds) Discoveries in Plant Biology. World Scientific Publishing Co., Hong Kong University of Science and Technology, Singapore

Lennon AM, Neueschwander UH, Ribas-Carbo M, Giles L, Ryals JA and Siedow JN (1997) The effects of salicylic acid and TMV infection upon the alternative oxidase of tobacco. Plant Physiol 115: 783–791

McCabe TC, Finnegan, PM, Millar AH, Day DA and Whelan J (1998) Differential expression of alternative oxidase genes in soybean cotyledons during postgerminative development. Plant Physiol 118: 675–682

Meeuse BJD (1975) Thermogenic respiration in aroids. Annu Rev Plant Physiol 26: 117–126

Millar AH and Day DA (1997) Nitric oxide inhibits the cytochrome oxidase but not the alternative oxidase of plant mitochondria. FEBS Lett 398: 155–158

Millar AH, Hoefnagel MHN, Day DA and Wiskich JT (1996) Specificity of the organic acid activation of the alternative oxidase in plant mitochondria. Plant Physiol 111: 613–618

Millar AH, Finnegan PM, Whelan J, Drevon J-J and Day DA (1997) Expression and kinetics of the mitochondrial alternative oxidase in nitrogen-fixing nodules of soybean roots. Plant Cell Environ 20: 1273–1282

Millar AH, Atkin OK, Menz RI, Henry B, Farquhar G and Day DA (1998) Analysis of respiratory chain regulation in roots of soybean seedlings. Plant Physiol 117: 1083–1093

Millenaar FF and Lambers H (2003) The alternative oxidase; in vivo regulation and function. Plant Biol 5: 2–15

Millenaar FF, Benschop J Wagner AM and Lambers H (1998) The role of the alternative oxidase in stabilizing the in vivo reduction state of the ubiquinone pool; and the activation state of the alternative oxidase. Plant Physiol 118: 599–607

Millenaar FF, Roelofs R, Gonzàlez-Meler MA, Siedow JN, Wagner AM and Lambers H (2000) The alternative oxidase during low-light conditions. Plant J 23: 623—632

Moore AL and Siedow JN (1991) The regulation and nature of the cyanide-resistant alternative oxidase of plant mitochondria. Biochim Biophys Acta 1059: 121–140

Moynihan MR, Ordentlich A and Raskin I (1995) Chilling-induced heat evolution in plants. Plant Physiol 108: 995–999

Palet A, Ribas-Carbo M, Argiles JM and Azcón-Bieto J (1991) Short-term effects of carbon dioxide on carnation callus cell respiration. Plant Physiol 96: 467–472

Palet A, Ribas-Carbo M, Gonzàlez-Meler MA, Aranda X and Azcón-Bieto J (1992) Short-term effects of CO_2/bicarbonate on plant respiration. In: H Lambers and LHW Van der Plas (eds) Molecular, Biochemical and Physiological Aspects of Plant Respiration, pp 597–602. SPB Academic Publishing, The Hague

Peñuelas J, Ribas-Carbo M and Giles L (1996) Effects of allelochemicals on plant respiration and oxygen isotope fractionation by the alternative oxidase. J Chem Ecol 22: 801–805

Poorter H, Van der Werf A, Atkin O and Lambers H (1991) Respiratory energy requirements depend on the potential growth rate of a plant species. Physiol Plant 83: 469--475

Purvis AC and Shewfelt RL (1993) Does the alternative pathway ameliorate chilling injury in sensitive plant tissues? Physiol Plant 88: 712--718

Raskin I, Ehmann A, Melander WR and Meeuse BJD (1987) Salicylic acid: A natural inducer of heat production in *Arum* lilies. Science 237: 1601–1602

Rennenberg H and Filner P (1983) Developmental changes in the potential for H_2S emission in cucurbit plants. Plant Physiol 71: 269–275

Ribas-Carbo M, Berry JA, Yakir D, Giles L, Robinson SA, Lennon AM and Siedow JN (1995) Electron partitioning between the cytochrome and alternative pathways in plant mitochondria. Plant Physiol 109: 829–837

Ribas-Carbo M, Lennon AM, Robinson SA, Giles L, Berry J and Siedow JN (1997) The regulation of the electron partitioning between the cytochrome and alternative pathways in soybean cotyledon and root mitochondria. Plant Physiol 113: 903–911

Ribas-Carbo M, Aroca R, Gonzalez-Meler MA, Irigoyen JJ and Sánchez-Díaz M (2000) The electron partitioning between the cytochrome and alternative respiratory pathways during chilling recovery in two cultivars of maize differing in chilling sensitivity. Plant Physiol 122: 199–204

Roberts JKM (1984) Study of plant metabolism in vivo using NMR spectroscopy. Annu Rev Plant Physiol 35: 375--386.

Roberts JKM, Wemmer D and Jardetzky O (1984) Measurements of mitochondrial ATPase activity in maize root tips by saturation transfer ^{31}P nuclear magnetic resonance. Plant Physiol 74: 632–639

Robinson SA, Yakir D, Ribas-Carbo M, Giles L, Osmond CB, Siedow JN and Berry JA (1992) Measurements of the engangement of cyanide-resistant respiration in the Crassulacean acid metabolism plant *Kalanchoë daigremontiana* with the use of on-line oxygen isotope discrimination. Plant Physiol 100: 1087–1091

Robinson SA, Ribas-Carbo M, Yakir D, Giles L, Reuveni Y and Berry JA (1995) Beyond SHAM and cyanide: Opportunities for studying the alternative oxidase in plant respiration using oxygen isotope discrimination. Aust J Plant Physiol 22: 487–496

Scheurwater I, Cornelissen C, Dictus F, Welschen R and Lambers H (1998) Why do fast- and slow-growing grass species differ so little in their rate of root respiration, considering the large differences in rate of growth and ion uptake? Plant Cell Environ 21: 995–1005

Scheurwater I, Clarkson DT, Purves J, Van Rijt G, Saker L, Welschen R and Lambers H (1999) Relatively large nitrate efflux can account for the high specific respiratory costs for nitrate transport in slow-growing grass species. Plant Soil 215: 123–134.

Seymour RS and Schultze-Motel P (1996) Thermoregulating lotus flowers. Nature 383: 305

Seymour RS, Schultze-Mote P and Lamprecht I (1998) Heat production by sacred lotus flowers depends on ambient temperature, not light cycle . J Exp Bot 49: 1213–1217

Shane MW, Cramer MD, Funayama-Noguchi S, Cawthray GR, Millar AH, Day DA and Lambers H (2004) Developmental physiology of cluster-root carboxylate synthesis and exudation in Harsh Hakea: Expression of phospho*enol*pyruvate carboxylase and the alternative oxidase. Plant Physiol 135: 459–560

Siedow, JN and Umbach AL (1995) Plant mitochondrial electron transfer and molecular biology. Plant Cell 7: 821–831

Siedow, JN and Umbach AL (2000) The mitochondrial cyanide-resistant oxidase: structural conservation amid regulatory diversity. Biochim Biophys Acta 1459: 432–439

Simons BH and Lambers H (1999) The alternative oxidase: is it a respiratory pathway allowing a plant to cope with stress? In: Lerner HR (ed) Plant Responses to Environmental Stresses: From Phytohormones to Genome Reorganization, pp 265–286. Plenum Press, New York

Simons BH, Millenaar FF, Mulder L, Van Loon LC and Lambers H 1999. Enhanced expression and activation of the alternative oxidase during infection of *Arabidopsis thaliana* with *Pseudomonas syringae* pv. tomato. Plant Physiol 120: 529–538

Stiles W and Leach W (1936) Respiration in Plants. Methuen and Co., London

Stewart CR, Martin BA, Reding L and Cerwick S (1990) Respiration and alternative oxidase in corn seedlings tissues during germination at different temperatures. Plant Physiol 92: 755–760

Van der Werf A, Welschen R and Lambers H (1992) Respiratory losses increase with decreasing inherent growth rate of a species and with decreasing nitrate supply: A search for explanations for these observations. In: H Lambers and LHW Van der Plas (eds) Plant Respiration. Molecular, Biochemical and Physiological Aspects, pp 421–432. SPB Academic Publishing, The Hague

Vani T and Raghavendra, S (1994) High mitochondrial activity but incomplete engagement of the cyanide-resistant alternative pathway in guard cell protoplasts of pea. Plant Physiol 105: 1263--1268

Vanlerberghe GC and McIntosh L (1992) Lower growth temperatures increase alternative oxidase protein in tobacco callus. Plant Physiol 100: 115–119

Vanlerberghe GC and McIntosh L (1996) Signals regulating the expression of the nuclear gene encoding alternative oxidase of plant mitochondria. Plant Physiol 111: 589–595

Vanlerberghe GC and McIntosch L (1997) Alternative oxidase: From gene to function. Annu Rev Plant Physiol Plant Mol Biol 48: 703–734

Vanlerberghe GC, Day DA, Wiskich JT, Vanlerberghe AE and McIntosh L (1995) Alternative oxidase activity in tobacco leaf mitochondria. Plant Physiol 109: 353–361

Wagner AM and Krab K (1995) The alternative respiration pathway in plants: Role and regulation. Physiol Plant 95: 318--325

Wagner AM, Van Emmerik WAM, Zwiers JH and Kaagman HMCM (1992) Energy metabolism of *Petunia hybrida* cell suspensions growing in the presence of antimycin A In: H Lambers and LHW Van der Plas (eds) Molecular, Biochemical and Physiological Aspects of Plant Respiration, pp 609--614. SPB Academic Publishing, The Hague

Whelan J, Millar AH and Day DA (1996) The alternative oxidase is encoded in a multigene family in soybean. Planta 198: 197–201

Chapter 2

Calorespirometry in Plant Biology

Lee D. Hansen* and Richard S. Criddle
Department of Chemistry and Biochemistry, Brigham Young University, Provo, UT 84602, U.S.A.

Bruce N. Smith
Department of Plant and Animal Sciences, Brigham Young University, Provo, UT 84602, U.S.A.

Summary

Calorespirometry is a means for understanding how plants adapt and acclimate metabolically to their environment. Analysis of the energetics of respiration shows that measurements of metabolic heat and CO_2 rates by calorespirometry, combined with estimates of substrate and biomass composition, are sufficient to calculate substrate carbon-conversion efficiencies, anabolic rates or rates of growth and development, and relative activities of metabolic paths. Calorespirometric measurements can thus be used to rapidly investigate the responses of plant growth and metabolism to varying conditions and to compare the responses of species and genotypes. Calorespirometric and calorimetric methods have been used to determine the temperature dependence of growth rate, temperature limits for growth, the kinetics of both chilling- and high-temperature responses, and the effects of toxins and nutrient deficiencies.

*Author for correspondence, email: Lee.Hansen@BYU.edu

H. Lambers and M. Ribas-Carbo (eds.), Plant Respiration, 17–30.

I. Introduction

Determining how plant metabolism adapts or acclimates to the diversity of plant habitats is a major challenge for plant physiologists. Some qualitative differences, e.g. C_3 and C_4 photosynthesis, are widely recognized, but the quantitative differences that exist even over short distances and small changes in environment have been difficult to measure. For example, a single hillside in northern Utah with less than 80 meters rise in elevation has two subspecies of sagebrush, *Artemesia tridentata* Nutt. ssp. *tridentata* at the low end of the hillside and *Artemesia tridentata* Nutt. ssp. *vaseyana* at the upper elevation with a stable hybrid swarm at mid elevations (Freeman et al., 1991; 1999; Wang et al., 1997; 1998). Plants or seeds taken from any of the three elevations and planted at other elevations fail to thrive and eventually die without a clear cause. Movement of soil among the gardens does not affect the result. Standard measures of the physiology of plants from the three areas, e.g. gas exchange, composition, predation, etc., have all failed to identify any correlation with abiotic characteristics of the three test sites that can account for the observations. Calorespirometric measurements however demonstrate clear differences among the three plant populations (Smith et al., 1999a, 2002) which are related to their survival at each site. The ability of calorespirometry to measure differences in respiratory properties indistinguishable by any other means thus can make a unique contribution to our understanding of adaptation and acclimation of plant metabolism to the environment.

Animal and microbial scientists have long relied on calorespirometric methods to provide a quantitative understanding of respiratory physiology (Gabriel, 1955). However, despite the unique contributions that calorimetric methods can and have made to understanding of plant respiration (Criddle et al., 1991; Hansen et al., 1995; Criddle and Hansen, 1999), plant scientists have been slow to adopt calorimetric methods. The first calorimetric measurements on plant respiration were done about 200 years ago, but until recently, calorimetric methods have seldom been used to study plants. One reason for this has been a focus on photosynthesis as the key property of plants, and calorimetric measurements on photosynthesizing systems are difficult because of the large background energy input from the light source (Johansson and Wadsö, 1997). Also, because of the equivalence between net photosynthesis and growth (Demetriades-Shaw et al., 1992; Hansen et al., 1998), photosynthesis has often been viewed as the controlling factor in plant growth rather than respiration-linked anabolic processes. This has led to widely accepted, respiration-based growth models that fit existing data (Amthor, 2000), but have little explanatory power in terms of the biochemistry and metabolic physiology of plants (Hansen et al., 1998, 2002a).

Respiration models have been limited by lack of a convenient method for measuring the substrate carbon-conversion efficiency. Large amounts of data are available on respiration rates, but very few data are available on directly measured efficiencies. Single measures of respiration rates, i.e. rates of CO_2 production and rates of O_2 uptake, have frequently been found to be correlated with growth rates (Amthor, 2000). However, the correlation is sometimes positive, sometimes negative, and sometimes nonexistent, and little is known about how the relation between metabolism and growth changes with experimental conditions. No single measure of respiration rate provides sufficient information to define the relation between metabolic and growth characteristics because growth rate depends on both the rate of respiratory metabolism and on the metabolic pathways (including futile pathways) that determine what fraction of the substrate carbon is incorporated into growth. When a large fraction of the substrate is lost as CO_2, i.e. when efficiency is low, a rapid respiration rate may still give slow growth rates. For rapid growth, both rate and efficiency must be high.

To fully understand the relation between metabolism and growth, both rate and efficiency must be measured. Calorespirometry provides a rapid and convenient means for simultaneously measuring rate and efficiency as functions of temperature. However, a model based on both plant biochemistry and thermodynamics is necessary. Otherwise, calorimetry would be just another method for measuring respiration rates, albeit a particularly useful one for measuring rates as a function of temperature. The model derived here shows that calorespirometric measurements are

Abbreviations: C_{AP} – anabolic products; C_S – substrate from a C-mole of substrate; q – heat; R_{AP} – specific rate of production of anabolic products; R_{CO_2} – specific rate of CO_2 production; R_q – specific rate of heat production; ΔH_B – enthalpy change for formation of a C-mole of anabolic products ; ΔH_{CO_2} – enthalpy change for combustion of C_s per mole of CO_2; ε – substrate carbon conversion efficiency; γ_S – oxidation state of carbon in C_S; μW – microwatt or microjoule s^{-1}

necessary to understand plant growth energetics, and hence the physiological ecology of plants.

The process of plant growth cannot be fully comprehended without a fundamental understanding of how energy from respiration is used in processing photosynthate and nutrients into new tissues. Such an understanding requires calorimetric measurements of energy losses, but the general unfamiliarity of plant physiologists with calorimetric methods and the thermodynamic models needed to extract meaningful information on plant physiology from calorespirometric data has made it difficult. However, the enthalpy and mass balance equations necessary for describing growth processes are readily derivable from the law of conservation of mass, the first law of thermodynamics and Thornton's rule. (This rule states that a near-constant amount of heat is produced per mole O_2 consumed during oxidation of organic substances (Battley, 1999; Hansen et al., 2004).) The model presented here consists of only four equations, but is an accurate description of the relation between plant respiratory metabolism and growth properties.

Calorimetry provides a means for measuring the energy lost as heat during respiratory metabolism in plant tissues, and thus enables quantitative examination of total energy balance. When calorimetric measurements of heat rate are combined with measurements of the rate of CO_2 production, the method is called calorespirometry, and energy use-efficiency and anabolic rates or growth rates can be calculated. When heat and CO_2 rates are measured for different plants or tissues over a range of temperatures and other growth conditions, responses of growth, efficiency, and activities of metabolic pathways to biotic and abiotic variables can be compared (Hansen et al., 2004). This chapter describes calorespirometric and data-analysis methods currently in use, and the unique information concerning plant respiration and growth that can be determined from calorespirometric measurements.

II. Methods for Calorespirometry Studies, Including an Illustrative Example

Any live plant material that can be fit into the measuring chamber of a heat conduction or power compensation calorimeter can be studied by calorespirometry. These two types of calorimeters measure instantaneous heat rates directly. Measuring chambers of calorimeters that have been used for determining metabolic rates of various organisms range from about 1 mL to room size (Kemp, 1999). Studies have been done on plant materials ranging from cells in culture to tissue sections to whole seedlings and whole heads of cauliflower. Because the time per measurement increases with sample size, most measurements have been done in small (1–2 mL), sealed-ampoule chambers.

Tissue culture studies require sterile conditions. Other tissues need not be sterile, but must be clean and relatively free from bacteria and fungi whose metabolism would interfere with measurement of tissue metabolism. Microbial contamination is generally not a problem in studies of leaf and stem tissue, but is more likely to be a problem with roots grown in media that cannot be cleanly removed. Sand culture of roots generally, but not always, produces tissue samples with acceptably low levels of contamination. Uncontaminated samples of most plant tissues produce a constant rate of heat output for several hours. Bacterial contamination is generally evident from an exponential increase in the heat rate as microbial cells replicate during an experiment. Loss of tissue viability or depletion of substrate is usually evident as a rapidly changing heat rate (Criddle and Hansen, 1999). Tissues of most plants are stable enough that cuttings can be taken and stored or transported for two to four days without any significant change in respiratory properties.

Because calorimeter ampoules must be sealed to prevent water loss, the build-up of toxic products or metabolic activators (Hansen and Criddle, 1989), tissue instability (Hemming, 1995), or temperature sensitivities (Smith et al., 1999b) can cause unstable heat rates. Wounding responses from cutting tissues are also evident as rapidly changing heat rates, but this response has been observed in only a very few cases. Plant species and tissues that have exhibited an unstable heat rate due to wounding include soybean (*Glycine max*) leaves (Hemming, 1995), cauliflower (*Brassica oleracea*) flowerettes (Hansen and Criddle, 1989), voodoo lily (*Sauromatum guttatum*) appendix tissue (Lytle, et al., 2000), potato tuber (*Solanum tuberosum*) tissue (Smith, et al., 2000) and a water fern (*Azolla mexicana*). Wound responses to cutting during preparation for loading into the calorimeter have not caused a measurable effect on metabolic activities of tissues from more than 100 other species studied to date (Criddle and Hansen, 1999 and references therein). The possibility of wound responses can be routinely checked by comparing a sample cut once with the same sample cut multiple times.

To illustrate the methods employed for plant metabolic calorimetry and the information that can be obtained by calorimetric methods, we selected, at random, a one-day study on growing shoots from two accessions of *Atriplex confertifolia*, one from Montana and one from North Dakota. The plants used in these experiments grew wild at their native sites. Young shoots were cut from the plants, cooled to about 5 °C, placed in open plastic bags, and transported to the calorimetry laboratory in Provo, Utah, USA. (Different accessions, cultivars, species, etc. can also be grown in chambers at various conditions to quantify acclimation effects or grown in common gardens to isolate genetic effects from acclimation.) Small segments of young growing leaves (100 to 150 mg fresh weight) were excised with a razor blade, weighed, and sealed into 1mL Hastelloy ampoules of commercially available calorimeters (Calorimetry Sciences Corporation, American Fork, Utah, USA Model 4100) for heat rate measurements at selected temperatures. Addition of water, buffer, or other reagents to the tissue samples was not necessary for these studies. Four calorimeters were used in the study, each holding three replicate samples of each accession in three separate measuring chambers. Two calorimeters were used to measure heat rates starting at 20 °C and proceeding upward at 5 °C intervals. Two were used for measurements from 20 °C downward at 5 °C intervals. At the end of the experiment, tissue samples were dried in a vacuum oven at 70 °C for at least 24 hours, then weighed. A vacuum oven is preferred for drying samples because less oxidation occurs and samples are less friable. At these conditions, weights of small tissue samples become constant after 24 hours.

Figure 1 is a schematic representation of the plant tissue sample, sample ampoule, calorimeter measuring chamber, and calorimeter used for these studies. References to, and construction details of, several sizes of calorimeters for measurements of metabolic heat rates may be found in Kemp (1999). Calorimeter ampoules must usually: a) be sealed to prevent the large heat effects of water evaporation, b) be capable of holding sufficient tissue to produce ample heat rate for accurate measurement, and c) have adequate head space for O_2 to support metabolism. An optimum calorimeter for most work on plant tissues is a heat conduction or power compensation calorimeter with multiple, removable, 1–2 cm^3 sample ampoules, baseline reproducibility of 3–5 μW, the ability to rapidly change temperature (~1 °C min^{-1}) and stabilize at the new temperature in a short time, and be capable of covering the necessary temperature range (0 to 50 °C) (Criddle and Hansen, 1999). Typical equilibration times for these types of calorimeters with 1–2 mL cells are 15–30 minutes. Baseline reproducibility does not need to be better than about 1% of the total expected heat rate, since that is about the reproducibility of tissue sample metabolism.

The calorimeters used in most of our experiments have three, 1-mL sample ampoules and one reference ampoule. Approximately 20 minutes are required for the ampoules and samples to reach thermal steady-state after insertion into the calorimeter. After this time, heat rates are recorded every minute or two for another 10-20 minutes (Fig. 2 shows an example of the data collected in the illustrative study). Then, the ampoules are removed from the calorimeter, the lids opened, a small, open vial containing 0.4 M NaOH is inserted, and the heat rate measured again. The CO_2 produced by respiration in the sealed ampoule reacts with the NaOH to form carbonate via the reaction

$$CO_2\,(g) + 2OH^-\,(aq) = CO_3^{2-}\,(aq) + H_2O \qquad (1)$$

which has an enthalpy change of –108.5 kJ mol^{-1} with 0.4 M NaOH (Criddle and Hansen, 1999). This concentration of NaOH has approximately the same water activity as plant tissues, thus minimizing transport of water between the tissue and the solution. A 40 μL volume of 0.4M NaOH is more than adequate to absorb all the CO_2 produced during measurements in 1 mL ampoules (there are 16 μmoles of NaOH in 40 μL of 0.4 M solution, and less than 4 μmole of CO_2 is produced during a typical measurement). The heat rates recorded under these conditions represent the sum of the rates of metabolic heat and heat from carbonate formation. Subtraction of the heat rate measured without NaOH yields the heat rate from carbonate formation in μW or μJ s^{-1}, and, when divided by the negative of the enthalpy change for reaction (1), i.e. 108.5 μJ $nanomole^{-1}$, the rate of CO_2 production. Once again, the sample ampoules are removed, the vials of NaOH are taken out, and heat rates measured for a third time. This repeat measurement tests the stability of the rate of heat production by the tissue over the time of the experiment and allows correction for any small changes in activity.

An instrument baseline constant determined with empty ampoules has been subtracted from the data in

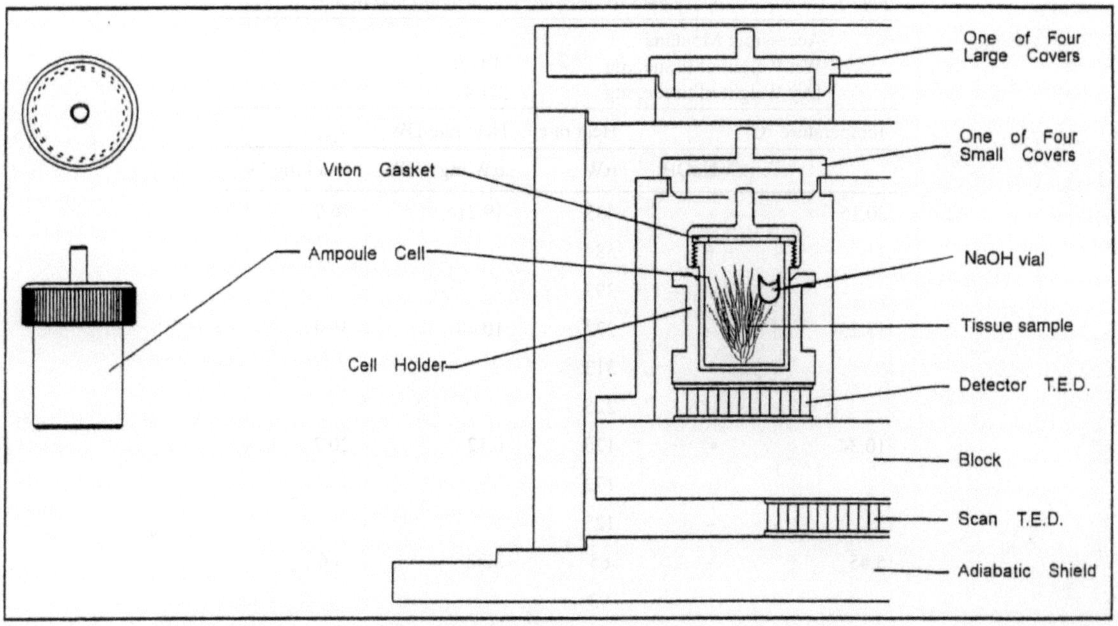

Fig. 1. Schematic of calorimeter with plant tissue and NaOH vial in ampoule. The diagram shows one of four measuring chambers in the Calorimetry Sciences Corp. (American Fork, UT) Model 4100 DSC. The T.E.D. (thermoelectric devices) are bismuth telluride thermopiles, the 1-mL cylindrical ampoule is made from Hastelloy, the vial is made of glass, the block and remainder of construction is nickel-plated aluminum. The scan T.E.D. is used to change the temperature of the block, to scan temperature up or down, or to hold the block at a constant temperature.

Fig. 2. Note that rates are measured directly, i.e. in μW or μJ s^{-1}. To analyze the data in Fig. 2, heat rates are taken at ca. 33 minutes (318 μW), at ca. 70 minutes (410 μw), and at ca. 111 minutes (293 μw). The heat rates at 33 and 111 minutes are averaged to give the metabolic heat rate at 70 minutes (306 μw), which is then subtracted from the heat rate measured at 70 minutes to give the heat rate (104 μW) for the CO_2 reaction. In studies of tissues cut from roots, shoots and leaves of hundreds of plant species, nearly all maintained relatively constant metabolic heat rates for the duration of the experiment, about 2 hours at each temperature for a total time of 6 to 8 hours. If metabolic activity of tissue in the calorimeter changes rapidly (i.e. from dying tissues, bacterial contamination or wounding response), the results are considered invalid. The heat rate data in Fig. 2 are decreasing somewhat faster than is usual, i.e. ca. 25 μW or 8% in 78 minutes, an indication that the tissue is stressed. To avoid tissue damage caused by stress, measurements are made beginning from a non-stress temperature and working toward stressful temperatures. Fresh samples could be used at each temperature, but this is usually not necessary and

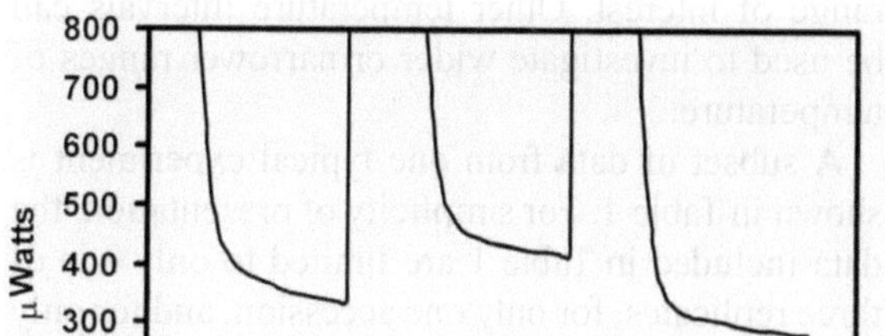

Fig. 2. A sample data set showing calorimetric determination of R_q and R_{CO_2}. Data points are at 2 minutes intervals. Data have been corrected for a constant instrument baseline measured with empty ampoules.

greatly increases the necessary effort.

Upon completion of one series of (- + - NaOH) measurements (e.g., as in Fig. 2) on a set of samples, the calorimeter temperature is changed, and measurements are repeated on the same samples. Temperature effects are routinely investigated at 5 °C intervals over

Table 1. Calorespirometry data for *Atriplex confertifolia* leaf tissues

Accession: Montana
Wet Weight of tissue, mg 148.0
Dry Weight of tissue, mg 21.4

Temperature °C		Heat rate	Heat rate/DW	R_{CO2}
	NaOH	μW	μW mg^{-1} DW	pmol mg^{-1} s^{-1}
20.15	–	433	19.21	76.7
	+	589		
	–	393		
15.25	–	232	10.68	36.4
	+	313		
	–	225		
10.34	–	125	6.12	20.2
	+	172		
	–	125		
5.45	–	65	3.32	8.6
	+	93		
	–	72		

the common growth temperature ranges encountered by the plants. Thus, a set of data from one calorimeter typically includes heat rates and CO_2 rates for three replicate samples over at least part of the temperature range of interest. Other temperature intervals can be used to investigate wider or narrower ranges of temperature.

A subset of data from one typical experiment is shown in Table 1. For simplicity of presentation, the data included in Table 1 are limited to only one of three replicates, for only one accession, and for only four temperatures. Though only a small fraction of the data collected in one day, this abbreviated data set illustrates the parameters that are directly measured by calorespirometry. Note that heat and CO_2 data are both measured directly as rates. Figures 3 and 4 show plots of all the data on heat rate and CO_2 rate with temperature taken in the one day study. As expected, R_q and R_{CO_2} change with temperature, but it is informative that these changes do not occur in parallel. Because the slopes differ, the ratio R_q/R_{CO_2} also changes with temperature (Fig. 5).

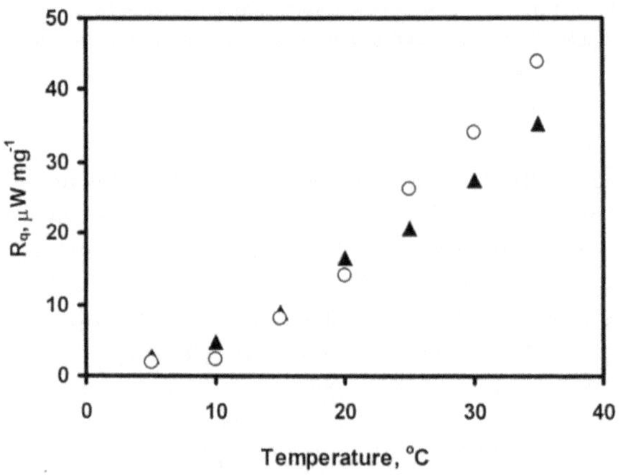

Fig. 3. Metabolic heat rate, R_q, versus temperature for two accessions of *Atriplex confertifolia,* circles are for the North Dakota accession and triangles are for the Montana accession.

III. Plant Growth Model

The ratio of heat rate (R_q) divided by the rate of CO_2 production (R_{CO_2}), or (R_q/R_{CO_2}), has units of energy (kJoules per mole), and is useful as an indication of efficiency. In simple descriptive terms, this ratio equals the amount of heat lost from the plant or tissue per mol of CO_2 produced. For plant tissues, R_q/R_{CO_2} values typically range from about 200 to 500 kJoules per mole^{-1}. Energy produced via catabolism in growing plants and used to drive anabolic reactions is largely lost as heat. Thus, small values for R_q/R_{CO_2} indicate that relatively little energy from catabolism is lost to the environment per unit of growth or of

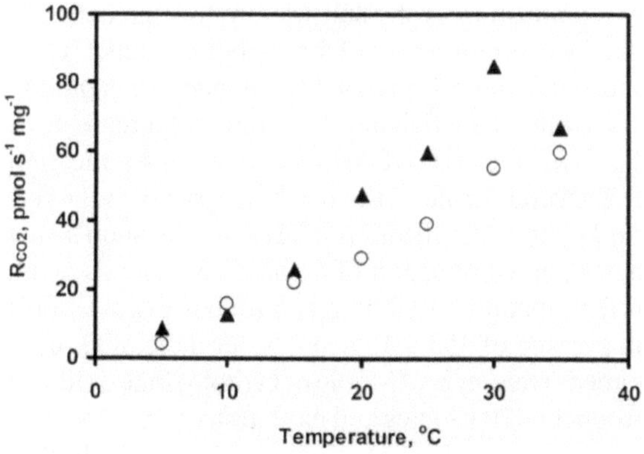

Fig. 4. Respiratory CO_2 rate, R_{CO_2}, versus temperature for two accessions of *Atriplex confertifolia*, triangles are for the North Dakota accession and circles are for the Montana accession.

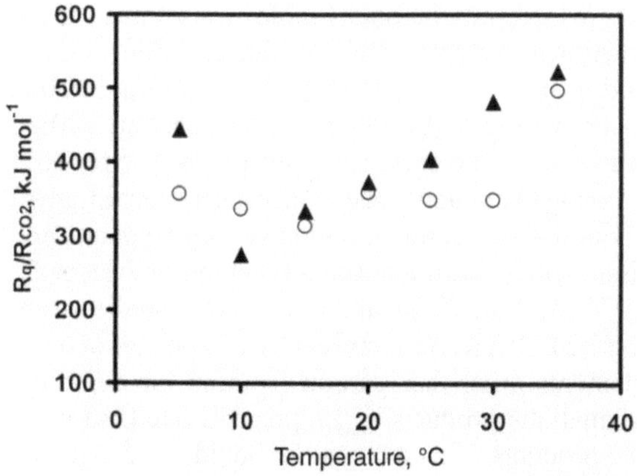

Fig. 5. The ratio, R_q/R_{CO_2}, versus temperature for two accessions of *Atriplex confertifolia*, triangles are for the North Dakota accession and circles are for the Montana accession.

anabolic reaction. In contrast, plants having a large R_q/R_{CO_2} lose much more energy per unit of growth. Clearly, an inverse relation exists between R_q/R_{CO_2} and efficiency in plant tissues with aerobic metabolism. In vitro heats of combustion measurements of carbohydrates yield 470 kJoules per mole CO_2 formed. Thus, a value of R_q/R_{CO_2} equal to 470 kJoules per mole CO_2 in aerobic metabolism of photosynthate would indicate zero efficiency, i.e. no growth, while values less than 470 kJoules per mole indicate a positive efficiency and values greater than 470 kJoules per mole indicate inclusion of reduced substances in substrate oxidized.

Furthermore, Fig. 5 includes plots of R_q/R_{CO_2} for two accessions of *Atriplex confertifolia* from different locations. Comparison of the data for the two accessions shows that R_q/R_{CO_2}, and thus efficiency, changes differently with temperature for the two accessions. Moreover, the temperatures of maximum efficiency of the two accessions differ. Several other studies (Criddle and Hansen, 1999; Smith, et al., 1999b) have shown that both cultivated and native species are adapted to maximize efficiency and growth rate at the temperature encountered most frequently during their growth season (the timing of which is often controlled by factors other than temperature, e.g., water).

These data show calorespirometric measurements can be used to identify conditions in which individual genotypes or phenotypes have low values of R_q/R_{CO_2} and thus optimum efficiency for conversion of photosynthate into plant biomass. Calorespirometry thus obviates the requirement for attempts to artificially partition respiration into ill-defined compartments such as growth and maintenance respiration, or to make growth measurements seeking empirical correlations that may predict effects of conditions or genotypes on growth. Moreover, since R_q/R_{CO_2} reflects the ratio of catabolism to anabolism, it also contains quantitative information on metabolic pathways (Hansen et al., 2004). Extracting this quantitative information requires a model of the biochemistry and thermodynamics of the growth processes.

The relation between growth and efficiency and calorespirometric data can be quantified with a relatively simple thermodynamic model of rates of energy and mass flows during plant growth. Development of this thermodynamic model for calorespirometric analysis of plant growth processes has been previously published (Criddle and Hansen 1999 and refs therein). Only four equations are needed to express plant growth rates quantitatively in terms measurable by, or calculated from, calorespirometric data. We divide respiration-linked growth metabolism into two parts, (a) the oxidative metabolism of a portion of the photosynthate via oxygen-consuming catabolic reactions to provide the energy or driving force for anabolic reactions, and (b) the anabolic reactions that result in growth. The rate of growth, broadly defined to include changes in composition as well as increases in mass or volume, is directly proportional to the rate of anabolism. The combined catabolic and anabolic processes of aerobic respiration are represented by the abbreviated chemical Eq. (2), the first of our four equations,

$$C_S + xO_2 + (N, P, K, etc.) \rightarrow \varepsilon\, C_{AP} + (1 - \varepsilon)CO_2 \quad (2)$$

where C_S represents one C-mole of substrate, ε is the fraction of substrate converted into anabolic products and (1 – ε) is therefore the fraction of substrate carbon lost as CO_2. ε is (by definition) the substrate carbon conversion efficiency. The value of ε is zero for non-growing plants and tissues and reaches a maximum of about 0.85 in rapidly growing vegetative plants and tissues under ideal conditions (Penning de Vries et al., 1974; Amthor, 2000, and unpublished studies from this lab). N, P, K, etc. represent the compounds of those elements other than carbon required for ε C-mole of anabolic products. C_{AP} represents one C-mole of the products of anabolism. It should be recognized that Eq. (2) is simply a statement of conservation of mass written in terms of plant substances. Equation (2) omits the details of the multiple processes involved in growth, but it is nonetheless accurate. The coefficients of Eq. (2) (i.e. the variables x and ε) depend on growth conditions (e.g., temperature, nutrients and water) and the nature of substrates and products incorporated into biomass. The dependence on conditions increases the complexity, but also allows quantifying the responses of growth processes to conditions by determination of ε and x under a variety of conditions.

From Eq. (2), the ratio of the rate of formation of anabolic products to the rate of CO_2 production is given by the ratio of the coefficients on C_{AP} and CO_2. Thus

$$R_{AP}/R_{CO_2} = \varepsilon/(1-\varepsilon) \text{ or } R_{AP} = R_{CO_2}[\varepsilon/(1-\varepsilon)] \quad (3)$$

where R_{AP} is the rate of formation of anabolic products and R_{CO_2} is the rate of production of CO_2. Equation 3 provides a relation giving the rate of formation of anabolic products, or rate of growth in terms of R_{CO_2}, and the substrate carbon conversion efficiency, ε.

From conservation of energy, the total rate of heat production, R_q, is the sum of the rates of heat produced by catabolism ($-R_{CO_2}\Delta H_{CO_2}$) and by anabolism ($-R_{AP}\Delta H_B$).

$$R_q = -R_{CO_2}\Delta H_{CO_2} - R_{AP}\Delta H_B \text{ or}$$

$$R_{AP} = (-R_{CO_2}\Delta H_{CO_2} - R_q)/\Delta H_B \quad (4)$$

The negative signs arise because of opposite sign conventions on heat (q) and enthalpy (H), H is negative and q is positive for exothermic processes. ΔH_{CO_2} is equal to the heat of combustion of the substrate (C_S) per C-mole and ΔH_B is equal to the enthalpy change for combustion of the substrate minus the enthalpy change for combustion of the anabolic product (C_{AP}), both per C-mole. Equation (4) thus allows calculation of the rate of anabolism, R_{AP}, from measurements of R_q, R_{CO_2}, and ΔH_B, if ΔH_{CO_2} is known. Application of Thornton's rule to the catabolic part of Eq. (2) is the key to quantification of ΔH_{CO_2}. Thornton's rule states that combustion of organic compounds (such as the substrates used for respiration in plants) yields an average of 455 ± 15 kJoules per mole of O_2 consumed. Thus, every C-mole of carbohydrate oxidized produces 470 kJoules and oxidation of a C-mole of a typical fatty acid to CO_2 produces about 650 kJoules. ΔH_{CO_2} thus can be approximated as $(-455)(1-\gamma_S/4)$ where γ_S is the average chemical oxidation state of carbon in the substrate (note that $\gamma_S = 0$ for carbohydrates, –1.8 for lipids, and –1.0 for proteins).

Combining Eqs. (3) and (4) to eliminate R_{AP} gives an equation for efficiency,

$$\varepsilon/(1-\varepsilon) = [(-R_q/R_{CO_2} - \Delta H_{CO_2})/\Delta H_B$$

$$\text{or } R_q/R_{CO_2} = -\Delta H_{CO_2} - \Delta H_B[\varepsilon/(1-\varepsilon)] \quad (5)$$

which allows calculation of ε from measurable or known quantities. R_q, R_{CO_2}, and ΔH_B are all measurable and, assuming the substrate is carbohydrate, $\Delta H_{CO_2} = -470$ kJoules per mole CO_2. Note that using the exact value for carbohydrates is usually more accurate than using the average value of Thornton's constant. However, approximating ΔH_{CO_2} in Eqs. (4) and (5) with $\gamma_S = 0$ generally causes negligible error except under highly stressful conditions or in tissues or plants where the substrate being oxidized includes compounds other than carbohydrates. Changes in efficiency as well as changes in substrate from carbohydrates are indicated by changes in R_q/R_{CO_2}.

Changes in R_q/R_{CO_2} from around 200 kJoules $mole^{-1}$ in young, rapidly growing, unstressed plants and tissues where a large fraction of the respiratory energy is stored in biomass, to values of 470 kJoules $mole^{-1}$ or greater in highly stressful conditions where the rate of energy accumulation is zero, represent very large changes in growth efficiency. Such changes are common in response to stresses from salt, hot and cold temperatures, toxins, tissue age, and other factors (e.g., see Fig. 5). R_q/R_{CO_2} is also increased by depletion of carbohydrates in tissues stored too long in the dark, but such increases are rarely seen in the 6–8 hours typically used to make calorespirometric measurements.

Four independent methods for determining ΔH_B (i.e. heats of combustion, elemental composition, growth rate, and substrate carbon conversion efficiency) have recently been shown to produce comparable results (Ellingson et al., 2002). These measurements of ΔH_B together with comparison of R_{AP} calculated from calorespirometric data with directly measured growth rates (Taylor et al., 1998) and temperature responses (Criddle et al., 1997, 2001; Smith et al., 1999b) have demonstrated the accuracy of the model embodied in Eqs. (2–5). Determination of ΔH_B requires specialized equipment, but fortunately it is not always necessary to measure ΔH_B. While growth rates and efficiencies may change markedly with small changes in growth conditions, plant composition, and therefore ΔH_B, generally does not. Moreover, ΔH_B values are not highly variable within classes of similar plants. Large numbers of published values of ΔH_B (i.e. heats of combustion) are available for many plants and tissues (Vertregt and Penning de Vries, 1987; Gary et al., 1995; Lamprecht, 1999). Thus, values of ΔH_B can be estimated, rather than measured, with acceptably small error for most studies. Also, for many purposes, simply assuming that ΔH_B is constant may be sufficient. For example, in studies of how ε and growth rates are changed by altered conditions, it may be sufficient to evaluate only (ΔH_B [ε /(1- ε)]) and (R_{AP} ΔH_B) (see Eqs. (4) and (5)) as functions of conditions. Table 2 and Fig. 6 show examples of each of these variables calculated from our illustrative data set. Note that variation with temperature of both efficiency and growth rate is different for the two accessions. The negative values in Fig. 6 indicate the assumption of $\gamma_S = 0$, i.e. carbohydrate as substrate, has become incorrect at these temperatures, and at the same time, that the temperature condition is highly stressful.

Analysis of Figs. 3 and 4 shows that the respiratory rates of the two accessions differ at temperatures at and above 20 °C. The specific heat rates of the North Dakota accession are greater than those of the Montana accession, and the same is true for the CO_2 rates. The CO_2 rate of the North Dakota accession decreases sharply between 30 and 35 °C, an indication of high-temperature stress. Note that the heat rate does not show a similar decrease. The data in figures 3 and 4 fit the Arrhenius equation (rate = A $e^{-\mu/T}$ where A and μ are accession specific constants and T is the Kelvin temperature) over most of the temperature range, but with different values of the Arrhenius temperature coefficient, μ (Criddle and Hansen, 1999).

Figure 5 shows the substrate carbon-conversion efficiency for the Montana accession is essentially constant from 5 to 30 °C and then abruptly decreases at 35 °C. Efficiency decreases continuously for the North Dakota accession from 10 to 35 ° and decreases

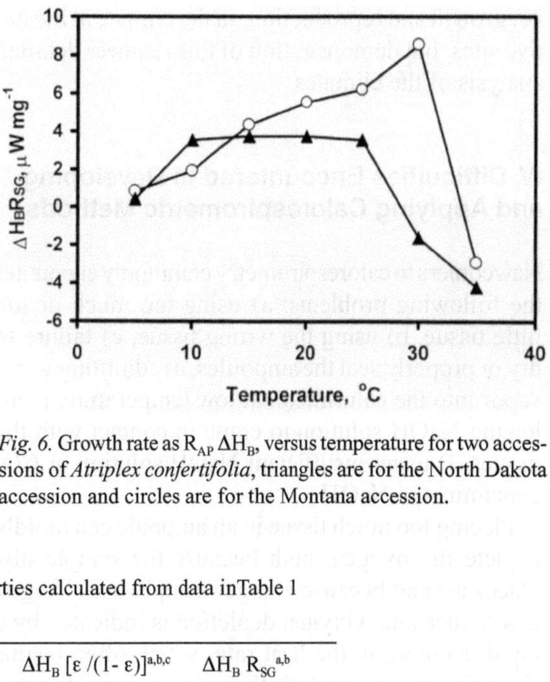

Fig. 6. Growth rate as R_{AP} ΔH_B, versus temperature for two accessions of *Atriplex confertifolia*, triangles are for the North Dakota accession and circles are for the Montana accession.

Table 2. Atriplex confertifolia metabolic properties calculated from data inTable 1

Accession: Montana

Temperature °C	$^aR_q/R_{CO_2}$ kJ mol^{-1}	ε ^{a,b,c}	ΔH_B [ε /(1- ε)]^{a,b,c}	ΔH_B R_{SG} ^{a,b} μW mg^{-1} DW
20.15	271 ± 29	0.80 ± 0.03	199 ± 29	11.9 ± 3.7
15.25	313 ± 30	0.76 ± 0.03	157 ± 30	4.3 ± 1.6
10.34	336 ± 35	0.73 ± 0.04	134 ± 35	1.9 ± 1.0
5.45	358 ± 57	0.69 ± 0.08	112 ± 57	0.8 ± 0.6

[a] Average and standard deviation of three samples. [b] Assuming ΔH_{CO_2} = –470 kJ mole^{-1}.
[c] Assuming ΔH_B = +50 kJ mole^{-1}

abruptly when temperature decreases from 10 to 5 °C. The R_q/R_{CO_2} values greater than 470 at high temperature in both accessions indicate the substrate for catabolism includes some compounds more reduced than carbohydrate. Together the variations in rate and efficiency produce the variations in growth rate seen in Fig. 6. Growth rate of the Montana accession increases steadily from 5 to 30 °C and then decreases abruptly at 35 °C. The increase from 5 to 30 °C is a consequence of an increasing rate at constant efficiency. The decrease at 35 °C is caused by both a loss of efficiency and a decrease in the CO_2 rate. Growth rate of the North Dakota accession increases from 5 to 10 °C, is constant from 10 to 25 °C, above which it drops to negative values. The curves show the low-temperature limit for growth of both accessions is about 5 °C, and the high-temperature limits are ca. 35 °C for the Montana accession and ca. 28 °C for the North Dakota accession. Note that the curves in Fig. 6 do not have the classic, textbook, bell shape. We expect these growth curves to optimize fitness, i.e. growth and reproduction, in the climate at the native sites, but demonstration of this requires detailed analysis of the climates.

IV. Difficulties Encountered in Developing and Applying Calorespirometric Methods

Newcomers to calorespirometry commonly encounter the following problems: a) using too much or too little tissue, b) using the wrong tissue, c) failure to dry or properly seal the ampoules, d) admitting water vapor into the calorimeter at low temperature, e) allowing NaOH solution to come in contact with the tissues, f) using insufficient NaOH solution or CO_2 contaminated NaOH.

Placing too much tissue in an ampoule can rapidly deplete the oxygen, both because the sample displaces air and because a larger sample uses oxygen at a higher rate. Oxygen depletion is indicated by a rapid decrease in the heat rate, which often begins abruptly. Placing too little tissue in an ampoule gives rates too small for accurate measurement, particularly of CO_2 by the method described here. The heat rate from the CO_2 reaction with NaOH is typically about 20% of the metabolic heat rate. For example, at a metabolic heat rate of 500 μW, the heat rate from the CO_2 reaction would be 100 μW which has an uncertainty of about 5% in our system. Metabolic heat rates above 500 μW use more than half the oxygen in a 1-mL ampoule during a measurement. Thus, heat rates around 400 μW in a 1-mL ampoule are a good compromise. Measurement of the same sample at different temperatures requires further compromise since rates roughly double every 10 °C temperature rise. Most rapidly-growing plant tissues produce 2 to 3 μW per mg fresh weight, so experience shows a good compromise is to use about 75 mg for measurements at 20 up to 40 °C and 150 mg for measurements at 20 down to 0 °C.

It must be kept in mind that the properties measured apply only to the tissue used in the measurement. Thus, growth rates and efficiencies should be calculated from calorespirometric data taken on the most rapidly growing tissue. Such rates are indicative of whole plant growth rates (Criddle and Hansen, 1999). Data taken on physiologically older tissue are not indicative of growth rate, but may be of interest for other reasons. Very little calorespirometric work has been done on older tissues.

Evaporation and condensation of water can cause heat effects that are much larger than metabolic heat. Ampoules must be kept clean and dry on the outside surfaces. Even the moisture in a fingerprint can cause significant error. Drying ampoule surfaces is conveniently done by wiping with an absorbent paper tissue just before sealing and inserting the ampoule into the calorimeter. Ampoules also must be properly sealed or water evaporation will cause large endothermic effects. This was a major problem in early efforts, but has been solved by better machining of ampoules and lids and choice of gasket materials by calorimeter manufacturers. Calorimeters constructed for use below room temperature always incorporate some means for keeping the atmosphere in the measuring chamber free from water vapor. Otherwise, water would condense at temperatures below the dew point with consequent large heat effects. Thus, the measuring chamber must not be opened, which admits room air, when the calorimeter temperature is below the dew point in the room.

If NaOH vials are not properly cleaned and filled, the solution can come in contact with the plant tissue causing a large, exothermic heat effect. Vials should be washed in dilute HCl (0.1 M) and thoroughly rinsed between uses. Filling can be done with a Pasteur pipette, but bubbles must be avoided in the vial. Bubbles can expand when temperature increases and push the solution out of the vial. The NaOH solution is normally held in the vial in any orientation by capillary action. Although no special care is necessary to keep CO_2 out of the NaOH solution, the presence of too much carbonate in the solution must be avoided. Carbonate reacts with CO_2, but with a smaller heat effect than hydroxide.

Many of the potential problems newcomers tend to worry about are usually not significant. Wounding responses of the tissue from cutting to fit in the ampoule are generally absent. The amount of water lost by the tissue to saturate the atmosphere in the ampoule is insignificant. The effects of decreasing oxygen and increasing CO_2, which can go from 20 to 10% and from near zero to 10%, respectively, during a typical measurement, have been found to be negligible. If necessary, pure oxygen can be used in the ampoule with little effect.

The major needs for improving current methods in plant calorespirometry are a faster, more precise method for measuring CO_2 rates and a compatible method for measuring O_2 rates simultaneously with the calorimetric measurement of heat rates. The standard deviations of the properties listed in Table 2 are typical of the reproducibility obtained by the methods described here. Much of the uncertainty is a consequence of the low precision of the CO_2 rate determination. Previous work employing Warburg methods, gas chromatography, mass spectrometry, and electrochemical methods (Criddle and Hansen, 1999) shows all of these to be much more difficult to employ than the calorimetric method for CO_2 described here. We are currently exploring the possibility of using optropes for both CO_2 and O_2 measurement. Although optrode methods are promising, we do not yet have sufficient data to fully evaluate their usefulness for calorespirometry.

V. Applications of Calorespirometry and Calorimetry

Curves of growth rate versus temperature are rapidly generated as illustrated in Fig. 6. These curves define the temperature for maximum growth rate and the temperature limits for growth of the tissue or plant at the time it was placed in the calorimeter. It must be kept in mind that these are very short-term measurements that allow no time for acclimation. Making such measurements on tissues from closely-related plants grown under a range of controlled conditions provides information on acclimation. Determination of growth curves for plants native to a range of climates but grown in a common garden provides information on genetic adaptation of plants as opposed to acclimation.

Because calorespirometry provides a direct way to measure the substrate carbon-conversion efficiency, the model described here gives the actual construction cost from measurements of R_q and R_{CO_2}. The construction cost estimated from measurements of composition or heats of combustion (Gary et al., 1995) provides only an upper limit for ε, and underestimates the total energy and hence carbon costs of biomass synthesis under real conditions. The only way to avoid such errors is to measure ε directly in living tissues and plants, and calorespirometry provides the only rapid way (and often the only feasible way) to measure ε in varying growth conditions.

Ecophysiologists have for many years sought tools to rapidly quantify growth rate and responses of plants to changing environmental conditions. The rates of developmental processes such as vegetative growth, development of flowers, stem growth, and seed development are proportional to their anabolic rates, R_{AP}, and these are all altered by conditions. R_{AP} can be determined by calorespirometry. Calorespirometry is thus a convenient means for determining the effects of different conditions on these processes (Criddle and Hansen, 1999).

Changing conditions always alter metabolism, but may or may not have an effect on the rate of development or growth. Equation 3 shows that anabolic rate is equal to the product of rate (R_{CO_2}) and efficiency ($\varepsilon/(1-\varepsilon)$). By definition, increasing stress must always reduce efficiency, but the catabolic rate R_{CO_2} may increase to compensate or partially compensate, with the result that the anabolic rate remains constant. Measurements of developmental or growth rates or catabolic rates (as R_{CO_2}) alone thus cannot reveal such stress effects, and can even be misleading. For example, a stress may increase R_{CO_2} while decreasing growth rate, giving an apparent inverse relation between the rate of respiration and growth. Or, what is thought to be an increasing stress may not be, as shown by an increasing efficiency. For example, the growth rate of most plants decreases with decreasing temperature, but measurements of ε show that efficiency actually increases with decreasing temperature for some plants (MacFarlane et al., 2002; Hemming et al., 1999). The *Atriplex confertifolia* data in Table 2 show the opposite trend with temperature.

Calorespirometric measurements can be made on tissues of different age, from different growth conditions, and with differing nutrient sources, or with tissues differing in biomass composition, to evaluate energy costs of differences in plant biosynthesis. It is only necessary to measure R_q and R_{CO_2} in control and test plants, calculate efficiency (or ΔH_B [ε /(1 – ε)]) and R_{AP} (or R_{AP} ΔH_B) with Eqs. (4) and (5). For example, comparison of plant energy efficiencies by calorespirometric measurements on plants that have been stressed with salt and unstressed controls can quantify the energy costs of the stress and subsequently be related to effects of stress on growth (Marcar et al., 2002). Additional measurements can investigate whether one genotype is more susceptible to a decrease in efficiency by a given stress than another. As another example of stress response data obtained by calorespirometry, the effects of temperature stress on respiratory metabolism in *Eucalyptus globulus* seedlings was published by Macfarlane et al. (2002). Measurements of R_{CO_2} and R_q as functions of temperature enabled calculation of growth rates, enthalpy conversion efficiencies, and even changes in P/O ratios for oxidative phosphorylation resulting from temperature changes. This study is thus an example of how calorespirometric studies can lead to information on activities of biochemical pathways beyond anabolism and catabolism.

The effects of toxins and nutrient stresses have also been studied by calorimetry (see refs. in Criddle and Hansen, 1999; Jones et al., 2000; Harris et al., 2001). Respiration rate can either increase or decrease in response to toxin and low-nutrient stresses. The effects on ε and R_{CO2} are nonlinear with concentrations of toxin or nutrient. Small increases in nutrient or toxin levels may cause large changes in respiration rate. Toxins or low levels of nutrients often decrease ε but simultaneously increase R_{CO_2} to compensate and maintain a near-constant growth rate (see Eq. 3). This may continue with increase in toxin or decrease in nutrient until a limit is reached where the ability to compensate is exceeded. Because such studies require direct measurement of ε, no method except calorespirometry could provide such information.

The use of calorimetry in many types of postharvest studies illustrates the breadth of application of calorimetric methods. The role of a symbiotic fungus in deterioration of quality of cut pineapple (*Ananas comosus*) was demonstrated and a protective treatment devised (Iverson et al., 1989). In a study on cauliflower (*Brassica oleracea*), a gas-phase, metabolic inhibitor was discovered that completely stopped any measurable heat production at temperatures below 20 °C. Although the inhibitor was never identified, it was found that cauliflower heads could be stored indefinitely with little loss in quality if the temperature was kept below 20 °C under the right atmospheric conditions in sealed containers (Hansen and Criddle, 1989).

Although contributions of calorespirometry to selection of crop plants have not been widely recognized (Thornley, 2002), the ability of calorespirometry to rapidly determine growth rate as a function of temperature (e.g., see Fig. 6) has proven useful for crop selection. Calorimetric methods have been used to speed selection of *Eucalyptus* provenances, families, and clones with appropriate temperature-growth properties for establishment of plantations in California and Mexico (Criddle et al., 1995; Anekonda et al., 1996; Criddle et al., 1996). *Eucalyptus* have also been identified for salt tolerance for saline regions in Australia (N. E. Marcar, R. S. Criddle, C. Macfarlane and L. D. Hansen, unpublished). Both Sierra and coast redwoods (*Sequoia gigantea* and *sempervirens*, respectively) have also been studied to match trees to specific environments (Anekonda et al., 1993, 1994). Planting times for lettuce (*Lactuca sativa*) varieties and cultivars to match climatic conditions have been established by the calorespirometric methods described here (R. S. Criddle, unpublished). Fungicide use in commercial banana (*Musa sapientum*) production was greatly reduced when it was discovered from calorimetric measurements that the older leaves that are subject to fungal infection were contributing nothing to production and could be removed without loss of yield (R. S. Criddle, unpublished).

Metabolic heat rates can be measured while scanning temperature with a temperature scanning calorimeter (Criddle et al., 1991), but it is not yet possible to make accurate, simultaneous measurements of CO_2 production while scanning temperature. Though the full potential of the calorespirometric method is thus not yet possible in continuous scanning studies, the response of heat rate to temperature alone can be very informative. The response of R_q to temperature shows upper and lower temperature limits for metabolism, temperature optima, and ranges where rates increase, decrease, or change abruptly with temperature (Smith et al., 1999b). Scans can be done with either increasing or decreasing temperature.

To summarize, measurement of respiratory heat rates provides a convenient and accurate means for measuring respiration rates as a function of temperature and other environmental variables. Simultaneous measurement of respiratory CO_2 and heat production rates provides information for determining anabolic rates, substrate carbon conversion efficiencies and growth rates over a range of conditions. Calorespirometry makes it possible to use short-term measurements to predict plant growth properties and to relate these properties to activities of metabolic pathways. Calorespirometry is thus a useful means for developing insight into adaptation and acclimation of plant respiratory metabolism to environmental variables.

Acknowledgments

We thank the many students, graduate and undergraduate, at Brigham Young University and University of California, Davis, for their assistance and contributions to the work summarized in this chapter. We thank both universities for their financial support of this effort.

References

Amthor JS (2000) The McCree-deWit-Penning de Vries-Thornley respiration paradigms: 30 years later. Ann Bot (London) 86:1–20

Anekonda TS, Criddle RS, Libby WJ and Hansen LD (1993) Spatial and temporal relationships between growth traits and metabolic heat rates in coast redwood. Can J For Res 23: 1793–1798

Anekonda TS, Criddle RS, Libby WJ, Breidenbach RW and Hansen LD (1994) Respiration rates predict differences in growth of coast redwood. Plant, Cell Environ 17: 197–203

Anekonda TS, Hansen LD, Bacca M and Criddle RS (1996) Selection for biomass production based on respiration parameters in eucalypts: Effects of origin and growth climates on growth rates. Can J For Res 26:1556–1568

Battley EH (1999) The thermodynamics of microbial growth. In: Kemp RB (ed) Handbook of Thermal Analysis and Calorimetry, Vol 4: From Macromolecules to Man, pp 219–265. Elsevier, Amsterdam

Criddle RS and Hansen LD (1999) Calorimetric methods for analysis of plant metabolism. In: Kemp RB (ed) Handbook of Thermal Analysis and Calorimetry, Vol. 4: From Macromolecules to Man, pp. 711–763. Elsevier, Amsterdam

Criddle RS, Breidenbach RW and Hansen LD (1991) Plant calorimetry: How to quantitatively compare apples and oranges. Thermochim Acta 193: 67–90

Criddle RS, Anekonda TS, Breidenbach RW and Hansen LD (1995) Site-fitness and growth-rate selection of *Eucalpytus* for biomass production. Thermochim Acta 251:335–349

Criddle RS, Anekonda TS, Sachs RM, Breidenbach RW and Hansen LD (1996) Selection for biomass production based on respiration parameters in eucalypts: Acclimation of growth and respiration to changing growth temperature. Can J For Res 26:1569–1576

Criddle RS, Smith BN and Hansen LD (1997) A respiration based description of plant growth rate responses to temperature. Planta 201: 441–445

Criddle RS, Smith BN, Hansen LD and Church JN (2001) Determination of plant growth rate and growth temperature range from measurement of physiological parameters. USDA Forest Service proceedings RMRS-P-21, pp 251–258

Demetriades-Shaw TH, Fuchs M, Kanemasu ET, Flitcroft I (1992) A note of caution concerning the relationship between accumulated intercepted solar radiation and crop growth. Agric Meterol 58:193–207

Ellingson D, Olson A, Matheson S, Criddle RS, Smith BN and Hansen LD (2002) Determination of the enthalpy change for anabolism by four methods. Thermochim Acta 400: 79–85

Freeman DC, Turner WA, McArthur ED and Graham JH (1991) Characterization of a narrow hybrid zone between two subspecies of big sagebrush (*Artemisia tridentata: Asteraceae*) Amer J Bot 78:805–815

Freeman DC, Wang H, Sanderson SC and McArthur ED, (1999) Characterization of a narrow hybrid zone between two subspecies of big sagebrush (*Artemisia tridentata: Asteraceae*. VII. Community and demographic analyses. Evol Ecol Res 1:487–502

Gabriel ML, (1955) Preface to Memoir on Heat. In: Gabriel ML and Fogel S (eds) Great Experiments in Biology, pp 85–86. Prentice Hall, Englewood Cliffs

Gary C, Frossard JS and Chenevard D (1995) Heat of combustion, degree of reduction and carbon content: 3 interrelated methods of estimating the construction cost of plant tissues. Agronomie 15:59–69

Hansen LD and Criddle RS (1989) Batch-injection attachment for the Hart DSC. Thermochim Acta 154:81–88

Hansen LD, Hopkin MS, Taylor DK, Anekonda TS, Rank DR, Breidenbach RW and Criddle RS (1995) Plant calorimetry. Part 2. Modeling the differences between apples and oranges. Thermochim Acta 250:215–232

Hansen LD, Breidenbach RW, Smith BN, Hansen JR and Criddle RS (1998) Misconceptions about the relation between plant growth and respiration. Bot Acta 111:255–260

Hansen LD, Criddle RS, Smith BN, MacFarlane C (2002a) Growth-maintenance component models are an inaccurate representation of plant respiration. Crop Sci 42:659

Hansen LD, MacFarlane C, McKinnon N, Smith BN, Criddle RS (2004) Use of calorespirometric ratios, heat per CO_2 and heat per O_2, to quantify metabolic paths and energetics of growing cells. Thermochim Acta, in press

Harris LC, Gul B, Khan A, Hansen LD and Smith BN (2001) Seasonal changes in respiration of halophytes in salt playas in the great basin. Wetlands Ecol and Manage 9: 463–468

Hemming DJB (1995) Calorimetric studies of plant respiration: Three applications of a thermodynamic model. M.S. Thesis, Brigham Young University, Provo, UT

Hemming DJB, Meyer SE, Smith BN and Hansen LD (1999) Respiration characteristics differ among cheatgrass (*Bromus tectorum L.*) populations. Great Basin Naturalist 59:355–360

Iverson E, Wilhelmsen E and Criddle RS (1989) Calorimetric examination of cut fresh pineapple metabolism. J Food Sci 54:1246–1249

Johansson P and Wadsö I (1997) A photo microcalorimetric system for studies of plant tissue. J Biochem Biophys Methods 35:103–114

Kemp RB (ed) (1999) Handbook of Thermal Analysis and Calorimetry, Vol. 4. Elsevier, Amsterdam

Jones AR, Lytle CM, Stone RL, Hansen LD and Smith BN (2000) Methylcyclopentyldienyl manganese tricarbonyl (MMT), plant uptake and effects on metabolism. Thermochim Acta 349:141–146

Lamprecht I (1999) Combustion Calorimetry. In: Kemp RB (ed) Handbook of Thermal Analysis and Calorimetry, Vol 4, pp 175–218. Elsevier, Amsterdam

Lytle CM, Smith BN, Hopkin MS, Hansen LD and Criddle RS (2000) Oxygen-dependence of metabolic heat production in the appendix tissue of the voodoo lily (*Sauromatum guttatum* schott) Thermochim Acta 349:135–140

Macfarlane C, Adams MA and Hansen LD (2002) Application of an enthalpy balance model of the relation between growth and respiration to temperature acclimation of *Eucalyptus globulus* seedlings. Proc Royal Soc London 269: 1499–1507

Marcar NE, Criddle RS, Guo JM, Zohar Y (2002) Analysis of respiratory metabolism correlates well with the response of *Eucalpytus camaldulensis* seedlings to NaCl and high pH. Funct Plant Biol 29:925–932

Penning de Vries FWT, Brunsting AHM and van Laar HH (1974) Products, requirements and efficiency of biosynthesis: A quantitative approach. J Theor Biol 45:339–377

Smith BN, Eldredge S, Moulton DL, Monaco TA, Jones AR, Hansen LD, McArthur ED and Freeman DC (1999b) Differences in temperature dependence of respiration distinguish subspecies and hybrid populations of big sagebrush: Nature versus nurture. USDA Forest Service Proceedings RMRS-P-11. pp 25–28

Smith BN, Jones AR, Hansen LD and Criddle RS (1999a) Growth, respiration rate, and efficiency responses to temperature. In: Pessarakli M (ed) Handbook of Plant and Crop Stress, 2nd Ed., pp 417–439. Marcel Dekker, New York

Smith BN, Hansen LD, Breidenbach RW, Criddle RS, Rank DR, Fontana AJ and Paige D (2000) Metabolic heat rate and respiratory substrate changes in aging potato slices. Thermochim Acta 349:121–124

Smith BN, Monaco TA, Jones C, Holmes RA, Hansen LD, McArthur ED and Freeman DC (2002) Stress induced metabolic differences between populations and subspecies of *Artemisia tridentata* (sagebrush) from a single hillside. Thermochim Acta 394:205–210

Taylor DK, Rank DR, Keiser DR, Smith BN, Criddle RS and Hansen LD (1998) Modelling temperature effects on growth-respiration relations of maize. Plant Cell Environ 21:1143–1151

Thornley JHM (2002) Response to growth-maintenance component models are an inaccurate representation of plant respiration. Crop Sci 42: 660

Vertregt N and Penning de Vries FWT (1987) A rapid method for determining the efficiency of biosynthesis of plant biomass J Theor Biology 128:109–119

Wang H, McArthur ED, Sanderson SC, Graham JH and Freeman DC (1997) Narrow hybrid zone between two subspecies of big sagebrush (*Artemisia tridentata:* Asteraceae). IV. Reciprocal transplant experiments. Evolution 51: 95–102

Wang H, Byrd DW, Howard DL, McArthur ED, Graham JH, and Freeman DC (1998) Narrow hybrid zone between two subspecies of big sagebrush (*Artemisia tridentata*: Asteraceae). V. Soil properties. Intl J Plant Sci 159:139–147

Chapter 3

The Application of the Oxygen-Isotope Technique to Assess Respiratory Pathway Partitioning

Miquel Ribas-Carbo[1]*, Sharon A. Robinson[1,2] and Larry Giles[3]

[1]*Departament de Biologia, Universitat de les Illes Balears, Ctra. Valldemossa Km. 7,5, 07122 Palma de Mallorca, Spain;* [2]*Institute for Conservation Biology, Department of Biological Sciences, University of Wollongong, Wollongong, NSW, 2522, Australia;* [3]*Carnegie Institution of Washington, Department of Global Ecology, 260 Panama St., Stanford, CA 94305-1297, U.S.A.*

Summary

The oxygen isotope technique is currently the only reliable method for studying relative electron partitioning between the cytochrome and alternative plant respiratory pathways. The theoretical background to this technique is described, as well as some of the difficulties that can complicate measurements. This chapter describes the development of systems over the last 15 years that currently allow measurement of respiration in both intact tissues and in the aqueous phase. Initially, the focus was on developing on-line systems for both gas and liquid phase measurements, but in recent years attention has shifted to the development of portable off-line systems which will allow measurements of respiratory electron partitioning in field studies. Measurements can now be made much more rapidly and accurately than a decade ago, however, the application of this technique is still limited by the availability of dedicated systems. Finally, a summary of data obtained with this technique is presented.

*Author for correspondence, email: mribas@uib.es

H. Lambers and M. Ribas-Carbo (eds.), Plant Respiration, 31–42.

I. Introduction

Plants have a cyanide-insensitive respiratory pathway in addition to the cytochrome pathway (Chapter 1, Lambers et al.). This alternative pathway draws electrons from ubiquinol to reduce oxygen to water. Unlike the cytochrome pathway, the transport of electrons through the alternative pathway is not linked to proton extrusion, and therefore not coupled to energy conservation. The alternative pathway is present in all plant species studied (Moore and Siedow, 1991).

For many years, studies of electron partitioning between the two respiratory pathways were performed with the use of specific inhibitors of the two pathways (i.e. cyanide for the cytochrome path, and SHAM for the alternative path). It was thought that electrons were only available to the alternative pathway when the cytochrome pathway was either saturated or inhibited (Dry et al., 1989). However, in 1995, it was shown that, under certain circumstances, both pathways compete for electrons under unsaturated or uninhibited conditions (Hoefnagel et al., 1995; Ribas-Carbo et al., 1995). It is now widely accepted that the only reliable technique to study electron partitioning between the cytochrome and alternative pathway is by using oxygen-isotope fractionation (Day et al., 1996). Although the methodology employed has changed dramatically in the last decade, the theoretical basis of the oxygen-isotope fractionation technique remains that described by Guy et al. (1989).

II. Theoretical Background

The origin of this methodology can be found in Bigeleisen and Wolfsberg (1958), Mariotti et al. (1981) and Hayes (1983). Oxygen-isotope fractionation is measured by examining the isotope fractionation of the substrate oxygen as it is consumed in a closed, leak-tight cuvette.

Since the energy needed to break the oxygen-oxygen bond of a molecule containing ^{18}O is greater than that to break the molecule $^{16}O{=}O^{16}$, both terminal oxidases of the plant mitochondrial electron transport chain react preferentially with $^{32}O_2$, rather than with $^{34}O_2$. However, because the two enzymes use different mechanisms to break that bond, they produce different isotope effects (Hoefs, 1987). This difference can be exploited to determine the relative flux through each terminal oxidase.

In general, the basis for measurement of fractionation is as follows. If α is the ratio of the rate of the reaction with ^{18}O to that with ^{16}O, then:

$$R_p = R\,\alpha \tag{1}$$

where R_p is the $^{18}O/^{16}O$ ratio of the product (H_2O), and R is that of the substrate (O_2). Since α generally differs from unity by only a few percent, fractionation is often given by D where:

$$D = (1 - \alpha) \times 1000 \tag{2}$$

and the units of D are parts per mil (‰). Generally, D is obtained directly from equation (1) by measurements of the isotope ratio of the substrate and product. However, since the product of an oxidase reaction is H_2O, and this is either the solvent for these reactions (liquid-phase) or very difficult to obtain (gas-phase), an alternative strategy has been adopted. Changes in the isotope ratio of the oxygen in the substrate pool are measured as the reaction proceeds in a closed system. If there is any isotopic fractionation during respiration, the oxygen-isotope ratio (R) of the remaining O_2 increases as the reaction proceeds. The respiratory isotope fractionation can be obtained by measuring R, and the fraction of molecular O_2 remaining at different times during the course of the reaction.

Therefore, if we define the following terms;

R_o = initial $^{18}O/^{16}O$

R = $^{18}O/^{16}O$ at time t

f = fraction of remaining oxygen at time t: f = $[O_2]/[O_2]_o$

the change in R through time would be:

$$\delta R/\delta t = (^{16}O\,(\delta^{18}O/\delta t) - {}^{18}O\,(\delta^{16}O/\delta t))/\,(^{16}O)^2 \tag{3}$$

and since

$$\delta^{18}O/\delta t = R\,\alpha\,(\delta^{16}O/\delta t) \tag{4}$$

we obtain:

Abbreviations: DMSO – dimethyl sulfoxide; DTT – dithiothreitol; R – $^{18}O/^{16}O$ ratio; SHAM – salicylhydroxamic acid; TCD – thermal conductivity detector; ΔD – isotopic fractionaltion factor

$$\delta R/R = \delta^{16}O/^{16}O\,(1-\alpha) \quad (5)$$

which, upon integration, yields:

$$\ln R/R_0 = -\ln {}^{16}O/{}^{16}O_0\,(1-\alpha) \quad (6)$$

Since only 0.4% of the O_2 contains ^{18}O, the ratio $^{16}O/^{16}O_0$ is a good approximation of O_2/O_{20} or f, and hence we may write,

$$D = \ln(R/R_0) / -\ln f \quad (7)$$

and D can be determined by the slope of the linear regression of a plot of ln R/R_0 vs –ln f, without forcing this line through the origin. (Henry et al., 1999).

The standard error of the slope is determined as (Neter and Wasserman, 1974):

$$SE = \frac{D\,(1-r^2)^{1/2}}{r\,(n-2)^{1/2}}$$

and indicates the precision of the measurement of isotopic fractionation (D). For accurate measurements this error should be less than 0.4‰, since the fractionation differential between the cytochrome pathway (18–20‰) and the alternative pathway (24–31‰) is between 6‰ and 12‰, for roots and green tissues, respectively (Robinson et al., 1995). In most cases, accurate determinations of D can be achieved with experiments consisting of six measurements, providing the r^2 of the linear regression is 0.995 or higher (Ribas-Carbo et al., 1995; Henry at al., 1999).

Since it is common practice in the botanical literature to express isotope fractionation in 'Δ' notation, the fractionation factors, D, are converted to Δ as described in Guy et al. (1993):

$$\Delta = \frac{D}{1-(D/1000)}$$

A. Calculation of the Electron Partitioning Between the Cytochrome and Alternative Pathways

The partitioning between the cytochrome and the alternative respiratory pathways (τ_a) is obtained as described by Ribas-Carbo et al. (1997):

$$\tau_a = \frac{\Delta_n - \Delta_c}{\Delta_a - \Delta_c}$$

where Δ_n is the oxygen-isotope fractionation measured in the absence of inhibitors, and Δ_c and Δ_a are the fractionation by the cytochrome and alternative pathway, respectively. These end-points for purely cytochrome or alternative pathway respiration are established for each experimental system using inhibitors of the alternative (SHAM, n-propylgallate) and cytochrome (usually KCN) oxidases, respectively. The cytochrome oxidase (Δ_c) consistently gives a fractionation between 18‰ and 20‰ (Table 1), while the fractionation of the alternative oxidase is much more variable, with values ranging from 24‰ to 31‰ (Table 2). The importance of these endpoints and some technical difficulties associated with their measurement are described below.

Residual respiration, which is any oxygen uptake in the presence of inhibitors of both the cytochrome (KCN) and alternative (SHAM) pathways, has been reported to have an isotopic fractionation between 19.6‰ and 21.0‰ (Guy et al., 1989; Ribas-Carbo et al., 1997). Because of the much lower fractionation by residual respiration, any significant residual respiration present in the tissue would decrease its alternative pathway fractionation (Δ_a), compared to isolated mitochondria, since the latter do not present residual respiration. Ribas-Carbo et al. (1997) showed that the oxygen-isotope fractionation by the alternative pathway was essentially the same in isolated mitochondria (30.9‰) and intact tissues (31.5‰) of green soybean (*Glycine max*) cotyledons. A similar result was also observed in etiolated soybean cotyledons. These results suggest that residual respiration maybe an artifact which only occurs in tissues in the presence of both inhibitors and therefore does not interfere with the oxygen isotope fractionation measurements (Ribas-Carbo et al., 1997).

B. Technical Difficulties Associated with the Isotopic-Fractionation Technique

Determination of the end-points (Δ_a and Δ_c) has not been a problem in the aqueous phase systems, which are used for mitochondrial or whole-cell studies. However, in whole-tissue studies poor infiltration of the inhibitors can cause difficulties in determining the two end-points, especially in dense or waxy tissues. In most organs, Δ_a is easy to obtain, since KCN penetrates tissues fairly easily; it can be applied by soaking samples in 1 mM KCN. However, some researchers have reported difficulties with infiltration of KCN into evergreen leaves, and had to resort to concentrations of 16 mM KCN to obtain full inhi-

Table 1. Summary of the discrimination values associated with respiration for a variety of plant tissues, measured in the presence and absence of KCN and SHAM and during uninhibited respiration. Partitioning of respiratory flux to the alternative pathway is determined, as a percentage of total flux, where end-point values are available (nd; not determined).

Species and tissue type	Δ_c (+SHAM) (‰)	Δ_a (+KCN) (‰)	Δ_n (control) (‰)	Partitioning (%)
Alocasia odora[A] (leaf disks)	19.9 ± 0.96	25.9 ± 2.5	19.6–20.4	0–8.9
Asparagus springeri[B] (intact mesophyll cells)	19.8[R]	26.1	20.6	0
Crassula argentea[C] (leaf discs)	nd	nd	21.8	nd
Crassula argentea[D] (leaf discs)	nd	nd	20.0 ± 1.34	nd
Giselina littoralis[D] (leaf halves)	15.7 ± 0.37	25.3 ± 0.53	18.5 ± 0.53	33
Gliricidia sepium[E] (leaves)	19.9 ± 0.2	30.7 ± 0.8	21.3–22.9	13–28
Glycine max[F/G] (etiolated cotyledons)	20.6 ± 0.6	25.5 ± 0.3	18.6 ± 0.8	0
Glycine max[F] (etiolated cotyledon mitochondria)	21.1 ± 0.5	25.4 ± 0.3	20.6–21.4	0–7
Glycine max[F/G] (green cotyledons)	20.0 ± 0.4	31.5 ± 0.3	25.2–27	45–61
Glycine max[F/H] (green cotyledon mitochondria)	19.9 ± 1.1	30.9 ± 0.6	20.2–25.5	0–51
Glycine max[I] (leaf discs)	nd	nd	23.4	nd
Glycine max[F/I] (roots)	20.8 ± 0.5	25.1 ± 0.6	19.7–22.2	0–33
Glycine max[J] (roots)	16.0–16.3	24.2–24.6	16.4–20.5	5–55
Glycine max[K] (roots)	17.05 ± 0.49	27.06 ± 0.02	16.6–18.6	0–15
Glycine max[F/H] (root mitochondria)	20.8 ± 0.3	25.0 ± 0.6	20.8–22.4	0–136
Kalanchoë daigremontiana[I] (leaf disks)	18.9	30.2	20.3–26.0	12–63
Medicago sativa[I] (whole etiolated seedlings)	18.7	25.7	20.0	0
Medicago sativa[B] (whole seedlings)	18.7	26.2	21.7	40
Nicotiana tabacum[L] (leaf disks)	20.1 ± 0.3	31.4 ± 0.2	23.2	27
Nicotiana tabacum[I] (leaves)	19.6 ± 0.2	29.8 ± 0.3	19.8–20.5	0–9
Phaseolus vulgaris[E] (leaf disks)	19.0 ± 2.2	26.7 ± 0.97	18.7–22.1	0–40
Phaseolus vulgaris[E] (leaves)	19.5 ± 0.5	30.3 ± 0.4	20.3–22.9	8–31
Philodendron[M] (roots)	nd	nd	11.9–20.2	nd
Poa annua[N] (roots)	19.5 ± 0.32	26.6 ± 0.10	21.6–23.6	30–60
Poa alpina[O] (roots)	19.16 ± 0.28	25.34 ± 0.15	20.49 ± 0.23	22 ± 4
Poa pratensis[O] (roots)	20.10 ± 0.06	26.33 ± 0.73	20.93 ± 0.16	13 ± 3
Poa compressa[O] (roots)	19.60 ± 0.22	25.17 ± 0.38	20.06 ± 0.50	11 ± 7
Poa trivali[O] (roots)	18.69 ± 0.19	26.06 ± 0.64	22.29 ± 0.48	49 ± 7
Ricinus communis[B] (endosperm mitochondria)	nd	nd	17.3	nd
Sauromatum guttatum[B] (appendix slices)	nd	nd	7.8 ± 2.7	nd
Spinacia oleracea[A] (leaf disks)	19.7 ± 0.96	28.9 ± 1.1	19.3–23.5	0–41
Symplocarpus foetidus[B] (spadix mitochondria)	17.4	24.1	22.6	78
Symplocarpus foetidus[B] (spadix slices)			10.9 ± 1.2	
Vigna radiata[K] (cotyledons)			18.9	
Vigna radiata[P] hypocotyls	20	30.8	20–21.5	0–15
Vigna radiata[P] (leaf disks)	20.2	30.1	21–24	10–40
Zea mays[Q] (leaf slices)	19.3 ± 0.3	29.9 ± 0.8	21.9–22.1	24–60

Data from; [A]Noguchi et al., 2001; [B]Guy et al., 1989 (recalculated as described in Robinson et al., 1995); [C]Robinson et al., 1995; [D]Nagel et al,. 2001; [E]Gonzàlez-Meler et al., 2001; [F]Ribas-Carbo et al., 1997; [G]Ribas-Carbo et al., 2000b; [H]Ribas-Carbo et al., 1995; [I]Robinson et al., 1992 (recalculated as described in Robinson et al., 1995); [J]Millar et al., 1998; [K]Ribas-Carbo unpublished, [L]Lennon et al., 1997; [M]Angert and Luz, 2001: [N]Millenaar et al,. 2000; [O]Millenaar et al., 2001; [P]Gonzàlez-Meler et al., 1999; [Q]Ribas-Carbo et al., 2000a; [R]data include results obtained with 200 μM disulfiram instead of SHAM.

bition (Nagel et al., 2001). The high concentration may have been required because infiltration and the subsequent 2 h evaporation period allowed the cyanide to volatilize (Nagel et al., 2001).

In many tissues the application of SHAM is more problematic; however, in most cases where SHAM

Table 2. Changes in discrimination values (Δ_n) measured during experimental manipulations of plant respiration. Data are means ± s.e except where ranges of means are shown.

Species by taxonomic group	Δ_n ($^o/_{oo}$)	n	Reference
Cotyledons			
Glycine max (green, control)	*24.3* ± 0.12	3–6	Penuelas et al., 1996
Glycine max (plus cinnamic acid)	*26.8* ± 0.36	3–6	Penuelas et al., 1996
Glycine max (plus α-pinene)	*25.2* ± 0.25	3–6	Penuelas et al.,1996
Glycine max (plus juglone)	*23.7* ± 0.16	3–6	Penuelas et al., 1996
Glycine max (plus quercetin)	*23.4* ± 0.15	3–6	Penuelas et al., 1996
Leaves			
Glycine max (90–95% RWC)	20.3 ± 0.4		Ribas-Carbo, unpublished
Glycine max (75–80% RWC)	23.0 ± 0.9		Ribas-Carbo, unpublished
Glycine max (60–70% RWC)	23.8 ± 0.6		Ribas-Carbo, unpublished
Kalanchoë daigremontiana acidification	22.89 ± 1.24	11	Robinson et al., 1992
Kalanchoë daigremontiana de-acidification	*25.6* ± 0.99	11	Robinson et al., 1992
Phaseolus vulgaris (high P)	20.3 ± 0.2	3–5	Gonzàlez-Meler et al., 2001
Phaseolus vulgaris (low P)	22.9 ± 0.4	3–5	Gonzàlez-Meler et al., 2001
Nicotiana tabacum (high P)	20.5 ± 0.2	3–5	Gonzàlez-Meler et al., 2001
Nicotiana tabacum (low P)	19.8 ± 0.2	3–5	Gonzàlez-Meler et al., 2001
Gliricidia sepium (high P)	21.3 ± 0.4	3–5	Gonzàlez-Meler et al., 2001
Gliricidia sepium (low P)	22.9 ± 0.3	3–5	Gonzàlez-Meler et al., 2001
Spinacia oleracea (high PPFD)	23.5 ± 0.39	2	Noguchi et al., 2001
Spinacia oleracea (low PPFD)	20.7 ± 1.3	2	Noguchi et al., 2001
Alocasia odora (high PPFD)	20.0 ± 0.36	2	Noguchi et al., 2001
Alocasia odora (low PPFD)	20.0	1	Noguchi et al., 2001
Zea mays (Penjalinan chill-sensitive cultivar)	25.5		Ribas-Carbo et al., 2000a
Zea mays (Z7 chill-tolerant cultivar)	23.0		Ribas-Carbo et al., 2000a
Roots			
Triticum aestivum (6–9d old)	12.6–14	16	Angert and Luz, 2001
Triticum aestivum (14–31d old)	15.0–15.6	21	Angert and Luz, 2001
Poa annua low sucrose	21.6–23.6	14	Millenaar et al., 2000
Poa annua high sucrose	20.9–22.1	7	Millenaar et al., 2000
Glycine max (4d old)	16.4 ± 0.07	3	Millar et al., 1998
Glycine max (7d old)	19.1 ± 0.45	3	Millar et al., 1998
Glycine max (17d old)	20.5 ± 0.4	3	Millar et al., 1998
Glycine max (control)	16.9	2	Gaston et al., 2003
Glycine max (plus imazethapyr)	19.9	*2*	Gaston et al., 2003
Glycine max (plus chlorosulfuron)	21.3	*2*	Gaston et al., 2003
Whole plants			
Lemna gibba plantlets 16 °C	20.6	1	Robinson et al., 1995
Lemna gibba plantlets 26 °C	21.4	1	Robinson et al., 1995

has been applied at a concentration of 2–10 mM dissolved in DMSO:H_2O (1:100), treatment for 20–30 minutes has been sufficient to ensure full inhibition of the alternative pathway (Robinson et al., 1992; Ribas-Carbo et al., 1997; Lennon et al., 1997). However, in some cases addition of the inhibitors leads to soaking of the tissues, which can cause diffusion problems with subsequent fractionation measurements.

Poor diffusion through tissues is perhaps the major limitation to the fractionation method. If diffusion limits the supply of oxygen to the terminal oxidases, then the discrimination values will be lower than those associated with respiration, and will vary depending on the rate of respiration. This is a problem with dense tissues, and was noted in the early measurements of Lane and Dole (1956) where carrot (*Daucus carota*) roots and potato (*Solanum tuberosum*) tubers gave Δ_n values of 9‰ compared with 25‰ for spinach (*Spinacea oleraceae*) leaves. Diffusion was also a problem with the voodoo lily (*Sauromatum guttatum*) and skunk cabbage (*Symplocarpus foetidus*) thermogenic tissues measured by Guy et al. (1989); however, these were measured in an aqueous-phase system, and improvements might be seen with similar tissues in a gas-phase system. This is particularly unfortunate since thermogenic tissues exhibit phenomenally fast respiration rates and alternative oxidase expression, and would make a fascinating area of study. Diffusion is less of a problem with leaves, but likely affects respiration measurements of thick tap-roots (Angert and Luz, 2001) and stems, and may be an issue with seeds. The implications of these diffusion limitations for atmospheric oxygen composition are discussed in Angert and Luz (2001).

The major problem of any isotope method of this type is the possibility of contamination from outside air in the form of leaks. The likelihood of this type of contamination increases with every additional connection. It also increases as the gradient between the closed cuvette and the ambient air increase. Many of the design improvements described below have been directed at reducing leakage.

III. Design Advances

Although the first measurements of respiratory ^{18}O discrimination were made by Lane and Dole as early as 1956, the first application of this technology for specific studies of electron partitioning between the cytochrome and the alternative respiratory pathways was more than thirty years later (Guy et al., 1989).

In the last decade the oxygen-isotope technique for plant respiration studies has improved enormously. The measuring systems have developed from the original design in which oxygen had to be purified and then converted to CO_2 (Guy et al. 1989), to the newest, in which air is directly injected from a cuvette into a dual-inlet mass-spectrometer system that measures the $^{18}O/^{16}O$ and O_2/N_2 ratios (Gaston et al., 2003). Along the way, many of the intermediate systems have been superceded, but several are still in use. The developments have mainly produced improvements in three aspects of this technique: a) sensitivity, b) reliability, and c) reproducibility. They have taken advantage of improvements in instrumentation to allow development of more efficient and flexible systems. Since the system developed by Guy et al. (1989) was an off-line system, the focus of many of the first improvements was to speed up measurements through the development of an online mass-spectrometry system. However, the future development of this technology will likely rely on the development of efficient, reliable off-line systems like those developed by Nagel et al. (2001) and Noguchi et al. (2001) which allow measurements to be made remotely from the mass spectrometer. This is essential if the technique is to be used in field studies.

Numerous variations of the on-line mass-spectrometry system have been developed, including liquid-phase systems for studies of algae, roots and isolated mitochondria, and gas-phase systems for studies with detached tissues like leaves, cotyledons and intact roots.

In this chapter we will detail the development of the different systems, and describe their basic features and limitations, along with examples of results obtained with them. Our goal is to provide a clear background to this technique and to assist future researchers in the field of plant respiration to apply the oxygen-isotope technique to their studies.

A The Original Off-Line System

The first system designed to study oxygen-isotope fractionation during plant respiration was developed at the Department of Plant Biology of the Carnegie Institution of Washington (Stanford, California, USA) in the late 1980s (Guy et al., 1989). Although it was groundbreaking in its day, this complicated and cumbersome liquid-phase system has been completely superseded.

This system featured a flexible and adjustable leak-tight liquid-phase cuvette (100–400 mL in volume) with an oxygen electrode at the bottom. Sample aliquots (10–150 mL) were withdrawn from the cuvette, and bubbled with high purity He to extract oxygen (water vapor and CO_2 were removed from the He stream by condensation in cold traps). The oxygen was first separated from other gas mixtures by chromatography, and then converted into CO_2 by reaction with graphite heated to 750°C. The CO_2 produced (10–20 μmol) was condensed and sealed in glass tubes. Isotope ratios of CO_2 were then analyzed at the Geophysical Laboratory of the Carnegie Institution of Washington, Washington, D.C. Isolation and purification of a single sample of oxygen thus took several hours.

B. On-line Gas Phase Systems

The second system dedicated to studying oxygen-isotope fractionation by plant respiration was developed at the Department of Botany at Duke University (Durham, North Carolina, USA) in the early 1990s (Robinson et al., 1992). Its major advance was that the mass spectrometer (VG-ISOTECH SIRA Series II, VG ISOGAS, Middlewich, UK) was able to measure the isotope ratio of oxygen ($^{18}O/^{16}O$). This was certainly a major technical leap forward, since the purification of oxygen and its conversion to CO_2 in a graphite furnace was no longer necessary.

In this system, respiration took place in a leaf-disk electrode unit (LD1 Hansatech Instruments Ltd, Kings Lynn, Norfolk, UK) from which, at regular intervals, small amounts of air were drawn into a 100 µL sample loop of a six-port Valco valve (Valco Instruments Co., Houston, Texas, USA) using a gas-tight syringe. This sample was then switched directly into the He flow. Water vapor and CO_2 were trapped, and O_2 and N_2 were then separated by gas chromatography, and identified by a Thermal Conductivity Detector (TCD). Since N_2 is not consumed during respiration, the O_2/N_2 ratio was used as a measure of total oxygen consumed by respiration. However, since O_2 and Ar elute together from the chromatographic column, a small correction was applied. The isotope ratio $^{18}O/^{16}O$ was determined directly from the ratio of masses 34 and 32 using an isotope-ratio mass spectrometer operated in a continuous flow mode.

The major disadvantage of this system was that it was prone to leaks, especially at the mixing syringe and oxygen electrode O-rings. The oxygen electrode itself may also discriminate against oxygen, and therefore might compromise results. Another limitation of this system is the intrinsic characteristics of the continuous flow mode of the mass spectrometer, which only allows one measurement of each sample limiting its accuracy. In practice, this means that at least 30% of all the oxygen present in the cuvette must be consumed during the experiment in order to obtain a reliable result, thus limiting the experimentation to tissues that have a relatively fast respiration rate. For slowly respiring tissues, this entails longer experiments than preferable. However, this system had several advantages over the original system, including its simple design, the possibility of rapid measurements (6–10 min per sample) and the reduced sample size.

This system has been further refined in the last decade with a series of minor modifications. Originally, the oxygen concentration was measured using an oxygen electrode; however, a more recent version of this system uses TCD detections of O_2 and N_2 to measure the total oxygen consumed. This meant that the oxygen electrode could be replaced by a commercial gas-tight syringe reducing the likelihood of leaks. In addition, as air was withdrawn from the syringe, helium was added into it with a mixing syringe, thus keeping the pressure constant (Robinson et al., 1995). In a further development, the gas tightness of the system was improved by replacing the sample syringe with a purpose-built, metal 3-mL cuvette with temperature control, and the mixing syringe with a plunger filled with Hg (González-Meler et al., 1999).

A similar system was developed at the Australian National University in Canberra, Australia (Millar et al., 1998). This system uses a water-jacketed 50-mL cuvette with adjustable volume (depending on tissue and respiration rate). The sample (125 µL), gas separation and isotope analysis are very similar to that described above (Robinson et al., 1992).

C. The On-Line Liquid Phase System

This system was also designed at the Department of Botany at Duke University, simultaneously with the gas-phase system (Ribas-Carbo et al., 1995). The reaction vessel is a water-jacketed 25-mL acrylic cuvette with a plunger that descends as samples are withdrawn during the course of the experiment. A small sample (3 mL) is withdrawn from the reaction vessel into a pre-evacuated sample chamber and flushed with high purity He until all the gases are purged. CO_2 and water vapor are then removed, and the remaining gases (O_2, N_2 and Ar) are adsorbed onto a coarse molecular sieve-5Å (500 µm; 20 mesh) trap at liquid N_2 temperatures. Thereafter, the trap is switched into the flow path of carrier gas for the gas chromatograph, and the inlet system of the continuous-flow isotope-ratio mass spectrometer, and the gases are released by warming the molecular sieve to 90 °C. Oxygen and N_2 gases are separated by chromatography and detected by TCD, and the oxygen-isotope composition is measured by mass

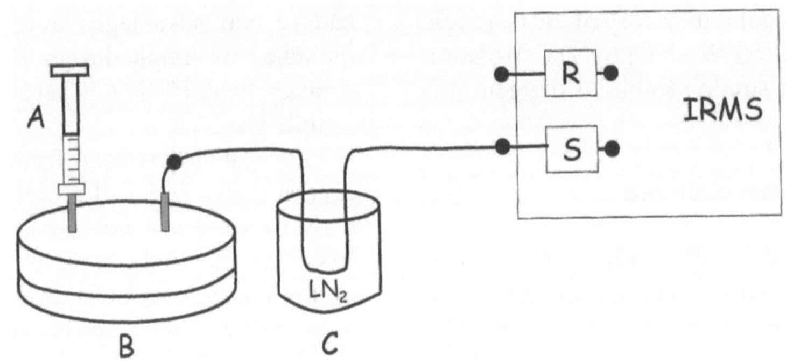

Fig. 1. Diagram of the most recent on-line oxygen isotope fractionation system (Gastón et al., 2003). A. 1 mL gas-tight syringe; B. 3 mL stainless steel cuvette; C. CO_2 and H_2O liquid nitrogen trap. IRMS, Dual Inlet Isotope Ratio Mass Spectometer; R; Reference bellows; S, Sample bellows; ● indicate valves.

spectrometry as described for the gas-phase system (Robinson et al., 1992).

The major problem with this system is, again, leakage, especially around the cuvette plunger and the multitude of valves. The entire line must be kept under positive pressure (up to 230 kPa) to avoid air being drawn into the system. Additionally, the valve through which the liquid sample passes has to be replaced often to avoid leaks. Another limitation is related to the minimum amount of oxygen that has to be consumed for accurate measurements (see above), although this is less of a problem when studying mitochondria, because their concentration can be adjusted to obtain a convenient oxygen-consumption rate.

The major advantages of this system are the easy application of inhibitors, the quickness of measurements, and the ability to work with isolated mitochondria, algae, cell-suspension cultures or enzymes.

D. The Dual-Inlet System

Figure 1 shows a diagram of the most recent system developed at the Department of Plant Biology of the Carnegie Institution of Washington (Gaston et al., 2003). Unlike the previous two systems, which were based on a continuous-flow mass spectrometer, this system is based on dual-inlet mass spectrometry.

The measuring system consists of a 3-mL stainless steel, closed cuvette from which 200 μL air samples are sequentially withdrawn and fed into the dual-inlet mass spectrometer sample bellows. The mass spectrometer (Finnigan Delta S, Thermo Finnigan LCC, San Jose, California, USA) simultaneously measures the m/z 34/32 ($^{18}O_2/^{16}O_2$) and m/z 32/28 (O_2/N_2) ratios of the sample gas. The sample is analyzed against standard air. Each measurement consists of four replicate cycles of each gas sample (cuvette sample and standard).

The respiration cuvette is equipped with two inlets: One connected to a 1-mL air-tight syringe, and the other connected to the mass spectrometer sample bellows through a 1-m long capillary tube (0.127 mm inside diameter). The bellow is first pre-evacuated and fully expanded to its maximum volume (≈30 mL). Then, the vacuum is closed, and the on-off pneumatically controlled micro-needle valve (Fig. 1) that connects the bellow with the cuvette, is opened. The sample air passes through a liquid N_2 trap to remove H_2O and CO_2, and into the bellows until the pressure reaches 500 Pa.

Before collection, the air is mixed by agitation with a gas-tight syringe initially containing 1 mL of air. Throughout the experiment, this syringe is used to both mix the air in the cuvette and to maintain the cuvette at constant pressure. The total time required for an entire respiration experiment (six data points) varies between 80 and 120 min, depending on the respiration rate of the tissue being studied. Respiration is measured from the decrease in the O_2/N_2 ratio, while the isotope fractionation is measured, as described above, from both O_2/N_2 and $^{18}O/^{16}O$ ratios.

The major advantage of this system is its amplified sensitivity (several-fold better than the systems described above), due to the inherent characteristics of the dual inlet mode, where each measurement is the average of several analyses (four in this setup) against a known standard. This sensitivity can be augmented by increasing the number of replicate cycles of each sample, as needed. This substantially decreases the minimum amount of oxygen that needs to be consumed throughout the experiment, from 30% to less than 5% of the total oxygen available. This improved sensitivity allows experiments to be performed using plant material with slow respiration rates. Another benefit of this system is its simplicity, which also diminishes the possibilities for leaks, although they must still be monitored conscientiously.

E. Off-Line Systems

Recently, several off-line systems have been developed to measure oxygen-isotope fractionation during plant respiration (Angert et al., 2001; Nagel et al., 2001; Noguchi et al., 2001). Despite having been developed for a similar purpose of collecting samples off-line, for subsequent analysis at a different location, the systems have very different designs.

Nagel et al. (2001) developed a sophisticated off-line system at the School of Life Sciences of the University of Dundee (Dundee, UK). This system consists of a large cuvette (approx. 150 mL) that contains a galvanic oxygen electrode, a thermocouple and an absolute pressure sensor, all sealed with a rubber lid, within which leaves are placed. The pressure inside the cuvette is maintained, as the samples are collected, by use of a rubber balloon filled with a salt solution (50 g L^{-1} NaCl) connected to a reservoir at atmospheric pressure. The whole system is placed in a water bath at controlled temperature. Air samples (10 mL) are extracted using gas-tight syringes, and transferred into pre-evacuated 10 mL flasks. These samples can then be sent to an isotope laboratory for analysis. In the isotope laboratory, the air samples are extracted into a vacuum line where the condensed gases are cryogenically removed, and oxygen is converted to CO_2 in a granite furnace. This CO_2 is then analyzed for $\delta^{18}O$.

This system has the advantage of an off-line system. However, there are several problems with this system, many of which can be resolved. These weaknesses, already identified in their manuscript (Nagel et al., 2001), can be summarized as follows: a) the experiments take a long time which increases the chance for leaks, and may reduce the physiological significance of the results; b) there is no mixing of air inside the cuvette; c) there is a chance of contamination of the gas sample during extraction with the gas-tight syringe; d) the conversion of oxygen to CO_2 in a graphite furnace is an additional complication of the system, and e) the measurement of oxygen inside the cuvette has to be compensated using the pressure measured inside the cuvette. Despite these problems, this system is another step towards measuring oxygen-isotope fractionation in the field.

The system developed by Noguchi et al. (2001) at Tsukuba University, Tsukuba Ibaraki, Japan is significantly simpler than the one developed by Nagel et al. (2001). The respiration cuvette is similar to the one described in Robinson et al. (1992). Air samples (100 μL) are extracted with a gas-tight syringe through a sample loop, and then collected into a glass tube, which is torch-sealed, after cryogenic removal of water and CO_2. Subsequently, the glass tubes containing the air samples are broken inside a flask, and analyzed using an autosampler, gas chromatography-mass spectrometry system, with gas chromatography to separate O_2 and N_2, and a continuous-flow isotope-ratio mass spectrometer to obtain the ratio $^{18}O/^{16}O$.

This system has the advantage of being rather fast, in the order of one hour per experiment, and allows laboratories to perform experiments even in the absence of an isotope-ratio mass spectrometer, since the samples obtained in one laboratory can be analyzed in another. This opens up the possibility of samples being routinely processed at a central facility (for example, on a commercial basis), as is the case for nitrogen- and carbon-isotopic analyses. The cost of this system is fairly low, and the costs of analyses can also be reasonable. This is certainly another step forward towards broadening the application of this technique in studies of plant respiration.

A third off-line system has been developed by Angert et al. (2001). Although this system was not specifically developed to study the electron partitioning between the cytochrome and alternative respiratory pathways, it is a feasible design to apply to these studies. This system can be arranged as a gas-phase as well as a liquid-phase system. Samples (4 mL of air or 100 mL of water) are collected in glass tubes. These samples are subsequently sent to a laboratory where the O_2/N_2 ratios and the $^{18}O/^{16}O$ are measured in an automated sampling system with a continuous-flow mode. This system has the advantage of allowing measurements of respiration of intact roots, either in soil or hydroponics.

The common disadvantages of all these off-line systems relate to the chance of contamination when transferring the sample into the flasks, and difficulties of transporting large numbers of flasks safely.

IV. Measurements Using the Isotopic Technique

The development of the oxygen-isotope technique has permitted its application to several studies of plant respiration pathways which are briefly summarized here and are also considered in more detail in the appropriate chapters herein (Chapter 1, Lambers et al.; Chapter 7, Atkin et al.; Chapter 13, González-Meler and Taneva). The studies performed since 1989 are summarized in Tables 1–2. Table 1 shows the discrimination values (end-points) associated with respiration for a variety of plant tissues, measured in the presence of KCN (Δ_a) and SHAM (Δ_c), as well as discrimination values (Δ_n) measured during uninhibited respiration. The partitioning of electrons to the alternative pathway is shown for those tissues where end-points are available. Table 2 details experiments where the flux through the alternative pathway has been measured in response to experimental treatments, including exposure to high or low light intensity, and low temperatures, in response to chemicals, and during ontogenetic changes.

Using the first system, Guy et al. (1989) determined the oxygen-isotope fractionation by the two pathways, and determined that the cytochrome pathway has a lower fractionation (18–20‰) than the alternative pathway (25‰), thus setting the benchmark for future experiments and developments. The early experiments concentrated on single determinations of fractionation during respiration; however, the development of simpler systems has allowed physiological measurements of fractionation and expanded our knowledge of the variability of the real fluxes through both the cytochrome and alternative respiratory pathways in plants. The first physiological application of this technique was to study the activity of the alternative pathway during the de-acidification of malate in the leaves of the CAM plant *Kalanchoë daigremontiana*, and to show that increased alternative pathway activity could account for all of the increased respiration during the de-acidification phase (Robinson et al., 1992). This study was also the first to show that an intrinsic difference exists between the discrimination of the alternative pathway in green versus non-green tissues (Table 1). Discrimination end-points for the cytochrome pathway (Δ_c) were 18.4–19.2‰ for almost all tissues measured; however, the end-point for the alternative pathway (Δ_a) was almost 5‰ higher in green (29.3–31.2‰) than in non-green tissues (25.6–25.7‰; Table 1; Robinson et al., 1995). These differences were subsequently verified by Ribas-Carbo et al. (1995), using mitochondria extracted from green and non-green tissues, thus excluding the possibility that they were caused by differential diffusion in the various tissues. The most likely explanation for this difference is that different alternative oxidase proteins are present in the different tissues (Finnegan et al., 1997; Saisho et al., 1997). Over a 36-h greening treatment, Δ_a increased from 27‰ to 32‰ in soybean cotyledons (Ribas-Carbo et al., 2000b), corresponding to a change in AOX gene expression from predominately AOX3 to AOX2 (Finnegan et al., 1997).

The uniformity of the oxygen-isotope fractionation values obtained for each pathway under a wide range of conditions (e.g. reduction status of the ubiquinone pool, addition of pyruvate and DTT), using different mitochondrial preparations confirmed the assumption that these values can be used as standard values for each pathway in subsequent experiments in which no inhibitors are added (Ribas-Carbo et al., 1995). However, these values might be slightly different between species.

V. Future Directions

The application and development of the oxygen-isotope technique will further improve as the number of laboratories that apply it increases. Since most laboratories do not possess an in situ mass spectrometer, the development and improvement of off-line systems that allow the collection of samples from laboratories that do not have a mass spectrometer and the establishment of laboratories prepared to analyze such samples is very important.

The off-line systems described in this chapter are clear examples of the potential of such systems. However, there are several issues that will have to be considered. The most important is the development of a leak-free system, in which the experiments are performed. This system should be easily portable, to allow measurements outside the laboratory, therefore economies of size and weight and the ability to operate without mains power are important. Another critical point is the collection of contaminant-free samples into containers that can be shipped safely from the collection site to the analysis laboratory. A third issue arises from the reliability of the analysis. It will certainly take time and dedication to design, build, test and develop systems that are reliable, affordable and finally applicable, in the many laboratories worldwide that are interested in this area of research. However, it will certainly be worthwhile, since research into the regulation of electron partitioning is currently limited by the availability of dedicated laboratory-based isotope systems, and there are especially severe restrictions on field research in this area.

References

Angert A and Luz B (2001) Fractionation of oxygen isotopes by root respiration: Implications for the isotopic composition of atmospheric O_2. Geochim Cosmochim Acta 65: 1695–1701

Bigeleisen J and Wolfsberg M (1959) Theoretical and experimental aspects of isotope effects in chemical kinetics. Adv Chem Phys 1: 15–76

Day DA, Krab K, Lambers H, Moore AL, Siedow JN, Wagner AM and Wiskich JT (1996) The cyanide-resistant oxidase: To inhibit or not to inhibit, that is the question. Plant Physiol 110: 1–2

Dry IB, Moore AL, Day DA and Wiskich JT (1989) Regulation of alternative pathway activity in plant mitochondria: Nonlinear relationship between electron flux and the redox poise of the quinone pool. Arch Biochem Biophys 273: 148–157

Finnegan PM, Whelan J, Millar AH, Zhang Q, Smith MK, Wiskich JT and Day DA (1997) Differential exprèssion of the multigene family encoding the soybean mitochondrial alternative oxidase. Plant Physiol 114: 455–466

Gaston S, Ribas-Carbo M, Busquets S, Berry JA, Zabalza A and Royuela M (2003) Changes in mitochondrial electron partitioning in response to herbicides inhibiting branched-chain amino acid biosynthesis in soybean. Plant Physiol 133: 1351–1359

Gonzàlez-Meler MA, Ribas-Carbo M, Giles L and Siedow JN (1999) The effect of growth and measurement temperature on the activity of the alternative respiratory pathway. Plant Physiol 120: 765–772

Gonzàlez-Meler M, Thomas RB, Giles L and Siedow JN (2001) Metabolic regulation of leaf respiration and alternative pathway activity in response to phosphate supply. Plant Cell Environ 24: 205–215

Guy RD, Berry JA, Fogel ML and Hoering TC (1989) Differential fractionation of oxygen isotopes by cyanide-resistant and cyanide-sensitive respiration in plants. Planta 177: 483–491

Hayes JM (1983) Practise and principles of isotopic measurements in inorganic geochemistry. In: W.G. Meinschein (ed) Organic Geochemistry of Contemporaneous and Ancient Sediments, pp 5-1–5-31, Great Lakes Section, Society of Economic Paleontologists and Mineralogists, Bloomington, Ind

Henry BK, Atkin OK, Day DA, Millar AH, Menz RI and Farquhar G (1999) Calculation of the oxygen isotope discrimination factor for studying plant respiration. Aust J Plant Physiol 26: 773–780

Hoefnagel MHN, Millar AH, Wiskich JT and Day DA (1995) Cytochrome and alternative respiratory pathways compete for electrons in the presence of pyruvate in soybean mitochondria. Arch Biochem Biophys 318: 394–400

Hoefs J (1987) Stable Isotope Geochemistry. Springer-Verlag, Berlin

Lane GA and Dole M (1956) Fractionation of oxygen isotopes during respiration. Science 123: 574–576

Lennon AM, Neueschwander UH, Ribas-Carbo M, Giles L, Ryals JA and Siedow JN (1997) The effects of salicylic acid and TMV infection upon the alternative oxidase of tobacco. Plant Physiol 115: 783–791

Mariotti A, Germon JC, Hubert P, Kaiser P, Letolle R, Tardieux A and Tardieux P (1981) Experimental determination of nitrogen kinetic isotope fractionation: Some principles; Illustration for the denitrification and nitrification processes. Plant Soil 62: 413–430

Millar AH, Atkin OK, Menz RI, Henry B, Farquhar G and Day DA (1998) Analysis of respiratory chain regulation in roots of soybean seedlings. Plant Physiol 117: 1083–1093

Millenaar FF, Roelofs R, Gonzàlez-Meler MA, Siedow JN Wagner, AM and Lambers H (2000). The alternative oxidase during low-light conditions. Plant J 23: 623–632

Millenaar FF, Gonzàlez-Meler MA, Fiorani F, Welschen R, Ribas-Carbo M, Siedow JN, Wagner AM and Lambers H (2001) Regulation of alternative oxidase activity in six wild monocotyledons species. An in vivo study at the whole root level. Plant Physiol 126:376–387

Moore AL and Siedow JN (1991) The regulation and nature of the cyanide-resistant alternative oxidase of plant mitochondria. Biochim Biophys Acta 1059: 121–140

Nagel OW, Waldron S and Jones HG (2001) An off-line implementation of the stable isotope technique for measurements of alternative respiratory pathway activities. Plant Physiol 127: 1279–1286

Neter J and Wasserman W (1974) Applied Linear Statistical Models: Regression, Analysis of Variance, and Experimental Designs. Richard D. Irwin Inc., Homewood, IL

Noguchi K, Go C-S, Terashima I, Ueda S and Yoshinari T (2001) Activities of the cyanide-resistant respiratory pathway in leaves of sun and shade species. Aust J Plant Physiol 28: 27–35

Peñuelas J, Ribas-Carbo M and Giles L (1996) Effects of allelochemicals on plant respiration and oxygen isotope fractionation by the alternative oxidase. J Chem Ecol 22: 801–805

Ribas-Carbo M, Berry JA, Yakir D, Giles L, Robinson SA, Lennon AM and Siedow JN (1995) Electron partitioning between the cytochrome and alternative pathways in plant mitochondria. Plant Physiol 109: 829–837

Ribas-Carbo M, Lennon AM, Robinson SA, Giles L, Berry J and Siedow JN (1997) The regulation of the electron partitioning between the cytochrome and alternative pathways in soybean cotyledon and root mitochondria. Plant Physiol 113: 903–911

Ribas-Carbo M, Aroca R, Gonzàlez -Meler MA, Irigoyen JJ and Sanchez-Diaz M (2000a) The electron partitioning between the cytochrome and alternative respiratory pathways during chilling recovery in two cultivars differing in chilling sensitivity. Plant Physiol 122:199–204

Ribas-Carbo M, Robinson SA, Meler-Gonzàlez MA, Lennon AM, Giles L, Siedow JN and Berry JA (2000b) Effects of light on respiration and oxygen isotope fractionation in soybean cotyledons. Plant Cell Environ 23: 983–989

Robinson SA, Yakir D, Ribas-Carbo M, Giles L, Osmond CB, Siedow JN and Berry JA (1992) Measurements of the engangement of cyanide-resistant respiration in the Crassulacean acid metabolism plant *Kalanchoë daigremontiana* with the use of on-line oxygen isotope discrimination. Plant Physiol 100: 1087–1091

Robinson SA, Ribas-Carbo M, Yakir D, Giles L, Reuveni Y and

Berry JA (1995) Beyond SHAM and cyanide: Opportunities for studying the alternative oxidase in plant respiration using oxygen isotope discrimination. Aust J Plant Physiol 22: 487–496

Saisho D, Nambara, E, Naito s, Tsutsumi N, Hirai A and Nakazono M. (1997). Characterization of the gene family for alternative oxidase from *Arabidopsis thaliana*. Plant Mol Biol 35: 585–596

Chapter 4

Respiration in Photosynthetic Cells: Gas Exchange Components, Interactions with Photorespiration and the Operation of Mitochondria in the Light

Vaughan Hurry*[1], Abir U. Igamberdiev[2], Olav Keerberg[3], Tiit Pärnik[3], Owen K. Atkin[4], Joana Zaragoza-Castells[4] and Per Gardeström[1]

[1]Umeå Plant Science Centre, Department of Plant Physiology, Umeå University, S-901 87 Umeå, Sweden; [2]Department of Plant Science, University of Manitoba, Winnipeg, MB, R3T 2N2, Canada; [3]Department of Plant Physiology, Institute of Experimental Biology, The Estonian Agricultural University, 76902 Harku, Estonia; [4]Department of Biology, University of York, York YO10 5YW, U.K.

*Author for correspondence, email: Vaughan.Hurry@plantphys.umu.se

H. Lambers and M. Ribas-Carbo (eds.), Plant Respiration, 43–61.

Summary

According to gas exchange measurements, mitochondrial oxygen consumption in the light is always fast, while respiratory CO_2 evolution is markedly decreased (compared with rates in darkness). We analyze the metabolic events that lead to such contrasting responses. In the light, the generation of NADH in mitochondria, both in the glycine decarboxylase reaction and in the tricarboxylic acid cycle, leads to increased NAD(P)H levels, which may increase the activity of the rotenone-insensitive NAD(P)H dehydrogenases. The resulting increase of the reduction level of ubiquinone activates the alternative oxidase. Stabilization of (photo)respiratory flux during the transition from darkness to light takes place at higher NADH/NAD^+ and ATP/ADP ratios. Maintenance of fast rates of mitochondrial electron transport in the light is facilitated by the import of oxaloacetate (OAA) from the cytosol to remove NADH, and by the export of citrate to the cytosol. This reduces the flow of metabolites in the tricarboxylic acid cycle, decreasing decarboxylation rates, while the rate of oxygen consumption reactions remain fast.

I. Introduction

In darkness, respiration provides for all of the cell's requirements for ATP and the metabolic intermediates needed for growth, maintenance, transport and nutrient assimilation. However in the light, ATP and reducing equivalents are formed in the chloroplasts by photosynthesis and these are used for carbon and nitrogen assimilation. The resulting products of these assimilatory reactions are transported to the cytosol where they are either exported to distant sinks or respired to produce energy and carbon skeletons through the operation of glycolysis and the TCA cycle and the coupled reactions of the electron transfer chain and oxidative phosphorylation. In C_3 plants, a fraction of photosynthetically assimilated carbon is also unavoidably dissipated through photorespiration, which begins with the oxygenation reaction of Rubisco. Some of the glycolate formed through photorespiration is converted back to 3-PGA through the combined activity of the chloroplasts, peroxisomes and mitochondria, resulting in the photorespiratory carbon and nitrogen cycles (Gardeström et al., 2002). In photorespiration, O_2 uptake is associated with the chloroplastic Rubisco and peroxisomal glycolate oxidase reactions, as well as with the oxidation of a part of photorespiratory NADH in the mitochondrial electron transport chain, whereas CO_2 release is due to mitochondrial decarboxylation of glycine. In the nitrogen cycle, NH_3 is released in the conversion of glycine to serine in the mitochondria, and re-assimilated into the primary amino acids glutamine and glutamate by the chloroplastic glutamine synthetase-glutamate synthase (GS-GOGAT) system (Keys et al., 1978; Bergmann et al., 1981; Wallsgrove et al., 1987). The cycle is closed when glutamate is used as the amino donor for the transamination of glyoxylate to glycine.

Respiratory and photorespiratory processes are well separated with respect to both carbon intermediates and compartmentation. However, the two pathways interact in the mitochondria, and the rate of respiration (often incorrectly referred to as a 'dark respiration') is strongly affected by both photosynthesis and photorespiration. Early isotopic studies demonstrated that leaves exposed to $^{14}CO_2$ also evolved radioactive CO_2 after long-term illumination (Zalensky et al., 1955; Doman, 1959; Goldsworthy, 1966), and in recent years the interactions between photosynthesis, respiration and photorespiration have become accepted as being major events in photosynthetic cells (Gardeström and Lernmark, 1995; Krömer, 1995; Hoefnagel et al., 1998; Atkin et al., 2000a, b; Gardeström et al., 2002; Igamberdiev and Lea, 2002). However, despite the increasing attention given to respiration in the light, considerable confusion has persisted in the literature over whether or not respiration in the light is slower than in the dark, and what role the photorespiratory carbon and nitrogen cycles play in the restriction of respiration in the light. There are two elements to this confusion. The first is that O_2 consumption is likely faster in the light than in darkness (Atkin et al., 2000b), whereas respiratory CO_2 release is likely to be slower in the light. The second is the extent to which respiratory

Abbreviations: AOX – alternative oxidase; C_i – internal CO_2; GDC – glycine decarboxylase complex; GO – glycolate oxidase; ICDH – isocitrate dehydrogenase; OAA – oxaloacetate; OG – 2-oxoglutarate; OPP – oxidative pentose phosphate pathway; PDC – pyruvate decarboxylase complex; PEP – phosphoenol pyruvate; PGA – 3-phosphoglycerate; *R* – respiration; TCA cycle – tricarboxylic acid cycle; Γ^* – CO_2 compensation point

CO_2 release is suppressed in the light; while many studies have indicated that mitochondrial CO_2 release is slower in the light (Brooks and Farquhar, 1985; Villar et al., 1994; Pärnik and Keerberg, 1995; Atkin et al., 1997, 2000a), several studies have reported no inhibition (Loreto et al., 1999; Loreto et al., 2001; Pinelli and Loreto, 2003) or even faster rates of CO_2 release in the light (Hurry et al., 1996). To address this confusion, in this chapter we discuss how respiration proceeds in the light, with an emphasis on substrate supply, and consider the major events that regulate mitochondrial function and respiration in photosynthetic tissues.

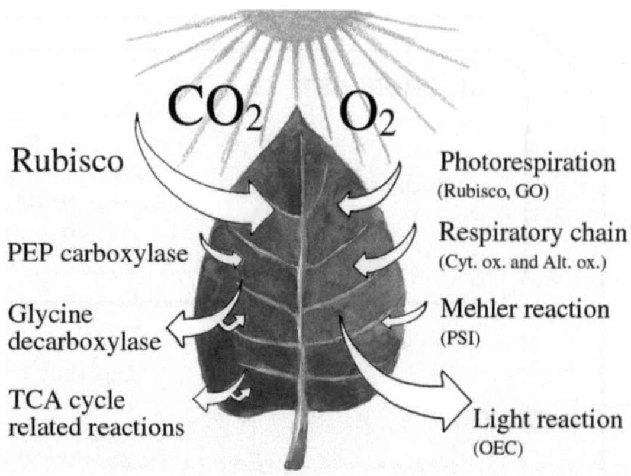

Fig. 1. Schematic representation of primary leaf gas exchange processes that take place in an illuminated leaf. Divided arrows indicate reassimilation.

II. Leaf Gas Exchange Components

In the literature, there has been some confusion about respiratory rates in the light, because there has been a lack of reference to whether it is the 'CO_2 component' or the 'O_2 component' that was considered in the various studies. To emphasize the differences that can be observed we have summarized the main gas fluxes occurring in an illuminated leaf (Fig. 1).

A. The CO_2 Component

CO_2 fixation by Rubisco is the dominant CO_2-uptake component. PEP carboxylase also fixes some CO_2 in the cytosol, but this is usually less than 5% of that fixed by Rubisco (Melzer and O'Leary, 1987; Raven and Farquhar, 1990). Respiration and photorespiration both release CO_2 in the mitochondria, and these two processes are the main pathways for decarboxylations in the light. Minor gas fluxes associated with the CO_2 released by specific biosynthetic reactions, mainly associated with OPP cycle, fatty-acid and amino-acid synthesis have not been included. Divided arrows in Fig. 1 symbolize intra- and intercellular refixation of respiratory and photorespiratory CO_2 by Rubisco (and PEP carboxylase). This flux is likely significant, but is difficult to quantify with current gas exchange methods. The total rate of leaf respiratory fluxes was calculated from PC_w curves, describing the dependence of the rate of photosynthesis (P) on the internal concentration of CO_2 (C_w) (Laisk and Oja, 1998). Using a model taking into account stomatal resistance, together with various gas exchange parameters, Gerbaud and Andre (1987) calculated that the actual rate of photorespiratory CO_2 release in C_3 plants could be 20 to 100% faster than the measured rate of CO_2 efflux from the leaf. Furthermore, using a radio-gasometric method, Pärnik et al. (1976) calculated that, depending on the stomatal status, the rate of leaf re-assimilation in C_3 plants was 10 to 65% of the total rate of decarboxylations (Pärnik and Keerberg, 1995; Pärnik et al., 2002). In C_3-C_4 intermediate and C_4 *Flaveria* species, 85 to 100% of the CO_2 was re-assimilated (Bauwe et al., 1987). Thus, there is considerable re-fixation of internally evolved CO_2 in the light, and this needs to be taken into account when assessing rates of *R* and photorespiration in the light.

B. The O_2 Component

During photosynthesis, several O_2-consuming reactions occur in parallel with O_2 production by Photosystem II. Two of these are the reactions related to photorespiratory O_2 uptake by Rubisco and glycolate oxidase. The terminal oxidases in the mitochondrial electron transport chain, cytochrome oxidase and alternative oxidase, represent additional O_2 uptake components (Fig. 2). Substrates for these oxidases can be derived from both respiration (TCA cycle related reactions) and photorespiration (glycine decarboxylation). Finally, the Mehler-reaction associated with PSI results in light-dependent O_2 uptake (Robinson, 1988). The quantitative significance of the Mehler reaction is often regarded as minor, but it has been proposed to function as a protective mechanism under high irradiances, and under such circumstances it may be more important (Osmond and Grace, 1995). Several approaches have been used to investigate the degree of mitochondrial O_2

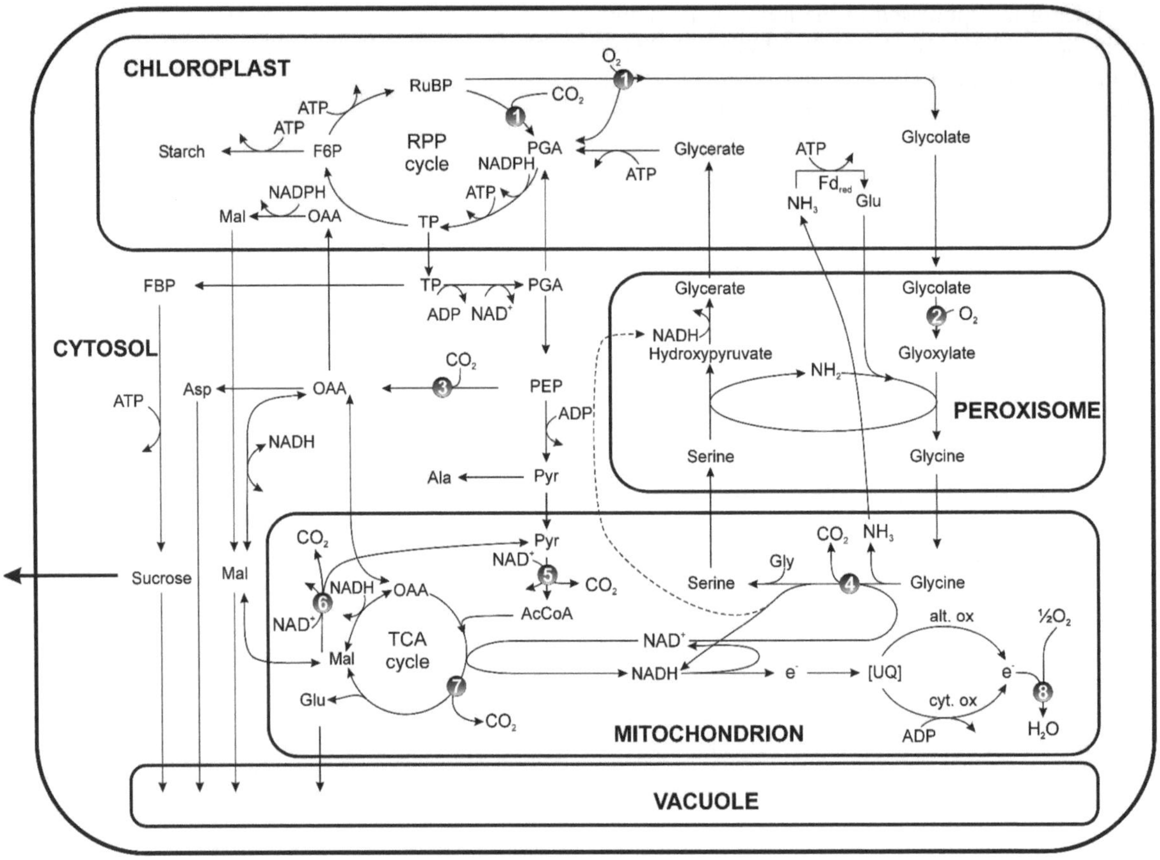

Fig. 2. Schematic representation of the main pathways of carbon metabolism in photosynthetic cells, showing the interactions between the different organelles and cellular compartments, and highlighting the key gas exchange points. The numbers indicate the various enzymes catalyzing the gas exchange steps: 1, Rubisco; 2, glycolate oxidase; 3, PEP carboxylase; 4, glycine decarboxylase; 5, pyruvate dehydrogenase; 6, NAD-malic enzyme; 7, decarboxylation reactions of the TCA cycle—isocitrate dehydrogenase and ketoglutarate dehydrogenase; 8, terminal oxidases of the mitochondrial electron transport chain—cytochrome oxidase and alternative oxidase.

uptake during photosynthesis. One is to measure oxygen exchange using mass spectrometry and the stable ^{16}O and ^{18}O isotopes. Estimates of respiratory activity in the light using this technique vary, with some investigators finding a decrease in the rate of mitochondrial oxygen uptake, and others do not (Raghavendra et al., 1994; Atkin et al., 2000b). These apparent discrepancies in the measured responses of mitochondria to light may reflect the variability in carbon supply from photosynthesis, the degree to which photorespiratory NADH is consumed in the mitochondria, and the degree to which excess photosynthetic redox equivalents are exported from the chloroplast (Atkin et al., 2000b). Mitochondrial O_2 uptake is thus likely to exceed that of dark O_2 uptake under high-light conditions that favor each of these components.

III. Respiratory and Photorespiratory Decarboxylations in the Light

As described above, the decarboxylation reactions coupled to respiration are mainly associated with the TCA cycle and related reactions, whereas the decarboxylation reactions linked to photorespiration are due to glycine decarboxylation in the glycolate pathway. To be able to determine the rate of respiration in the light these components must be separated.

A. Respiratory Decarboxylations in the Light

Estimates of respiratory decarboxylation in the light, and thus TCA cycle activity, have been made by several groups, by feeding ^{14}C-labeled glycolytic or TCA cycle substrates in pulse-chase experiments, using

$^{14}CO_2$ (Pärnik and Keerberg, 1995) or $^{13}CO_2$ (Loreto et al., 1999; Loreto et al., 2001), and regular gas exchange approaches such as the Laisk (1977) and Kok (1948) methods. A description of these approaches is given in Atkin et al. (2000b), with recent advances in the use of $^{13}CO_2$-insensitive infrared gas analyzers being described by Loreto et al. (1999, 2001). A review of the literature shows that mitochondrial CO_2 release continues in the light, but the effect of light on respiratory flux is variable. In some cases, respiratory CO_2 release continues in the light at rates that are similar (Marsh et al., 1965; Chapman and Graham, 1974) or even faster (Hurry et al., 1996) than in darkness. Similarly, in algae and photoautotrophic cells under low photorespiratory conditions, mass spectrometric studies using stable $^{12}C/^{13}C$ isotopes have shown that light has little effect on respiratory CO_2 release (Weger et al., 1988; Turpin et al., 1990; Avelange et al., 1991). In contrast, substantial inhibition of respiratory CO_2 release in the light has been reported under conditions that favor photorespiration (low internal CO_2 concentrations and/or high temperatures). This includes numerous studies using the Kok (1948) and Laisk (1977) methods (Brooks and Farquhar, 1985; Kirschbaum and Farquhar, 1987; Villar et al., 1994, 1995; Atkin et al., 1997, 1998ab, 2000a). Moreover, using a ^{14}C-labeling technique that takes into account re-fixation of respiratory CO_2 by Rubisco, Pärnik and Keerberg (1995) demonstrated that respiration *(R)* is lower in the light than in darkness in several species. Light inhibited leaf *R* by 14, 46 and 55% in wheat (*Triticum aestivum*), tobacco (*Nicotiana tobacum*) and barley (*Hordeum vulgare*), respectively (Pärnik and Keerberg, 1995). Previous experiments using $^{14}CO_2$ confirm that *R* is inhibited by light in wheat (McCashin et al., 1988). In contrast, using the ^{14}C method, Hurry et al. (1996) found *R* was stimulated in the light in winter rye (*Secale cereale*), showing that *R* is not repressed in the light in all species under photorespiratory conditions. Moreover, although Loreto et al. (1999, 2001) found *R* to be lower in the light than in the dark, the difference in rates appeared to reflect re-fixation of respiratory CO_2 by Rubisco in the light, rather than actual light inhibition of respiratory flux. However, Loreto et al. (2001) did find that light suppresses leaf *R* per se in salt- and water-stressed *Zea mays* leaves. Thus, the response of leaf *R* to irradiance is highly variable among plant species, and is affected by environmental conditions.

B. Photorespiratory Decarboxylations

Rubisco ultimately determines the input of carbon into the glycolate pathway. The flux of carbon through the photorespiratory glycolate cycle has been experimentally determined to be in the range of 15 to 30% of the rate of photosynthetic CO_2 fixation (Keerberg and Viil, 1988; Keerberg et al., 1989). As photorespiration is as dynamic a process as photosynthesis, the rate can be modulated by changing the light intensity, lowering the internal CO_2/O_2 concentrations or changing the leaf temperature (Ogren, 1984). These conditions also modulate CO_2 fixation, leading to changes in the production of primary photosynthates. Using the radiogasometric method, it has been demonstrated that photorespiratory decarboxylations in leaves of wheat, tobacco and barley account for 80–90% of total decarboxylations in normal air and at saturating irradiances (Pärnik and Keerberg, 1995). Similarly, estimations using the CO_2-compensation point method also show that photorespiration is three to five times faster than respiratory rates in aspen (*Populus tremula*) leaves (Laisk, 1977) and eight to nine times in cotton (*Gossypium hirsutum*) leaves (Rasulov and Oja, 1982). In C_3 species, photorespiration is thus the prevailing component of the total decarboxylations in the light under normal physiological conditions. However, there is no fixed ratio between photorespiratory and respiratory rates of decarboxylation. For example, during leaf development in cotton plants, both photorespiration and respiration decreased per unit leaf area, and the ratio remains unchanged (Rasulov et al., 1983). However, following the removal of sink organs, leaves of cotton plants showed an increase only of the respiratory component, with a concomitant decrease in the photorespiration/respiration ratio (Rasulov, 1986). Similarly, in barley leaves measured at 26, 13 and 7 °C, lowering the temperature resulted in a decrease in the rate of photorespiration (2.3, 0.9 and 0.5 $\mu mol\ CO_2\ m^{-2}\ s^{-1}$, respectively), with a concomitant increase of respiration in the light [0.3, 1.0 and 1.2 $\mu mol\ CO_2\ m^{-2}\ s^{-1}$, respectively (Pärnik and Keerberg, 1995)], again resulting in a decrease in the ratio of photorespiration to respiration. Furthermore, in transgenic potato (*Solanum tuberosum*) with reduced glycine decarboxylase, slower rates of photorespiration were compensated by faster rates of respiration, resulting in total decarboxylations being equal to those in wild-type plants (Keerberg et al., 1999). These data indicate that in the light the cellular requirements for ATP, reducing equivalents and carbon skeletons are met by both photorespiration and respiration. Together, they function to balance the needs of the cell depending on the environmental and developmental condition of the plant, and can respond plastically to varying metabolic demands.

C. Estimates of Respiratory Decarboxylation in the Light Using Different Methods

A factor that can contribute to the variable effects of light on leaf respiration are the different methodologies used to estimate respiratory CO_2 release in the light. Differences in steady-state conditions during measurements (e.g., at ambient or low atmospheric CO_2 concentrations) and whether or not re-fixation of respiratory CO_2 by Rubisco is taken into account could affect estimates of how much respiratory CO_2 is released into the atmosphere by leaves in the light. Here we compare two methods (Laisk, 1977; Pärnik and Keerberg, 1995) to assess whether variation in the degree of inhibition in the light reflects methodological differences. We begin by briefly describing the two methods, and then show data comparing rates of respiratory CO_2 release in the light obtained using the two methods.

The radiogasometric method (Pärnik and Keerberg, 1995) provides estimates of mitochondrial CO_2 release in the light under ambient atmospheric conditions (e.g., 360 µL L^{-1} CO_2 and 21% O_2). This method distinguishes the substrates (primary or stored photosynthates) for, and the rates of, photorespiratory and respiratory carbon fluxes on the basis of different labeling kinetics in leaves exposed to $^{14}CO_2$ and different concentrations of oxygen. Re-fixation of respiratory CO_2 by Rubisco is accounted for by recording the initial rate of $^{14}CO_2$ efflux after $^{14}CO_2$-labeled leaves are transferred to a very high (30 mL L^{-1}) $^{12}CO_2$ concentration. This results in a reduced probability of the re-fixation of labeled $^{14}CO_2$, allowing the $^{14}CO_2$ released by mitochondrial respiration to be measured accurately. As a result, actual rates of mitochondrial non-photorespiratory CO_2 release are determined, rather than just the rate of respiratory CO_2 release from the leaf surface into the surrounding atmosphere (which might be substantially less than the actual rate of mitochondrial CO_2 release if substantial re-fixation is taking place). The method assumes that respiration in the light is not affected by decreasing the concentration of O_2 from 21 to 1.5%, which appears reasonable considering the affinity of the oxidases involved.

The Laisk method (Laisk, 1977) analyzes the rate of net CO_2 exchange at low internal CO_2 concentrations (c_i) and varying irradiances. Net CO_2 exchange is related to leaf *R* in the light (R_{light}) according to $A = v_c - 0.5v_o - R_{light}$, where v_c and $0.5v_o$ are the rates of carboxylation and oxygenation of Rubisco, respectively. At a low c_i value (Γ^*; typically around 40 µL L^{-1} for leaves measured at 25°C), CO_2 fixed by Rubisco is matched by the CO_2 that is released by glycine decarboxylation (i.e. $v_c = 0.5v_o$), resulting in net CO_2 exchange being equal to R_{light}. The Laisk method assumes that R_{light} is the same at Γ^* as it is at 360 µL L^{-1} CO_2. Although earlier studies suggested that mitochondrial respiration was inhibited by elevated concentrations of CO_2, and was therefore higher at low CO_2 concentrations (Drake et al., 1999), several recent studies have demonstrated that mitochondrial respiration is largely CO_2 insensitive (Amthor, 2000; Jahnke, 2001; Bruhn et al., 2002; Chapter 13, Gonzalez-Meler and Taneva). It seems likely, therefore, that the assumption that R_{light} is the same at Γ^* as it is at 360 µL L^{-1} CO_2, is correct. The method also assumes that R_{light} is not substrate limited during prolonged exposure to low c_i values; Atkin et al. (1998a) used a rapid gas exchange system to show that this assumption is valid.

Care must be taken, however, when using many of the commercially available clamp-on leaf chambers (e.g., Licor 6400 and Parkinson chambers) to determine R_{light} at Γ^* as CO_2 can diffuse through the gasket material that seals the leaf from the surrounding atmosphere (Bruhn et al., 2002). When CO_2 concentrations in the chamber are low (e.g., close to Γ^*) substantial inward diffusion of CO_2 from the surrounding atmosphere into the chamber is likely; this can result in substantial overestimates of respiratory CO_2 release by leaves (Fig. 3). Inward diffusion of respiratory CO_2 from darkened leaf material under the gasket into the chamber will also likely result in over-estimates of respiratory CO_2 release (Pons and Welschen, 2002), particularly in chambers where the area under the gasket is high in comparison with the area inside the cuvette. For example, in the Licor 6400 the gasket area (7.45 cm^2) is greater than the area inside the chamber (6 cm^2). We recently assessed the impact of gasket diffusion and inward diffusion of CO_2 from leaf material under the Licor 6400 gasket (using the 6 cm^2 chamber) on measured rates of leaf respiration in *Plantago lanceolata* (Joana Zaragoza Castells, Olav Keerberg, Tiit Pärnik and Owen Atkin, unpublished data). In addition to substantial gasket

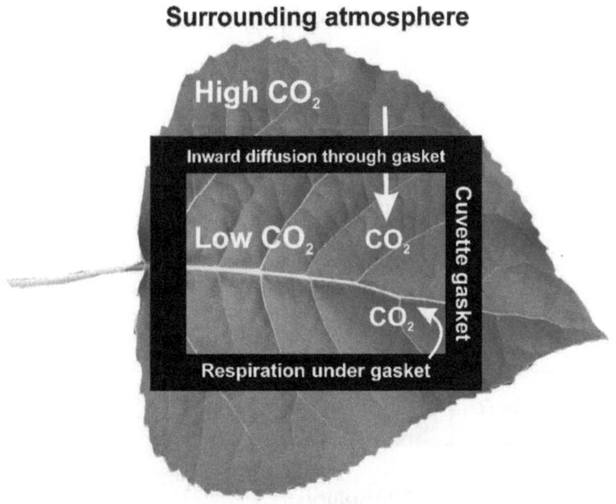

Fig. 3. Diagrammatic illustration of how CO_2 diffuses from the surrounding atmosphere into the chamber of many commercial gas analyzer systems (whenever the CO_2 concentration in the atmosphere is substantially higher than that within the chamber) and how CO_2 released from respiration by leaf material under the gasket can diffuse inward into the cuvette. Both factors result in the measured rate of CO_2 release (either in the light or dark) being an overestimate of the actual rate of respiration taking place within the chamber itself.

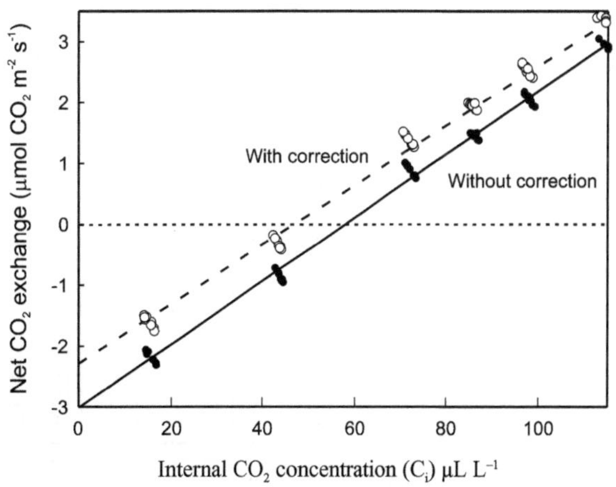

Fig. 4. Plot of net CO_2 exchange in the light versus internal CO_2 concentration for uncorrected (open symbols) and corrected (closed symbols) data. In this example, corrections were made for diffusion of CO_2 through the gasket of a Licor 6400 leaf chamber and for inward diffusion of CO_2 released by respiration taking place under the gasket. Rates of leaf respiration in the light are taken as the rate of net CO_2 release at Γ^* (approximately 40 μL L^{-1} in this case). Data are from a single *Plantago lanceolata* leaf (Joana Zaragoza-Castells and Owen Atkin, unpublished).

diffusion when CO_2 concentrations in the chamber were lower than that of the surrounding atmosphere (Fig. 3), approximately 45% of the CO_2 released by leaves under the gasket entered the chamber. To assess the impact of both factors on estimated rates of leaf respiration in the light, we corrected the rates of net CO_2 exchange in the light measured at several internal CO_2 concentrations. These data show that correction for CO_2 diffusion through the gasket from the atmosphere and inward diffusion of respired CO_2 from leaf material under the gasket increased net CO_2 release at low internal CO_2 concentrations (Fig. 4), and thus decreased estimates of leaf respiration in the light (Fig. 5). From these experiments, the estimates of leaf respiration in the light were nearly two-fold lower following correction for CO_2 diffusion under the gasket compared with estimates that are not corrected for diffusion, and the correction substantially increased the apparent inhibition of leaf respiratory CO_2 release (Fig. 5).

To assess whether the estimates of leaf respiratory CO_2 release in the light [obtained using a Licor 6400 system and Laisk (1977) method] were accurate, we compared the corrected rates with respiration rates in the light measured using the radiogasometric method (Pärnik and Keerberg, 1995). The corrected rates of respiration were similar to those obtained using the radiogasometric method (Fig. 5). Given that the latter method takes into account re-fixation of CO_2 and measures respiratory CO_2 release under ambient CO_2 conditions (rather than at Γ^*), it seems likely that accurate estimates of mitochondrial CO_2 release can be obtained using the Laisk method (Laisk, 1977), provided corrections for CO_2 diffusion from the atmosphere and from the darkened portions of the leaf under the gasket are made. However, failure to correct for CO_2 diffusion under the gasket will result in substantial overestimates of leaf respiratory CO_2 release in the light. Such errors could be even more substantial at high temperatures, where rates of respiration in darkness under the gasket are high (and thus contribution of inward diffusion of gasket respiratory CO_2 is high), but where actual rates of respiration in the light are relatively slow (due to the greater inhibition of respiration that occurs at high temperatures (Atkin et al., 2000a)).

IV. Availability of Substrates for Cellular Decarboxylations

A. Respiratory Decarboxylations

The substrates for respiratory decarboxylations are

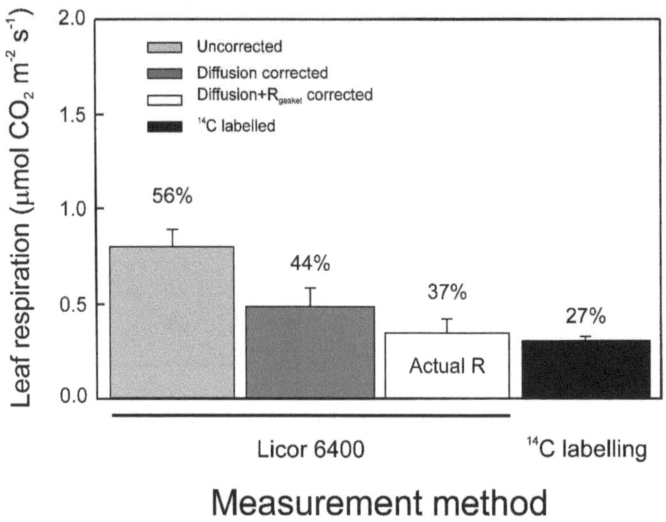

Fig. 5. Calculated rates of leaf respiration in the light using the Laisk (1977) and the ^{14}C radiogasometric (Pärnik and Keerberg 1995) methods. A Licor 6400 system was used to estimate rates of R_{light} according to the Laisk method (see Fig. 4), with corrections being made for diffusion of CO_2 through the gasket material and inward diffusion of CO_2 released by respiration taking place under the gasket (R_{gasket}). The percentage values represent the extent to which respiration continued in the light relative to rates measured in darkness. Error bars are standard errors (n=3).

supplied by glycolysis, either from the products of sucrose and starch breakdown or by diverting primary photosynthetic products such as triose phosphates directly to respiration (Fig. 2). Depending on tissue type and developmental stage, other carbohydrates, organic acids, seed lipids, and protein could also contribute to respiratory carbon flow.

The involvement of primary photosynthates in respiration in the light was shown in radio-labeling experiments where leaves were exposed to $^{14}CO_2$ for different intervals, and kinetics of ^{14}C incorporation into various compounds was analyzed (Ivanova et al., 1993; Keerberg et al., 1999). In different plant species, 25–80% of respiratory substrates are derived from primary photosynthates (Pärnik et al., 2002). However, respiratory decarboxylations of stored photosynthates are suppressed in the light, and this suppression is most severe in starch-accumulating species (e.g., *Arabidopsis* and tobacco) where utilization of stored photosynthates is 4- to 14-fold lower in the light than in darkness, and to a lesser extent in sucrose-accumulating cereals (e.g., wheat, barley and winter rye), where utilization of stored photosynthates is only about 2-fold lower in the light than in darkness (Pärnik et al., 2002). These findings suggest that the consumption of starch in respiratory decarboxylations might be inhibited in the light. This conclusion was checked experimentally using *Arabidopsis*, where the catabolism of labeled photosynthates was followed in the light and in the dark (Keerberg et al., 2001). It was found that the degradation of starch was blocked in the light (600 µmol m^{-2} s^{-1}) and the only stored substrate pools that were available as substrates for respiratory decarboxylations in the light were soluble compounds, mainly sucrose. Suppression of starch degradation in the light has also been observed in leaves of pea (*Pisum sativum*) (Kruger et al., 1983), sugar beet (*Beta vulgaris*) (Fox and Geiger, 1984) and in anaerobic cultures of *Chlamydomonas reinhardtii* (Gfeller and Gibbs, 1984). Inhibition was not observed in sugar beet leaves under low concentrations of CO_2 (Fox and Geiger, 1984), in *Chlorella* cells at high temperatures (Nakamura and Miyachi, 1982) or in tobacco transformants with repressed triose phosphate translocator (Hausler et al., 1998). However, no respiration measurements were performed in these studies. The mechanisms of light regulation of starch metabolism are not known but inhibition of starch degradation in the light appears to depend on several environmental and endogenous factors, and it may be that the degradation of starch takes place only when the rate of photosynthesis drops below some threshold value (Fondy et al., 1989). The net result of this difference in starch- vs. sucrose-accumulating plants is that in sucrose-accumulating species such as the cereals, the total rate of respiratory decarboxylations (primary + stored photosynthates) in the light is similar to or even higher than in darkness, while in leaves of starch-accumulating species such as Arabidopsis and tobacco, total respiratory decarboxylation is generally lower in the light.

B. Photorespiratory Decarboxylations

It is generally agreed that photorespiratory glycine is the main substrate for leaf mitochondria in the light (Krömer, 1995). The amount of glycine available to the mitochondria is determined by the rate of the oxygenation reaction in the chloroplast stroma. The rate of glycine decarboxylation, calculated from the rate of carbon flux through the glycolate cycle, correlates with the rate of photorespiratory decarboxylations of newly fixed products of photosynthesis (Kumarasinghe et al., 1977; Keerberg et al., 1989; Ivanova et al., 1993). However, substrates for photorespiratory decarboxylations could also be derived from stored products of photosynthesis (Goldsworthy, 1970). After incubating illuminated leaves with labeled CO_2, photorespiratory $^{14}CO_2$ efflux into CO_2-free medium continued for a long time after the $^{14}CO_2$ was removed (Laisk, 1977). Mahon et al. (1974) also noted that after a 15 min exposure of sunflower (*Helianthus annuus*) leaves to $^{14}CO_2$ the specific radioactivity of PGA, glycine and serine was significantly lower than the specific radioactivity of the $^{14}CO_2$ originally fed to the leaves. This indicated a flow of carbon into the reductive pentose phosphate cycle from some unlabeled source. Using the method of Pärnik and Keerberg (1995) it has been shown that in several species 18 to 27% of the substrates for the oxygenase reaction (or photorespiratory decarboxylations) are derived from stored photosynthates (Hurry et al., 1996). Yamauchi and Yamada (1985) have also shown that glycolate can be synthesized from storage material at high O_2/CO_2 ratios. At low concentrations of CO_2, near to the CO_2 compensation point where most of gasometric measurements of respiration are performed, the substrates for photorespiration must be primarily derived from stored photosynthates (the contribution of primary photosynthates must be low when the net photosynthetic rate is close to zero).

V. Metabolic Fluxes in Plant Mitochondria in the Light

For an analysis of the mechanisms causing the reported gas exchange rates, we will turn to a description of the actual metabolic fluxes through mitochondria in the light. This will allow us to estimate the biochemical background of the O_2 and CO_2 components of gas exchange in the light, and to explain the observed light effects on respiration.

A. Estimation of Carbon Fluxes in vivo

In the absence of photorespiration, at saturating light and high temperature (30 °C), photosynthetic rates can reach a maximum of around 7 μmol (O_2 evolved) mg^{-1} (Chl) min^{-1} (calculated from Edwards and Walker (1983)). At 25 °C, photosynthetic rates can be as high as 5 μmol O_2 mg^{-1} (Chl) min^{-1}. The rate of respiration in darkness during the post-illumination period (which is 2–2.5 times faster than in prolonged darkness) is about ten times slower than the photosynthetic rate (Byrd et al., 1992). Thus, at 25 °C respiration can be as high as 0.5 μmol O_2 consumed mg^{-1} (Chl) min^{-1}, and in prolonged darkness near 0.2 μmol (O_2 consumed) mg^{-1} (Chl) min^{-1}. Taking the mitochondrial volume to be 4 μl mg^{-1} Chl [determined to be approximately the same for barley, spinach (*Spinacia oleracea*) and potato (Winter et al., 1993, 1994)], we can calculate maximal respiratory rates of 120 nmol (O_2 consumed) mg^{-1} (mitochondrial protein) min^{-1} after illumination, and near 50 nmol (O_2 consumed) mg^{-1} (mitochondrial protein) min^{-1} during prolonged darkness. The latter value corresponds to the maximum capacity of isocitrate oxidation, the bottleneck in the TCA cycle (Day and Wiskich, 1977).

The rate of photorespiration depends on the CO_2 concentration near the sites of carboxylation. Even at atmospheric CO_2 levels of 370 μL L^{-1} the pCO_2 is around 260 μL L^{-1} in the sub-stomatal cavities and 180 μL L^{-1} near Rubisco (von Caemmerer and Evans, 1991), and may be as low 90 μL L^{-1} during drought or in xeromorphic plants (Di Marco et al., 1990). The pO_2 in the interior of C_3 leaves is practically constant (corresponding to atmospheric pO_2), both in the dark and during illumination, due to effective ventilation (Ligeza et al., 1997). Calculations of the ratio of oxygenation to carboxylation of ribulose-1,5-bisphosphate (V_o/V_c) and of the rate of photorespiration, based on the O_2/CO_2 ratio in the air, suggests that the photorespiratory rate is about 20–25% the rate of carboxylation at 370–300 μL L^{-1} CO_2 (Sharkey, 1988). This value is in good agreement with the experimental estimations of 15–30% (Keerberg and

Viil, 1988; Keerberg et al., 1989), increasing up to 50% during CO_2 depletion (e.g., glacial CO_2 levels or following stomatal closure due to water stress). The latter value corresponds to a rate of 1200 nmol (CO_2 evolved) mg^{-1} (mitochondrial protein) min^{-1} or 600 nmol O_2 consumed mg^{-1} (mitochondrial protein) min^{-1} by the glycine decarboxylase reaction (assuming that all NADH is oxidized in the electron transport chain). The glycine decarboxylase complex (GDC) accounts for approximately 50% of the total protein found in the matrix of leaf mitochondria (Day et al., 1985; Oliver, 1994), and inhibition of GDC by as little as 30–50% leads to the accumulation of glycine, increased susceptibility to water stress, and to formate accumulation (Wingler et al., 1999a; Wingler et al., 1999b; Heineke et al., 2001). These data indicate that the GDC operates at sub-saturating substrate levels at current atmospheric concentrations of CO_2, and that the amount of GDC protein is close to a possible maximum in order to support a very fast rate of photorespiratory flux.

B. Comparison to the Rates in Isolated Mitochondria

The rates for oxidation of glycine in isolated mitochondria are usually in the range of 200 nmol (O_2 consumed) mg^{-1} (mitochondrial protein) min^{-1}, and for isocitrate, they are not more than 50 nmol (O_2 consumed) mg^{-1} (mitochondrial protein) min^{-1} (Wiskich and Dry, 1985; Bykova et al., 1998). In the presence of malate, the rate of glycine oxidation increases to 500–600 nmol (O_2 consumed) mg^{-1} (mitochondrial protein) min^{-1} (Wiskich and Dry, 1985; Bykova et al., 1998), so it is close to the maximal possible photorespiratory fluxes in vivo. As yet, there is no clear explanation for the activation of glycine oxidation by malate; however, OAA formation during malate oxidation may recycle NADH formed in the GDC reaction, and facilitate NADH oxidation in the mitochondrial electron transport chain (Wiskich and Dry, 1985).

VI. Change of Mitochondrial Electron Transport in the Light

At low rates of mitochondrial NADH production (usually in unstressed non-photosynthetic tissues) complex I is the major NADH dehydrogenase, and the cytochrome pathway fulfils the energetic demands of the cell. In the light, when ATP is formed photosynthetically in the chloroplasts, the role of the mitochondria changes, and the production of carbon skeletons and cellular redox regulation becomes more important (Gardeström et al., 2002). To some extent this is supported by a high flux of electrons through the mitochondrial electron transport chain, supported by the operation of pathways that are not coupled to ATP synthesis. These pathways include the rotenone-resistant NADH and NADPH dehydrogenases, and the cyanide-insensitive alternative oxidase.

A. NAD(P)H Dehydrogenases on the Mitochondrial Membranes

Three NAD(P)H dehydrogenases on the inner side of the inner mitochondrial membrane are responsible for the oxidation of the reducing equivalents formed in the mitochondrial matrix. The *Km* for NAD(P)H of complex I is about 7 μM, the *Km* of the rotenone-resistant NADH dehydrogenase is 80 μM, and the *Km* of the NADPH dehydrogenase (which is Ca^{2+}-dependent) is 25 μM (Møller, 2001). There are also two external dehydrogenases, one NADH and the other NADPH dependent, on the inner mitochondrial membrane; both are Ca^{2+}-dependent. The NADH dehydrogenase on the outer mitochondrial membrane may also be connected to the electron transport chain of mitochondria (Møller and Lin, 1986). The conclusion that there are separate dehydrogenases specific for NADH or NADPH, with different membrane localization, is based on currently available kinetic data (Melo et al., 1996; Bykova and Møller, 2001; Møller, 2001).

B. Consequences of the NADH Produced from Glycine Oxidation

Plant mitochondria contain a pool of NAD(H) of around 1.5–2 mM, and a much smaller pool of NADP(H) of around 0.15–0.2 mM (Igamberdiev et al., 2001). In darkness, and during photosynthesis at high pCO_2, the NAD(H) pool is highly oxidized with an $NADH/NAD^+$ ratio of around 0.05. Based on estimations of metabolic fluxes, during photorespiratory glycine oxidation, NADH will be produced at rates that are several-fold higher than could be produced by the TCA cycle alone. Consequently, the NADH/NAD ratio in mitochondria increases from around 0.05 to 0.2–0.3 in photorespiratory conditions (Wigge et al., 1993; Igamberdiev and Gardeström, 2003).

This increase in the NADH/NAD ratio corresponds to a total NADH concentration of around 0.4 mM in photorespiratory conditions, compared to 0.15 mM in non-photorespiratory conditions (Igamberdiev and Gardeström, 2003). Such concentrations of NADH would inhibit GDC, which has a *Ki* for NADH of 15 μM (Oliver, 1994). However, the actual concentration of free NADH will be lower than this, especially in green tissue where the GDC concentration is about 0.2 mM, because the GDC will bind the majority of the available NADH (Møller, 2001). Even so, measurements of an increased total NADH content in mitochondria in photorespiratory conditions, relative to darkness and non-photorespiratory conditions, would reflect a higher redox state of the mitochondrial NAD(H) pool.

The NADH generated in photorespiratory conditions has important consequences for the metabolism of photosynthetic cells. As pointed out above, the redox state of the mitochondrial NAD(H) pool is increased, which will result in the engagement of the rotenone-insensitive NADH dehydrogenase (Igamberdiev et al., 1997; Bykova et al., 1998). Under maximal rates of glycine plus malate oxidation in isolated pea leaf mitochondria, complex I was responsible for the oxidation of almost 50% of the internal NAD(P)H, with the rotenone-insensitive NADH dehydrogenase and the NADPH dehydrogenase (which can reach near 15% of complex I activity) accounting for the remaining activity (Bykova et al., 1998). Thus, the induction/activation of the non-coupled pathways of electron transport in the light prevents the depletion of ADP and $NAD(P)^+$ which is necessary for the turnover of metabolic cycles (Igamberdiev, 1999; Igamberdiev and Kleczkowski, 2003). It also minimizes the risk that the levels of reduced pyridine nucleotides, reduced ubiquinone and the ATP/ADP ratio increase to the point that they are harmful to cellular metabolism (Purvis, 1997; Maxwell et al., 1999).

Another important consequence of the increasing NADH level in the light is its connection to an increase in the reduction level of NADPH. In darkness, the NADP(H) pool of pea leaf protoplasts is about 20% reduced which increases to about 50% in the light during non-photorespiratory conditions, and to about 75% in photorespiratory conditions (Igamberdiev and Gardeström, 2003). This increase can be achieved via trans-hydrogenation between NADH and $NADP^+$ (Bykova and Møller, 2001). Plant mitochondria lack a proton-translocating transhydrogenase (Bykova et al., 1998), but possess two non-energy-linked transhydrogenase activities, one belonging to a side reaction of complex I, and the other to a soluble (possibly weakly attached to the membrane) transhydrogenase-like enzyme. The high mitochondrial NADPH level will lead to the engagement of the internal NADPH dehydrogenase, as shown in experiments that revealed the involvement of an internal NADPH dehydrogenase during glycine oxidation (Bykova and Møller, 2001). The increased NADPH levels inside mitochondria under photorespiratory conditions will facilitate the reduction of glutathione, activate the alternative oxidase (possibly via a thioredoxin system), and also result in isocitrate oxidation (see below and Fig. 6).

Interestingly, studies of gene expression of complex I and two homologues of bacterial and yeast NADH dehydrogenases in potato leaves have shown that expression of the internal rotenone-insensitive NADH-dehydrogenase is completely light dependent, and shows a diurnal rhythm with a sharp maximum just after dawn (Svensson and Rasmusson, 2001; Møller, 2002; Michalecka et al., 2003). However, the importance of this finding for the in vivo activity of the complexes needs to be established. Despite the involvement of non-phosphorylating bypasses, the mitochondrial ATP/ADP ratio increases in the light under photorespiratory conditions, and this increase is directly coupled to the oxidation of photorespiratory glycine (Gardeström and Wigge, 1988). Furthermore, many enzymes participating in the TCA cycle can be phosphorylated, including aconitase, succinyl-CoA ligase, NAD-isocitrate dehydrogenase, NAD-malic enzyme, and two subunits of malate dehydrogenase as well as several complexes of the mitochondrial electron transport chain (Bykova et al., 2003). This represents a potentially very important mechanism for regulating the activity of the TCA cycle, but it is not yet clear how phosphorylation is linked to regulation of the activity of these proteins.

C. The Role of Alternative Oxidase

Electron transport to the alternative oxidase is not coupled to ATP production, resulting in a very flexible coupling between electron transport and oxidative phosphorylation. For a long time the alternative oxidase was regarded as a more or less passive overflow mechanism. Recent progress in understanding the regulation of the alternative oxidase has changed this view (Ribas-Carbo et al., 1997, 2000). It is now clear that the alternative oxidase can play a very significant

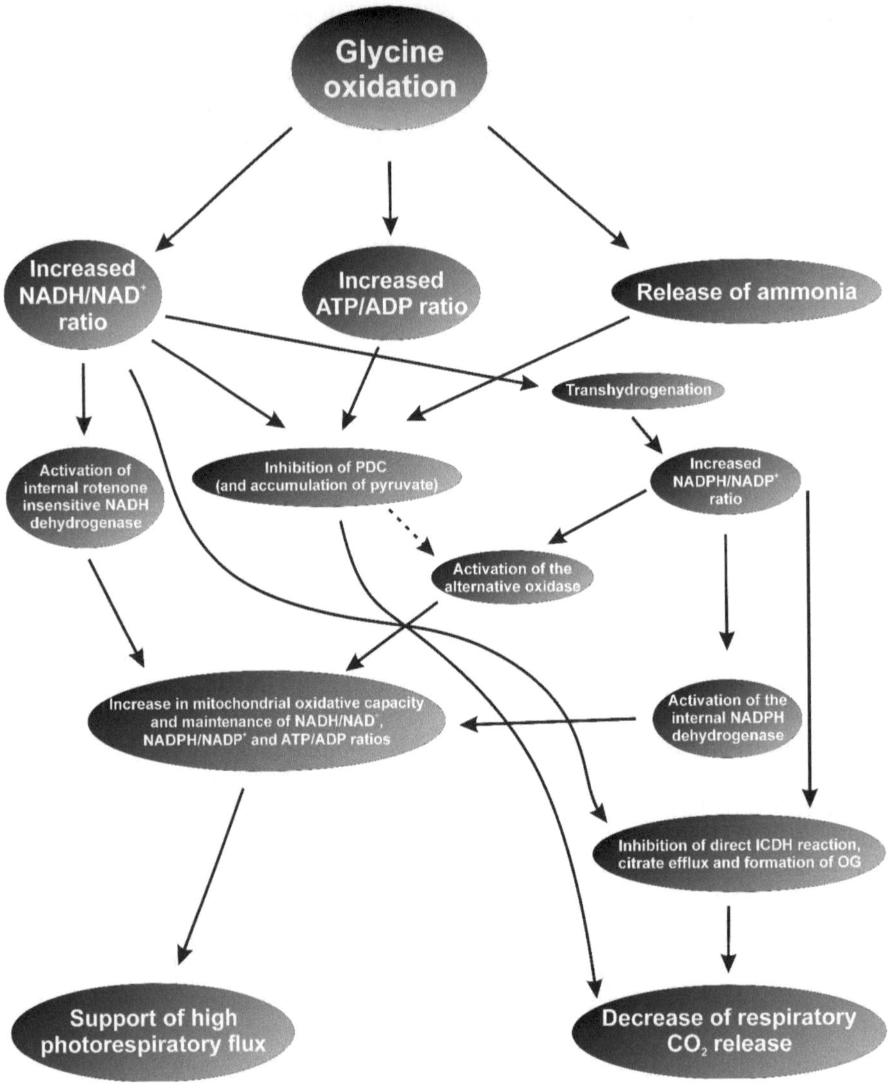

Fig. 6. Schematic representation of the events during photorespiration and mitochondrial respiration. Activation of the pathways not coupled to ATP synthesis support a high flux through the mitochondria and switching to a partial TCA cycle decreases respiratory CO_2 release.

role in the coupling between electron transport and oxidative phosphorylation, and thus for the energy- and redox-balance in the cell. During photorespiration, NADH/NAD and NADPH/NADP ratios increase nearly three-fold (Wigge et al., 1993; Igamberdiev et al., 2001; Igamberdiev and Gardeström, 2003). The increased NADPH can activate AOX via reduction of a disulfide bond, probably via the thioredoxin system. The reduced form may then be activated by pyruvate, which can increase in the light due to inhibition of the mitochondrial PDC (see below and Fig. 6). Citrate also accumulates in the light, and this leads to the activation of AOX gene expression (Vanlerberghe et al., 1995; Vanlerberghe and McIntosh, 1996). Cytochrome oxidase is inhibited by an increase in the ATP/ADP ratio, and under these conditions the engagement of AOX significantly increases. Thus, it is likely that AOX is in a more activated state in the light than in the dark, particularly in photorespiratory conditions. In support of this, direct evidence for the involvement of the AOX during respiration in the light has been obtained using oxygen-isotope-fractionation techniques (Ribas-Carbo et al., 1997, 2000) and by data showing that phytochrome may play a role in altering electron partitioning between the cytochrome and alternative pathways (M. Ribas-Carbo, unpublished).

VIII. Operation of the Tricarboxylic Acid (TCA) Cycle in the Light

The increase in the reduction level of NADH/NAD and NADPH/NADP and an increased ATP/ADP ratio in mitochondria in the light have important consequences for the operation of the TCA cycle. As we set out below, the TCA cycle is reorganized in the light so that it changes from being the main source of energy in the cell to become a flexible mechanism that enables the cell to sustain the photosynthetic process, both through the production of carbon skeletons and by contributing to the redox homeostasis of the cell.

A. Inhibition of the Pyruvate Dehydrogenase Complex in the Light

An important step for regulation of the TCA cycle is the entry of carbon via PDC. This enzyme complex is subject to regulation by metabolites linked to photorespiration. The amount of PDC is also developmentally regulated, with a high activity in young developing leaves, and lower activity in mature leaves (Luethy et al., 2001). The mechanisms for regulation are both allosteric and via reversible phosphorylation, where the activity is decreased by phosphorylation (Tovar-Méndez et al., 2003). The phosphorylation level is modulated by the relative activities of a specific kinase/phosphatase pair, where the kinase activity is stimulated by ammonium and ATP. PDC activity is allosterically inhibited by NADH and acetyl-CoA. Thus, high concentrations of ATP, NADH and ammonium which result from photorespiratory reactions, tend to inhibit PDC. Increased content of these metabolites can thus result in a limitation of carbon flux into the TCA cycle in the light. However, the enzyme is also stimulated by pyruvate, which acts as inhibitor of PDC kinase. Steady state mtPDC activity has been reported to decrease in the light in pea leaves (Budde and Randall, 1990), and photorespiratory inhibitors abolished this light-dependent decrease, suggesting that photorespiratory metabolism was responsible (Gemel and Randall, 1992). On the other hand, in a study with barley leaf protoplasts, Krömer et al. (1993) observed no change in the activation state for PDC in darkness and in the light under photorespiratory and non-photorespiratory conditions. The in vivo activity in the light will therefore depend on the balance between stimulating and inhibitory factors.

B. Partial Tricarboxylic Acid (TCA) Cycle in the Light

The possibility that a partial TCA cycle may operate in the light to supply carbon skeletons for biosynthetic purposes has been discussed on several occasions (Chen and Gadal, 1990; Hanning and Heldt, 1993). One point of debate has been over which compound was the most likely to be exported from mitochondria to the cytosol in such circumstances. Citrate has been suggested as the most likely candidate, in preference to 2-oxoglutarate (OG). In isolated pea leaf mitochondria citrate export was more than ten times faster than the export of OG during oxidation of OAA and nearly two-fold faster for spinach mitochondria (Hanning and Heldt, 1993). Recent measurements of subcellular pyridine nucleotide redox status, and the kinetic properties of the key enzymes involved, also support the conclusion that citrate is the exported species (Igamberdiev and Gardeström, 2003).

In darkness, when both the mitochondrial NADP- and NAD- pools are oxidized, both isocitrate dehydrogenases will operate in the forward direction, and the partitioning of metabolic flux between them will depend on the concentration of isocitrate. During photosynthesis, in non-photorespiratory conditions, NAD-ICDH will retain some ability to oxidize isocitrate because the reduction level in the mitochondria is not too high. In these conditions, if mitochondrial concentrations of isocitrate and OG are in the same range, the mitochondrial NADP-ICDH will be near equilibrium between the forward and reverse reactions, because of the similar levels of $NADP^+$ and NADPH. In photorespiratory conditions, when the NADH/NAD and NADPH/NADP ratios are high, only the reverse reaction of NADP-ICDH may take place and NAD-ICDH is inhibited.

The partial TCA cycle is important for maintaining the OG concentration necessary for glutamate biosynthesis. The glutamate formed is readily transaminated, with the formation of other amino acids and the release of OG. Thus, flux through the TCA cycle in the light is necessary to provide OG for net nitrogen assimilation. PDC and isocitrate oxidation will be important steps for the control of this flux. It

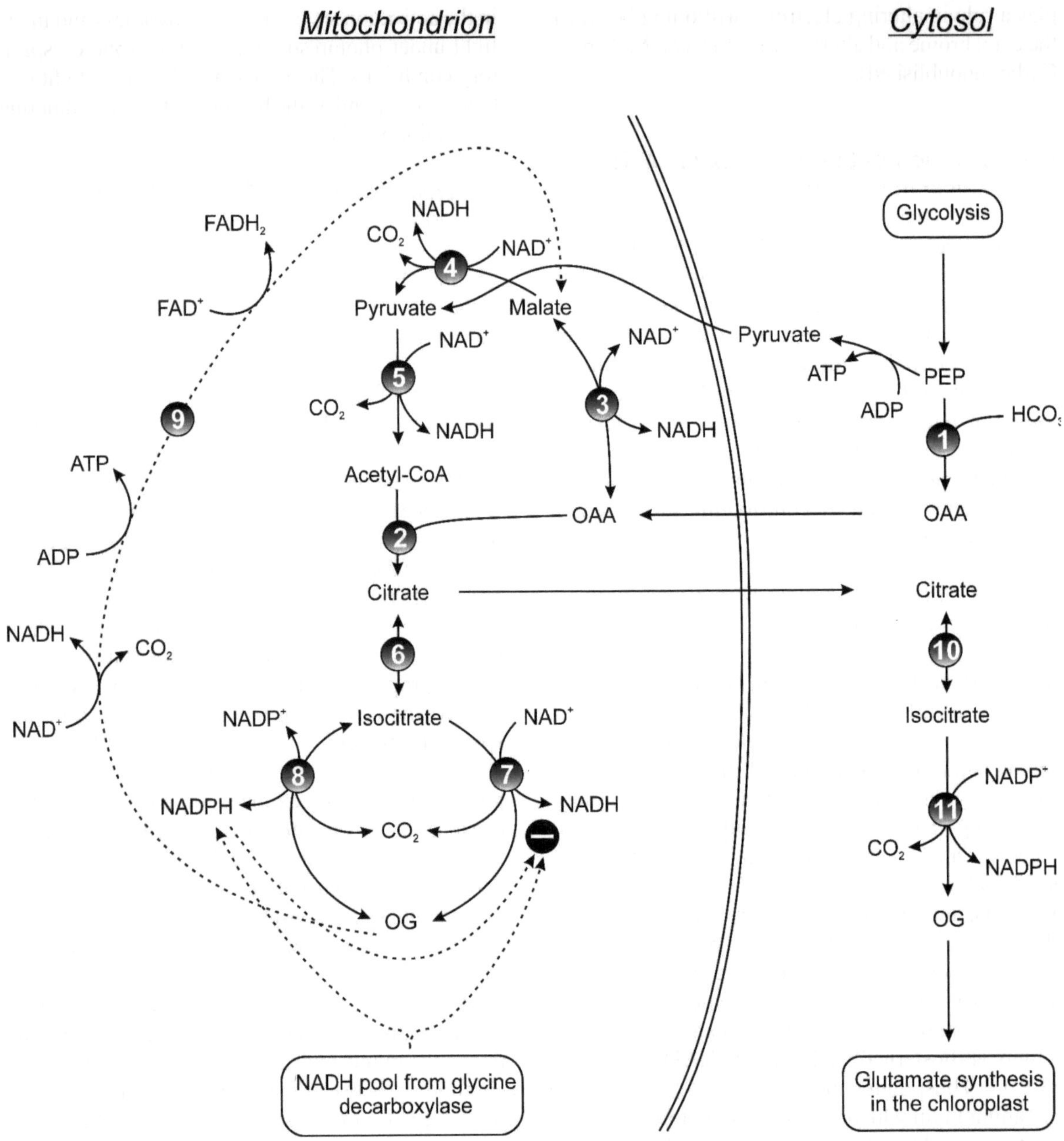

Fig. 7. Operation of the partial tricarboxylic acid cycle in the light. In the light, the main glycolytic substrate for mitochondria is likely OAA formed by PEP-carboxylase (1). One OAA molecule is condensed with acetyl-CoA in the citrate synthase reaction (2). Another OAA molecule is reduced by malate dehydrogenase (3) to malate, which is decarboxylated by NAD-malic enzyme (4) to pyruvate entering the reaction catalyzed by the pyruvate dehydrogenase complex (5). Citrate is converted to isocitrate by the mitochondrial aconitase (6). An increase in the intra-mitochondrial NADH pool caused by glycine oxidation inhibits NAD-ICDH (7) and stimulates the reverse reaction of NADP-ICDH (8) via transhydrogenation with $NADP^+$. Because the aconitase equilibrium favors citrate formation, this will lead to citrate efflux from mitochondria via citrate/OAA transporter or other tricarboxylate transporter. The sequence of reactions of the tricarboxylic acid cycle from 2-oxoglutarate (OG) to malate (9) is suppressed under these conditions. Citrate in the cytosol is converted by cytosolic aconitase (10) to isocitrate, which turns to OG by the cytosolic NADP-ICDH (11). Dashed lines indicate the inhibition of NAD-ICDH by NADH and NADPH (shown by the minus sign) and show connections with the NADH pool from the glycine decarboxylase reaction.

should be noted that refixation of photorespiratory ammonia proceeds without net input of OG. The OG needed for this will be recycled by transamination in the peroxisomes, linked to the glycolate cycle. Citrate export from mitochondria may also be important for maintaining the cytosolic NADPH/NADP ratio in the light (see Fig. 7).

The TCA cycle can fulfill several functions, most commonly considered are to supply reducing equivalents for oxidative phosphorylation and carbon skeletons for biosynthetic reactions. Although not absolute, it is reasonable to assume that a complete TCA cycle will be more important in darkness than in the light to support oxidative phosphorylation. However in the light, a partial TCA cycle might dominate to supply carbon skeletons and to maintain redox homeostasis, while oxidation of photorespiratory glycine may take over the role to support electron transport for ATP formation. In the extreme case, the difference between darkness and light might also include a switch between inactive and active PEP carboxylase. This is supported by evidence that the enzyme is more active in the light (Krömer et al., 1996).

The importance of the ICDH enzymes in switching between a complete TCA cycle and a partial TCA cycle is shown in Fig. 7. Two extreme cases are considered: (1) a partial TCA cycle with PEP carboxylase serving the role as terminal glycolytic enzyme and export of citrate from mitochondria with OG formed in the cytosol, and (2) a complete TCA cycle involving pyruvate kinase and assumed to prevail in darkness.

(1) In the PEP carboxylase reaction, OAA is formed in the cytosol and enters the mitochondria in exchange for citrate via the citrate/OAA transporter or a separate tricarboxylate uniporter (Picault et al., 2002; De Palma et al., 2003). The OAA in mitochondria is either reduced to malate, or converted to citrate via condensation with acetyl-CoA. The latter is formed from pyruvate by the pyruvate dehydrogenase complex. The pyruvate is a product of malate decarboxylation by NAD-malic enzyme, which is relatively insensitive to high reduction levels of the pyridine nucleotides (Pascal et al., 1990). At high reduction levels in the mitochondria, particularly during active glycine oxidation, NAD-ICDH is inhibited and mNADP-ICDH operates in the reverse direction. Under these conditions, citrate is exported to the cytosol, where it is converted by the cytosolic aconitase and NADP-isocitrate dehydrogenase to 2-oxoglutarate (OG). The stoichiometry of one OG molecule formation from two molecules of PEP can be presented by the following equation:

$$2\ \text{PEP} + 2\text{HCO}_3^- + \text{NADH} \rightarrow \text{OG} + 3\text{CO}_2 + 2\text{NADH} + \text{NADPH}$$

In summary: $$2\ \text{PEP} \rightarrow \text{OG} + \text{CO}_2 + \text{NADH} + \text{NADPH} \quad (1)$$

This total balance follows from the consumption of two bicarbonate-anions in the PEP-carboxylase reaction, the formation of two CO_2 molecules in the mitochondria during the decarboxylation of malate and pyruvate, and the formation of one CO_2 molecule in the cytosol in the NADP-ICDH reaction. In the mitochondria, one NADH is consumed and two are formed, and in the cytosol one NADPH is formed.

(2) The complete TCA cycle, starting from PEP and including PK, can be presented as:

$$\text{PEP} \rightarrow 3\ \text{CO}_2 + 4\ \text{NADH} + \text{FADH}_2 + 2\ \text{ATP} \quad (2)$$

All three carbons of PEP are oxidized to CO_2 and 4 NADH are formed in one turn of the TCA cycle, FAD is reduced by SDH and one ATP is formed in the PK reaction and another by substrate level phosphorylation in the TCA cycle.

By comparing Eqs. (1) and (2) we see that relative to the complete TCA cycle, the partial TCA cycle evolves only one sixth of the CO_2, forms one fifth of the reducing power (but half of this amount is the anabolic reducing power, NADPH in the cytosol) and no substrate phosphorylation occurs. This means that there is a substantial reduction in decarboxylation of TCA cycle intermediates during respiration in the light. In fact, if the switch is complete, as indicated in the example above, a switch to a partial TCA cycle will decrease respiratory decarboxylations by ⅔ without decreased PDC activity. However, the normal situation in a leaf would probably be some mixture between a total and a partial TCA cycle in most metabolic situations.

C. Malate and Citrate Valves

We propose that there are two redox valves in operation in the photosynthetic plant cell; one driven by the chloroplasts, and another driven by mitochondria. The malate valve, driven by NADPH formed by photosynthetic electron transport, prevents over-reduction of the chloroplasts and increases the NADH/NAD ratio in different cellular compartments (Krömer and Scheibe, 1996). Another valve, the 'citrate valve,' is driven by the increased reduction level in mitochondria linked to photorespiratory glycine oxidation, reduces NADP pools and supplies OG for glutamate biosynthesis. The active operation of the citrate valve corresponds to the transition from the full to the partial TCA cycle in plant mitochondria. The partial TCA cycle maintains the operation of the citrate valve, supplying the anabolic reduction power (NADPH) via oxidation of isocitrate in the cytosol and peroxisomes. In photorespiratory conditions, part of the NADPH pool in the cytosol is used for the reduction of glyoxylate and hydroxypyruvate exported from peroxisomes (Krömer and Heldt, 1991). NADPH/NADP turnover may be provided by the participation of cNADP-ICDH and NADP-dependent hydroxypyruvate reductase (Igamberdiev and Kleczkowski, 2001). Operation of the modified TCA cycle and the citrate valve also maintains the concentrations of OG, OAA and pyruvate in the cytosol and mitochondria which is important for nitrogen assimilation in the light. Studies with a barley mutant deficient in mitochondrial GDC showed that, under photorespiratory conditions, the chloroplasts were over-reduced and over-energized (Igamberdiev et al., 2001). This gives support to a function for photorespiration as an effective redox-transfer mechanism from chloroplasts involving mitochondrial reactions.

IX. Conclusion

The reorganization of mitochondrial metabolism in the light, including the provision of high photorespiratory fluxes, switching to non-coupled pathways of electron transport and the transition to a partial TCA cycle, helps to explain the variable changes in the gas exchange components that have been observed for leaf respiration in the light. The operation of a partial TCA cycle in the light would result in a decrease in the decarboxylation rates and oxygen consumption, while photorespiration is characterized by a fast rate of oxygen consumption and decarboxylation occurs only in one reaction catalyzed by the glycine decarboxylase complex. Thus, with some input from CO_2 refixation, the net decarboxylation rate in the light is decreased, while (photo)respiratory oxygen consumption remains high.

Acknowledgments

Cooperation between the authors' groups is supported by travel grants from the NorFA network on Temperature Stress and Acclimation in Plants.

References

Amthor JS (2000) Direct effect of elevated CO_2 on nocturnal in situ leaf respiration in nine temperate deciduous tree species is small. Tree Physiol 20: 139–144

Atkin OK, Westbeek MHM, Cambridge ML, Lambers H and Pons TL (1997) Leaf respiration in light and darkness: A comparison of slow- and fast-growing Poa species. Plant Physiol 113: 961–965

Atkin OK, Evans JR and Siebke K (1998a) Relationship between the inhibition of leaf respiration by light and enhancement of leaf dark respiration following light treatment. Aust J Plant Physiol 25: 437–443

Atkin OK, Schortemeyer M, McFarlane N and Evans JR (1998b) Variation in the components of relative growth rate in ten Acacia species from contrasting environments. Plant Cell Environ 21: 1007–1017

Atkin OK, Evans JR, Ball MC, Lambers H and Pons TL (2000a) Leaf respiration of snow gum in the light and dark. Interactions between temperature and irradiance. Plant Physiol 122: 915–923

Atkin OK, Millar AH, Garderström P and Day DA (2000b) Photosynthesis, carbohydrate metabolism and respiration in leaves of higher plants. In: RC Leegood, TD Sharkey, S von Caemmerer (eds) Photosynthesis: Physiology and Metabolism, pp 153–175. Kluwer Academic Publishers, Dordrecht

Avelange MH, Thiery JM, Sarrey F, Gans P and Rebeille F (1991) Mass-spectrometric determination of O_2 and CO_2 gas exchange in illuminated higher plant cells—evidence for light inhibition of substrate decarboxylations. Planta 183: 150–157

Bauwe H, Keerberg O, Bassuner R, Pärnik T and Bassuner B (1987) Reassimilation of carbon dioxide by Flaveria (Asteraceae) species representing different types of photosynthesis. Planta 172: 214–218

Bergmann A, Garderström P and Ericson I (1981) Release and refixation of ammonia during photorespiration. Physiol Plant 53: 528–532

Brooks A and Farquhar GD (1985) Effect of temperature on the CO_2-O_2 specificity of ribulose-1,5-bisphosphate carboxylase/oxygenase and the rate of respiration in the light. Estimates from gas exchange measurements on spinach. Planta 165: 397–406

Bruhn D, Mikkelsen TN and Atkin OK (2002) Does the direct effect of atmospheric CO_2 concentration on leaf respiration vary with temperature? Responses in two species of *Plantago* that differ in relative growth rate. Physiol Plant 114: 57–64

Budde RJA and Randall DD (1990) Pea leaf mitochondrial pyruvate dehydrogenase complex is inactivated in vivo in a light-dependent manner. Proc Natl Acad Sci USA 87: 673–676

Bykova NV and Møller IM (2001) Involvement of matrix NADP turnover in the oxidation of NAD^+-linked substrates by pea leaf mitochondria. Physiol Plant 111: 448–456

Bykova NV, Igamberdiev AU and Møller IM (1998) Contribution of respiratory NAD(P)H dehydrogenases to glycine and glycine plus malate oxidation by pea leaf mitochondria. In: IM Møller, P Garderström, K Glimelius and E Glaser (eds) Plant Mitochondria: From Gene to Function. Backhuys Publishers, Leiden, pp 347–351

Bykova NV, Egsgaard H and Møller IM (2003) Identification of 14 new phosphoproteins involved in important plant mitochondrial processes. FEBS Lett 540: 141–146

Byrd GT, Sage RF and Brown RH (1992) A comparison of dark respiration between C_3 and C_4 plants. Plant Physiol 100: 191–198

Chapman EA and Graham D (1974) The effect of light on the tricarboxylic acid cycle in green leaves. I. Relative rates of the cycle in the dark and in the light. Plant Physiol 53: 879–885

Chen RD and Gadal P (1990) Structure, functions and regulation of NAD- and NADP-dependent isocitrate dehydrogenases in higher plants and in other organisms. Plant Physiol Biochem 28: 411–427

Day D and Wiskich JT (1977) Factors limiting respiration by isolated cauliflower mitochondria. Phytochemistry 16: 1499–1502

Day D, Neuburger M and Douce R (1985) Interactions between glycine decarboxylase, the tricarboxylic acid cycle and the respiratory chain in pea leaf mitochondria. Aust J Plant Physiol 12: 119–130

De Palma A, Scalera V, Bisaccia F and Prezioso G (2003) Citrate uniport by the mitochondrial tricarboxylate carrier: A basis for a new hypothesis for the transport mechanism. J Bioenerg Biomembr 35: 133–140

Di Marco G, Manes F, Tricoli D and Vitale E (1990) Fluorescence parameters measured concurrently with net photosynthesis to investigate chloroplastic CO_2 concentration in leaves of Quercus ilex L. J Plant Physiol 136: 538–543

Doman NG (1959) On the interaction between photosynthesis and respiration in plants. Biochemistry (Moscow) 24: 19–24

Drake BG, Azcon-Bieto J, Berry J, Bunce J, Dijkstra P, Farrar J, Gifford RM, Gonzalez-Meler MA, Koch G, Lambers H, Siedow J and Wullschleger S (1999) Does elevated atmospheric CO_2 concentration inhibit mitochondrial respiration in green plants? Plant Cell Environ 22: 649–657

Edwards GE and Walker D (1983) C_3, C_4: Mechanisms, and Cellular and Environmental Regulation of Photosynthesis. Blackwell, Oxford

Fondy BR, Geiger DR and Servaites JC (1989) Photosynthesis, carbohydrate-metabolism and export in *Beta vulgaris* L and *Phaseolus vulgaris* L during square and sinusoidal light regimes. Plant Physiol 89: 396–402

Fox TC and Geiger DR (1984) Effects of decreased net carbon exchange on carbohydrate metabolism in sugar beet source leaves. Plant Physiol 76: 762–768

Gardeström P and Lernmark U (1995) The contribution of mitochondria to energetic metabolism in photosynthetic cells. J Bioenerg Biomembr 27: 415–421

Gardeström P and Wigge B (1988) Influence of photorespiration on ATP/ADP ratios in the chloroplasts, mitochondria, and cytosol, studied by rapid fractionation of barley (*Hordeum vulgare*) protoplasts. Plant Physiol 88: 69–76

Gardeström P, Igamberdiev AU and Raghavendra AS (2002) Mitochondrial functions in the light and significance to carbon-nitrogen interactions. In: Foyer CH and Noctor G (eds) Photosynthetic Nitrogen Assimilation and Associated Carbon and Respiratory Metabolism, pp 151–172. Kluwer Academic Publishers, Dordrecht

Gemel J and Randall DD (1992) Light regulation of leaf mitochondrial pyruvate dehydrogenase complex. Plant Physiol 100: 908–914

Gerbaud A and Andre M (1987) An evaluation of the recycling in measurements of photorespiration. Plant Physiol 83: 933–937

Gfeller RP and Gibbs M (1984) Fermentative metabolism in *Chlamydomonas reinhardtii*. 1. Analysis of fermentative products from starch in dark and light. Plant Physiol 75: 212–218

Goldsworthy A (1966) Experiments on the origin of CO_2 released by tobacco leaf segments in the light. Phytochemistry 5: 1013–1019

Goldsworthy A (1970) Photorespiration. Bot Rev 36: 321–340

Hanning I and Heldt HW (1993) On the function of mitochondrial metabolism during photosynthesis in spinach (*Spinacia oleracea* L.) leaves. Plant Physiol 103: 1147–1154

Hausler RE, Schlieben NH, Schulz B and Flugge UI (1998) Compensation of decreased triose phosphate translocator activity by accelerated starch turnover and glucose transport in transgenic tobacco. Planta 204: 366–376

Heineke D, Bykova N, Gardeström P and Bauwe H (2001) Metabolic response of potato plants to an antisense reduction of the P-protein of glycine decarboxylase. Planta 212: 880–887

Hoefnagel MHN, Atkin OK and Wiskich JT (1998) Interdependence between chloroplasts and mitochondria in the light and the dark. Biochim Biophys Acta 1366: 235–255

Hurry V, Keerberg O, Pärnik T, Öquist G and Gardeström P (1996) Effect of cold hardening on the components of respiratory decarboxylation in the light and in the dark in leaves of winter rye. Plant Physiol 111: 713–719

Igamberdiev AU (1999) Foundations of metabolic organization: Coherence as a basis of computational properties in metabolic networks. BioSystems 50: 1–16

Igamberdiev AU and Gardeström P (2003) Regulation of NAD- and NADP-dependent isocitrate dehydrogenases by reduction levels of pyridine nucleotides in mitochondria and cytosol of pea leaves. Biochim Biophys Acta 1606: 117–125

Igamberdiev AU and Kleczkowski LA (2001) Implications of adenylate kinase-governed equilibrium of adenylates on contents of free magnesium in plant cells and compartments. Biochem J 360: 225–231

Igamberdiev AU and Kleczkowski LA (2003) Membrane potential, adenylate levels and Mg^{2+} are interconnected via adenylate kinase equilibrium in plant cells. Biochim Biophys Acta 1607: 111–119

Igamberdiev AU and Lea PJ (2002) The role of peroxisomes in the integration of metabolism and evolution of land plants. Phytochemistry 60: 651–674

Igamberdiev AU, Bykova NV and Gardeström P (1997) Involvement of cyanide-resistant and rotenone-insensitive pathways of mitochondrial electron transport during oxidation of glycine in higher plants. FEBS Lett 412: 265–269

Igamberdiev AU, Bykova NV, Lea PJ and Gardeström P (2001) The role of photorespiration in redox and energy balance of photosynthetic plant cells: A study with a barley mutant deficient in glycine decarboxylase. Physiol Plant 111: 427–438

Ivanova H, Keerberg O and Pärnik T (1993) Influence of oxygen concentration on the rates of carbon fluxes in the biochemical system of CO_2 assimilation. Proc Estonian Acad Sci Chem 42: 185–197

Jahnke S (2001) Atmospheric CO_2 concentration does not directly affect leaf respiration in bean or poplar. Plant Cell Environ 24: 1139–1151

Keerberg O, Drozdova IS, Pärnik TR, Keerberg HI and Voskresenskaya NP (1989) Components of photosynthetic and respiratory CO_2 exchange and photosynthetic carbon metabolism in barley seedlings grown under red and blue light. Russ J Plant Physiol 36: 642–652

Keerberg O, Ivanova H, Keerberg H and Pärnik T (1999) CO_2 exchange of potato transformants with reduced activities of glycine decarboxylase. In: GE de Vries, K Metzlaff (eds) Phytosfere '99—Highlights in European Plant Biotechnology, pp 215–219. Elsevier, Amsterdam

Keerberg O and Viil J (1988) Quantitative characteristics of photosynthetic carbon metabolism. In: AA Nichiporovich (ed) Photosynthesis and Production Processes, pp 40–53. Nauka, Moscow

Keerberg O, Ivanova H, Keerberg H and Pärnik T (2001) Contribution of primary and stored photosynthates to photorespiration. In: PS2001 Proceedings: 12th International Congress on Photosynthesis, S15–002. CSIRO, Melbourne (CD/ROM)

Keys AJ, Bird IF, Cornelius MJ, Lea PJ, Wallsgrove RM and Miflin BJ (1978) Photorespiratory nitrogen cycle. Nature 275: 741–743

Kirschbaum MUF and Farquhar GD (1987) Investigation of the CO_2 dependence of quantum yield and respiration in *Eucalyptus pauciflora*. Plant Physiol 83: 1032–1036

Kok B (1948) A critical consideration of the quantum yield of *Chlorella* photosynthesis. Enzymologia 13: 1–56

Krömer S (1995) Respiration during photosynthesis. Annu Rev Plant Physiol Plant Mol Biol 46: 47–70

Krömer S and Heldt HW (1991) Respiration of pea leaf mitochondria and redox transfer between the mitochondrial and extramitochondrial compartment. Biochim Biophys Acta 1057: 42–50

Krömer S and Scheibe R (1996) Function of the chloroplastic malate valve for respiration during photosynthesis. Biochem Soc Trans 24: 761–766

Krömer S, Malmberg G and Gardeström P (1993) Mitochondrial contribution to photosynthetic metabolism—a study with barley (*Hordeum vulgare* L) leaf protoplasts at different light intensities and CO_2 concentrations. Plant Physiol 102: 947–955

Krömer S, Gardeström P and Samuelsson G (1996) Regulation of the supply of oxaloacetate for mitochondrial metabolism via phosphoenolpyruvate carboxylase in barley leaf protoplasts. II. Effects of metabolites on PEPC activity at different activation states of the protein. Biochim Biophys Acta 1289: 351–361

Kruger NJ, Bulpin PV and ap Rees T (1983) The extent of starch degradation in the light is pea leaves. Planta 157: 271–273

Kumarasinghe KS, Keys AJ and Whittingham CP (1977) The flux of carbon through the glycolate pathway during photosynthesis in leaves. J Exp Bot 28: 1247–1257

Laisk A and Oja V (1998) Dynamics of Leaf Photosynthesis. CSIRO Publishing, Collingwood

Laisk AH (1977) Equipment and methods for the measurement of PC_w curves. In: AA Nichiporovich, ed, Kinetika Fotosinteza i Fotodykhaniya u C_3 Rastenij (Kinetics of photosynthesis and Photorespiration of C_3 Plants). Nauka, Moscow, pp 48–71

Ligeza A, Tikhonov AN and Subczynski WK (1997) In situ measurements of oxygen production and consumption using paramagnetic fusinite particles injected into a bean leaf. Biochim Biophys Acta 1319: 133–137

Loreto F, Delfine S and Di Marco G (1999) Estimation of photorespiratory carbon dioxide recycling during photosynthesis. Aust J Plant Physiol 26: 733–736

Loreto F, Velikova V and Di Marco G (2001) Respiration in the light measured by (CO_2)-C^{12} emission in (CO_2)-C^{13} atmosphere in maize leaves. Aust J Plant Physiol 28: 1103–1108

Luethy MH, Gemel J, Johnston ML, Mooney BP, Miernyk JA and Randall DD (2001) Developmental expression of the mitochondrial pyruvate dehydrogenase complex in pea (*Pisum sativum*) seedlings. Physiol Plant 112: 559–566

Mahon JD, Fock H and Canvin DT (1974) Changes in specific radioactivities of sunflower leaf metabolites during photosynthesis in $^{14}CO_2$ and $^{12}CO_2$ at normal and low oxygen. Planta 120: 123–134

Marsh HVJ, Galmiche JM and Gibbs M (1965) Effect of light on the tricarboxylic acid cycle in *Scenedesmus*. Plant Physiol 40: 1913–1922

Maxwell DP, Wang Y and McIntosh L (1999) The alternative oxidase lowers mitochondrial reactive oxygen production in plant cells. P Natl Acad Sci USA 96: 8271–8276

McCashin BG, Cossins EA and Canvin DT (1988) Dark respiration during photosynthesis in wheat leaf slices. Plant Physiol 87: 155–161

Melo AMP, Roberts TH and Møller IM (1996) Evidence for the presence of two rotenone-insensitive NAD(P)H dehydrogenases on the inner surface of the inner membrane of potato tuber mitochondria. Biochim Biophys Acta 1276: 133–139

Melzer E and O'Leary MH (1987) Anapleurotic CO_2 fixation by phosphoenolpyruvate carboxylase in C_3 plants. Plant Physiol 84: 58–60

Michalecka AM, Svensson AS, Johansson FI, Agius SC, Johanson U, Brennicke A, Binder S and Rasmusson AG (2003) *Arabidopsis* genes encoding mitochondrial type II NAD(P)H dehydrogenases have different evolutionary origin and show distinct responses to light. Plant Physiol 133: 642–652

Møller IM (2001) Plant mitochondria and oxidative stress: electron transport, NADPH turnover, and metabolism of reactive oxygen species. Annu Rev Plant Physiol Plant Mol Biol 52: 561–591

Møller IM (2002) A new dawn for plant mitochondrial NAD(P)H dehydrogenases. Trends Plant Sci 7: 235–237

Møller IM and Lin W (1986) Membrane-bound NAD(P)H dehydrogenases in higher plant cells. Annu Rev Plant Physiol 37: 309–334

Nakamura Y and Miyachi S (1982) Effect of temperature on starch degradation in *Chlorella vulgaris* 11h cells. Plant Cell Physiol 23: 333–341

Ogren WL (1984) Photorespiration: Pathways, regulation, and modification. Annu Rev Plant Physiol 35: 415–442

Oliver DJ (1994) The glycine decarboxylase complex from plant mitochondria. Annu Rev Plant Physiol Plant Mol Biol 45: 323–337

Osmond CB and Grace SC (1995) Perspectives on photoinhibition and photorespiration in the field—quintessential inefficiencies of the light and dark reactions of photosynthesis. J Exp Bot 46: 1351–1362

Pärnik T and Keerberg O (1995) Decarboxylation of primary and end-products of photosynthesis at different oxygen concentrations. J Exp Bot 46: 1439–1447

Pärnik T, Keerberg O and Viil J (1976) Estimation of photorespiration in bean and maize leaves. Newslett Appl Nucl Meth Biol Agr 6: 5–7

Pärnik TR, Voronin PY, Ivanova HN and Keerberg OF (2002) Respiratory CO_2 fluxes in photosynthesizing leaves of C_3 species varying in rates of starch synthesis. Russ J Plant Physiol 49: 821–827

Pascal N, Dumas R and Douce R (1990) Comparison of the kinetic behavior toward pyridine nucleotides of NAD^+-linked dehydrogenases from plant mitochondria. Plant Physiol 94: 189–193

Picault N, Palmieri L, Pisano I, Hodges M and Palmieri F (2002) Identification of a novel transporter for dicarboxylates and tricarboxylates in plant mitochondria—bacterial expression, reconstitution, functional characterization, and tissue distribution. J Biol Chem 277: 24204–24211

Pinelli P and Loreto F (2003) (CO_2)-C^{12} emission from different metabolic pathways measured in illuminated and darkened C_3 and C_4 leaves at low, atmospheric and elevated CO_2 concentration. J Exp Bot 54: 1761–1769

Pons TL and Welschen RAM (2002) Overestimation of respiration rates in commercially available clamp-on leaf chambers. Complications with measurement of net photosynthesis. Plant Cell Environ 25: 1367–1372

Purvis AC (1997) Role of the alternative oxidase in limiting superoxide production by plant mitochondria. Physiol Plant 100: 165–170

Raghavendra AS, Padmasree K and Saradadevi K (1994) Interdependence of photosynthesis and respiration in plant cells—interactions between chloroplasts and mitochondria. Plant Sci 97: 1–14

Rasulov BH (1986) Photosynthetic parameters of cotton leaves at various source-sink relations within the whole plant system. Russ J Plant Physiol 33: 922–929

Rasulov BH and Oja V (1982) Estimation of components of respiration in the light considering the presence of residual oxygen. Russ J Plant Physiol 30: 616–622

Rasulov BH, Laisk AH and Asrorov KA (1983) Photosynthesis and photorespiration in some cotton species during ontogeny. Russ J Plant Physiol 30: 616–645

Raven JA and Farquhar GD (1990) The influence of N-metabolism and organic acid synthesis on the natural abundance of isotopes of carbon in plants. New Phytol 116: 505–529

Ribas-Carbo M, Lennon AM, Robinson SA, Giles L, Berry JA and Siedow JN (1997) The regulation of electron partitioning between the cytochrome and alternative pathways in soybean cotyledon and root mitochondria. Plant Physiol 113: 903–911

Ribas-Carbo M, Robinson SA, Gonzalez-Meler MA, Lennon AM, Giles L, Siedow JN and Berry JA (2000) Effects of light on respiration and oxygen isotope fractionation in soybean cotyledons. Plant Cell Environ 23: 983–989

Robinson JM (1988) Does O_2 photoreduction occur within chloroplasts in vivo? Physiol Plant 72: 666–680

Sharkey TD (1988) Estimating the rate of photosynthesis in leaves. Physiol Plant 73: 147–152

Svensson AS and Rasmusson AG (2001) Light-dependent gene expression for proteins in the respiratory chain of potato leaves. Plant J 28: 73–82

Tovar-Méndez A, Miernyk JA and Randall DD (2003) Regulation of pyruvate dehydrogenase complex activity in plant cells. Eur J Biochem 270: 1043–1049

Turpin DH, Botha FC, Smith RG, Feil R, Horsey AK and Vanlerberghe GC (1990) Regulation of carbon partitioning to respiration during dark ammonium assimilation by the green alga *Selenastrum minutum*. Plant Physiol 93: 166–175

Vanlerberghe GC and McIntosh L (1996) Signals regulating the expression of the nuclear gene encoding alternative oxidase of plant mitochondria. Plant Physiol 111: 589–595

Vanlerberghe GC, Day DA, Wiskich JT, Vanlerberghe AE and McIntosh L (1995) Alternative oxidase activity in tobacco leaf mitochondria — dependence on tricarboxylic acid cycle-mediated redox regulation and pyruvate activation. Plant Physiol 109: 353–361

Villar R, Held AA and Merino J (1994) Comparison of methods to estimate dark respiration in the light in leaves of 2 woody species. Plant Physiol 105: 167–172

Villar R, Held AA and Merino J (1995) Dark leaf respiration in light and darkness of an evergreen and a deciduous plant species. Plant Physiol 107: 421–427

von Caemmerer S and Evans JR (1991) Determination of the average partial pressure of CO_2 in chloroplasts from leaves of several C_3 plants. Aust J Plant Physiol 18: 287–305

Wallsgrove RM, Turner JC, Hall NP, Kendall AC and Bright SWJ (1987) Barley mutants lacking chloroplast glutamine synthetase. Biochemical and genetic analysis. Plant Physiol 83: 155–158

Weger HG, Birch DG, Elrifi IR and Turpin DH (1988) Ammonium assimilation requires mitochondrial respiration in the light—a study with the green alga *Selenastrum minutum*. Plant Physiol 86: 688–692

Wigge B, Krömer S and Gardeström P (1993) The redox levels and subcellular distribution of pyridine nucleotides in illuminated barley leaf protoplasts studied by rapid fractionation. Physiol Plant 88: 10–18

Wingler A, Lea PJ and Leegood RC (1999a) Photorespiratory metabolism of glyoxylate and formate in glycine-accumulating mutants of barley and *Amaranthus edulis*. Planta 207: 518–526

Wingler A, Quick WP, Bungard RA, Bailey KJ, Lea PJ and Leegood RC (1999b) The role of photorespiration during drought stress: An analysis utilizing barley mutants with reduced activities of photorespiratory enzymes. Plant Cell Environ 22: 361–373

Winter H, Robinson DG and Heldt HW (1993) Subcellular volumes and metabolite concentrations in barley leaves. Planta 191: 180–190

Winter H, Robinson DG and Heldt HW (1994) Subcellular volumes and metabolite concentrations in spinach leaves. Planta 193: 530–535

Wiskich JT and Dry IB (1985) The tricarboxylic acid cycle in plant mitochondria: Its operation and regulation. In: R Douce, D Day (eds) Higher Plant Cell Respiration, Vol 18, pp 281–313. Springer-Verlag, Berlin

Yamauchi M and Yamada Y (1985) Glycolate synthesis in tomato leaf disks: Involvement of storage material in photorespiration. J Fac Agr Kyushu Univ 30: 135–147

Zalensky OV, Voznesensky VL, Ponomaryeva MM and Shtanko TP (1955) Effect of temperature on the metabolism of carbon fixed in the process of photosynthesis. Bot J (Russia) 40: 347–358

Chapter 5

Effects of Light Intensity and Carbohydrate Status on Leaf and Root Respiration

Ko Noguchi*

Department of Biology, Graduate School of Science, Osaka University, Machikaneyama-cho 1-1, Toyonaka, Osaka, Japan

Summary

A positive correlation has been observed between dark respiration and carbohydrate status/light intensity during prior illumination in both leaves and roots of many species. This correlation is often ascribed to an indirect effect: changes in carbohydrate status/light intensity are thought to influence various ATP-consuming processes (growth, maintenance and ion uptake), and adenylate demands for these processes are thought to restrict respiration rates. However, some data clearly indicate that this correlation is partly caused by a direct

*author for correspondence, email: knoguchi@bio.sci.osaka-u.ac.jp

H. Lambers and M. Ribas-Carbo (eds.), Plant Respiration, 63–83.

effect of carbohydrate as substrates for respiration both in leaves and in roots. In leaves of some species, in vivo activity of the alternative oxidase (AOX) in mitochondria is high when carbohydrate status is high (e.g., leaves after illumination), and AOX would have an important role as an energy-overflow pathway, while this correlation between carbohydrate status and in vivo AOX activity does not exist in leaves of other species. These different responses to carbohydrate status among plant species may be related to their ecological traits. However, the significance and physiological mechanisms of these different responses are still unknown. Day respiration (non-photorespiratory mitochondrial CO_2 production or O_2 consumption in the light) also depends on light intensity, although measurements of day respiration are still hard to make. High-light intensity induces fast rates of O_2 uptake in the light which would support fast rates of photosynthesis; rates of CO_2 production in the light also depend on light intensities under low irradiances. Growth light intensity also has a direct influence on dark respiration, especially at photo-oxidative light intensities. If excess light intensity overwhelms avoiding and scavenging systems in leaves, photoinhibition in photosystems occurs in leaves. Under these conditions, non-phosphorylating pathways, such as AOX and uncoupling protein, would consume reducing equivalents efficiently, and prevent the over-reduction in the electron transfer of chloroplasts and mitochondria.

I. Introduction

A positive correlation between dark respiration and light intensity during a prior light period has been found in many cases, e.g., *Trifolium repens* (white clover) plants (McCree and Troughton, 1966), a *Solanum tuberosum* (potato) canopy (Sale, 1974), *Zea mays* (maize) roots (Massimino et al., 1981), maize shoots (André et al., 1982), and *Triticum aestivum* (wheat) leaves (Azcón-Bieto and Osmond, 1983). This correlation has often been ascribed to a variation in the amounts of carbohydrates, which form respiratory substrates. This is because a similar positive correlation between carbohydrate levels and dark respiratory rates has also been observed in many studies, such as *Vicia faba* (field bean) plants (Breeze and Elston, 1978), wheat, *Lolium perrene* (ryegrass), and maize plants (Penning de Vries et al., 1979), wheat leaves (Azcón-Bieto and Osmond, 1983; Fig. 1), *Spinacia oleracea* (spinach) leaves (Noguchi et al., 1996), and leaves of five boreal tree species (Tjoelker et al., 1999).

Abbreviations: AOX – alternative oxidase; APX – ascorbate peroxidase; CAT – catalase; c_i – internal CO_2 concentration; COX – cytochrome *c* oxidase; DHAP – dihydroacetone phosphate; GDC – glycine decarboxylase; HNE – 4-hydroxy-2-nonenal; LEDR – light-enhanced dark respiration; LMA – leaf mass per area; ME – malic enzyme; O_2^- – superoxide anion; OAA – oxaloacetate; PDC – pyruvate dehydrogenase; PFK – ATP-dependent phosphofructokinase; PGA – phosphoglycerate; P_i – inorganic phosphate; PIB – post-illumination burst; PK – pyruvate kinase; PS I – Photosystem I; PS II – Photosystem II; R_{area} – respiratory rate per unit leaf area; R_d – day respiration; R_{mass} – respiratory rate per unit dry weight; ROS – reactive oxygen species; RQ – respiratory quotient; SHAM – salicylhydroxamic acid; SLA – specific leaf area; SOD – superoxide dismutase; SPS – sucrose phosphate synthase

Why is there a correlation between carbohydrate status and respiratory rates? Does carbohydrate status directly determine dark respiratory rates, because carbohydrates are used as substrates? Plant respiration rate is thought to be limited by these factors: substrate availability, the rates of processes that use respiratory products (e.g., ATP), and/or the capacity of respiratory pathway (e.g., amounts of respiratory enzymes) (Amthor, 1995; Lambers et al., 1998). Adenylate control can restrict several steps in respiratory pathway, such as the cytochrome pathway in the mitochondrial respiratory chain via oxidative phosphorylation, ATP-dependent phosphofructokinase (PFK) and pyruvate kinase (PK) in glycolysis, and glucose-6-phosphate

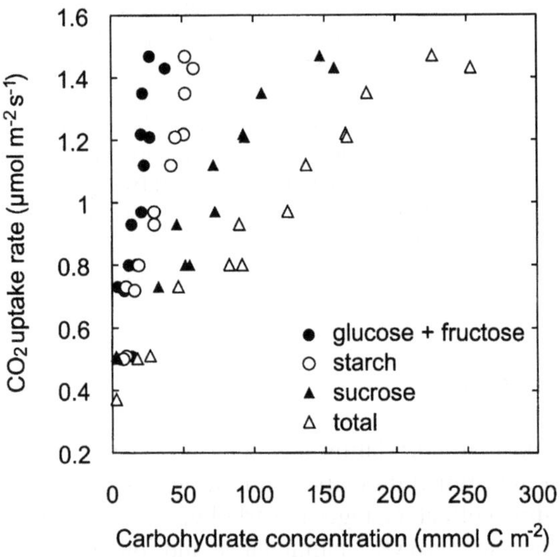

Fig. 1. Relationship between CO_2 uptake rate and several carbohydrate fractions in mature *Triticum aestivum* (wheat) leaves. Based on Azcón-Bieto and Osmond (1983).

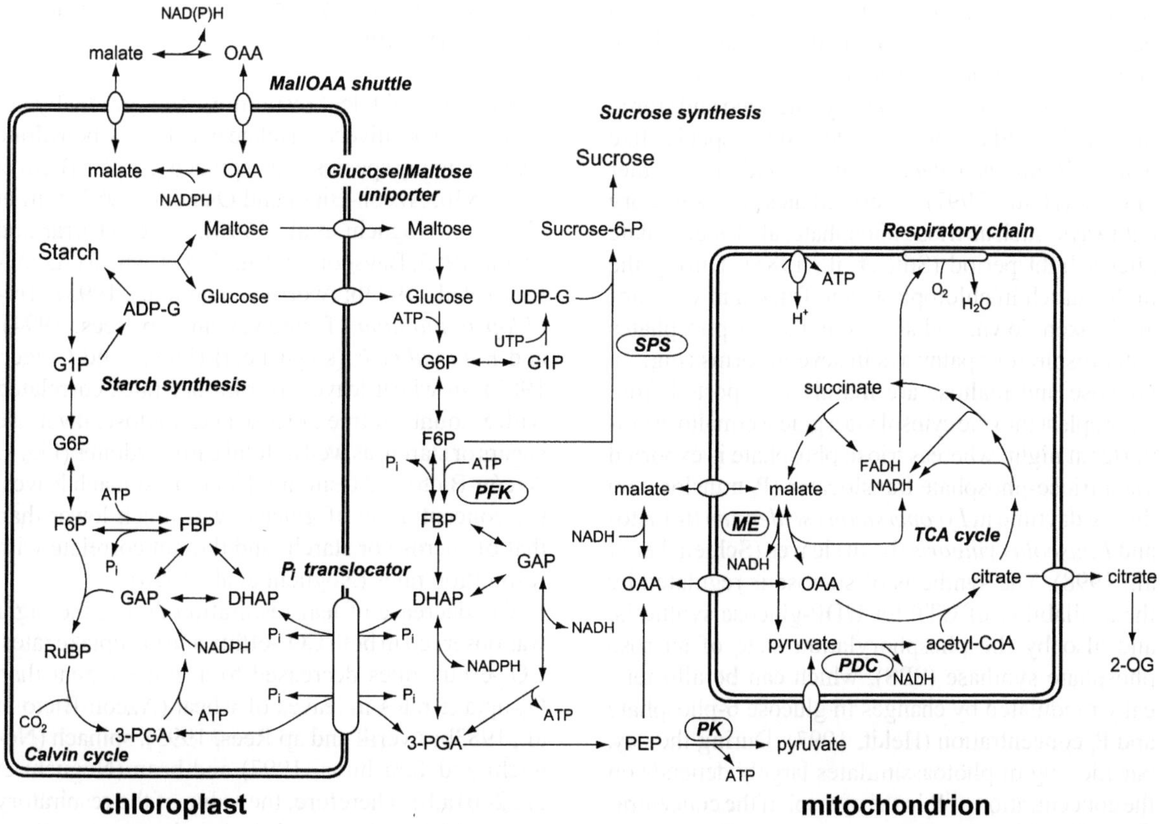

Fig. 2. Metabolic pathways in mitochondria and chloroplasts of plant cells. Abbreviations: G1P: glucose 1-phosphate; G6P: glucose 6-phosphate; ADP-G: ADP-glucose; F6P: fructose 6-phosphate; FBP: fructose 1,6-bisphosphate; GAP: glyceraldehyde 3-phosphate; DHAP: dihydroacetone phosphate; RuBP: Ribulose 1,5-bisphosphate; 3-PGA: 3-phosphoglycerate; UDP-G: UDP-glucose; Sucrose 6-P: sucrose 6-phosphate; PEP: phosphoenolpyruvate; OAA: oxaloacetate; CoA: coenzyme A; PFK: ATP-dependent phosphofructokinase; PK: pyruvate kinase; SPS: sucrose phosphate synthase; ME: malic enzyme; PDC: pyruvate dehydrogenase.

dehydrogenase in the pentose phosphate pathway (Fig. 2). Several reviews suggested that, in young plant cells, carbohydrate availability or capacity of respiratory components would determine respiratory rates, whereas in mature plant cells, a respiration rate is controlled by rates of processes that use respiratory products, such as growth, transport, nutrient uptake and assimilation, and maintenance, via adenylate control (Beevers, 1974; Farrar, 1985; ap Rees, 1988; Amthor, 1989, 1994, 1995; Farrar and Williams, 1991; Amthor et al., 2000).

What mechanisms can account for the correlation between carbohydrate status and respiratory rates in leaves, roots and whole plants? Does the carbohydrate status directly influence respiratory rates? Is there any difference between organs and between conditions? This chapter, firstly, reviews carbohydrate metabolism and respiratory pathways in leaf and root cells, and then summarizes some (eco)physiological studies on correlations between carbohydrate status and respiratory rates. It then examines the underlying mechanisms of the correlation between carbohydrate status and respiration in leaves and roots. Finally, a direct effect of strong light intensity on leaf respiration in the light of stress physiology is presented.

II. Relationship between Leaf Respiration and Carbohydrate Status

A. Carbohydrate and Respiratory Metabolisms in Mature Leaves

During daytime, carbohydrates accumulate in mature leaves as a result of photosynthesis. There is variation in accumulated carbohydrates among plant species (Huber, 1989). In *Cucumis sativus* (cucumber) and *Glycine max* (soybean), leaves mainly store starch, whereas sucrose is mostly stored in leaves of field bean and *Pisum sativum* (pea). Starch is an insoluble polyglucose formed in plastids, whereas fructans are soluble polyfructoses that are synthesized and stored in vacuoles and can accumulate in other species like wheat, *Hordeum vulgare* (barley) and some other grasses (Heldt, 1997). Intermediates of respiratory pathways, such as triose phosphate, also accumulate after a light period (Stitt et al., 1985). During the night, starch in chloroplasts and fructan in vacuoles are broken down, and sugars and sugar phosphates enter respiratory pathways in several forms (Fig. 2). Glucose and maltose are mainly transported from chloroplast into the cytosol via a glucose/maltose uniporter at night, whereas triose phosphate is exported via a triose-phosphate translocator (P_i translocator) during daytime in *Lycopersicon esculentum* (tomato) and *Phaseolus vulgaris* (bean) leaves (Schleucher et al., 1998). The synthesis of sucrose is regulated by the availability of UTP for UDP-glucose synthesis, and also by the phosphorylation state of sucrose phosphate synthase (SPS), which can be allosterically modulated by changes in glucose 6-phosphate and P_i concentration (Heldt, 1997). During the day, partitioning of photoassimilates largely depends on the concentration of P_i in the cytosol. If the concentration of P_i is high, triose phosphate is rapidly exported from chloroplasts, and converted to sucrose. However, if the P_i concentration is too low, triose phosphate is used for starch synthesis in chloroplasts. Eighty percent of sucrose is accumulated in vacuoles, where its concentration is 26–120 mM in barley leaves (Farrar and Farrar, 1986). Sucrose transport across the tonoplast is not energized (Pollock and Farrar, 1996). At night, in mature leaves, most of the carbohydrates are translocated to sink organs (e.g., stem, root, and young leaves). In potato leaves, 67–84% of carbohydrates were translocated, and in bean leaves about 70% (Bouma et al., 1995). In spinach leaves, 65–70% were translocated (K. Noguchi, unpublished data) and 71% and 76% in leaves of bean and *Alocasia odora*, a sun and shade species, respectively (Noguchi et al., 2001b). The remainders of the carbohydrates are used for respiratory substrates in leaves during the night, to produce respiratory energy (ATP) and carbon skeletons. About 70 to 95% of all hexose phosphate is degraded via PFK and aldolase in glycolysis, whereas the remainder are oxidized to triose phosphate via the oxidative pentose phosphate pathway (ap Rees, 1980).

B. Correlation between Carbohydrate Status and Respiration

During the night, leaf respiratory rates gradually decrease, and positively correlate with leaf carbohydrate levels in many species, e.g., spinach (Azcón-Bieto et al., 1983b; Azcón-Bieto and Osmond, 1983; Stitt et al., 1990; Noguchi et al., 1996), barley (Farrar and Farrar, 1985; Baysdorfer et al., 1987), wheat (Azcón-Bieto et al., 1983b; Averill and ap Rees, 1995), *Arabidopsis thaliana* (Trethewey and ap Rees, 1994), and *Beta vulgaris* (sugar beet) (Fondy and Geiger, 1982). In wheat leaves, respiratory rates correlated with amounts of free glucose and fructose, invertase sugars or starch, as well as total carbohydrates (Fig. 1, Azcón-Bieto and Osmond, 1983). In spinach leaves, the concentration of glucose was much lower than that of sucrose or starch, and did not correlate with respiration rates (Noguchi et al., 1996).

The decrease of leaf respiration during the night was observed in both CO_2-efflux and O_2-uptake rates. CO_2-efflux rates decreased to a grater extent than O_2-uptake rates in leaves of wheat (Azcón-Bieto et al., 1983b; Averill and ap Rees, 1995), spinach (Noguchi and Terashima, 1997) and bean (Noguchi et al., 2001a,b). Therefore, the value of the respiratory quotient (RQ) decreased during the night in these species. The RQ of wheat leaves after six hours of illumination was 1.80, whereas that at the end of the night was 0.93 (Azcón-Bieto et al., 1983b). This is probably because at the beginning of the night, NADH is consumed in several processes, such as nitrate assimilation and biosynthesis of reduced components in the leaves, and later in the night, NADH is predominantly consumed by the respiratory chain (Lambers et al., 1998).

The correlation between carbohydrate status and respiration can be ascribed, in part, to a direct effect of carbohydrates as respiratory substrates in spinach and wheat leaves (Azcón-Bieto et al., 1983b; Trethewey and ap Rees, 1994; Noguchi and Terashima, 1997). This was demonstrated by an addition of exogenous sugars or an uncoupler to leaf segments. At the end of the night, respiratory rates of leaves were largely enhanced by exogenous sugars, but, after illumination, respiratory rates were less enhanced by sugars (Fig. 3). In wheat and spinach leaves, addition of an uncoupler did not enhance the respiratory rates throughout the night. Therefore, in these leaves, carbohydrate status would directly control respiratory rates during the night.

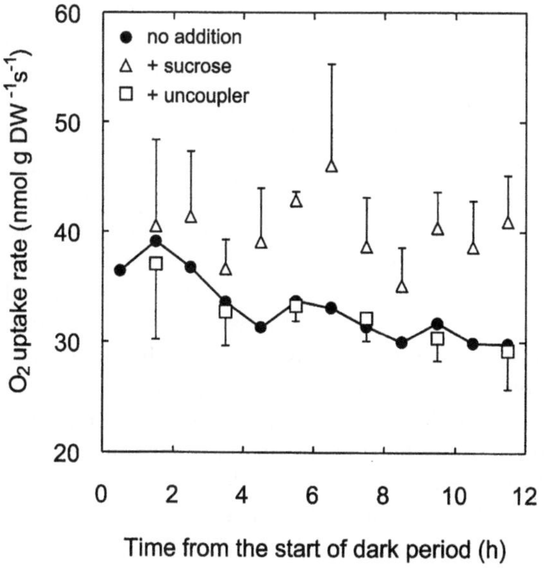

Fig. 3. Effects of the addition of sucrose or uncoupler on the rate of O_2 uptake in *Spinacia oleracea* (spinach) leaves. In the presence of sucrose (triangles) or uncoupler (squares), each datum point represents the mean of the rates and the bars denote standard deviation. Based on Noguchi and Terashima (1997).

Does carbohydrate availability, rather than adenylate demand, mainly restrict the respiratory rates in all leaves that show a positive correlation between respiration and carbohydrate status, as it does in spinach and wheat? A good correlation between respiration and carbohydrate status might reflect other physiological mechanisms. In mature leaves, respiratory ATP is used for maintenance of existing components (e.g., proteins and lipid membranes), carbohydrate export (starch degradation and phloem loading), and maintenance of ion gradients across membranes (Amthor, 1994; Lambers et al., 1998; Noguchi et al., 2001b). If ATP consumption for these processes decreased at night, respiratory rates would also decrease via adenylate control. Estimations of the rate of ATP consumption in each process, and studies of the relationships between respiration and ATP consumptions for these processes as well as a direct addition of sugars or uncouplers to leaves are needed. Estimation of total respiratory ATP production rate is also important, in order to understand the effect of adenylate demand on respiratory rates. The estimation of respiratory ATP production depends on the respiratory pathways that are used, especially in the respiratory electron transport in mitochondria.

C. Relationship between Carbohydrate Status and Alternative Oxidase in Mature Leaves

Plant respiratory chains have two quinol-oxidizing pathways, the cytochrome pathway and the alternative pathway. The alternative pathway consists of one enzyme, the alternative oxidase (AOX), which is sensitive to salicylhydroxamic acid (SHAM) (Vanlerberghe and McIntosh, 1997). Since electron transport from ubiquinol to AOX is not coupled to proton export, engagement of AOX will largely influence ATP production rates (Amthor, 1994; Noguchi et al., 2001b). If all electrons enter into AOX, respiratory ATP production would be about one-third of maximal production. Also, since AOX is not restricted by adenylate demands, the engagement of AOX will make the relationship between carbohydrate status and respiration rates complicated. If there is a good correlation between carbohydrate status and respiratory rates of leaves during the night, can both pathways respond to carbohydrate status in the leaves? SHAM-sensitive O_2 uptake of wheat and spinach leaves was fast after an illumination period, and decreased during the night (Azcón-Bieto et al., 1983b). At the end of the night, enhanced respiratory rates by exogenous sucrose or glycine were sensitive to SHAM (Azcón-Bieto et al., 1983a). This suggested that AOX is engaged early during the night, and AOX would consume a large proportion of carbohydrates when carbohydrate status is high (energy-overflow hypothesis, Lambers, 1985). However, SHAM-sensitive respiration cannot be interpreted to indicate in vivo AOX activity (in vivo contribution of AOX to total respiration) (Day et al., 1996). Studies using an inhibitor such as SHAM may underestimate AOX activity, especially at the end of the night. Studies with the ^{18}O-fractionation technique can solve this problem, but only a few studies have been conducted. In spinach leaves and bean primary leaves (Noguchi et al., 2001a), in vivo AOX activities were, indeed, high early during the night when carbohydrate concentrations were high, while rates were slow later during the night when carbohydrate status was low. In these leaves, AOX would consume a large fraction of carbohydrates early during the night. However, in *Nicotiana tabacum* (tobacco) and bean leaves under high phosphate condition, there was no correlation between concentration of soluble carbohydrate and in vivo AOX activity, whereas there was a significant correlation between concentration of soluble carbohydrates and activities of cytochrome *c* oxidase

(COX) or total respiratory rates (Gonzàlez-Meler et al., 2001). In soybean cotyledons, in vivo AOX activity decreased during the dark period, but there was no correlation between in vivo AOX activity and carbohydrate levels (Ribas-Carbo et al., 2000). Direct effects of carbohydrates to in vivo AOX activity will solve the contradictions between results. If AOX responds to carbohydrate status in leaves of some species, how do carbohydrate levels influence in vivo AOX activity in the leaves? AOX activity is usually determined by the redox state of the ubiquinone pool, the redox state of the AOX protein and the allosteric effects due to α-keto acid, such as pyruvate and α-keto glutarate (Vanlerberghe and McIntosh, 1997). How can these factors be influenced by carbohydrate status during the night? Changes in carbohydrate levels would influence substrate levels in mitochondria, and then redox levels of the ubiquinone pool. The redox state of AOX protein is thought to be affected by intramitochondrial NADPH levels via the thioredoxin pathway in mitochondria (Vanlerberghe and McIntosh, 1997). Intramitochondrial NADPH is generated by isocitrate dehydrogenase and malate dehydrogenase. Carbohydrate levels would affect levels of TCA-cycle intermediates, and then NADPH levels. α-keto acids also respond to carbohydrate levels. Pyruvate and α-keto glutarate concentrations were 34% and 130% higher at the start of night than at the end of night in wheat leaves, respectively (Averill and ap Rees, 1995). Although the pyruvate concentration at the cellular level is high enough to activate AOX (Millenaar et al., 1998, Siedow and Umbach, 2000), most of pyruvate might occur outside mitochondria and its mitochondrial concentration is unknown. Gastón et al. (2003) observed that both pyruvate concentration and in vivo AOX activity increased after inhibiting branched chain amino acid biosynthesis in soybean roots. Thus, carbohydrate status may influence in vivo activity of AOX via the above activating systems in mature leaves. In roots, however, the carbohydrate status would not influence in vivo activity of AOX (Millenaar et al. 2000; see details in Section III.D).

D. A Lack of Correlation between Carbohydrate Status and Respiratory Rates in Mature Leaves of Some Species

In the examples discussed above, leaves showed a good correlation between respiration rates and carbohydrate status. However, in leaves of other species, there is no correlation between carbohydrate status and respiration (pea, Azcón-Bieto et al., 1983b; shade-tolerant species, *Alocasia odora*, Noguchi et al., 1996; alpine perennial species, McCutchan and Monson, 2001a). These species showed almost constant respiratory rates during the night. In leaves of the shade-tolerant *A. odora*, both CO_2-efflux and O_2-uptake rates were slow and constant throughout the night (Noguchi et al., 1996, Noguchi and Terashima, 1997). This slow rate of leaf respiration would be useful for a positive carbon balance under deep-shade environments. Exogenous sucrose and glycine did not enhance O_2-uptake rates, even at the end of the night. However, an uncoupler did enhance O_2-uptake rates of *A. odora* leaves throughout the night (Fig. 4). The ADP concentration of *A. odora* leaves was lower, and both ATP/ADP and energy charge were higher than those of spinach leaves throughout the night (Noguchi and Terashima, 1997). Thus, respiratory rates of *A. odora* leaves were strictly limited by adenylate demand for ATP-consuming processes. When ATP-consuming processes (carbohydrate export and protein turnover) were compared between *A. odora* leaves and bean primary leaves, these rates were lower in *A. odora* leaves (Noguchi et al., 2001b). Thus, these results suggested that the low ATP demand for cellular processes restricted respiration rates via oxidative phosphorylation in *A. odora* leaves. Adenylate control on respiration has also been observed in leaves of other species. An uncoupler enhanced O_2-uptake rates after a period in the light in barley leaves (Farrar and Rayns, 1987), *Lolium perenne* (Day et al., 1985) and bean leaves (Azcón-Bieto et al., 1983c). An uncoupler relieves not only the cytochrome pathway but also substrate supply to mitochondria, because PK in glycolysis is also restricted by adenylates (Pasteur effect; Fig. 2). In *L. perenne* leaves, exogenous malate and glycine enhanced O_2-uptake rates, but exogenous sucrose only stimulated respiration when an uncoupler was also present. Thus, adenylates restricted the flow of substrates into mitochondria via glycolysis in *L. perenne* leaves. Does AOX activity respond to carbohydrate status in leaves of these species like in spinach? Studies with the ^{18}O-discrimination technique showed that *A. odora* leaves had a low in vivo activity of AOX, irrespective of carbohydrate status throughout the night (Noguchi et al., 2001a), whereas *A. odora* leaves showed fast rates of KCN-insensi-

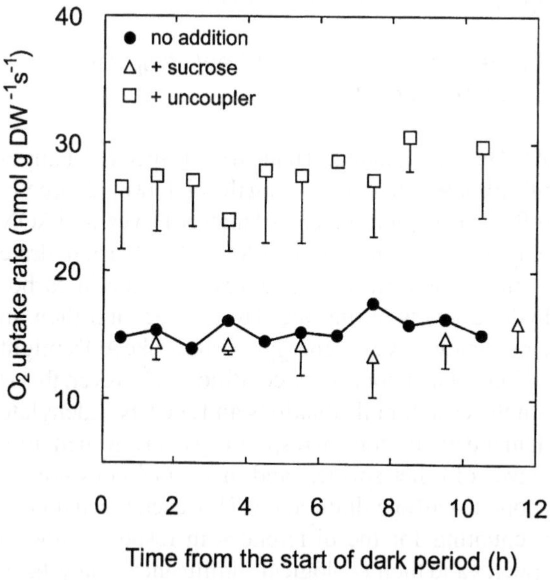

Fig. 4. Effects of the addition of sucrose or uncoupler on the rate of O_2 uptake in leaves of *Alocasia odora*, a shade-tolerant species. In the presence of sucrose (triangles) or uncoupler (squares), each datum point represents the mean of the rates and the bars denote standard deviation. Based on Noguchi and Terashima (1997).

tive O_2 uptake (Noguchi, 1995). This suggested that neither AOX nor COX activities in *A. odora* leaves responded to carbohydrate status, and that a high efficiency of ATP production was achieved.

In some cases, enhanced respiration occurred late in the night or early in the morning (called 'morning rise') in mature leaves of *Oryza sativa* (rice) (Akita et al., 1993; Lee and Akita, 2000), whole plants of 14 crop species, such as rice, soybean, wheat (Yamagishi et al., 1990), *Medicago sativa* (alfalfa) shoots (Pearson and Hunt, 1972), and shoots and roots of wheat (Gerbaud et al., 1988). Gerbaud et al. (1988) found that respiration of young wheat shoots deceased throughout the night, but older shoots showed an enhanced respiration at the end of night. Akita et al. (1993) observed a similar tendency in rice. Rates of both CO_2 efflux and O_2 uptake showed a 'morning rise,' and RQ values of rice leaves did not change during the night (M. Oda and S. Akita, personal communication). The underlying mechanisms of the morning rise have not yet been clarified, but the carbohydrate status would not influence this enhancement in mature leaves, because carbohydrate levels did not increase late at night in most cases.

E. Relationship between Leaf Respiration and Carbohydrate Status during Leaf Development and Senescence

Leaf respiration changes during development in most species (Tetley and Thimann, 1974; Jurik et al., 1979; Azcón-Bieto et al., 1983c; Collier and Thibodeau, 1995; Oleksyn et al., 2000; McCutchan and Monson, 2001b). Respiration rates were high in young expanding leaves, and decreased rapidly as leaves showed maximum photosynthetic rates. Thereafter, respiration rates were almost constant until senescence. A senescence-induced respiratory burst has been observed in some species (Tetley and Thimann, 1974; McCutchan and Monson, 2001b), but not in others (Jurik et al., 1979; Azcón-Bieto et al., 1983c, Collier and Thibodeau, 1995). Does the effect of carbohydrate status on leaf respiration also change with age?

Leaf respiration of bean correlated with free fructose and glucose, but not with invertase sugars (mostly sucrose) or starch (Azcón-Bieto et al., 1983c). Leaves of *Betula pendula*, a deciduous tree, showed gradually increased respiratory rates and soluble carbohydrates during summer and autumn (Oleksyn et al., 2000). This increased respiration rate would be related to acclimation to low temperatures. A similar trend was observed in leaves of evergreen trees (Kozolowski and Pallardy, 1997). During senescence, in leaves of two deciduous tree species, *Populus tremuloides* and *Quercus rubra*, both chlorophyll and nitrogen concentration gradually declined for two months before defoliation, but respiratory rates decreased quickly for three weeks before defoliation in parallel with soluble sugar contents (Fig. 5, Collier and Thibodeau, 1995). In *B. pendula* leaves, both respiratory rates and soluble carbohydrate levels decreased (Oleksyn et al., 2000). However, these correlation studies do not always indicate a direct influence of carbohydrate status on respiratory rates during leaf development. In bean leaves, exogenous sucrose did not enhance respiration at any ages, but an uncoupler enhanced respiration rates, especially in young leaves (Azcón-Bieto et al., 1983c).

There are no data on in vivo activities of AOX and COX with leaf developmental stage using the ^{18}O-discrimination technique, except for soybean cotyledons (Ribas-Carbo et al., 2000). In parallel with greening, total respiration rates of soybean cotyledons showed a peak, and then gradually decreased. In vivo

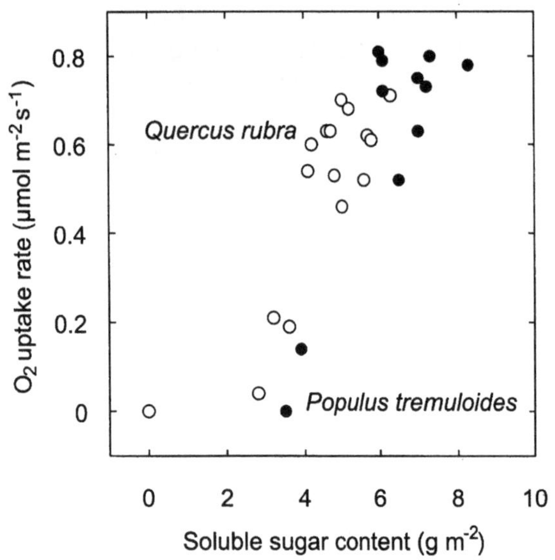

Fig. 5. Relationship between respiration rate and soluble sugar content in leaves of *Populus tremuloides* (open symbols) and *Quercus rubra* (filled symbols) during autumnal senescence. Based on Collier and Thibodeau (1995).

activity of COX increased during the first three hours of exposure to light, and then decreased, whereas in vivo activity of AOX peaked after 24 hours of greening and gradually decreased. After 50 hours of greening, both oxidases showed a similar activity. In young leaves, the capacity of the cytochrome pathway was high, and then it decreased. However, the capacity of the alternative pathway did not change in leaves of bean (Azcón-Bieto et al., 1983c) and of two deciduous tree species, *P. tremuloides* and *Q. rubra* (Collier and Thibodeau, 1995). In pea leaf mitochondria, the capacity of the cytochrome pathway also decreased, but the activity of the alternative pathway did not change during the same measurement period (Azcón-Bieto et al., 1983c). The calculated P/O ratio (the efficiency of oxidative phosphorylation) also decreased with leaf age in pea mitochondria (Geronimo and Beevers, 1964). A similar tendency was observed in senescent mung bean (Wen and Liang, 1993) and soybean cotyledons (Azcón-Bieto et al., 1989). During senescence, the COX capacity in leaves of two deciduous tree species, *P. tremuloides* and *Q. rubra,* decreased more quickly than the AOX capacity (Collier and Thibodeau, 1995). Thus, these data suggest that AOX would be active, even in senescent leaves, although it is not know whether in vivo activity of AOX relates to carbohydrate status with leaf developmental stage from the available data.

F. Different Responses of Leaf Respiration to Carbohydrate Status

All these results suggest that, in some species, changes of carbohydrate status during the night would directly influence respiration rates of mature leaves, and AOX may act as an energy overflow when carbohydrate status is high. In mature leaves, variation in carbohydrate levels during the day is large, and then an emergency valve for energy overflow like AOX might be important under stress conditions. However, this is not the case for all situations and species. Adenylate demand would control respiratory pathway in mature leaves of other species, and then AOX does not respond to carbohydrate status. However, mechanisms accounting for the differences in response among species and their ecological significance, if any, have not yet been clarified. It is relatively easy to measure dark respiration of mature leaves with a portable gas-exchange system or an oxygen electrode, and to sample leaf materials even under field conditions. Even though there is a large collection of data on respiration rates and carbohydrate levels in leaves of various plant species (Amthor, 1989), these data is not sufficient to gain an understanding of the mechanisms explaining the relationships between leaf respiration and carbohydrate levels. Further analysis of the correlation between respiration rates and carbohydrate levels under various conditions and with various plant species, as well as to examine in vivo AOX activities and changes of respiratory rates by the addition of sugar/uncoupler to leaves, in order to understand the effect of carbohydrate status on leaf respiration pathway in vivo. Understanding the relationship between levels of carbohydrates and metabolites, such as respiratory intermediates, reducing equivalents, and adenylates, and estimate fluxes of carbohydrates between tissues and between subcellular components is also necessary (Farrar, 1985; ap Rees, 1988; Winter et al., 1994). Different responses of respiration to carbohydrate status may be understood better after a comparison of plant functional groups and different species. Reich et al. (1998) examined leaves of 69 species from four functional groups (forbs, broad-leafed trees and shrubs, and conifers) in six biomes. They predicted dark respiratory rates of leaves by combinations of leaf traits (leaf life span, nitrogen concentration, specific leaf area). If detailed physiological experiments will be added to their results, this would help to understand different responses of respiration to carbohydrate status in mature leaves, and also the ecological significance of different responses.

III. Relationship between Root Respiration and Carbohydrate Status

A. Carbohydrate Metabolism in Roots

Is the relationship between respiration and carbohydrate status in roots similar to that in leaves? Between one-third to two-thirds of all photosynthates are respired per day, and between 8 and 52% of the daily produced photosynthates are consumed in the roots (Lambers et al., 2002). This percentage is influenced by several factors, such as maximum growth rate of the species, nutrient availability, and plant age. Slower potential growth rate and lower nutrient supply are accompanied with greater partitioning of photoassimilates to root respiration (Van der Werf et al., 1992). Moreover, this percentage tends to decrease with increased plant age (Van der Werf et al., 1988). These changes in carbohydrate partitioning would be partly influenced by changes in cellular demands for respiratory energy (growth, maintenance, and ion uptake) in roots.

Photosynthates are transported from shoot to roots via the phloem. Sucrose is a major transported form in phloem of species with an apoplasmic loading type. In the species that have a symplasmic loading type, oligosaccharides are major forms (Van Bel, 1993). Transported sucrose is unloaded from sieve elements of phloem into root parenchyma cells. There are also two pathways in phloem unloading, symplasmic and apoplasmic (Heldt, 1997). In symplasmic unloading, sucrose is directly transported into root parenchyma cells via plasmodesmata. In apoplasmic unloading, sucrose is exported from sieve elements into an apoplasmic region. Sucrose is partly hydrolyzed into glucose and fructose by invertase in the apoplast, and sucrose and hexoses are taken up into root cells via disaccharide and monosaccharide transporters, respectively (Williams et al., 2000). Both transporters are energy-dependent symporters. In roots, symplasmic unloading is a major process, while in storage sinks (e.g., tubers) predominates the apoplasmic pathway (Heldt, 1997). In barley roots, 45% of sucrose entering the root was respired (Farrar, 1985). In roots, imported sugars are compartmented into three pools: cytosolic, vacuolar, and apoplasmic, and stored as fructan and starch (Fig. 6, Farrar and Williams, 1991). Although, the cytosolic pool size is small, it is stable, because this pool is supported by import supply from shoot and other pools.

B. Root Respiration and its Relationship with Carbohydrate Status

Root respiration supports growth, maintenance, and active uptake of nutrients from soil or culture solution. Root respiration also supports nutrient assimilation in roots, although a considerable fraction of nitrate assimilation may occur in shoots during the day (Amthor, 1995). Like leaf respiration, root respiration may show a diurnal fluctuation in many species, for example *Helianthus annuus* (sunflower) (Hatrick and Bowling, 1973; Casadesús et al., 1995), *Lolium multiflorum* (Hansen, 1980), cucumber (Lambers et al., 2002, based on data from Challa, 1976), *Holcus lanatus* (Lambers et al., 2002, based on data from Scheurwater et al., 1998), barley (Farrar, 1981), alfalfa (Pearson and Hunt, 1972). In most cases, respiratory rates in roots gradually decreased during night and then increased during daytime (e.g., barley root, Far-

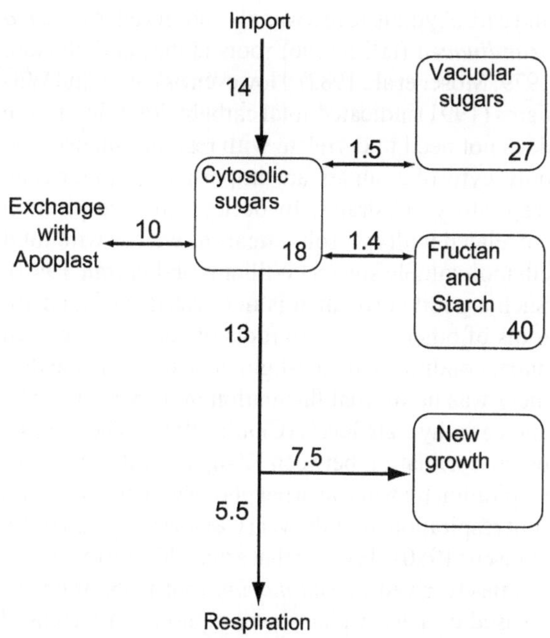

Fig. 6. Compartmentation and fluxes of carbohydrates in seminal roots of young *Hordeum vulgare* (barley). Numbers in boxes are pool sizes (mg g^{-1}) and next to arrows are fluxes (mg g^{-1} h^{-1}). Based on Farrar and Williams (1991).

rar, 1981). However, respiratory rates in cucumber roots increased during the night and showed minimum rates during midday (Lambers et al., 2002, based on data from Challa, 1976, Fig. 7). As seen in leaf respiration, CO_2-efflux rates showed larger diurnal fluctuation than O_2-uptake rates in roots of *Holcus lanatus* (Lambers et al., 2002, based on data from Scheurwater et al., 1998) and *Derris elliptica* (Huck et al., 1962), and thus RQ values changed throughout the day. Possibly, the larger variation in CO_2-efflux rates is associated with variation in nitrate reduction in the roots, because an additional two molecules of CO_2 are produced per molecule of nitrate reduced to ammonium. RQ value can also be high when organic acids (e.g., malate) are used as respiratory substrates. Malate, which is produced during nitrate reduction in leaves, would be translocated to roots via the phloem, and then decarboxylated in the roots (Ben Zioni et al., 1971; Lambers et al., 1998).

There was a diurnal fluctuation in the level of soluble sugars, and a good correlation between soluble sugar levels and respiratory rates in roots of cucumber, grown under conditions of low light intensity and short day (Fig. 7). A good correlation between root respiration and translocation from the shoot was observed in sunflower (Hatrick and Bowling, 1973). Another good correlation between respiration rates and carbohydrate levels was also observed in *Festuca arundinacea* (tall fescue) roots (Jones and Nelson, 1979; Moser et al., 1982). However, as Farrar and Williams (1991) indicated, total carbohydrates in a tissue does not need to correlate with respiration, because only cytosolic sugars are important as immediate respiratory substrates. In barley roots, respiration correlated with cytosolic sugars as well as with total ethanol-soluble sugars (Williams and Farrar, 1990). Such a good correlation is not invariably found for roots of other species. In roots of cucumber, grown under conditions of high-light intensity and long day, there was no diurnal fluctuation in respiratory rates and carbohydrate levels (Challa, 1976). There was a better correlation between K^+-uptake rates and root respiration than between net photosynthetic rates and root respiration in sunflower (Casadesús et al., 1995). Hansen (1980) observed that after addition of NO_3^- to nitrate-deprived *L. multiflorum* root respiration increased, even when plants were exposed to a decreased light intensity. There was no correlation between root respiration and soluble carbohydrate content in roots of barley, after being placed under prolonged darkness (Farrar, 1981), and in which leaves or roots were pruned (Williams and Farrar, 1990). Therefore, a diurnal fluctuation in root respiration was found in most species, although this fluctuation cannot be simply related to a diurnal change in carbohydrate status in the roots.

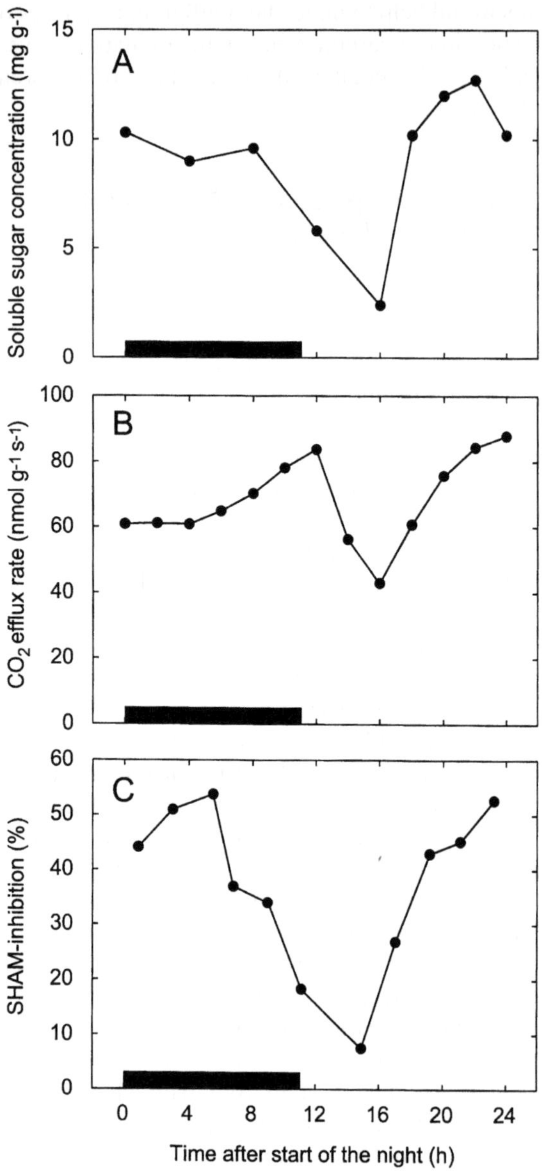

Fig. 7. Diurnal change of the soluble sugar concentration (A), CO_2 efflux rate (B), the percentage of respiration that is sensitive to SHAM (C) in the roots of *Cucumis sativus* (cucumber). A and B are based on Challa (1976), and C is based on Lambers (1980).

C. Fine and Coarse Controls on Root Respiration by Carbohydrate Status

Does a different relationship between carbohydrate status and root respiration mean a different regulation of root respiration? Also, does a good correlation simply mean a direct effect of carbohydrate levels on root respiration as a substrate? Or, alternatively, is the correlation due to an indirect effect of carbohydrate status via ATP demand? These questions have been addressed in many experiments using addition of sugars or uncouplers to roots. Roots have frequently been chosen for respiratory studies, because they can easily absorb exogenous sugars or uncouplers, and avoid effects of concurrent photosynthesis and photorespiration (reviewed by Farrar and Williams, 1991; Lambers et al., 2002). Farrar and co-workers elaborately examined the relationships between respiration and carbohydrate status with barley and wheat plants, for which the source:sink ratio was changed by selective pruning or shading (Farrar and Jones, 1986, Bingham and Farrar, 1988, Williams and Farrar, 1990, Bingham and Stevenson, 1993). They suggested that root respiration rate is restricted by adenylates for short time periods (minutes to hours, fine control), while carbohydrate status influences the capacity of respiratory rate after longer time periods (several hours to days, coarse control). They also suggested that short-term supply of sugars to roots only stimulates respiration as a substrate when carbohydrate contents in cells is low, such as after prolonged dark treatment and at the end of the night (Farrar and Williams, 1991). The good correlation observed between respiration and carbohydrate levels is supposedly induced by a long-term effect of carbohydrate status on respiration. Many other data support these views. An uncoupler enhanced root respiration in spinach, maize and bean (Day and Lambers, 1983), barley (Farrar, 1981), wheat (Bingham and Stevenson, 1993) and *L. perenne* (Day et al., 1985). In roots of spinach, maize and *L. perenne*, adenylate demand determines respiratory rates via an effect through limitation of the flux of electrons through the respiratory chain, because the uncoupler enhanced root respiration in the presence of SHAM. In contrast, in bean roots, adenylate control limits respiration via the substrate supply to mitochondria via glycolysis. Other indirect evidence suggesting the adenylate control in root respiration is that an increase in the demand for ATP results in an increased rate of respiration, such as 'salt respiration' (the rise in respiration rates shortly after adding anions; Willis and Yemm, 1955). In excised maize root tips, respiration rates gradually decreased, and exogenous glucose restored the decreased O_2 uptake (Saglio and Pradet, 1980; Brouquisse et al., 1991; Williams and Farrar, 1992). However, stimulation by exogenous glucose was not rapid (after more than 1 hour) (Saglio and Pradet, 1980), and an uncoupler also enhanced respiration rates of maize roots throughout the starvation period (Brouquisse et al., 1991; Williams and Farrar, 1992).

Farrar and Williams (1991) suggested that changes in carbohydrate status induce long-term changes of various cellular processes due to gene transcription, followed by *de novo* protein synthesis (Farrar, 1985; Farrar and Williams, 1991). The protein pattern in the roots of shoot-pruned barley was affected within 24 hours (Williams et al., 1992). Mitochondria isolated from the roots of shaded barley plants showed changes in respiratory properties, such as decreased maximal activities of COX (McDonnell and Farrar, 1992). Specific proteins increase, while others decrease in sucrose-fed roots of *Pennisetum americanum* (millet) (Baysdorfer and Van der Woude, 1988). The expression of the *sus1* gene increased with the concentration of glucose supplied to excised maize root (Koch et al., 1992). An increase in sugars may stimulate root nitrate uptake and assimilation in bean (Hänisch ten Cate and Breterler, 1981), and enhance cell division and differentiation in barley (Williams and Farrar, 1990). In contrast, accumulated carbohydrates in source leaves led to changes of photosynthetic rates in leaves (Neals and Incoll, 1968; Azcón-Bieto, 1983; Pollock and Farrar, 1996). There was a positive correlation between carbohydrate contents and suppression of photosynthetic rates in wheat leaves (Azcón-Bieto, 1983). Feeding sugar to detached leaves of spinach and cold-girdling, which decreased export of photoassimilates, reduced expression of specific photosynthetic genes, such as *rbsS* (Krapp et al., 1993; Krapp and Stitt, 1995). Aldolase, triose-phosphate isomerase, and phosphoglucomutase, which have a dual role in photosynthesis and respiration, were not subject to carbohydrate repression. Activities of glycolytic enzymes either increased or remained unaltered (Krapp and Stitt, 1994). Addition of sucrose to tobacco seedlings also enhanced maximal activities of sucrose synthase, pyrophosphate-dependent phosphofructokinase and PFK in roots (Paul and Stitt,

1993). Further details of carbohydrate repression of gene expression and sugar-sensing mechanisms in higher plants can be found in Sheen (1994), Koch (1996), Graham and Martin (2000), and Hellmann et al. (2000).

In some cases, exogenous carbohydrates stimulated root respiration directly and rapidly. In pea excised roots, exogenous sucrose (20–100 mM) enhanced respiration rates (Fig. 8, Bryce and ap Rees, 1985). The effect of exogenous sucrose on pea root respiration increased with increasing time of exposure, and with increasing concentration of exogenous sucrose. In pea roots, exogenous sucrose enhanced respiration before endogenous sugars were depleted. Exogenous sugars enhanced respiratory rates of excised roots in maize and soybean (Huck et al., 1962), maize and pea (Crawford and Huxter, 1977), and bean (Hänisch ten Cate and Breteler, 1981). Since the energy cost of taking up sugars would be very small (about 2% of total respiration rates in the case of barley roots; Farrar and Williams, 1991), root respiratory rates would be restricted by availability of carbohydrates in these cases.

D. Relationship between Carbohydrate Status and Alternative Oxidase (AOX) in Roots

AOX is not restricted by adenylate demand, and in spinach and bean primary leaves, AOX appears to act as an energy overflow pathway when carbohydrate levels were high at the beginning of the night (Azcón-Bieto et al., 1983b; Noguchi et al., 2001a). How do the cytochrome and alternative pathway respond to carbohydrate status in roots? SHAM-sensitive respiration rates correlated with soluble carbohydrate levels in cucumber roots (Fig. 7C, Lambers, 1980), although in vivo AOX activity might be underestimated when SHAM-sensitive respiration rates is measured (Day et al., 1996). Millenaar et al. (2000) transferred *Poa annua* plants from high- to low-light intensity, and at the same time from long-day to short-day conditions. They measured respiration rates, carbohydrate levels, maximal activity of COX, and in vivo AOX activity, which was measured with the ^{18}O-discrimination technique, in roots. Both respiration rates and maximal activity of COX decreased in parallel with carbohydrate levels, but in vivo AOX activity did not change for 64 hours after transfer (Fig. 9). They also

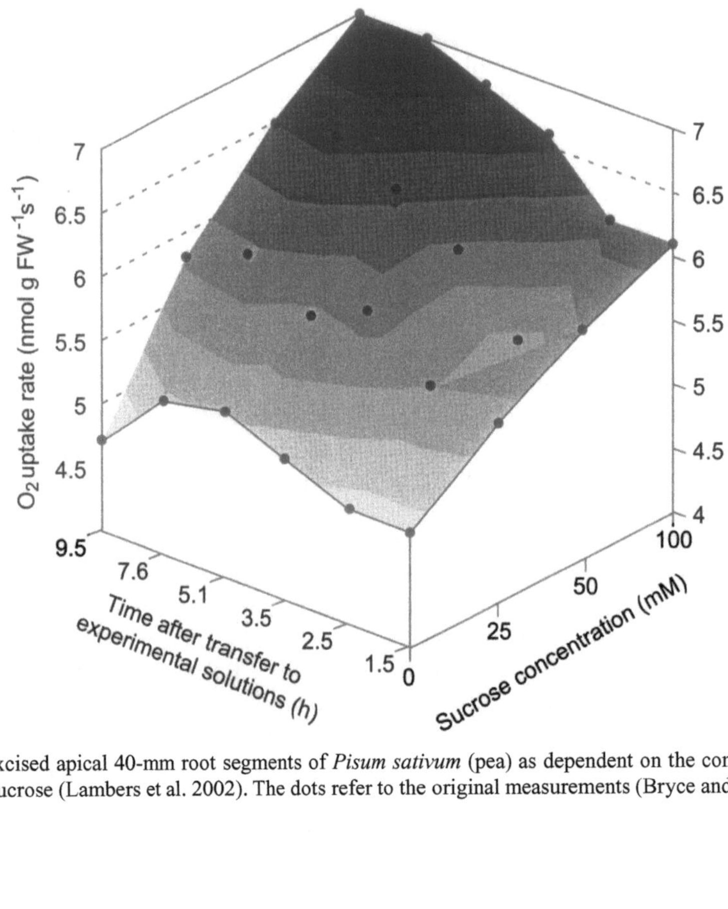

Fig. 8. Respiration of excised apical 40-mm root segments of *Pisum sativum* (pea) as dependent on the concentration and time after addition of exogenous sucrose (Lambers et al. 2002). The dots refer to the original measurements (Bryce and ap Rees 1985).

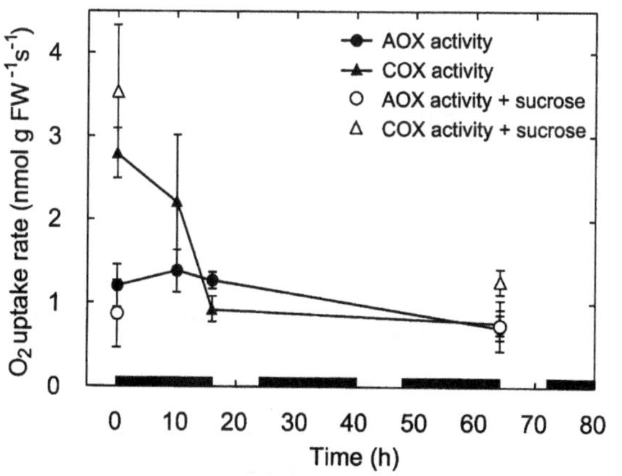

Fig. 9. Activities of the alternative oxidase and the cytochrome pathway with sucrose or without sucrose in *Poa annua* roots. Measurements were conducted with the ^{18}O-discrimination technique, and at different times after the transfer to low-light conditions. Error bars represent standard error. Based on Millenaar et al. (2000).

measured in vivo activities of AOX and COX in the presence of exogenous sucrose with *P. annua* roots. In vivo activity of COX was enhanced by exogenous sucrose, whereas that of AOX was repressed. Total respiration rates could not be restored to the level of high-light condition by exogenous sucrose. AOX protein under the low-light condition showed only the reduced form, which is more active than the oxidized form, and this was the same under high-light condition. Exogenous glucose and sucrose did not increase AOX protein content 24 hours after sugar addition (Millenaar et al., 2002). These results indicate that AOX did not act as an energy overflow, even when carbohydrate levels were high in *P. annua* roots. Under low carbohydrate status, what role does AOX have? The authors suggested that AOX stabilizes the redox state of the respiratory chain, when the cytochrome pathway is restricted (Millenaar et al., 1998). A different response of AOX to carbohydrate status may exist between leaves and roots, and we need detailed investigations on such differences.

IV. Relationship between Light Intensity and Respiration

A. Effect of Light Intensity on Dark Respiration of Leaves

1. Growth Light Intensity and Leaf Respiration

Light intensity varies in a natural habitat; plants show acclimation and adaptation to the habitat light intensity. For example, shade leaves show a higher specific leaf area (SLA) and slower respiratory rates per unit dry weight (R_{mass}), thus optimizing their carbon balance and compensating for slow photosynthetic rates (Björkman, 1981). Fast respiratory rates in sunny conditions would be useful for rapid translocation of carbohydrates from leaves, and maintenance of large amounts of proteins of the photosynthetic apparatus. Six *Piper* species that consisted of generalist, gap and shade species showed slower respiration rates per unit leaf area (R_{area}) under lower daily light intensity (Fredeen and Field, 1991). Lusk and Reich (2000) examined respiratory rates of leaves in 11 cold-temperate tree species, and a response of R_{area} to light availability was dominated by that of R_{mass} for conifers, and by those of leaf mass per area (LMA) and R_{mass} for most angiosperms. They also found that shade-tolerant angiosperms tend to have lower R_{mass} than shade-intolerant ones.

2. Effects of Growth Light Intensity on Leaf Respiration via Carbohydrates

How does light intensity influence dark respiration in leaves? Reduced daily light intensity means reduced production of photosynthates, i.e. reduced carbohydrate status in leaves. Conversely, a reduced light intensity means reduced costs for maintenance of photosynthetic components and carbohydrate export. Does light intensity influence respiratory rates of leaves via carbohydrate status as substrate or via adenylate demands? As discussed above (see Section II.B), carbohydrate status can determine leaf respiratory rates in some species, such as spinach and wheat. Spinach leaves under shade conditions showed slower respiration rates and lower carbohydrate levels (Noguchi et al., 1996). They also showed slower rates of KCN-resistant respiration (Noguchi, 1995) and lower in vivo activity of AOX (Noguchi et al., 2001a). Lower carbohydrate status would influence respiration rates and in vivo AOX activity of spinach leaves under lower light intensity. Lower activity of AOX in spinach leaves provides a highly efficient respiratory ATP production under low-light environments. Lusk and Reich (2000) found that leaf respiratory rates per unit leaf nitrogen (Rd/N) were correlated with light availability in 11 tree species, whereas leaf R_{mass} did not correlate with leaf nitrogen content. This suggests that maintenance cost of nitrogen compounds, which mainly consist of photosynthetic enzymes, would not determine leaf respiratory rates, and that the correlation between Rd/N and light availability reflected the influence of leaf carbohydrate status on respiratory rates and/or variation in the proportion of leaf nitrogen allocation to protein.

However, in other species, carbohydrate status did not explain slower respiration rates under lower light intensity. In *A. odora*, a shade species, lower light intensity caused lower carbohydrate levels in the leaves, but had little effect on R_{mass} and in vivo activity of AOX (Noguchi et al., 1996, 2001a).

3. Transfer Experiments between High- and Low-Light Intensity

When *Alocasia macrorrhiza*, a shade species, was transferred between low- and high-light intensities, R_{area} of *A. macrorrhiza* acclimated to a new light condition within a few days (Sims and Pearcy, 1991), whereas light-saturated photosynthetic rates did not. Naidu and DeLucia (1997) also observed respiratory acclimation to new light environments in leaves of two deciduous trees, *Acer saccharum* and *Quercus rubra*. In *A. odora*, both R_{area} and R_{mass} changed (Fig. 10, Noguchi et al., 2001c), and, comparing CO_2-efflux and O_2-uptake rates, CO_2-efflux rates changed more quickly than O_2-uptake rates; that is, RQ changed after transfer between light environments. In *A. odora* leaves, only uncoupler enhanced leaf respiration while exogenous sugar did not. Thus, changes in growth light intensity modified the adenylate demand for some cellular processes, which influenced leaf respiration rates of *A. odora*. In this species, maximal rates of O_2-uptake in the presence of sugar and uncoupler, also acclimated to the new light intensity. This indicates that the abundance of the respiratory components changed after transfer. In *A. odora*, maximal activity of NAD-isocitrate dehydrogenase also acclimated in parallel with maximal rates of respiration. Bunce et al. (1977) also showed a positive correlation between respiratory rates and maximal activities of malate dehydrogenase in soybean leaves after transfer between high- and low-light conditions. However, in *A. odora* leaves, maximal activity of COX did not acclimate to new light environments (Noguchi et al., 2001c). AOX activity might change after transfer, and ^{18}O-discrimination experiments would clarify any relationships between respiratory compounds and ATP-consuming processes.

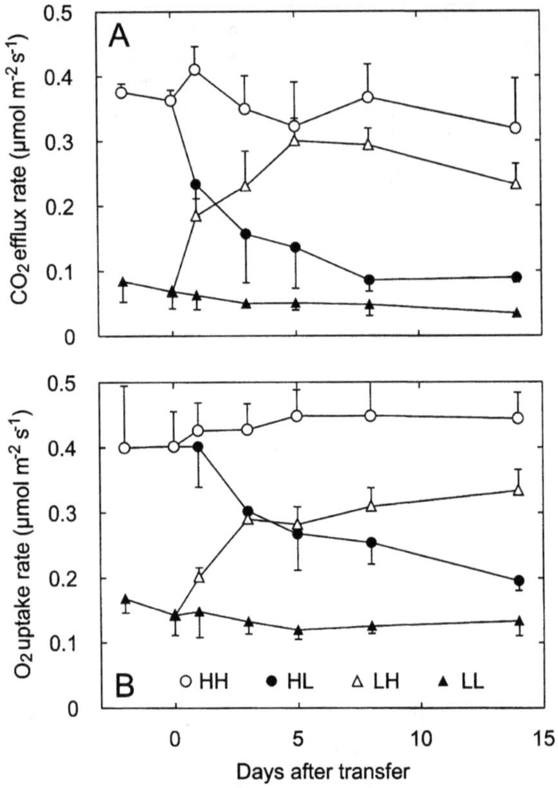

Fig. 10. Changes in the rate of CO_2 efflux (A) and O_2 uptake (B) of *Alocasia odora* leaves, expressed on a leaf area basis. Plants originally grown in high light (circles) or low light (triangles) were transferred to high light (open symbols) or low light (closed symbols) on day 0. Error bars represent standard deviation. Based on Noguchi et al. (2001c).

B. Effect of Light Intensity on Day Respiration of Leaves

Day respiration refers to non-photorespiratory mitochondrial respiration (CO_2 efflux and O_2 uptake) during the daytime (Hoefnagel et al., 1998; Atkin et al., 2000). Measurement of day respiration is much more complicated than that of dark respiration, because in the light O_2 efflux from photosystem and O_2 uptake by Rubisco oxygenation and the Mehler reaction occur as well as mitochondrial O_2 consumption. CO_2 exchange also includes CO_2 uptakes by Rubisco and PEP carboxylase and CO_2 efflux by glycine decarboxylation, and the activities of the TCA cycle and the oxidative pentose pathway. Several methods, such as using stable isotopes of oxygen, ^{14}C-labeling pulse-chase, the Laisk method and the Kok method, have been devised for estimation of day respiration, but all these methods have their limitations (Atkin et al., 2000). Studies with these methods suggest that mitochondrial O_2 consumption can be maintained or even stimulated in the light, and that non-photorespiratory mitochondrial CO_2 efflux is inhibited in the light. Does light intensity influence day respiration? Is there a different response between O_2-uptake and CO_2-efflux rates to light intensity? The effects of light intensity on O_2-uptake and CO_2-efflux rates are reviewed in the following sections.

1. Effect of Light Intensity on Mitochondrial O_2 Uptake in the Light

In the light, plant mitochondria provide carbon skeletons of the TCA cycle for nitrogen assimilation and ATP for additional demands, such as sucrose synthesis, protein synthesis, and nitrogen assimilation (Hoefnagel et al., 1998; Padmasree and Raghavendra, 1998). In chloroplasts, ATP is provided both by non-cyclic electron transport, and also by cyclic electron transport and the Mehler reaction. Although these two pathways provide ATP as the sole product and supply extra ATP to demand (Heldt, 1997), mitochondrial oxidative phosphorylation maintains most of the cytosolic ATP pool, and is essential for maximal rates of photosynthesis (Krömer, 1995). The degree to which mitochondrial ATP production is required for cellular processes in the light, is determined by environmental factors and plant conditions. Hurry et al. (1995) reported that mitochondria supplied ATP in illuminated non-hardened leaves of *Secale cereale* (winter rye), but not in cold-hardened leaves.

Plant mitochondria also have other roles in the light. One is the oxidation of NADH produced by glycine decarboxylase (GDC) in photorespiration, and another is the removal of excess photosynthetic reducing equivalents, which can lead to damage of the photosynthetic electron transport system (Hoefnagel et al., 1998). NADH produced by GDC is exported to the cytosol via the malate/oxaloacetate shuttle (Mal/OAA shuttle, Fig. 2) and oxidized in the peroxisome, although Krömer and Heldt (1991) suggested that only 25 to 50% of the NADH produced by GDC is exported, while the remainder would be oxidized in the mitochondria. When the chloroplast NADPH/NADP ratio is too high, photosynthetic electron transport components will be highly reduced, increasing the chances for photoinhibition. There are several mechanisms to avoid the over-reduction of the chloroplast, such as state transition, heat dissipation, photorespiration and nitrogen assimilation. However, if the over-reduction of the chloroplast exceeds the consumption by these mechanisms, excess reducing equivalents can be exported from chloroplasts via the Mal/OAA shuttle and PGA/DHAP shuttle (P_i translocator) (Fig. 2). In order to regenerate OAA for transport back to the chloroplast, malate must be oxidized in the cytosol, peroxisome and mitochondria. Malate oxidation in mitochondria provides reducing equivalents (NADH). Thus, high light intensity would increase mitochondrial O_2 consumption. In the green alga *Chlamydomonas reinhardtii*, the rate of mitochondrial O_2 consumption in the light is faster at high light intensity than at low light intensity (Xue et al., 1996). The excess reducing equivalents in the chloroplasts would be exported to the mitochondria in an irradiance-dependent manner. In addition to producing ATP for cellular processes and oxidizing excess photosynthetic reducing equivalents, mitochondria have another role in the light: production of carbon skeletons for biosynthesis. During illumination, a partial TCA cycle operates in cells and mitochondria taking up OAA from the cytosol via an OAA translocator. OAA is converted to citrate via the TCA cycle, and citrate is exported to the cytosol and converted to 2-oxoglutalate, which becomes the major source of carbon skeletons for amino acid biosynthesis in chloroplasts (Hanning and Heldt, 1993, Fig. 2).

In the light, is AOX used as a sink for NADH? Under conditions where ATP demand is low, the recycling of ADP would limit the rate of oxidation via the cytochrome pathway. However, AOX allows electron flow to continue when the demand for ATP is limited. In vivo AOX activity of soybean cotyledon decreased after transfer to continuous darkness. However, this decrease did not correlate with carbohydrate status, because 5 min of illumination every 12 hours was sufficient to keep AOX engaged to the same extent as in plants grown in the light (Ribas-Carbo et al., 2000). Thus, AOX may serve as a sink for NADH in the mitochondria of soybean cotyledon in the light. High light intensity may increase the electron flow via AOX under illumination.

2. Effect of Light Intensity on Mitochondrial CO_2 Efflux in the Light

In the light, does non-photorespiratory mitochondrial CO_2 efflux (R_d) also depend on light intensity? In the Laisk method (Atkin et al., 2000), R_d is estimated as the rate of CO_2 release at Γ^*, where the CO_2 fixed by

Rubisco is matched by the CO_2 released by GDC.

$$A = v_c - 0.5\ v_o - R_d,$$

where A is the rate of net CO_2 exchange, and v_c and v_o are the rates of carboxylation and oxygenation of Rubisco, respectively. To determine both Γ^* and R_d, regression lines of A versus internal CO_2 concentration (c_i) under different light intensities must be constructed, and both parameters can be estimated as the node of the regression lines. The values of light intensity used to construct regression lines should be at least 80 μmol photons $m^{-2}\ s^{-1}$, because R_d decreases sharply at lower irradiance values (Brooks and Farquhar, 1985; Atkin et al., 1998, 2000). The Laisk method assumes that R_d is constant across a wide range of c_i and light intensity values above 80 μmol photons $m^{-2}\ s^{-1}$. Thus, we could not know whether light intensity influences R_d with this method, but estimated R_d changed with light intensity in the range of less than 80 μmol photons $m^{-2}\ s^{-1}$ (Brooks and Farquhar, 1985). Moreover, not only the light inhibition of R_d (dark respiration minus day respiration) but also light-enhanced dark respiration (LEDR) is equally sensitive to increasing light intensity in the light period (Atkin et al., 1998). LEDR is a second rise in dark respiration, following a rapid increase in respiration, the post-illumination burst (PIB), after darkness (Atkin et al., 2000). Several studies suggested that PIB represents CO_2 release by GDC and LEDR represents an increased supply of non-photorespiratory substrates (pyruvate and malate), which accumulate during photosynthesis (Raghavendra et al., 1994). The mechanism by which R_d is inhibited in the light is still unclear, but the model for LEDR may provide information for inhibition of R_d in the light (Hill and Bryce, 1992). According to this model, inhibition of R_d in the light is caused by the low activities of malic enzyme (ME) and pyruvate decarboxylase (PDC) in the light (Fig. 2). Examination of the dependence of ME and PDC activities on light intensity will provide a mechanism of the dependence of R_d on irradiance.

C. Effect of Light Intensity on Root Respiration

Daily light intensity also affects root respiration. Under a low light intensity, slow rates of root respiration were observed in *Lolium multiflorum* (Hansen, 1980), maize (Massimino et al., 1981), barley (Bingham and Farrar, 1988; Williams and Farrar, 1990), and *Plantago major* (Kuiper and Smid, 1985), *Calamagrostis epigejos*, a perennial rhizomatous grass (Gloser et al., 1996). When plants were transferred from a high to a low light intensity, root respiratory rates decreased in *P. annua* (Millenaar et al., 2000), and maize (Massimino et al., 1981). In the low-light environments, these slow rates of root respiration would be associated with low levels of carbohydrates, which would be influenced by low levels of translocation from shoot. Decreases of carbohydrate levels after shading were observed in roots of *P. annua* (Millenaar et al., 2000) and barley (Williams and Farrar, 1990). However, root respiratory rates can be influenced by differences in metabolic activities in shoot and root at different growth light intensities. Hansen (1980) observed that root respiratory rates in *L. multiflorum* decreased after withdrawal of nitrogen from the nutrient solution, even under a high light intensity. After re-supply of nitrate, high respiratory rates were observed under both high- and low-light conditions. Since carbohydrate accumulation increases at low supply of nitrate, nitrate uptake would mainly influence root respiration of *L. multiflorum*. High-light intensity also increases nitrate assimilation in the shoot, which increases ion uptake in roots. Thus, under a high light intensity, root respiration would be influenced by high demands for ATP for ion uptake, and by high levels of translocated carbohydrates. Examining diurnal changes of these processes and temporary pattern after transfer will clarify what mechanisms mainly influence root respiration.

D. Effect of Excess Light Intensity on Leaf Respiration

Under an excess light intensity, the over-reduction of the electron transport chain in chloroplasts is prevented by several energy-dissipating systems, such as state transition, non-photochemical quenching, photorespiration, and Mal/OAA and DHAP/PGA shuttles, which export reducing equivalents from chloroplasts (Fig. 2, Hoefnagel et al., 1998). However, when the excess reducing equivalents overwhelm the consumption by these systems, the excess reducing equivalents can lead to the over-reduction of electron transport in the chloroplasts, followed by the generation of reactive oxygen species (ROS). ROS are detoxified by scavenging systems, including superoxide dismutase (SOD), ascorbate peroxidase (APX) and catalase (CAT). However, especially when the excess light intensity is associated with other stress factors (e.g., low temperature and drought), these scavenging systems cannot detoxify ROS, and ROS causes photoinhibition of Photosystem I (Sonoike, 1996) and Photosystem II (Osmond, 1994).

In leaves, an excess light intensity can also lead to a high level of substrates in mitochondria, leading to the over-reduction of the respiratory chain, especially of the ubiquinone pool. The over-reduction of the ubiquinone pool induces O_2^- at sites of Complex I and III (Møller, 2001). ROS production can be decreased by engagement of AOX in the mitochondria. Decreasing the amount of mitochondrial AOX increased the sensitivity of plants to oxidative stress, which was induced by an inhibitor of the cytochrome pathway (Maxwell et al., 1999). A similar role has been proposed for the mitochondrial uncoupling protein.

Under photo-oxidative conditions, does in vivo activity of AOX increase and can AOX prevent the redox level of the ubiquinone pool? Transfer from shade to sun conditions increased the leaf respiration rate, which was accompanied with PS II photoinhibition in leaves of two deciduous trees, *A. saccharum* and *Q. rubra* (Naidu and De Lucia, 1997) and of *A. odora*, a shade species (K. Noguchi, A.H. Millar, H. Lambers, D.A. Day, unpublished). In *A. odora*, the redox state of AOX was oxidized (less inactive form) under a low light intensity, but most of AOX protein was reduced (more active form) under photo-oxidative conditions (K. Noguchi, A.H. Millar, H. Lambers, D.A. Day, unpublished). Under oxidative conditions where GDC was inhibited, AOX protein was induced in pea leaves (Taylor et al., 2002). AOX protein and other non-phosphorylating pathways may be activated and decrease the redox level of the ubiquinone pool under such conditions. Recently, the thioredoxin and the NADPH-dependent thioredoxin reductase have been described in plant mitochondria (Laloi et al., 2001, Balmer and Buchanan, 2002). This thioredoxin system could be involved in a redox control of mitochondrial enzymes, including AOX. If the high-light intensity activates this thioredoxin system, AOX would be indirectly activated by the light intensity. We need to investigate what mechanisms enhance NADPH levels in mitochondria after the high-light illumination and to examine a detail relationship between the regulation of AOX and NADPH levels in mitochondria.

V. Concluding Remarks

The above results suggest that carbohydrate status influences respiratory rates in both leaves and roots of some species, and this is more common in leaves than in roots. Is there any difference in response between leaves and roots to carbohydrate status? In leaves, carbohydrates are produced and consumed in the same cell, and diurnal variations in carbohydrate levels are larger than that in roots. A balance between production and consumption in leaves would be stable in most situations. However, this balance can be lost in some situations, and accumulation of carbohydrates can reduce photosynthate production. Thus, respiratory rates of leaves depend on carbohydrate status, and AOX would work as an energy-overflow pathway, especially under stress conditions. However, in roots, carbohydrates are imported from the shoot, and the amount of carbohydrates is fairly tightly regulated. The translocation from shoot to roots would be controlled by a range of variables distributed between shoot and roots (Farrar and Jones, 2000). Thus, respiratory rates of roots would be determined by adenylate demands rather than carbohydrate status in most cases, and adenylate demands in roots (e.g., ion uptake, growth) are influenced by metabolic activities in shoots (e.g., photosynthesis, nitrogen assimilation). However, these linkages between shoot and root have mostly been studied in crop species, herbaceous plants and seedlings of tree species, whereas the bulk of global-level plant respiration occurs in large trees. Rigid relationships between shoot and roots in large trees may be loose, because of a long distance of transport between sources and sinks, and a slower turnover of the carbohydrate pool (Amthor, 1995). Moreover, in large trees, net primary production is a small fraction of gross primary production, especially in the tropics (Kozolowski and Pallardy, 1997). In this situation, the effects of light intensity/carbohydrate status on leaf respiration are important to plant productivity, because leaves contain a large percentage of living tissues. Non-phosphorylating pathways, such as AOX, will also be important in this situation. The role of AOX has not been clearly identified for non-thermogenic tissues, but several roles have been suggested (Moore et al., 2002). AOX and other non-phosphorylating pathways, such as rotenone-insensitive NAD(P)H dehydrogenase and uncoupling protein, determine the respiratory efficiency. Thus, it is important to clarify how environmental and genetic factors, including light intensity/carbohydrate status, influence the in vivo activities of non-phosphorylating pathways, and whether non-phosphorylating pathways are engaged to maximize growth rate with various plant species.

References

Akita S, Ishikawa T, Lee BW and Katayama K (1993) Variation of the dark respiration rate of tissues and organs of rice (*Oryza sativa* L.) with differential physiological age and causal factors. Jap J Crop Sci 62: 73–80

Amthor JS (1989) Respiration and Crop Productivity. Springer-Verlag, Berlin

Amthor JS (1994) Respiration and carbon assimilate use. In: Boote KJ, Bennett JM, Sinclair TR and Paulsen GM (eds) Physiology and Determination of Crop Yield, pp 221–250. American Society of Agronomy, Madison

Amthor JS (1995) Higher plant respiration and its relationships to photosynthesis. In: Schulze ED and Caldwell MM (eds) Ecophysiology of Photosynthesis, pp 71–101. Springer-Verlag, Berlin

Amthor JS (2000) The McCree-de Wit-Penning de Vries-Thornley respiration paradigms: 30 years later. Ann Bot 86: 1–20

André M, Massimino J, Daguenet A, Massimino D and Thiery J (1982) The effect of a day at low irradiance of a maize crop. II. Photosynthesis, transpiration and respiration. Physiol Plant 54: 283–288

ap Rees T (1980) Assessment of the contributions of metabolic pathways to plant respiration. In: Davies DD (ed) The Biochemistry of Plants, Vol 2, Metabolism and Respiration, pp 1–29. Academic Press, New York

ap Rees T (1988) Hexose phosphate metabolism by nonphotosynthetic tissues of higher plants. In: Preiss J (ed) The Biochemistry of Plants, Vol 14, Carbohydrates, pp 1–33. Academic Press, New York

Atkin OK, Evans JR and Siebke K (1998) Relationship between the inhibition of leaf respiration by light and enhancement of leaf dark respiration following light treatment. Aust J Plant Physiol 25: 437–443

Atkin OK, Millar AH, Gardeström P and Day DA (2000) Photosynthesis, carbohydrate metabolism and respiration in leaves of higher plants. In: Leegood RC, Sharkey TD and von Caemmerer S (eds) Photosynthesis: Physiology and Metabolism, pp 154–175. Kluwer Academic Publishers, Dordrecht

Averill RH and ap Rees T (1995) The control of respiration in wheat (*Triticum aestivum* L.) leaves. Planta 196: 344–349

Azcón-Bieto J (1983) Inhibition of photosynthesis by carbohydrates in wheat leaves. Plant Physiol 73: 681–686

Azcón-Bieto J and Osmond CB (1983) Relationship between photosynthesis and respiration. Plant Physiol 71: 574–581

Azcón-Bieto J, Day DA and Lambers H (1983a) The regulation of respiration in the dark in wheat leaf slices. Plant Sci Lett 32: 313–320

Azcón-Bieto J, Lambers H and Day DA (1983b) Effect of photosynthesis and carbohydrate status on respiratory rates and the involvement of the alternative pathway in leaf respiration. Plant Physiol 72: 598–603

Azcón-Bieto J, Lambers H and Day DA (1983c) Respiratory properties of developing bean and pea leaves. Aust J Plant Physiol 10: 237–245

Azcón-Bieto J, Salom CL, Mackie ND and Day DA (1989) The regulation of mitochondrial activity during greening and senescence of soybean cotyledons. Plant Physiol Biochem 27: 827–836

Balmer Y and Buchanan BB (2002) Yet another plant thioredoxin. Trend Plant Sci 7: 191–193

Baysdorfer C and van der Woude WJ (1988) Carbohydrate responsive proteins in the roots of *Pennisetum americanum*. Plant Physiol 87: 566–570

Baysdorfer C, Sicher RC and Kremer DF (1987) Relationship between fructose 2,6-bisphosphate and carbohydrate metabolism in darkened barley primary leaves. Plant Physiol 84: 766–769

Beevers H (1974) Conceptual developments in metabolic control, 1924-1974. Plant Physiol 54: 437–442

Ben Zioni A, Vaadia Y and Lips H (1971) Nitrate uptake by roots as regulated by nitrate reduction products of the shoot. Physiol Plant 24: 288–290

Bingham IJ and Farrar JF (1988) Regulation of respiration in roots of barley. Physiol Plant 73: 278–285

Bingham IJ and Stevenson EA (1993) Control of root growth: Effects of carbohydrates on the extension, branching and rate of respiration of different fractions of wheat roots. Physiol Plant 88: 149–158

Björkman O (1981) Responses to different quantum flux densities. In: Lange OL, Nobel PS, Osmond CB and Ziegler H (eds) Physiological Plant Ecology, pp 57–107. Springer-Verlag, Berlin

Bouma TJ, de Visser R, van Leeuwen PH, de Kock MJ and Lambers H (1995) The respiratory energy requirements involved in nocturnal carbohydrate export from starch-storing mature source leaves and their contribution to leaf dark respiration. J Exp Bot 46: 1185–1194

Breeze V and Elston J (1978) Some effects of temperature and substrate content upon respiration and the carbon balance of field beans (*Vicia faba* L.). Ann Bot 42: 863–876

Brooks A and Farquhar GD (1985) Effect of temperature on the CO_2/O_2 specificity of ribulose-1,5-bisphosphate carboxylase/oxygenase and the rate of respiration in the light. Planta 165: 397–406

Brouquisse R, James F, Raymond P and Pradet A (1991) Study of glucose starvation in excised maize root tips. Plant Physiol 96: 619–626

Bryce JH and ap Rees T (1985) Effects of sucrose on the rate of respiration of the roots of *Pisum sativum*. J Plant Physiol 120: 363–367

Bunce JA, Patterson DT, Peet MM and Alberte RS (1977) Light acclimation during and after leaf expansion in soybean. Plant Physiol 60: 255–258

Casadesús J, Tapia L and Lambers H (1995) Regulation of K^+ and NO_3^- fluxes in roots of sunflower (*Helianthus annuus*) after changes in light intensity. Physiol Plant 93: 279–285

Challa H (1976) An analysis of the diurnal course of growth, carbon dioxide exchange and carbohydrate reserve content of cucumber. Agricultural Research Reports 861. Centre for Agricultural Publishing and Documentation, Wageningen

Collier DE and Thibodeau BA (1995) Changes in respiration and chemical content during autumnal senescence of *Populus tremuloides* and *Quercus rubra* leaves. Tree Physiol 15: 759–764

Crawford RMM and Huxter TJ (1977) Root growth and carbohydrate metabolism at low temperature. J Exp Bot 28: 917–925

Day DA and Lambers H (1983) The regulation of glycolysis and electron transport in roots. Physiol Plant 58: 155–160

Day DA, de Vos OC, Wilson D and Lambers H (1985) Regulation of respiration in the leaves and roots of two *Lolium perenne* populations with contrasting mature leaf respiration rates and crop yields. Plant Physiol 78: 678–683

Day DA, Krab K, Lambers H, Moore AL, Siedow JN, Wagner AM and Wiskich JT (1996) The cyanide-resistant oxidase: To inhibit or not to inhibit, that is the question. Plant Physiol 110: 1–2

Farrar JF (1981) Respiration rate of barley roots: Its relation to growth, substrate supply and the illumination of the shoot. Ann Bot 48: 53–63

Farrar JF (1985) The respiratory source of CO_2. Plant Cell Environ 8: 427–438

Farrar JF and Jones DL (1986) Modification of respiration and carbohydrate status of barley roots by selective pruning. New Phytol 102: 513–521

Farrar JF and Jones DL (2000) The control of carbon acquisition by roots. New Phytol 147: 43–53

Farrar JF and Rayns FW (1987) Respiration of leaves of barley infected with powdery mildew: Increased engagement of the alternative oxidase. New Phytol 107: 119–125

Farrar JF and Williams JHH (1991) Control of the rate of respiration in roots: Compartmentation, demand and the supply of substrate. In: Emes MJ (ed) Compartmentation of Plant Metabolism in Non-photosynthetic Tissues, pp 167–188. Cambridge University Press, Cambridge

Farrar SC and Farrar JF (1985) Carbon fluxes in leaf blades of barley. New Phytol 100: 271–283

Farrar SC and Farrar JF (1986) Compartmentation and fluxes of sucrose in intact leaf blades of barley. New Phytol 103: 645–657

Fondy BR and Geiger DR (1982) Diurnal pattern of translocation and carbohydrate metabolism in source leaves of *Beta vulgaris* L. Plant Physiol 70: 671–676

Fredeen AL and Field CB (1991) Leaf respiration in *Piper* species native to a Mexican rainforest. Physiol Plant 82: 85–92

Gastón S, Ribas-Carbo M, Busquets S, Berry JA, Zabalza A and Royuela M (2003) Changes in mitochondrial electron partitioning in response to herbicides inhibiting branched-chain amino acid biosynthesis in soybean. Plant Physiol 133: 1351–1359

Gerbaud A, André M and Richaud C (1988) Gas exchange and nutrition patterns during the life cycle of an artificial wheat crop. Physiol Plant 73: 471–478

Geromino J and Beevers H (1964) Effects of aging and temperature on respiratory metabolism of green leaves. Plant Physiol 39: 786–793

Gloser V, Scheurwater I and Lambers H (1996) The interactive effect of irradiance and source of nitrogen on growth and root respiration of *Calamagrostis epigejos*. New Phytol 134: 407–412

Gonzàlez-Meler MA, Giles L, Thomas RB and Siedow JN (2001) Metabolic regulation of leaf respiration and alternative pathway activity in response to phosphate supply. Plant Cell Environ 24: 205–215

Graham IA and Martin T (2000) Control of photosynthesis, allocation and partitioning by sugar regulated gene expression. In: Leegood RC, Sharkey TD and von Caemmerer S (eds) Photosynthesis: Physiology and Metabolism, pp 233–248. Kluwer Academic Publishers, Dordrecht

Hänisch ten Cate CH and Breteler H (1981) Role of sugars in nitrate utilization by roots of dwarf bean. Physiol Plant 52: 129–135

Hanning I and Heldt HW (1993) On the function of mitochondrial metabolism during photosynthesis in spinach (*Spinacia oleracea* L.) leaves. Plant Physiol 103: 1147–1154

Hansen GK (1980) Diurnal variation of root respiration rates and nitrate uptake as influenced by nitrogen supply. Physiol Plant 48: 421–427

Hatrick AA and Bowling DJF (1973) A study of the relationship between root and shoot metabolism. J Exp Bot 24: 607–613

Heldt H-W (1997) Plant Biochemistry and Molecular Biology. Oxford University Press, New York

Hellmann H, Barker L, Funck D and Frommer WB (2000) The regulation of assimilate allocation and transport. Aust J Plant Physiol 27: 583–594

Hill SA and Bryce JH (1992) Malate metabolism and light-enhanced dark respiration in barley mesophyll protoplasts. In: Lambers H and van der Plas LHW (eds) Molecular, Biochemical and Physiological Aspects of Plant Respiration, pp 221–230. SPB Academic Publishing, The Hague

Hoefnagel MHN, Atkin OK and Wiskich JT (1998) Interdependence between chloroplasts and mitochondria in the light and the dark. Biochim Biophys Acta 1366: 235–255

Huber SC (1989) Biochemical mechanism for regulation of sucrose accumulation in leaves during photosynthesis. Plant Physiol 91: 656–662

Huck MG, Hageman RH and Hanson JB (1962) Diurnal variation in root respiration. Plant Physiol 37: 371–375

Hurry VM, Tobiæson M, Krömer S, Gardeström P and Öquist G (1995) Mitochondria contribute to increased photosynthetic capacity of leaves of winter rye (*Sacale cereale* L.) following cold-hardening. Plant Cell Environ 18: 69–76

Jones RJ and Nelson CJ (1979) Respiration and concentration of water soluble carbohydrate in plant parts of contrasting tall fescue genotypes. Crop Sci 19: 367–372

Jurik TW, Chabot JF and Chabot BF (1979) Ontogeny of photosynthetic performance in *Fragaria virginiana* under changing light regimes. Plant Physiol 63: 542–547

Koch KE (1996) Carbohydrate-modulated gene expression in plants. Annu Rev Plant Physiol Plant Mol Biol 47: 509–540

Koch KE, Nolte KD, Duke ER, McCarty DR and Avlgne WT (1992) Sugar levels modulate differential expression of maize sucrose synthase genes. Plant Cell 4: 59–69

Kozolowski TT and Pallardy SG (1997) Physiology of Woody Plants. Academic Press, New York

Krapp A and Stitt M (1994) Influence of high carbohydrate content on the activity of plastidic and cytosolic isoenzyme pairs in photosynthetic tissues. Plant Cell Environ 17: 861–866

Krapp A and Stitt M (1995) An evaluation of direct and indirect mechanisms for the 'sink-regulation' of photosynthesis in spinach: Changes in gas exchange, carbohydrates, metabolites, enzyme activities and steady-state transcript levels after cold-girdling source leaves. Planta 195: 313–323

Krapp A, Hofmann B, Schäfer C and Stitt M (1993) Regulation of the expression of *rbcS* and other photosynthetic genes by carbohydrates—a mechanism for the 'sink regulation' of photosynthesis? Plant J 3: 817–828

Krömer S (1995) Respiration during photosynthesis. Annu Rev Plant Physiol Plant Mol Biol 46: 45–70

Krömer S and Heldt HW (1991) Respiration of pea leaf mitochondria and redox transfer between the mitochondrial and extramitochondrial compartment. Biochim Biophys Acta 1057: 42–50

Kuiper D and Smid A (1985) Genetic differentiation and phenotypic plasticity in *Plantago major* ssp. *major*: I. The effect of differences in level of irradiance on growth, photosynthesis, respiration and chlorophyll content. Physiol Plant 65: 520–528

Laloi C, Rayapuram N., Chartier Y, Grienenberger J-M, Bonnard G and Meyer Y (2001) Identification and characterization of a mitochondrial thioredoxin system in plants. Proc Natl Acad Sci USA 98: 14144–14149

Lambers H (1980) The physiological significance of cyanide-resistant respiration in higher plants. Plant Cell Environ 3: 293–302

Lambers H (1985) Respiration in intact plants and tissues: Its regulation and dependence on environmental factors, metabolism and invaded organisms. In: Douce R and Day DA (eds) Higher Plant Cell Respiration, pp 418–473. Springer-Verlag, Berlin.

Lambers H, Chapin III FS and Pons TL (1998) Plant Physiological Ecology. Springer-Verlag, Berlin.

Lambers H, Atkin OK and Millenaar FF (2002) Respiratory patterns in roots in relation to their functioning. In: Waisel Y, Eshel A and Kafkafi K (eds) Plant Roots. The Hidden Half, pp 521–552. Marcel Dekker Inc, New York

Lee K and Akita S (2000) Factors causing the variation in the temperature coefficient of dark respiration in rice (*Oryza sativa* L.). Plant Prod Sci 3: 38–42

Lusk CH and Reich PB (2000) Relationships of leaf dark respiration with light environment and tissue nitrogen content in juveniles of 11 cold-temperate tree species. Oecologia 123: 318–329

Massimino D, André M, Richaud C, Daguenet A, Massimino J and Vivoli J (1981) The effect of a day at low irradiance of a maize crop. I. Root respiration and uptake of N, P and K. Physiol Plant 51: 150–155

Maxwell DP, Wang Y and McIntosh L (1999) The alternative oxidase lowers mitochondrial reactive oxygen production in plant cells. Proc Nat Acad Sci 96: 8271–8276

McCree KJ and Troughton JH (1966) Prediction of growth rate at different light levels from measured photosynthesis and respiration rates. Plant Physiol 41: 559–566

McCutchan CL and Monson RK (2001a) Night-time respiration rate and leaf carbohydrate concentration are not coupled in two alpine perennial species. New Phytol 149: 419–430

McCutchan CL and Monson RK (2001b) Effects of tissue-type and development on dark respiration in two herbaceous perennials. Ann Bot 87: 355–364

McDonnell E and Farrar JF (1992) Substrate supply and its effect on mitochondrial and whole tissue respiration in barley roots. In: Lambers H and van der Plas LHW (eds) Molecular, Biochemical and Physiological Aspects of Plant Respiration, pp 455–462. SPB Academic Publishing, The Hague

Millenaar FF, Benschop JJ, Wagner AM and Lambers H (1998) The role of the alternative oxidase in stabilizing the in vivo reduction state of the ubiquinone pool and the activation state of the alternative oxidase. Plant Physiol 118: 599–607

Millenaar FF, Roelofs R, Gonzàlez-Meler MA, Siedow JN, Wagner AM and Lambers H (2000) The alternative oxidase in roots of *Poa annua* after transfer from high-light to low-light conditions. Plant J 23: 623–632

Millenaar FF, Gonzàlez-Meler MA, Siedow JN, Wagner AM and Lambers H (2002) Role of sugars and organic acids in regulating the concentration and activity of the alternative oxidase in *Poa annua* roots. J Exp Bot 53: 1081–1088

Møller IM (2001) Plant mitochondria and oxidative stress: Electron transport, NADPH turnover, and metabolism of reactive oxygen species. Annu Rev Plant Physiol Plant Mol Biol 52: 561–591

Moore AL, Albury MS, Crichton PG and Affourtit C (2002) Function of the alternative oxidase: Is it still a scavenger? Trend Plant Sci 7: 478–481

Moser LE, Volenec JJ and Nelson CJ (1982) Respiration, carbohydrate content, and leaf growth of tall fescue. Crop Sci 22: 781–786

Naidu SL and DeLucia EH (1997) Acclimation of shade-developed leaves on saplings exposed to late-season canopy gaps. Tree Physiol 17: 367–376

Neals TF and Incoll LD (1968) The control of leaf photosynthesis rate by the level of assimilate concentration in the leaf: A review of the hypothesis. Bot Rev 34: 107–124

Noguchi K (1995) Comparative ecophysiological study of the respiratory acclimation of leaves to different light environments in *Spinacia oleracea* L., a sun species, and *Alocasia macrorrhiza* (L.) G. Don., a shade species. Ms Thesis. Tokyo University, Tokyo

Noguchi K and Terashima I (1997) Different regulation of leaf respiration between *Spinacia oleracea*, a sun species, and *Alocasia odora*, a shade species. Physiol Plant 101: 1–7

Noguchi K, Sonoike K and Terashima I (1996) Acclimation of respiratory properties of leaves of *Spinacia oleracea* L., a sun species, and of *Alocasia macrorrhiza* (L.) G. Don., a shade species, to changes in growth irradiance. Plant Cell Physiol 37: 377–384

Noguchi K, Go C-S, Terashima I, Ueda S. and Yoshinari T (2001a) Activities of the cyanide-resistant respiratory pathway in leaves of sun and shade species. Aust J Plant Physiol 28: 27–35

Noguchi K, Go C-S, Miyazawa S-I, Terashima I, Ueda S and Yoshinari T (2001b) Costs of protein turnover and carbohydrate export in leaves of sun and shade species. Aust J Plant Physiol 28: 37–47

Noguchi K, Nakajima N and Terashima I (2001) Acclimation of leaf respiratory properties in *Alocasia odora* following reciprocal transfers of plants between high- and low-light environments. Plant Cell Environ 24: 831–839

Oleksyn J, Zytkowiak R, Reich PB, Tjoelker MG and Karolewski P (2000) Ontogenetic patterns of leaf CO_2 exchange, morphology and chemistry in *Betula pendula* trees. Trees 14: 271–281

Osmond CB (1994) What is photoinhibition? Some insights from comparisons of shade and sun plants. In: Baker NR and Bowyer JR (eds) Photoinhibition of Photosynthesis, pp 1–24. Bios Scientific Publishers Limited, Oxford

Padmasree K and Raghavendra AS (1998) Interaction with respiration and nitrogen metabolism. In: Raghavendra AS (ed) Photosynthesis—A Comprehensive Treatise, pp 197–211. Cambridge University Press, London

Paul MJ and Stitt M (1993) Effects of nitrogen and phosphorus deficiencies on levels of carbohydrates, respiratory enzymes and metabolites in seedlings of tobacco and their response to exogenous sucrose. Plant Cell Environ 16: 1047–1057

Pearson CJ and Hunt LA (1972) Studies on the daily course of carbon exchange in alfalfa plants. Can J Bot 50: 1377–1384

Penning de Vries FWT, Witlage JM and Kremer D (1979) Rates of respiration and of increase in structural dry matter in young wheat, ryegrass and maize plants in relation to temperature, to water stress and to their sugar content. Ann Bot 44: 595–609

Pollock CJ and Farrar JF (1996) Source-sink relations: The role of sucrose. In: Baker NR (ed) Photosynthesis and the Environment, pp 261–279. Kluwer Academic Publishers, Dordrecht

Raghavendra AS, Padmasree K and Saradadevi K (1994) Interdependence of photosynthesis and respiration in plant cells: Interactions between chloroplasts and mitochondria. Plant Sci 97: 1–14

Reich PB, Walters MB, Ellsworth DS, Vose JM, Volin JC, Gresham C and Bowman WD (1998) Relationships of leaf dark respiration to leaf nitrogen, specific leaf area and leaf life-span: A test across biomes and functional groups. Oecologia 114: 471–482

Ribas-Carbo M, Robinson SA, Gonzàlez-Meler MA, Lennon AM, Giles L, Siedow JN and Berry JA (2000) Effects of light on respiration and oxygen isotope fractionation in soybean cotyledons. Plant Cell Environ 23: 983–989

Saglio PH and Pradet A (1980) Soluble sugars, respiration, and energy charge during aging of excised maize root tips. Plant Physiol 66: 516–519

Sale PJM (1974) Productivity of vegetable crops in a region of high solar input. III. Carbon balance of potato crops. Aust J Plant Physiol 1: 283–296

Scheurwater I, Cornelissen C, Dictus F, Welschen R and Lambers H (1998) Why do fast- and slow-growing grass species differ so little in their rate of root respiration, considering the large differences in rate of growth and ion uptake? Plant Cell Environ 21: 995–1005

Schleucher J, Vanderveer PJ and Sharkey TD (1998) Export of carbon from chloroplasts at night. Plant Physiol 118: 1439–1445

Sheen J (1994) Feedback control of gene expression. Photosynth Res 39: 427–438

Siedow JN and Umbach AL (2000) The mitochondrial cyanide-resistant oxidase: structural conservation amid regulatory diversity. Biochim Biophys Acta 1459: 432–439

Sims DA and Pearcy RW (1991) Photosynthesis and respiration in *Alocasia macrorrhiza* following transfers to high and low light. Oecologia 86: 447–453

Sonoike K (1996) Photoinhibition of Photosystem I: Its physiological significance in the chilling sensitivity of plants. Plant Cell Physiol 37: 239–247

Stitt M, Wirtz W, Gerhardt R, Heldt HW, Spencer C, Walker D and Foyer C (1985) A comparative study of metabolite levels in plant leaf material in the dark. Planta 166: 354–364

Stitt M, von Schaewen A and Willmitzer L (1990) 'Sink' regulation of photosynthetic metabolism in transgenic tobacco plants expressing yeast invertase in their cell wall involves a decrease of the Calvin-cycle enzymes and an increase of glycolytic enzymes. Planta 183: 40–50

Taylor NL, Day DA and Millar AH (2002) Environmental stress causes oxidative damage to plant mitochondria leading to inhibition of glycine decarboxylase. J Biol Chem 277: 42663–42668

Tetley RM and Thimann KV (1974) The metabolism of oat leaves during senescence. I. Respiration, carbohydrate metabolism, and the action of cytokinins. Plant Physiol 54: 294–303

Tjoelker MG, Reich PB and Oleksyn J (1999) Changes in leaf nitrogen and carbohydrates underlie temperature and CO_2 acclimation of dark respiration in five boreal tree species. Plant Cell Environ 22: 767–778

Trethewey RN and ap Rees T (1994) A mutant of *Arabidopsis thaliana* lacking the ability to transport glucose across the chloroplast envelop. Biochem J 301: 449–454

Van Bel AJE (1993) Strategies of phloem loading. Annu Rev Plant Physiol Plant Mol Biol 44: 253–281

Van der Werf A, Kooijman A, Welschen R and Lambers H (1988) Respiration energy costs for the maintenance of biomass, for growth and for ion uptake in roots of *Carex diandra* and *Carex acutiformis*. Physiol Plant 72: 483–491

Van der Werf A, Welschen R, Lambers H (1992) Respiratory losses increase with decreasing inherent growth rate of a species and with decreasing nitrate supply: A search for explanations for these observations. In: Lambers H and van der Plas LHW (eds) Molecular, Biochemical and Physiological Aspects of Plant Respiration, pp 421–432. SPB Academic Publishing, The Hague

Vanlerberghe GC and McIntosh L (1997) Alternative oxidase: From gene to function. Annu Rev Plant Physiol Plant Mol Biol 48: 703–734

Wen J-Q and Liang H-G (1993) Studies on energy status and mitochondrial respiration during growth and senescence of mung bean cotyledon. Physiol Plant 89: 805–810

Williams JHH and Farrar JF (1990) Control of barley root respiration. Physiol Plant 79: 259–266

Williams JHH and Farrar JF (1992) Substrate supply and respiratory control. In: Lambers H and Van der Plas LHW (eds) Molecular, Biochemical and Physiological Aspects of Plant Respiration, pp 471–475. SPB Academic Publishing, The Hague

Williams JHH, Winters AL, and Farrar JF (1992) Sucrose: A novel plant growth regulator. In: Lambers H and Van der Plas LHW (eds) Molecular, Biochemical and Physiological Aspects of Plant Respiration, pp 463–469. SPB Academic Publishing, The Hague

Williams LE, Lemoine R and Sauer N (2000) Sugar transporters in higher plants—a diversity of roles and complex regulation. Trend Plant Sci 5: 283–290

Willis AJ and Yemm EW (1955) The respiration of barley plants VIII. Nitrogen assimilation and the respiration of the root system. New Phytol 54: 163–181

Winter H, Robinson DG and Heldt HW (1994) Subcellular volumes and metabolite concentrations in spinach leaves. Planta 193: 530–535

Xue X, Gauthier DA, Turpin DH and Weger HG (1996) Interaction between photosynthesis and respiration in the green alga *Chlamydomonas reinhardtii*. Plant Physiol 112: 1005–1014

Yamagishi J, Akita S and Takanashi J (1990) Time-course of respiration and its relation to water condition in various plant species. Jap J Crop Sci 59: 169–173

Chapter 6

The Effects of Water Stress on Plant Respiration

Jaume Flexas*, Jeroni Galmes, Miquel Ribas-Carbo and Hipólito Medrano
Departament de Biologia; Grup de Recerca en Biologia de les Plantes en Condicions Mediterrànies; Universitat de les Illes Balears, Carretera de Valldemossa Km 7.5, 07122 Palma de Mallorca, Spain

Summary

Plant growth can be limited by several factors, among which a lack of water is considered of major importance. Despite the vast knowledge of the effect of water stress on photosynthesis, there is much less known about its effect on respiration. Respiration, unlike photosynthesis, never halts, and it reflects the overall metabolism. However, the data available on the effect of water stress on respiration show large variation, from inhibition to stimulation under different water-stress conditions. This chapter combines a review of the latest studies of the effect of water stress on plant respiration with the compilation of data from different authors and recent results to develop a working hypothesis to explain how respiration is regulated under water stress. Leaf respiration shows a biphasic response to Relative Water Content (RWC), decreasing in the initial stages of water stress (RWC > 60%), and increasing as RWC decreases below 50%. Under this hypothesis, the initial decrease in respiration would be related to the immediate inhibition of leaf growth and, consequently, the growth respiration component. The increase of respiration at lower RWC would relate to an increasing metabolism as the plant triggers acclimation mechanisms to resist water stress. These mechanisms would increase the maintenance component of respiration, and, as such, the overall respiration rate. This hypothesis aims to give a metabolic explanation for the observed results, and to raise questions that can direct future plant respiration experiments.

*author for correspondence, email: jaume.flexas@uib.es

H. Lambers and M. Ribas-Carbo (eds.), Plant Respiration, 85–94.

I. Introduction

Water stress is considered to be a major environmental factor limiting plant productivity world-wide (Boyer, 1982; 1996). Equally, it is well recognized that plant productivity largely depends on the balance between photosynthesis and respiration (Lambers et al., 1998). The effects of water stress on photosynthesis have been studied and debated elaborately (Hsiao, 1973; Boyer, 1976; Chaves, 1991; Lawlor, 1995; Cornic and Massacci, 1996; Flexas and Medrano, 2002; Lawlor and Cornic, 2002). However, since photosynthesis is limited to favorable environmental conditions (including light) and to green biomass, whereas respiration occurs continuously in every cell of every plant organ, the latter may be the more important factor controlling productivity, particularly when photosynthesis is largely suppressed, such as under water-stress conditions. A recent large-scale eddy-correlation study further illustrates the importance of ecosystem respiration in determining productivity (Valentini et al,. 1999). In a transect across Europe, large differences in annual net primary production were independent of gross primary assimilation, which was relatively similar among different ecosystems, but strongly dependent on ecosystem respiration (Valentini et al., 1999). It is now also well documented that ecosystem respiration is strongly affected by water availability (Bowling et al., 2002).

However, in spite of its highly recognized importance, the effects of water stress on respiration at the physiological level are largely unknown, partly because only a limited number of studies are available, and partly because of the apparent contradictions among these studies. Certainly, the available experimental evidence does not support a clear pattern of respiration response to water stress, different studies showing either increased, unaffected or decreased rates of respiration (Hsiao, 1973; Amthor, 1989).

This chapter does not seek to exhaustively review all the literature concerning the response of plant respiration to water stress. This would result merely in a list of contrasting examples. Rather, the aims of the present chapter will be: (i) to summarize the information on effects of water stress on plant respiration, focusing mainly on the literature published after the most recent reviews (Amthor, 1989; Amthor and McCree, 1990); (ii) to search for a theoretical background to reconcile the apparent contradictions about respiration response to water stress; and (iii) to highlight the research priorities in this field for the near future.

Abbreviations: CAM – Crassulacean acid metabolism; PEPCK – phosphoenol pyruvate carboxylase kinase; RWC – relative water content.

II. The Effects of Water Stress on Respiration Rates of Different Plant Organs

Previous reviews concerning plant respiration responses to water stress highlight the apparent discrepancies among studies (Hsiao, 1973; Hanson and Hitz, 1982; Amthor, 1989; Amthor and McCree, 1990). Among the earlier studies, several described a water-stress-induced decreased respiration rate in leaves (Brix, 1962; Brown and Thomas, 1980), shoots (Boyer, 1970), roots (Rice and Eastin, 1986), flower apices (Pheloung and Barlow, 1981) or whole plants (Penning de Vries et al., 1979; Wilson et al., 1980; McCree et al., 1984; McCree, 1986). Others showed almost unaffected (Lawlor, 1976) or even increased respiration rates in water-stressed plants (Upchurch et al., 1955; Shearman et al., 1972). More recent studies, using a diversity of techniques to determine respiration rates, have not resolved these apparent contradictions. While several studies have again shown decreased respiration rates under water stress in different plant organs (Palta and Nobel, 1989; González-Meler et al., 1997; Escalona et al., 1999; Ghashghaie et al., 2001; Haupt-Herting et al., 2001), others have again shown unaffected rates (Loboda, 1993), or an increased respiration rate under water stress (Zagdanska, 1995). Ghashghaie et al. (2001) showed in sunflower (*Helianthus annuus*) that leaf respiration decreased at early stages of water stress, and then increased even above control values at later stages.

Moreover, recent studies have shown the influence of respiratory acclimation in the response to water stress. Collier and Cummins (1996), for instance, in a study with *Saxifraga cernua*, showed a progressive decline in total leaf respiration as water stress developed slowly in plants grown on an organic substrate. In contrast, in plants grown in vermiculite where water stress developed more rapidly, total leaf respiration initially increased, and then declined steeply. Zagdanska (1995) showed that pre-acclimation to water stress resulted in higher total respiration in wheat (*Triticum aestivum*) leaves. However, similar responses of respiration to subsequent water stress were observed in both pre-acclimated and non-acclimated plants. Also, Palta and Nobel (1989) showed in *Agave deserti* that root respiration declined as soil water potential decreased, but the precise response was age-dependent and different in established and rain roots. Moreover, the respiration rates of established roots never reached zero, and recovered rapidly and completely upon rewatering. In contrast, the respiration rates of rain roots rapidly reached zero, and did not recover upon rewatering. Bryla et al. (1997) have shown in citrus (*Citrus volkameriana*) seedlings that water stress induced a progressive decline in root respiration.

All these studies have been performed using a single or a few plant species or genotypes, under particular environmental conditions, and using different techniques to assess respiration, thereby making direct comparison difficult. Therefore, the confusion might arise from the difficulties of directly comparing different experiments. In this sense, at least three possible causes for the above-mentioned contradictions are apparent: (i) that the discrepancies in the response of respiration rates to water stress among several studies are simply due to the different species, organs and techniques used; (ii) that different responses to water stress arise from complex interactions with other environmental factors, e.g., ambient temperature; and (iii) that a change in the response of respiration to water stress occurs at a certain threshold of water stress intensity.

In order to discard the first two possible causes, Gulías et al. (2002) compared six species developing water stress under the same conditions in the field. Total leaf respiration rate was determined from light-response curves of CO_2 assimilation. One species (*Rhamnus ludovici-salvatoris*) showed a progressive decrease of leaf respiration during water stress. Another (*Quercus humilis*) showed an initial increase at mild water stress, followed by a large decrease at severe water stress. Leaf respiration was unaffected by water stress in the other four species. To further confirm that interspecific differences do occur, irrespective of the environmental conditions and the technique used to assess respiration rates, J. Galmés et al. (unpublished) have recently analyzed six additional species, growing in a cabinet under identical conditions (800 μmol photons m^{-2} s^{-1}; 50% RH; 25 °C), all with a similar total leaf area and subjected to identical water stress treatments (withholding water for 15 consecutive days). In this study, respiration during a water-stress cycle, was monitored polarographically with a liquid-phase oxygen electrode, after dark-adapting the leaves for 30 minutes. The response of respiration to water stress turned out to be very variable (Fig. 1). While *Mentha aquatica* and *Pistacia lentiscus* exhibited almost constant values throughout the water-stress cycle, respiration in *Phlomis italica* showed an initial decline until day 8, and a sustained slow respiration rate thereafter. *Lysimachia minoricensis* showed an initial decline, but respiration increased above initial values by day 15. Finally, *Cistus albidus* and *Hypericum balearicum* showed a very irregular response of respiration to water stress. These results clearly demonstrate inter-specific differences, and show that environmental and methodological differences do not fully account for the observed discrepancies.

III. The Relationship between Leaf Respiration and Relative Water Content

The fact that respiration rates decrease under water stress in some studies, while they increase in others may be due to differences in the species tolerance to water stress, as well as to different water stress intensities. Clearly, different species develop different degrees of water stress under similar water shortage. Following the idea of Lawlor and Cornic (2002) that differences in tissue relative water content (RWC) may account for most of the observed metabolic responses to water stress, we have attempted to make a comparison of different studies using RWC as a reference parameter for the intensity of water stress (Fig. 2). We have pooled data from several studies (Zagdańska, 1995; Ghashghaie et al., 2001; Gulías et al., 2002; J. Galmés et al., unpublished), covering a total of 14 species, including herbs, shrubs and trees, and both crop and wild species. The relationship between dark respiration and RWC showed a biphasic response. Initially, as RWC decreased to ca. 70%, there was a decreasing trend of respiration. For RWC values between 70% and 55%, there was a remarkably consistent slow respiration. At RWC below 55% the respiration rate eventually increased, sometimes even above control values (Zagdańska, 1995; Ghashghaie et al., 2001).

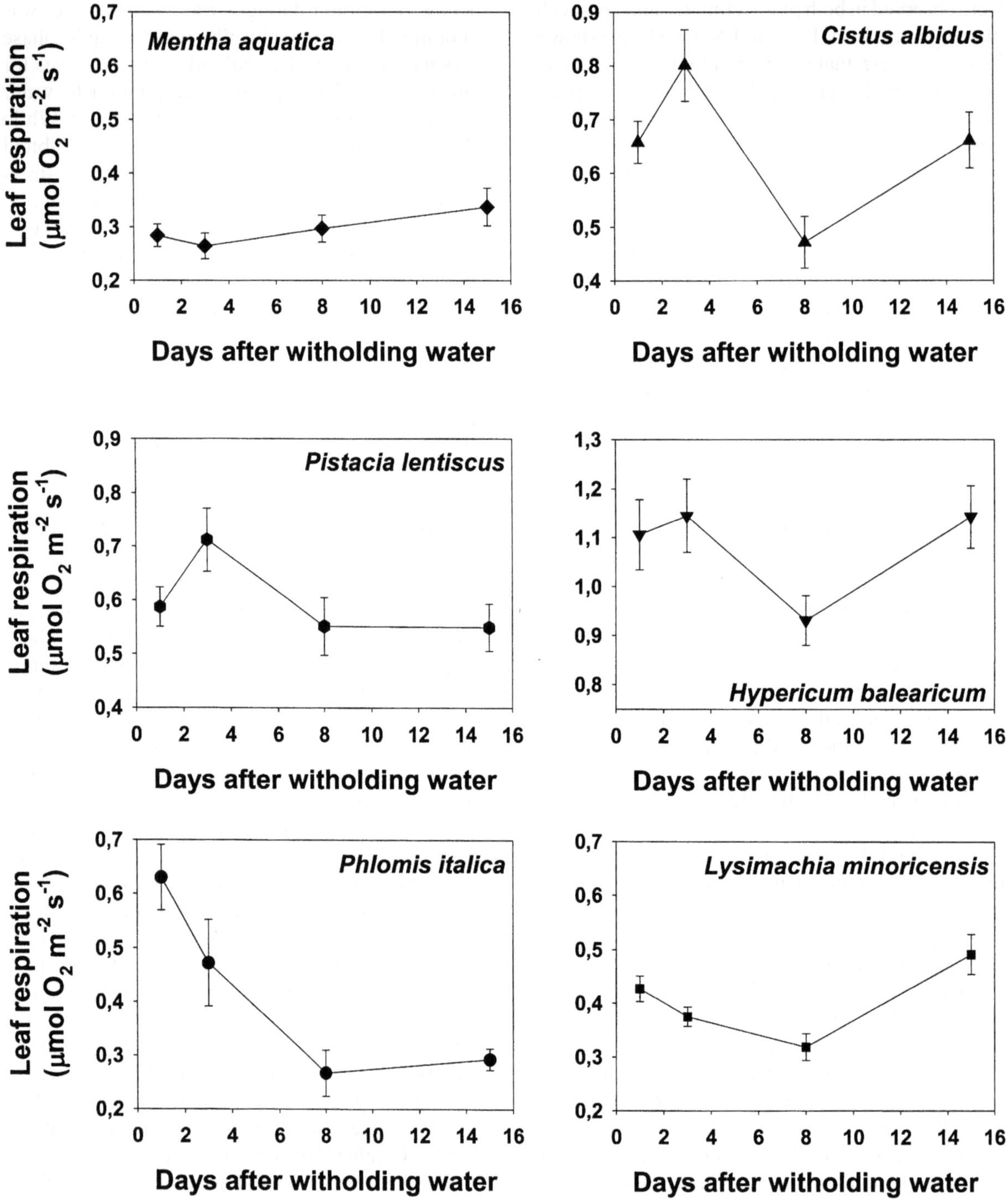

Fig. 1. Different responses of total leaf respiration to water stress in six Mediterranean species. The six species were grown in a cabinet under identical conditions (800 µmol photons m^{-2} s^{-1}; 50% RH; 25°C), and had a similar total leaf area at the onset of the experiment, they were then subjected to identical water-stress treatments (withholding water for 15 consecutive days). In this study, respiration was measured polarographically with a liquid-phase oxygen electrode, after dark-adjusting the leaves for 30 minutes (from J. Galmés et al., unpublished).

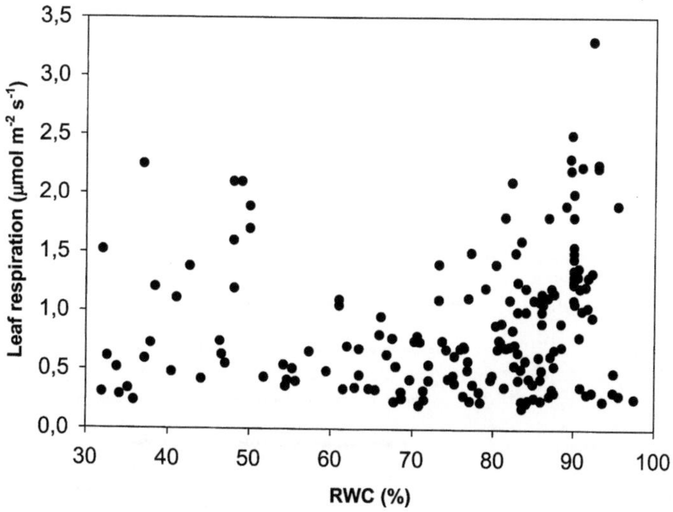

Fig. 2. The relationship between total leaf respiration (μmol O_2 m^{-2} s^{-1} or μmol CO_2 m^{-2} s^{-1}, assuming a 1:1 O_2:CO_2 relationship during respiration) and relative water content (RWC). The set includes data from Zagdańska (1995), Ghashghaie et al. (2001), Gulías et al. (2002) and J. Galmés et al. (unpublished). The species included are: *Triticum aestivum, Helianhus annuus, Nicotiana sylvestris, Rhamnus alaternus, Rhamnus ludovici-salvatoris, Pistacia lentiscus, Pistacia therebinthus, Quercus humilis, Quercus ilex, Hypericum balearicum, Lysimachia minoricensis, Mentha aquatica, Cistus albidus* and *Phlomis italica.*

IV. Possible Causes for the Biphasic Response of Respiration to Relative Water Content

The biphasic response of respiration to RWC may be explained by the differences in sensitivity of different physiological processes to water stress (Hsiao, 1973, Fig. 3). The first physiological consequences of a mild water stress consist of a decrease in cell expansion, cell-wall synthesis, protein synthesis, stomatal closure and photosynthesis. These decreases will result in reduced plant growth, and, therefore, the growth component of respiration would be decreased (R_G; see Chapter 10, Bouma). One, or both of these two factors, may induce a progressive down-regulation of respiration as water stress becomes more intense. However, although respiration in well watered plants depends on photosynthetic rates (Azcón-Bieto and Osmond, 1983; Noguchi, 2004) this may not necessarily be true under water stress. Early studies have shown that both photosynthesis and respiration are affected by water stress; however, photosynthesis is much more affected than respiration (Upchurch et al., 1955; Brix, 1962; Boyer, 1970). While some authors have observed a good correspondence between photosynthetic rates and respiration during a water-stress cycle (Lee Chung et al., 1994), others have not. An example of the latter situation is shown in Fig. 4 (J. Galmés et al., unpublished). Leaf respiration rates during a water-stress cycle were independent of photosynthesis in *Pistacia lentiscus* and *Hypericum balearicum*. While photosynthesis was strongly suppressed in both species, respiration was almost constant during the cycle. Moreover, respiration was always faster in *Hypericum* than in *Pistacia*. Other evidence suggests that the direct relationship between photosynthesis and respiration is impaired by water stress. For instance, Collier and Cummins (1996) showed a good correspondence between leaf respiration rates and soluble sugar content, but the precise relationship differed strongly between plants that were stressed rapidly or slowly. Moreover, ^{13}C-fractionation studies suggest that the patterns of use of recent photoassimilates and reserve substances to drive respiration may well change under water-stress conditions (Duranceau et al., 1999; Ghashaghaie et al., 2001). Taken together, the evidence suggests that decreased rates of photosynthesis, and, consequently, decreased availability of photosynthates, is not the main cause for decreased respiration rates at early stages of water stress.

Recent studies have not differentiated between the responses of the growth and maintenance components of total respiration to water stress. An early study by Wilson et al. (1980) in sorghum (*Sorghum bicolor)* plants subjected to slowly developing water

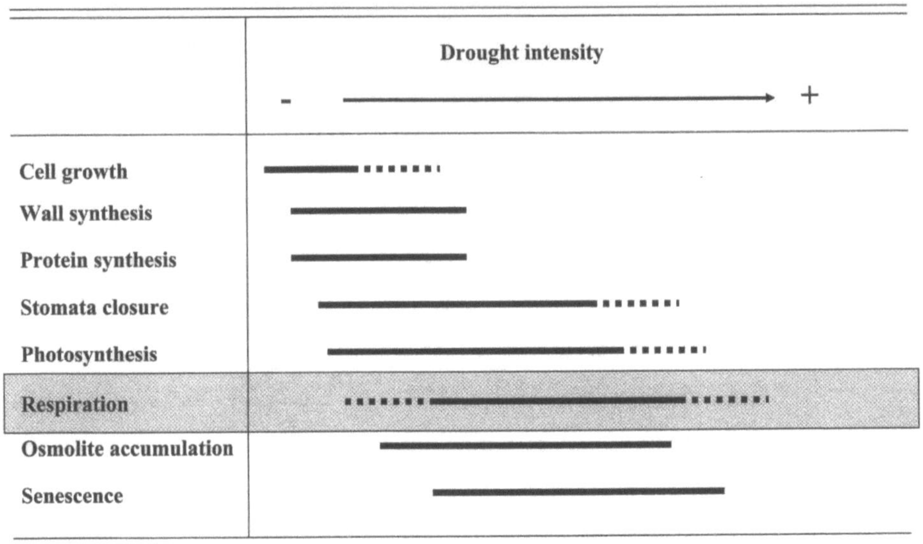

Fig. 3. Theoretical sequence of metabolic processes, including respiration, as affected at different levels of water stress. Modified from Hsiao (1973).

stress showed that whole plant respiration decreased linearly with leaf water potential. However, although both growth and maintenance components of respiration decreased, the slope of the response of growth respiration was three times steeper than that of the maintenance component. This result supports the hypothesis that decreased plant growth is the main reason for the initial decrease in respiration in plants subjected to water stress. Slow-growing species show slower respiration rates, even at high RWC (Lambers et al., 1998). If decreased growth is indeed the main cause for decreased respiration under water stress, then the respiration rates of these species are expected to be less affected by water stress. Certainly, this was the case in slow-growing species such as *Pistacia lentiscus*, *Rhamnus alaternus*, *Quercus ilex* and *Mentha aquatica* (Gulías et al., 2002; J. Galmés et al., unpublished; Fig. 1).

Another possibility to explain the initial decrease of respiration rates would be that water stress might impair some enzymes involved in respiration. However, these seem to be quite insensitive to water stress. The effect of water stress on mitochondrial activity of several key respiratory enzymes and oxidative capacities has been studied in two CAM species, *Prenia sladeniana* (ME-type) and *Crassula lycopodioides* (PEPCK-type) (Herppich and Peckmann, 2000), with the results showing that cytochrome *c* oxidase, NADH-malic enzyme, malate dehydrogenase, fumarase and citrate synthase were unaffected by mild to moderate water stress.

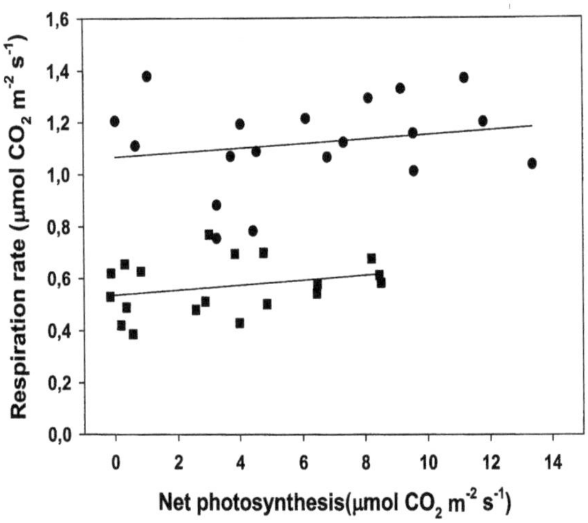

Fig. 4. The relationship between leaf respiration and photosynthesis during a water-stress cycle (see Fig. 1) in *Pistacia lentiscus* (squares) and *Hypericum balearicum* (circles). Data from J. Galmés et al. unpublished.

At more severe levels of water stress, by contrast, the situation can be reversed, i.e. respiration rates may eventually increase (Fig. 2). After hormonal changes take place, there is an accumulation of proline and other compatible solutes, and a general change in metabolism (Fig. 3). All these changes might induce an increase in the maintenance component of respiration under moderate to severe water stress. This was indeed shown by Moldau et al. (1980), Hitz et al. (1982) and Moldau and Rahi (1983). Moreover, since ATP production in the chloroplasts is decreased under water stress, it can be expected that the excess reducing equivalents would be used in mitochondria to overcome an excessive reduction state (and, consequently, susceptibility to oxidative stress) of the system (Lawlor, 1995; Wagner and Krab, 1995). Water-stress-induced senescence and its associated metabolism could also imply higher needs for respiration. Therefore, a biphasic response of respiration to water stress could be expected, consisting of progressive depression of respiration rates at initial stages of water stress, followed by subsequent increases below the threshold water-stress intensity.

An important point to consider is how electron partitioning between the cytochrome pathway and the cyanide-resistant alternative pathway changes under water stress. The role of the alternative pathway is still under debate (Chapter 1, Lambers et al.), but it has been proposed that the activity of this pathway is somehow related to stress and to oxidative stress alleviation (Wagner and Krab, 1995; Lambers et al., 1998; Ribas-Carbo et al., 2000). In this case, it might be hypothesized that the activity of the cytochrome pathway would progressively decrease during water stress, because of the reduced demand for ATP, while the activity of the alternative pathway may eventually increase during the second phase of water stress, to alleviate over-reduction of the electron transport chain. However, there is a general lack of knowledge on the effect of water stress on the electron partitioning between the cytochrome and alternative pathway, and on the role the alternative pathway might play under water stress. A few studies have addressed this subject (Zagdańska, 1995; Collier and Cummins, 1996; González-Meler et al., 1997), but these have used specific inhibitors for the cytochrome (KCN) and alternative (SHAM) respiratory pathways and used the assumption of the 'overflow theory' (Moller et al., 1988) which we now know to be invalid (Day et al., 1996; Chapter 1, Lambers et al.).

Zagdańska (1995) showed in wheat leaves that, while the SHAM-resistant respiration increased in both acclimated and non-acclimated plants, cyanide-resistant respiration was strongly stimulated in acclimated plants and slightly inhibited in non-acclimated plants. Collier and Cummins (1996) studying *Saxifraga cernua* leaves observed that cyanide-resistant respiration decreased as water deficit increased, while SHAM-resistant respiration was differentially sensitive to the rate of development of leaf water deficit. When water stress was applied slowly, SHAM-resistant respiration was initially unchanged, while during a fast-developing water stress its activity initially increased up to 40% with a sharp decrease as the leaves lost turgor. González-Meler et al. (1997) showed that water stress decreased SHAM-resistant respiration in both *Phaseolus vulgaris* and *Capsicum annuum* leaves, but SHAM-sensitive respiration was very slow without any variation in cyanide-resistant respiration. The combination of these analyses suggests that there is variation among species, tissues and conditions.

Recently we addressed the lack of published experiments by studying the effect of water stress on electron partitioning using the oxygen-isotope discrimination technique, which is now known to be the most reliable (Day et al., 1996; Chapters 1, Lambers et al.; and 3, Ribas-Carbo et al.). When soybean (*Glycine max*) plants were subjected to progressive dehydration, the cytochrome pathway decreased progressively, while the activity of the alternative oxidase pathway increased (Fig. 5; M. Ribas-Carbo et al., unpublished results). These results agree with our hypothesis, but further studies would be needed using other species to confirm this trend.

VI. Concluding Remarks

The general knowledge on respiration responses to water stress has increased little since the 1980s. Different studies have reached opposite conclusions regarding the response to water stress of total respiration rates of different plant organs, of various respiratory enzymes or of the partitioning of electron transport between the cytochrome and the alternative pathways. Given the diversity of the results obtained, it seems that the regulation of respiration under water stress reflects a complex metabolic regulation, rather than simply being a consequence of decreased photosynthesis and/or inhibition of one or a few enzymes. Moreover, the recent use of stable-isotope techniques suggests that the pattern of respiratory use of recent photoassimilates versus accumulated reserves may well change under water stress, in a species-dependent manner.

The present evidence indicates a biphasic response of respiration to decreasing relative water content. The initial tendency is for the rate of respiration to

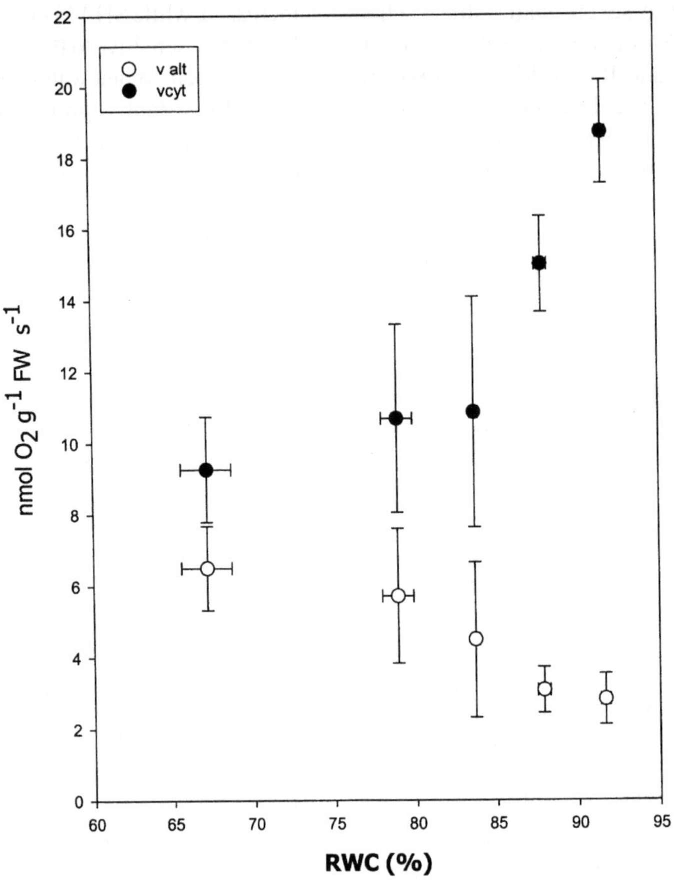

Fig. 5. Activities of the cytochrome and alternative pathway in soybean leaves at different relative water contents (RWC). Values were obtained by oxygen-isotope fractionation measurements (Chapter 3, Ribas-Carbo et al.). Water stress was induced by monitoring and reducing water soil availability. Values are the mean±SEM of five to seven measurements.

decrease, probably as a consequence of decreased energy demand for growth. A second trend that appears at severe water stress is the increase of respiration rates, possibly as a consequence of enhanced metabolism (osmoregulation, water-stress-induced senescence processes). However, the pattern is not yet completely clear. It seems, for instance, that fast-growing species show a more pronounced biphasic response than slow-growing species.

Some research priorities for the near future would be:

(i) To test the possible occurrence of a biphasic response of respiration to water stress in a larger number of species, all grown under similar environmental conditions.

(ii) To analyze the possible metabolic factors underlying the respiratory response to different degrees of water stress. For this, it would be desirable to undertake studies in which a large number of parameters could be analyzed, including any possible water-stress factor that might trigger the response of respiration (soil and plant water potential, relative water content, stomatal conductance, abscisic acid content), and also as many potential targets as possible (activity of respiratory enzymes, mitochondrial structure, content and patterns of use of different respiratory substrates, photosynthesis).

(iii) To prioritize carbon balance studies at the whole plant level. It would be important to determine both the pattern of response of respiration rates of different plant organs to developing water stress, and to discern the response of the growth versus the maintenance components of respiration.

(iv) To further analyze, using appropriate isotope techniques, the effects of water stress on electron partitioning between the cytochrome and the alternative pathways in different species.

All this knowledge would be necessary to fully understand the importance of respiration to plant carbon balance during water stress which is the first step to enable prediction and management of crop growth and yields in water-stress-prone areas.

References

Amthor J (1989) Respiration and crop productivity. Springer-Verlag, New York

Amthor J and McCree KJ (1990) Carbon balance of stressed plants: A conceptual model for integrating research results. In Alscher RG and Cumming JR (eds) Stress Responses in Plants: Adaptations and Acclimation Mechanisms. pp 1–15 Wiley-Liss, Inc. New York

Azcón-Bieto J and Osmond CB (1983) Relationship between photosynthesis and respiration. Plant Physiol 71: 574–581

Bell DT, Koeppe DE and Miller RJ (1971) The effects of drought stress on respiration of isolated corn mitochondria. Plant Physiol 48: 413–15

Bowling DR, McDowell NG, Bond BJ, Law BE and Ehleringer JR (2002) C-13 content of ecosystem respiration is linked to precipitation and vapor pressure deficit. Oecologia 131:113–124

Boyer JS (1970) Leaf enlargement and metabolic rates in corn, soybean, and sunflower at various leaf water potentials. Plant Physiol 46: 233–235

Boyer JS (1976) Water deficit and photosynthesis. In: Kozlowski TT (ed) Water Deficit and Plant Growth, Vol 4, pp 153–190. Academic Press, New York

Boyer JS (1982) Plant productivity and environment. Science 218: 443–448

Boyer JS (1996) Advances in drought tolerance in plants. Adv Agron 56: 187–218

Brix H (1962) The effect of water stress on the rates of photosynthesis and respiration in tomato plants and loblolly pine seedlings. Physiol Plant 15: 10–20

Brown KW and Thomas JC (1980) The influence of water stress preconditioning on dark respiration. Physiol Plant 49: 205–209

Bryla DR, Bouma TJ and Eissenstat DM (1997) Root respiration in citrus acclimates to temperature and slows during drought. Plant Cell Environ 20: 1411–1420

Chaves MM (1991) Effects of water deficits on carbon assimilation. J Exp Bot 42: 1–16

Collier DE and Cummins WR (1996) The rate of development of water deficits affects *Saxifraga cernua* leaf respiration. Physiol Plant 96: 291–297

Cornic G and Massacci A (1996) Leaf photosynthesis under drought stress. In: Baker NR (ed) Photosynthesis and the Environment, pp 247–266. Kluwer Academic Publishers, Dordrecht

Day DA, Krab K, Lambers H, Moore AL, Siedow JN, Wagner AM and Wiskich JT (1996) The cyanide-resistant oxidase: To inhibit or not to inhibit, that is the question. Plant Physiol. 110: 1–2

Duranceau M, Ghasghaie J, Badeck FW, Deleens E and Cornic G (1999) $\delta^{13}C$ of CO_2 respired in the dark in relation to $\delta^{13}C$ of leaf carbohydrates in *Phaseolus vulgaris* L. under progressive drought. Plant Cell Environ 22: 515–523

Escalona JM, Flexas J and Medrano H (1999) Stomatal and non-stomatal limitations of photosynthesis under water stress in field grown grapevine. Aust J Plant Physiol 26: 421–433

Flexas J and Medrano H (2002) Photosynthetic responses of C_3 plants to drought. In: Hemantaranjan A (ed) Advances in Plant Physiology, Vol 4, pp 1–56. Scientific Publishers, Jodhpur

Ghashghaie J, Duranceau M, Badeck FW, Cornic G, Adeline M-T and Deleens E (2001) $\delta^{13}C$ of CO_2 respired in the dark in relation to $\delta^{13}C$ of leaf metabolites: Comparison between *Nicotiana sylvestris* and *Helianthus annuus* under drought. Plant Cell Environ 24: 505–515

Gonzalez-Meler MA, Matamala R and Peñuelas J (1997) Effects of prolonged drought stress and nitrogen deficiency on the respiratory O_2 uptake of bean and pepper leaves. Photosynthetica 34: 505–512

Gulías J, Flexas J, Abadía A and Medrano H (2002) Photosynthetic responses to water deficit in six Mediterranean sclerophyll species: Possible factors explaining the declining distribution of *Rhamnus ludovici-salvatoris*, an endemic Balearic species. Tree Physiol 22: 687–697

Hanson AD and Hitz WD (1982) Metabolic responses of mesophytes to plant water deficits. Annu Rev Plant Physiol 33: 163–203

Haupt-Herting S, Klug K and Fock HP (2001) A new approach to measure gross CO_2 fluxes in leaves. Gross CO_2 assimilation, photorespiration, and mitochondrial respiration in the light in tomato under drought stress. Plant Physiol 126: 388–396

Herppich WB and Peckmann K (2000) Influence of drought on mitochondrial activity, photosynthesis, nocturnal acid accumulation and water relations in the CAM plant *Prenia sladeniana* (ME-type) and *Crassula lycopodioides* (PEPCK-type) Ann Bot 86: 611–620

Hitz WE, Ladyman JAR and Hanson AD (1982) Betaine synthesis and accumulation in barley during field water stress. Crop Sci 22: 47–54

Hsiao T (1973) Plant responses to water stress. Annu Rev Plant Phys 24: 519–570

Lambers H, Chapin III FS and Pons TL (1998) Plant physiological ecology. Springer-Verlag, New York

Lawlor DW (1976) Water stress induced changes in photosynthesis, photorespiration, respiration and CO_2 compensation concentration of wheat. Photosynthetica 10: 378–387

Lawlor DW (1995) Effects of water deficit on photosynthesis. In: Smirnoff N (ed), Environment and Plant Metabolism, pp 129–160. BIOS Scientific Publishers, Oxford

Lawlor DW and Cornic G (2002) Photosynthetic carbon assimilation and associated metabolism in relation to water deficits in higher plants. Plant Cell Environ 25: 275–294

Lee Chung Y, Tsuno Y, Nakano J and Yamaguchi T. (1994) Ecophysiological studies on the drought resistance of soybean: I. Changes in photosynthesis, transpiration and root respiration with soil moisture deficit. Jap J Crop Sci 63: 215–222

Loboda T (1993) Gas exchange of different spring cereal genotypes under normal and drought conditions. Photosynthetica 29: 567–572

McCree KJ, Kallsen CE and Richardson SG (1984) Carbon balance of sorghum plants during osmotic adjustment to water stress. Plant Physiol 76: 898–902

McCree KJ (1986) Measuring the whole-plant daily carbon balance. Photosynthetica 20: 82–93

Moldau H and Rakhi M (1983) Enhancement of maintenance respiration under water stress. In: Marcelle R, Clijsters H and van Poucke M (eds) Effects of Stress on Photosynthesis, pp 121–131. Martinus Nijhoff/Dr W Junk Publishers, The Hague

Moldau KS, Syber YK and Rakhi M (1980) Components of dark respiration in bean under conditions of water deficit. Sov Plant Physiol 27: 1–6

Moller IM, Berczi A, Van der Plas LHW and Lambers H (1988) Measurement of the activity and capacity of the alternative pathway in intact plant-tissues: Identification of problems and possible solutions. Physiol Plant 72: 642–649

Palta JA and Nobel P (1989) Root respiration for Agave deserti: Influence of temperature, water status and root age on daily patterns. J Exp Botany 40: 181–186

Penning de Vries FWT, Witlage J and Kremer D (1979) Rates of respiration and of increase in structural dry matter in young wheat, ryegrass and maize plants in relation to temperature, to water stress and to their sugar content. Ann Bot 44: 595–609

Pham Thi AT and Vieira da Silva J (1980) Effects of water stress on mitochondrial ultrastructure of cotton leaves: Some metabolic consequences. Z Pflanzenphysiol 100: 351–358

Pheloung P and Barlow EWR (1981) Respiration and carbohydrate accumulation in the water-stressed wheat apex. J Exp Bot 32: 921–931

Ribas-Carbo M, Aroca R, Gonzalez-Meler MA, Irigoyen JJ and Sanchez-Diaz M (2000) The electron partitioning between the cytochrome and alternative respiratory pathways during chilling recovery in two cultivars differing in chilling sensitivity. Plant Physiol 122: 199–204

Rice JR and Eastin JD (1986). Grain sorghum root responses to water and temperature during reproductive development. Crop Sci 26:547–551

Shearman LL, Esatin JD, Sullivan CY and Kinbacher EJ (1972) Carbon dioxide exchange in water-stressed maize sorghum. Crop Sci 12: 406–409

Upchurch RP, Peterson ML and Hagan RM (1955) Effect of soil-moisture content on the rate of photosynthesis and respiration in ladino clover (*Trifolium repens* L.). Plant Physiol 30:297–303

Valentini R, Matteucci G, Dolman AJ, Schulze E-D, Rebmann C, Moore EJ, Granier A, Gross P, Jensen NO, Pilegaard K, Lindroth A, Grelle A, Bernhofer G, Grünwald T, Aubinet M, Ceulemans R, Kowaslki AS, Vesala T, Rannik Ü, Berbigier P, Loustau D, Guömundsson J, Thorgeirsson H, Ibrom A, Mongerstern K, Clement R, Moncrieff J, Montagnani L, Minerbi S and Jarvis PG (1999) Respiration as the main determinant of carbon balance in European forests. Nature 404: 861–865

Wagner AM and Krab K (1995) The alternative respiration pathway in plants: Role and regulation. Physiol Plant 95: 318–325

Wilson DR, Van Bavel CHM and McCree KJ (1980) Carbon balance of water deficit grain sorghum plants. Crop Sci 20: 153–159

Zagdańska B (1995) Respiratory energy demand for protein turnover and ion transport in wheat leaves upon water demand. Physiol Plant 95: 428–436

Chapter 7

Response of Plant Respiration to Changes in Temperature: Mechanisms and Consequences of Variations in Q_{10} Values and Acclimation

Owen K. Atkin*
Department of Biology, The University of York, PO Box 373, York YO10 5YW, U.K.

Dan Bruhn
Research School of Biological Sciences, Australian National University, GPO Box 475, Canberra, ACT 2601, Australia

Mark G. Tjoelker
Department of Forest Science, Texas A&M University, 2135 TAMU, College Station, TX 77843-2135, U.S.A.

*Author for correspondence, email: OKA1@york.ac.uk

H. Lambers and M. Ribas-Carbo (eds.), Plant Respiration, 95–135.

Summary

The effects of short- and long-term changes in temperature on plant respiration (*R*) are reviewed. We discuss the methods available for quantifying the short- and long-term temperature-dependence of *R*. The extent to which the Q_{10} (the proportional change in *R* with a 10 °C increase in temperature) and the degree of thermal acclimation (change in the temperature-response curve of *R* following a long-term change in growth temperature) vary within and amongst plant species are assessed. We show that Q_{10} values are highly variable (e.g., being affected by measuring and growth temperature, irradiance and drought), but most plant species exhibit similar Q_{10} values (in darkness) when grown and measured under identical conditions (i.e. little evidence of inherent differences in the Q_{10} of plant *R*). The possible mechanisms responsible for variability in the Q_{10} are discussed; high Q_{10} values occur in tissues where respiratory flux is substrate saturated (i.e. capacity limited). This is illustrated using plots of reduced ubiquinone versus O_2 uptake in isolated mitochondria. The degree of acclimation is also highly variable amongst plant species. This variability is due, in part, to some studies exposing pre-existing roots/leaves to a new growth temperature, whereas others compare roots/leaves that develop at different temperature. In most cases, maximal acclimation requires that new leaves and/or roots be developed following a change in growth temperature. In addition to its link with development, acclimation is also often associated with changes in the Q_{10}, particularly in pre-existing leaves/roots transferred from one environment to another. The importance of acclimation in determining annual rates of *R* as a component of net primary productivity and net ecosystem CO_2 exchange is discussed. The importance of acclimation for future atmospheric CO_2 concentrations is highlighted, including a positive feedback effect of climate warming on the carbon cycle. This review shows that the assumptions of coupled global circulation models (that Q_{10}

values are constant and that *R* does not acclimate to long-term changes in temperature) are incorrect, and this may lead to overestimation of the effects of climate warming on respiratory CO_2 flux.

I. Introduction

Plant respiration (*R*) plays a critical role in a wide range of ecological phenomena, from the performance of individual plants to global atmospheric CO_2 concentrations. *R* couples the production of energy and carbon skeletons (necessary for biosynthesis and cellular maintenance) to the release of large amounts of CO_2; up to two thirds of daily photosynthetic carbon gain is released into the atmosphere by plant *R* (Poorter et al., 1990; Van der Werf et al., 1994; Atkin et al., 1996; Loveys et al., 2002). At the ecosystem level, plant *R* contributes up to 65% of the total CO_2 released into the atmosphere (Xu et al., 2001; Reichstein et al., 2002) with the remaining CO_2 release coming from heterotrophic soil *R* (R_h). Recently, Valentini et al. (2000) concluded that total ecosystem *R* is the main determinant of net ecosystem CO_2 exchange in northern European forests. Globally, terrestrial plant *R* releases approximately 64 Gt C year^{-1} into the atmosphere (Raich and Schlesinger, 1992; Amthor, 1997; Field, 2001). This is a large flux, compared with the relatively small release of CO_2 from the use of fossil fuels and cement production (totals 5.5 Gt year^{-1}) and changing land use (about 1.6 Gt C year^{-1}) (Schimel, 1995). Clearly, increases in plant *R* in response to climate warming could have a substantial impact on atmospheric CO_2 concentrations. In this chapter, we discuss the effects of both short- and long-term changes in temperature on plant *R* and underlying respiratory metabolism. Unless otherwise stated, *R* refers to specific rates of respiration per unit leaf or root dry mass. The respiratory Q_{10} values reported in this chapter have been calculated using rates of respiratory CO_2 release in most cases, and O_2 uptake in others. It is not known whether Q_{10} values differ depending on the way rates of *R* were measured; consequently, readers should be cautious when comparing Q_{10} values obtained from studies that used contrasting measuring units.

It has long been known that *R* is sensitive to short-term changes in temperature (e.g., Fig. 1) in measurements lasting minutes to a few hours (Wager, 1941; James, 1953; Forward, 1960) with many studies assuming that the relationship between *R* and temperature is exponential with a constant Q_{10} values (i.e. proportional change in *R* with a 10 °C increase in temperature) of 2.0–2.3 (e.g., Atkin and Day, 1990; Ryan, 1991; Raich and Schlesinger, 1992; Cox et al., 2000). This review shows that the common use of a Q_{10} value in the range 2.0–2.3 is probably appropriate only over a restricted temperature range (ca. 15 to 25 °C) and for measurements conducted over short time periods. However, further analysis suggests that even within this temperature range, large deviations in Q_{10} from values of 2.0–2.3 can occur.

The effect of prolonged exposure to different temperatures on *R* depends on the degree to which *R* acclimates to a change in temperature. Acclimation to contrasting growth temperatures is associated with a change in the temperature-response curve of *R* in many plant species (Fig. 1). Thermal acclimation may result in changes in the intercept, slope (shape of the curve), and/or temperature optimum of the short-term temperature response function of *R*. For example, shifting warm-grown plants to a lower growth temperature (e.g., for several days) increases the rate of *R* measured at common measurement temperature (Rook, 1969; Chabot and Billings, 1972; Pisek et al., 1973; Larigauderie and Körner, 1995; Körner,

Abbreviations: Acclim$_x$ – acclimation ratio of *R* calculated using various (*x*) methods; AOX – alternative oxidase; ATP – adenosine triphosphate; CI – complex I; CII – complex II (succinate dehydrogenase); CIII – complex III (cytochrome bc_1 complex); CIV complex IV (cytochrome oxidase); CV – complex V (ATP synthase); Cyt – cytochrome; E_a – activation energy; ExtNDH – external NAD(P)H dehydrogenase; F1,6bP – fructose 1,6 bisphosphate; GCMs – global circulation models; GPP – gross primary production; F_v/F_m – ratio of variable to maximal fluorescence; k – temperature coefficient of *R*; LTER – Long-term Ecological Research; LTR_{10} – long-term Q_{10} of *R*; ME – malic enzyme; NADH – nicotinamide-adenine dinucleotide, reduced form; NADPH – nicotinamide-adenine dinucleotide phosphate, reduced form; NEE – net ecosystem CO_2 exchange; NPP – net primary productivity; OAA – oxaloacetate; PDC – pyruvate dehydrogenase complex; PEP – phosphoenolpyruvate; PFK – phosphofrucokinase; PK – pyruvate kinase; *R* – plant respiration per unit dry mass; R_a – autotrophic *R*; R_e – ecosystem *R*; R_h – heterotrophic respiration; $R_{RefTemp}$ – *R* measured at a reference temperature; R_T – *R* at any given T; R_g – universal gas constant; RIB – internal, rotenone-insensitive NADH dehydrogenase; SDH – succinate dehydrogenase; TEM – terrestrial ecosystem model; T – temperature; T_{growth} – growth temperature; T_m – transition temperature; UQ_r – ubiquinol; ubiquinone (UQ); UQ_r/UQ_t – proportion of UQ in reduced form

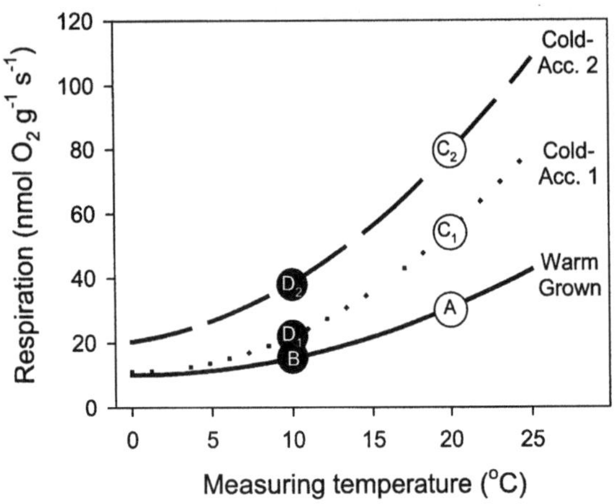

Fig. 1. Diagrammatic representation of temperature response of respiration for warm-grown and cold-acclimated (i.e. plants exposed to low growth temperatures for extended periods). In Cold-Acc. 1, respiration rates are similar to those of the warm-grown plants when measured at low temperatures. However, large differences are observed in rates of *R* at higher measuring temperatures (i.e. the Q_{10} increases following cold acclimation). In Cold-Acc. 2, differences in rates of *R* are observed at both low and high measuring temperatures. The letters refer to rates of *R* at specific temperatures as used in Eqs. 10–13.

1999; Atkin et al., 2000b; Covey-Crump et al., 2002). Differences in the rate of *R* at standard measuring temperatures are also exhibited by plants that grew and developed under contrasting temperature regimes (either in the lab or in the field) (e.g., Fig. 2; Billings and Mooney, 1968; Chabot and Billings, 1972; Körner and Larcher, 1988; Collier and Cummins, 1990; Semikhatova et al. 1992; Goldstein et al., 1996; Collier, 1996; Arnone and Körner, 1997). In some cases, cold-acclimation is associated with an increase in the rate of *R* only at moderate to high measuring temperatures, with no change in *R* at low measuring temperatures (i.e. Q_{10} increases) (Fig. 1). In other cases, acclimation is associated with an increase in the rate of *R* over a wide range of measuring temperatures (Fig. 1). Acclimation of *R* can occur within a 1–2 day period following a change in ambient temperature (Rook, 1969; Billings et al., 1971; Chabot and Billings, 1972; Atkin et al., 2000b; Covey-Crump et al., 2002; Bolstad et al. 2003), raising the possibility that plant *R* may dynamically acclimate to changes in thermal environment with an onset of the acclimation processes within perhaps one hour.

Acclimation can also result in respiratory homeostasis [i.e. identical rates of *R* in plants grown and measured in contrasting temperatures (Körner and Larcher, 1988; Semikhatova et al., 1992; Goldstein et al., 1996; Arnone and Körner, 1997; Körner, 1999; Atkin et al., 2000b)] (Fig. 1). Such changes can result in annual respiratory CO_2 release being substantially reduced in leaves and roots that exhibit a high degree of thermal acclimation of *R* compared with that of tissues that do not acclimate (Atkin et al., 2000a). As a result, respiratory CO_2 release under natural climatic conditions may not be accurately predicted using information from a single short-term temperature response function of *R* alone in leaves and roots that acclimate to temperature. Moreover, failure to take into account acclimation may result in an over-estimate of the effects of global warming on respiratory CO_2 release over long periods (Luo et al., 2001), particularly in coupled global circulation models (GCMs) that assume a positive feedback of global warming on *R* (Cox et al., 2000). In fact, there are also examples where cold-grown plants exhibit higher rates than warm-grown plants, when each are measured at their respective growth temperatures [i.e. *R* 'over-acclimates' (Loveys et al., 2003)].

This chapter begins by discussing the methods and analysis procedures used to assess the effect of short-term changes in temperature on plant *R*. It then assesses the extent to which Q_{10} varies within and among plant species. The probable mechanisms responsible for *R* being temperature sensitive are discussed (including the effect of temperature on regulation of *R* by substrate supply, demand for respiratory energy and the maximum catalytic activity of respiratory enzymes). It also assesses which of these factors are probably responsible for variation in Q_{10} values. Acclimation to long-term changes in temperature is then discussed, starting with a discussion of what is meant by acclimation, followed by the methods used to quantify the degree of acclimation and the extent to which acclimation varies within and among plant species. The mechanisms responsible for acclimation are discussed (in particular the role of plant development in determining the magnitude of thermal acclimation); the link between acclimation and changes in the Q_{10} is also highlighted. The importance of acclimation in determining annual rates of *R* as a component of net primary productivity (NPP) and net ecosystem CO_2 exchange (NEE) is discussed. We highlight the importance of acclimation for future atmospheric CO_2 concentrations, including the impact on coupled GCMs that incorporate a positive-feedback effect of climate warming on the carbon cycle (e.g., Cox et al., 2000).

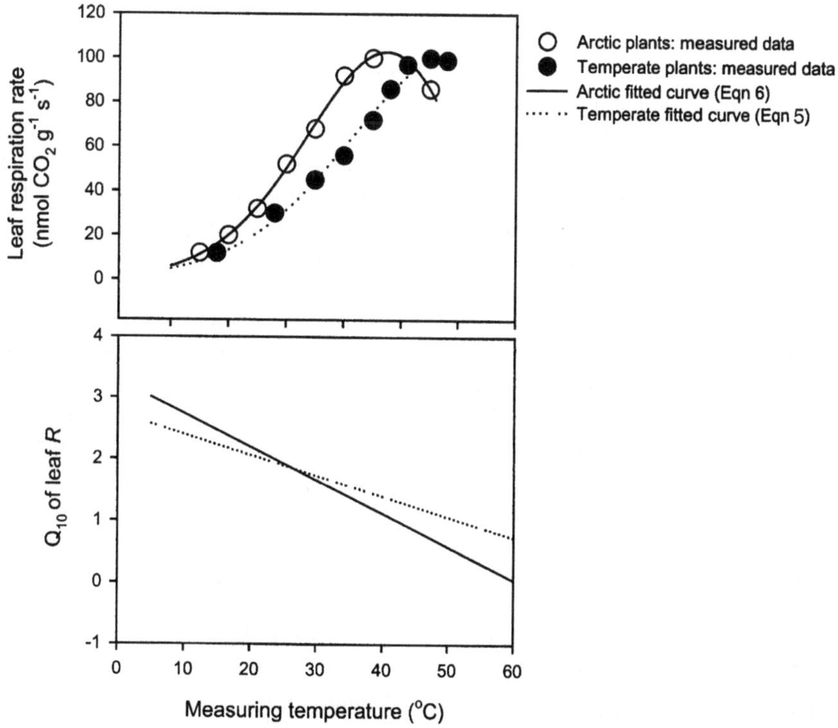

Fig. 2. Temperature-response curves of leaf *R* (A) and temperature dependence of the Q_{10} (B) of arctic and temperate plants grown under field conditions. Data in (A) were reported by Semikatova et al. (1992), with curves being fitted using Eq. 6 (R_0 and the constants c and b were estimated via iteration/non-linear regression). Data shown in (B) were calculated from (A) [using log-transformed values of *R* and Eq. 4 (Section II.A.3)].

II. Short-term Changes in Temperature

A. Characterizing the Temperature Response of Respiration

The methodology used in studies of plant respiration responses to temperature can have an important impact on the obtained results. This section critically discusses several of the commonly used approaches, and emphasizes the fact that many studies use Arrhenius theory incorrectly when analyzing the temperature response of *R* in intact leaves and roots. Finally, the review highlights the advantages and limitations of the Q_{10} approach when predicting rates of *R*.

1. Methodology

Respiration rates in intact plants or individual leaves or roots may be determined in enclosed cuvettes using infrared gas analyzers that measure net CO_2 efflux or using oxygen electrodes that measure net O_2 uptake. The latter method is typically used to determine rates of root respiration in roots bathed in a buffered hydroponic medium and temperature-controlled water bath (Burton et al., 1996). Since net CO_2 efflux rates can be small, especially at low measurement temperatures, care should be taken to minimize leaks in infrared-based gas exchange systems that may occur where gaskets contact plant tissues (Bruhn et al., 2002). When using an open system, enclosing more tissue within the chamber may increase the measured CO_2 differentials. This may be accomplished by measuring entire leaves or roots in a larger cuvette, rather than the smaller leaf area sampled in cuvettes typically designed for photosynthesis measurements. Examples of custom-built chambers for leaves include a portable device using a peltier cell for temperature control (Hubbard et al., 1995) and for roots a chamber using an extended heat sink for insertion into soil (Burton and Pregitzer, 2002). To generate a temperature response curve, a controlled-environment (such as a growth chamber) provides a readily controllable and reproducible set of temperatures and that for statistical reasons could be applied in random order.

To generate a temperature-response curve on intact leaves (e.g., Fig. 1) attached to the rest of the plant, the entire plant should be subjected to step changes in temperature, as recent evidence suggests that different temperature response functions result when leaf temperature is uncoupled from the ambient temperature of the plant. Both Atkin et al. (2000b) and Griffin et al. (2002) found that individual leaves exhibit greater Q_{10} values when their temperature tracks that of the surrounding air/whole shoot compared with when the leaf temperature response is measured with the surrounding air/shoot temperature constant. For example, the Q_{10} of leaf *R* in *Eucalyptus pauciflora* (between a measurement temperature range of 10–22 °C) was 2.6 when the plant and leaf temperatures were the same, and 2.1 when the plant was kept at 10 °C (with only the temperature of the measured leaf being altered) (Atkin et al., 2000b). Similarly, in *Populus deltoides* the Q_{10} of leaf *R* was 2.1 when the temperatures of the measured leaves matched the rest of the stand, and 1.7 when the leaf temperature was manipulated alone (with stand temperature being kept constant) (Griffin et al., 2002). Griffin et al. (2002) calculated that total night-time CO_2 release over five days was 21% lower when using the Q_{10} estimates made using the traditional leaf manipulation compared with that using the stand-level manipulation. Thus, accurate estimates of total daily leaf *R* need to be made using measurements carried out with the attached leaf at the same temperature as that of the surrounding air/shoot to avoid physiological uncoupling. For root respiration measurements, it is often not practical to measure rates of *R* using attached, intact roots. However, several investigators have developed cuvettes that enclose roots, and are placed back in the soil for measurements for extended periods of time (Rakonczay et al., 1997; Bryla et al., 2001). Typically, either detached whole roots or root pieces are used. Excision does not have a major effect on root respiration when assessed within 30 mins after severing of the shoot (Lambers et al., 1993). However, repeated measurements of *R* at different temperatures using detached roots do lead to a progressive reduction of *R* at a standard temperature (Loveys et al., 2003). Alternatively, separate roots may be used for each measuring temperature with care taken to control for root order, size, and age. The same sampling constraints are likely to apply to the measurement of O_2 uptake in leaf segments.

Another issue to consider when generating temperature-response curves of *R* of leaves attached to the rest of the plant is whether the same leaves should be used for all temperatures. When using individual plants, measurements may take several hours per plant (investigators normally let the plant material adjust for approximately 30–60 mins at each new measurement temperature). As the short-term temperature response can change within hours (Lawrence and Holaday, 2000), generation of temperature-response curves and Q_{10} values using single plants can lead to an underestimate of the actual in situ Q_{10} value. Therefore, in studies where the instantaneous Q_{10} value is needed, it may be best to conduct temperature-response curve measurements using different plant material for each measurement temperature (Bruhn et al., 2002). However, use of individual plants is adequate where the aim is to describe the response of *R* to diurnal shifts in temperature (e.g., under field conditions).

2. Using Arrhenius plots

Many studies have used Arrhenius theory to characterize the response of *R* to short-term changes in temperature (Crawford and Palin, 1981; Sowell and Spomer, 1986; Lloyd and Taylor, 1994; Turnbull et al., 2001; Griffin et al., 2002a,b); in Arrhenius plots, log *R* is plotted against the reciprocal of the absolute temperature [1/T (K)]. If *R* exhibits an exponential relationship with 1/T (K) then the Arrhenius plot should yield a straight line of slope E_a/R_g where R_g is the universal gas constant (8.314 J mol^{-1} K^{-1}) and *E* is the activation energy (J mol^{-1}) for the reaction in question (Forward, 1960; Berry and Raison, 1981). This theory is typically used as a standard of reference for reactions in physical chemistry; it has also been used to characterize the temperature response of *R* in several studies (see references above). However, its use in a complex physiological process such as *R* needs to be treated with caution. There are four reasons for this. First, Arrhenius plots normally assume that the reaction is substrate saturated; as discussed in Section III.A, *R* in intact tissues is rarely substrate saturated, particularly at moderate temperatures. Second, application of Arrhenius theory to the temperature response of *R* implies that there is biological meaning to the value of E_a [i.e. E_a is the amount of excess energy (activation energy) that must be acquired by participating molecules before the reaction will proceed]. Third, no single E_a will describe the temperature response of the series of highly regulated reactions that take place in the respiratory chain. In such cases, the E_a derived from Arrhenius plots of log R against 1/T (K) merely describes the overall *temperature coefficient* of the respiratory metabolism. Fourth, over the non-exponential regions of the temperature response function, estimates of E_a for plant R will continually change with measuring temperature (such changes should not be confused with break points in Arrhenius plots where there is a sharp transition from one constant E_a to another). For these reasons, we suggest that the temperature response of R to short-term changes in temperature should not be described using Arrhenius theory (although the outcomes from using Arrhenius theory are unlikely to differ substantially from that when the Q_{10} is used over limited temperature ranges). In addition, in the rare cases in which R has been determined over its entire temperature range, rates appear to increase as a sigmoid function of measurement temperature to an optimum, and then abruptly decline towards zero at a maximum temperature. Neither the Arrhenius nor the Q_{10} model (described below) adequately account for the complex shape of the observed temperature responses of R. Nevertheless, there are available equations that provide adequate fits to observed temperature response data over a wide range of measuring temperatures (e.g., Eq. 6 in Section II.A.3.b). In addition, Cannell and Thornley (1998) describe a cubic function of an appropriate sigmoid shape that may be useful in modeling the temperature response of R over its entire temperature range.

3. Using Q_{10} values

a. Calculating the Q_{10}

An alternative to Arrhenius theory with a long tradition in plant physiology is to use Q_{10} to describe the short-term temperature response of R. Berry and Raison (1981) suggested that Q_{10} offers an important advantage over Arrhenius theory (derived from physical chemistry) when interpreting the temperature dependence of R; that being that Q_{10} does not imply a mechanistic explanation (whereas the use of E_a does). The Q_{10} is simply the ratio of R at one temperature to that at 10 °C lower. Although Q_1 values (i.e. the proportional increase in respiration per degree increase in temperature) could also be used, the dominance of Q_{10} as a descriptor of the temperature sensitivity of respiration suggests that its use will continue. The Q_{10} can be calculated according to:

$$Q_{10} = \left(\frac{R_2}{R_1}\right)^{\left[\frac{10}{(T_2-T_1)}\right]} \quad (1)$$

where R_1 is the respiration rate measured at a colder temperature (T_1) and R_2 is respiration measured at a warmer temperature (T_2). In Eq. 1, T_1 and T_2 do not have to be 10 °C apart. A rearrangement of Eq. 1 provides the following formula:

$$R_2 = R_1 Q_{10}^{[(T_2-T_1)/10]} \quad (2)$$

in which R may be predicted as a function of the Q_{10} and the measurement temperature (T).

When rates of R have been determined over a range of measurement temperatures (but below the optimum temperature), a simple exponential function will often adequately describe the temperature response. R at any given T (R_T) can be predicted using a model of the form:

$$R_T = R_{RefTemp}(e^{kT}) \quad (3)$$

where $R_{RefTemp}$ is R at 0 °C and k is a temperature coefficient. Equation 3 may be fitted using standard non-linear regression techniques. Alternatively, k may be determined by linear regression of log-transformed R plotted against measurement temperature (T) (derived by log-transforming both sides of Eq. 3) (note: this value of k may differ from that obtained from Eq. 3). Q_{10} may be then estimated from k using the following formula:

$$Q_{10} = e^{10k} \quad (4)$$

Equations. 1 through 4 do not, however, provide information on the extent to which the temperature coefficient of R changes with measuring temperature (rather, they provide an estimate of the average Q_{10} for the temperature range T_1 to T_2).

The short-term response of R to temperature is not strictly exponential, except perhaps over a limited temperature range below the optimum temperature. In other words, the temperature sensitivity of R may change with measurement temperature, implying that Q_{10} is temperature-dependent (e.g., Figs 2 and 4; see Section II.B.2). This may be revealed by a lack of fit of R against measurement temperature when using exponential temperature-response functions (Eqs. 2 or 3). To overcome these limitations, an estimate of

Q_{10} at each temperature is needed. If the regression slope of log-transformed *R* against measurement temperature (k in Eqs. 3 and 4) is linear, then a single Q_{10} value can be used across all temperatures over a defined measurement temperature interval. However, the slope may not be constant, as would be evident by the lack of linearity in the regression fit, and a significant polynomial fit to the log-transformed *R* versus temperature data. This fitted polynomial equation can then be differentiated to get the slope (i.e. k) at each temperature; these slopes may then used to calculate Q_{10} values at each temperature (Eq. 4). It should be noted that the use of log-transformed *R* versus temperature plots to determine the temperature dependence of Q_{10} requires the analysis to be based on a large number of measurements conducted at several temperatures. Reliance on too few replicates and/or measurement temperatures may result in inadequate statistical power to adequately distinguish between a linear or polynomial fit to the log-transformed *R* versus temperature plots, and perhaps erroneous conclusions being made about the temperature dependence of the Q_{10}.

When the above analysis clearly shows that Q_{10} varies with measuring temperature, the extent of that temperature dependence can be approximated via linear regression of the Q_{10} values plotted against T to yield a formula:

$$Q_{10} = c - bT \qquad (5)$$

where c is the Q_{10} at 0 °C and b is the slope of the Q_{10} versus T plot. Section II.B.2 provides generalized equations (Eqs. 8 and 9) to describe the approximate temperature dependence of leaf and root Q_{10} across biomes and contrasting plant taxa. It should be noted that if the data range includes *R* measured at values lower and higher than the optimum temperature, then Q_{10} may in fact exhibit a non-linear decline with increasing measurement temperature throughout a broad temperature range.

b. Fitting curves to measured data

Curve fitting to measured *R* data can be accomplished in several ways. Here, we describe the use of standard non-linear regression techniques to fit curves by iteration to existing data using Eq. 2. This equation may be written as a function with R_2 as the predicted value of *R* at any temperature by substituting T for $T_2 - T_1$ and when R_1 represents *R* at 0 °C (i.e. $T_1 = 0$). In cases where Q_{10} is temperature-independent, estimates of R_1 (i.e. *R* at 0 °C) and the average Q_{10} over the temperature range are estimated parameters, based on non-linear regression. The resulting expression may then be used to predict *R* and plot the fitted curves.

Whenever Q_{10} is temperature-dependent (e.g., data are measured over a broad temperature range that approaches or includes the temperature optimum of *R*; Figs 2 and 4), curve fitting to measured data requires Eq. 2 to be modified. By substituting Eq. 5 into Eq. 2, the following expression is obtained:

$$R_T = R_0(c - bT)^{[T/10]} \qquad (6)$$

where R_0 is *R* at 0 °C, and c and b are constants that describe the intercept and slope of Q_{10} versus temperature plots, respectively). Again, standard non-linear regression methods can be used to estimate R_0, c and b. These values will then be used to predict R_T and plot the fitted curves. Examples of such fits are shown in Fig. 2A.

c. Predicting R in the Absence of a Measured Temperature-Response

In cases where the temperature response of *R* has not been determined experimentally, *R* at different temperatures (R_T) can be modeled. Assuming that the Q_{10} value is temperature insensitive, then the rates of R_T at any given temperature (T) can be predicted using Eq. 2, with R_2 representing R_T and R_1 being *R* measured at a reference temperature. However, as stated above, Q_{10} is often temperature sensitive. Then, an equation that takes into account the temperature dependence of the Q_{10} is needed to successfully predict rates of R_T at given temperatures. Intuitively, one might replace the single Q_{10} value in Eq. 2 with a term that describes the temperature dependence of the Q_{10} (e.g., using Eqs. 8 or 9 in Section II.B.2). However, this approach fails to accurately predict rates of R_T, particularly at measuring temperatures that are much higher than the reference temperature (an example is shown in Fig. 3), owing to the fact the Q_{10} describes the proportional change in *R* across a 10 °C interval. To predict rates of R_T when using a temperature-dependent Q_{10}, the Q_{10} at the midpoint between the reference (T_1) and prediction temperature (T_2) should be used as shown in the following equation:

$$R_2 = R_1[x - y((T_2 + T_1)/2)]^{[(T_2 - T_1)/10]} \qquad (7)$$

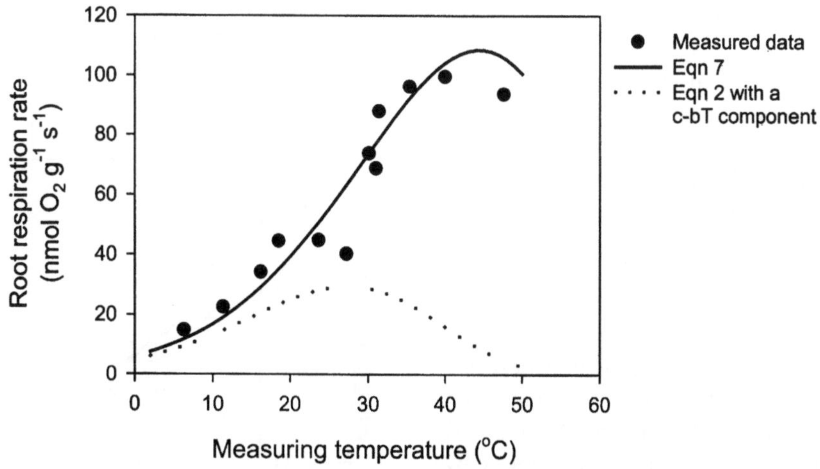

Fig. 3. Comparison of rates of R predicted using Eq. 2 (with a c-bT term to describe the temperature dependence of the Q_{10}) and Eq. 7 against observed data (Covey-Crump et al., 2002). Assumed values in both equations: R_1 was taken as the rate of R at 0 °C (6.01 nmol O_2 g^{-1} s^{-1}), c = 3.00 and b = 0.045 (as per Eq. 9). See text for further details.

where x and y are constants that describe the temperature dependence of the Q_{10} (see Eq. 8 for leaves and Eq. 9 for roots in Section II.B.2). R_1 and R_2 are rates of R at temperatures T_1 and T_2, respectively, where R_1 can be either higher or lower than R_2. To use Eq. 7, an initial R_1 value at T_1 is either measured or obtained from published literature. Rates of R at new temperature (R_2) can then be predicted using a temperature-dependent Q_{10} equation (e.g., Eq. 5). Equation 7 may be used in predicting temporal changes (e.g., diurnal) in R with changing ambient temperatures by solving for a new R_2 for each successive temperature-measurement interval. An example of this approach is shown in Fig. 3.

B. Variations in the Q_{10} of Respiration

This section addresses the variability of leaf and root Q_{10} values. It shows that Q_{10} values are often highly variable; Q_{10} depends upon both measurement temperature and growth temperatures, and differs between thermally contrasting biomes. The Q_{10} of leaf R is reduced in the light compared with the Q_{10} of leaf R in the dark. Q_{10} is also lower in water-stressed plants than in their fully watered counterparts, and in roots compared with that in leaves. Leaves in the upper canopy of trees are also more temperature sensitive than their lower-canopy counterparts. Large seasonal variations in Q_{10} have also been reported. However, despite this variability, most plant species exhibit similar Q_{10} values (leaves or roots) of R (in darkness) when grown and measured under identical conditions.

1. Inter-specific Variation

The *range* of Q_{10} values reported for leaves and roots of contrasting plant species is substantial [leaves: 1.36 to 4.2 (Azcón-Bieto, 1992; Larigauderie and Körner, 1995; Tjoelker et al., 2001); roots: 1.1 to 4.6 (Higgins and Spomer, 1976; Boone et al., 1998; Tjoelker et al., 1999a, 2001)]. Despite this, most species exhibit Q_{10} values that fall within a narrow range, especially when values are compared at a common measurement temperature interval. For example, Ivanova et al. (1989) found that the mean Q_{10} for leaf R in 34 temperate and arctic plant species was 2.45 (between measurement temperatures of 10 and, 20 °C) with the upper and lower 95% confidence intervals being 2.62 and 2.27, respectively. Similarly, the mean leaf Q_{10} of 65 species (15 °C midpoint) reported by Tjoelker et al. (2001) was 2.50, with the upper and lower 95% confidence intervals of 2.62 and 2.39, respectively. In a review with published values of 125 species, Larigauderie and Körner (1995) show that the majority of species exhibited leaf Q_{10} values between 2.0 and 2.5, with the overall mean being 2.3. Thus, despite large differences between the highest and lowest Q_{10} exhibited by contrasting species, most species show relatively similar Q_{10} values over a given measurement temperature range.

2. Measurement Temperature

There is ample evidence that the Q_{10} is not constant over the entire range of measurement temperatures in a temperature-response function, nor are values necessarily close to 2.0. Rather, Q_{10} typically declines with increasing measurement temperature (e.g., Fig. 2B; Wager, 1941; James, 1953; Forward, 1960; Ivanova et al., 1989; Palta and Nobel, 1989; Bruhn et al., 2002; Covey-Crump et al., 2002). Although the temperature dependence of the Q_{10} is not strictly linear; a linear relationship does adequately describe the mean temperature dependence of 56 species from the Arctic to the tropics over the range of 5 to 35 °C (Tjoelker et al. 2001; Fig. 4). Using data reported by Ivanova et al. (1989) for Arctic, sub-arctic, temperate and desert environments (Table 1), we have expanded the Tjoelker et al. (2001) leaf *R* Q_{10} data set to include a total of 116 species [see Atkin and Tjoelker (2003 for details]; the measurement temperature dependence of the mean Q_{10} of leaf *R* is described by:

$$\text{Leaf } Q_{10} = 3.09 - 0.043T \quad (8)$$

for species from the arctic, boreal, temperate, and tropical biomes which shared a common regression slope. The Q_{10} response for the 14 desert species (Table 1) to measurement temperature appeared to differ from that of the other biomes, perhaps owing to a higher optimum temperature.

A survey of published Q_{10} values (5 to 35 °C) of root *R* of 21 boreal and temperate tree species (Tjoelker et al., unpublished data) reveals that the temperature dependence of Q_{10} of root *R* is:

$$\text{Root } Q_{10} = 3.00 - 0.045T \quad (9)$$

The intercept and slope values in Eq. 9 are similar to those of Eq. 8, suggesting that the temperature dependence of the mean Q_{10} of root *R* may parallel that of leaf *R*. Literature evidence (Eqs. 8 and 9, Table 1) demonstrates that the Q_{10} of leaf and root *R* declines near-linearly with increasing short-term measurement temperature in a consistent manner across diverse plant taxa.

Equations 8 and 9 may accurately predict the mean temperature dependence of leaf and root Q_{10} (which decline linearly with increasing short-term measurement temperature), when the fits are applied to data from numerous species and growth environments. The general relationships (Eqs. 8 and 9) are probably useful in modeling plant/ecosystem responses to temperature and global warming, particularly over a moderate measurement temperature range (10–30 °C) below the optimum temperature of *R*.

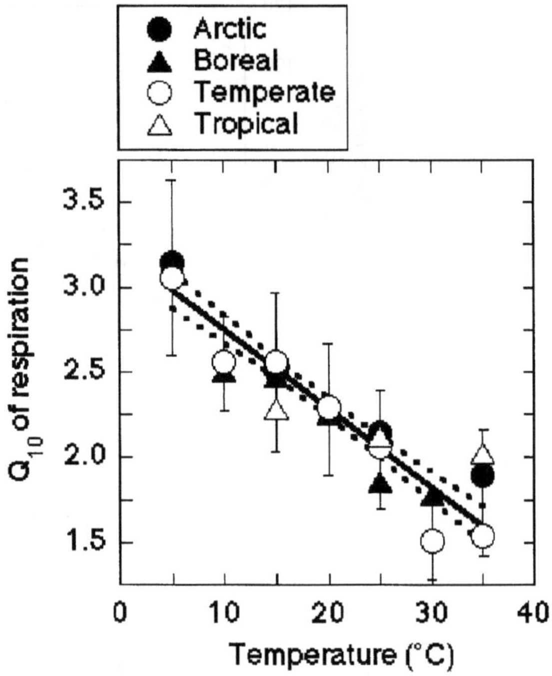

Fig. 4. Q_{10} of leaf respiration as a function of measuring temperature across biomes. Symbols are the mean Q_{10} of Arctic (16 species), boreal (6 species), temperate (31 species) and tropical (3 species) biomes (measured under field conditions) plotted against the midpoints in 5 °C classes. Error bars indicate ±1 SD of the class means of all observations. Data are of 56 species from 23 studies (from Tjoelker et al., 2001).

3. Growth Temperature

Although changing growth temperature can affect the Q_{10} of leaf and root *R* (e.g., schematically shown in Fig. 1; Wager, 1941; Atkin et al., 2000b; Covey-Crump et al., 2002; Loveys et al., 2003), the response to growth temperature is not consistent. Differences in Q_{10} (for a common measurement temperature interval) between plants grown in contrasting growth temperatures ultimately arises from changes in the shape, elevation (intercept), or the optimum temperature of the short-term temperature-response function as a result of temperature acclimation. Moreover, as we address below, growth-temperature effects on Q_{10} appear to differ between plants exposed to a new thermal regime for several days and leaves/roots

Table 1. Mean Q_{10} of leaf dark respiration for plant species of various biomes measured in their growth environments at a range of measurement temperatures (adapted from Ivanova et al., 1989; Table 4). *n* is the number of species sampled.

			Measurement temperature interval (°C)							
Biome	Location	n	5–15	10–20	15–25	20–30	25–35	30–40	35–45	45–55
Arctic	Wrangel Island	18	2.95	2.42	2.13	1.94	1.58	.	.	.
Sub-arctic	Hibbins Mts	17	.	1.86	1.96	1.92	1.73	1.58	1.39	.
Boreal	St. Petersberg	18	.	2.48	2.05	1.95	.	1.71	1.46	.
Temperate	Caucasus	19	.	.	2.44	2.17	1.91	1.77	.	.
Desert	Karakum	14	.	.	.	1.80	.	1.84	2.18	2.05

that develop under contrasting temperatures. Covey-Crump et al. (2002) found that the Q_{10} (between 15–23 °C) of *Plantago lanceolata* root *R* was greater at low measurement temperatures in plants exposed to 15 °C for 7 days than in plants kept at 23 °C. Similarly, Rook (1969) found that the Q_{10} (between 15–30 °C) of leaf *R* in *Pinus radiata* seedlings grown at 33/28 °C increased following 2 days exposure to 15/10 °C (rates of *R* measured at 15–30 °C increased significantly, whereas there was no change in *R* measured at 8 °C). Conversely, shifting of 15/10 °C grown plants to 33/28 °C resulted in a decrease of Q_{10} within two days (again, no change in *R* at 8 °C was observed). However, shifting from a growth temperature of 25/20 °C to 15/10 °C for 7 days had no effect on the Q_{10} of root *R* of several species (calculated using rates of *R* measured at 15 and 25 °C; Loveys et al., 2003). Moreover, no growth-temperature-dependent differences in Q_{10} values were apparent in plant leaves or roots that had *developed* at different temperatures (i.e. were not shifted; Tjoelker et al., 1999a,b; Loveys et al., 2003).

A factor that contributes to the differences in Q_{10} values among differing growth temperatures is the effect of growth temperature on the optimum temperature of *R* (e.g., Fig. 2A). The optimum temperature is the high temperature point where *R* stops increasing and Q_{10} equals 1.0. With further increases in temperature, *R* begins to decline, and the Q_{10} falls below 1.0. Typically, the optimum temperature of *R* is lower in cold-grown plants than in their warm-grown counterparts. For example, Covey-Crump et al. (2002) found that the optimum temperature of *Plantago lanceolata* roots was approximately 44 °C in plants grown and maintained at 23 °C, whereas it was approximately 31 °C in plants exposed to 15/10 °C for 7 days. Similarly, in the study by Semikhatova et al. (1992), the optimum temperature differed between plants growing at Wrangel Island (Arctic; 34–40 °C), Kola (Sub-Arctic; 36–45 °C), St. Petersburg (cold temperate; 41–47 °C) and in the Karakum Desert (40–54 °C).

Of the 20 species assayed at the Arctic site by Semikhatova et al. (1992), seven exhibited optimum temperatures between 34 and 35 °C, three between 36 and 37 °C, four at 38 °C and four at 40 °C. This demonstrates that considerable inter-species differences exist in the optimum temperature of leaf *R*. By extrapolating from Eq. 8, a mean temperature optimum of 50 °C for leaf *R* for all species and growth environments can be predicted. Whilst this is correct for comparisons of multiple species from several biomes, substantial variability exists in the optimum temperature of individual species and between contrasting environments.

Using data reported by Semikhatova et al. (1992), the temperature dependence of the Q_{10} of Arctic and temperate-grown plants (Fig. 2) was calculated. Over the 15 and 25 °C measurement temperature range, there was little difference in the Q_{10} of the Arctic and temperate-grown plants. However, below 15 °C, the Arctic plants exhibited substantially higher Q_{10} values than their temperate counterparts. Above 25 °C, Arctic plants exhibited lower Q_{10} values than the temperate plants. Thus, while growth temperature has little effect on Q_{10} values in the moderate temperature range (e.g., 15–25 °C; as found by Loveys et al., 2003), it does alter the overall temperature dependency of Q_{10}, when a broader range of measurement temperatures is considered.

4. Inter-biome Variation

When considering the extent to which Q_{10} values differ between biomes, three questions need to be considered. Firstly, are there in situ differences in the

Q_{10} of plants growing in separate biomes? Secondly, are such differences simply the result of Q_{10} values not being corrected for differences in air and soil temperatures among the biomes (an effect of measurement temperature)? And thirdly, are there inherent differences in the Q_{10} of plant species characteristic of the contrasting biomes?

In the literature analysis conducted by Tjoelker et al. (2001) where the Q_{10} values of 56 species were compared (Fig. 4), mean Q_{10} values that were not temperature-corrected (i.e. they were for the most part determined at measurement temperatures experienced by plants in the field) were 2.14 for tropical, 2.26 for temperate, 2.20 for boreal, and 2.56 for Arctic biomes (i.e. higher in the coldest climates). Why was this? Clearly, measurement-temperature-dependent changes in the Q_{10} will have played a role, because the mean Q_{10} values of the temperate, boreal and Arctic biomes in the Tjoelker et al. (2001) study were similar (2.31, 2.22 and 2.42, respectively) when Q_{10} values were compared at a common temperature. Moreover, Burton et al. (2002) found that the Q_{10} of fine root respiration of forest trees was similar in North American biomes when compared at a common temperature (Table 2).

Are there inherent differences in the Q_{10} of contrasting biomes? Criddle et al. (1994) found that temperature-corrected Q_{10} values varied with climate of origin among woody species, but not among annuals or herbaceous perennials. However, in a controlled-environment experiment, Larigauderie and Körner (1995) found no evidence that temperature-corrected leaf Q_{10} values depend on altitude of plant origin. There was also no systematic variation in leaf or root Q_{10} values among species characteristic of alpine, temperate and arid environments in the study by Loveys et al. (2003). Thus, differences in Q_{10} values among biomes are unlikely to reflect inherent differences in the species characteristic of each biome.

5. Leaves Versus Roots

A survey of the literature shows that Q_{10} values (typically within the 10–30 °C measurement temperature range) range between 1.4 to 4.2 for leaves (Azcón-Bieto, 1992; Larigauderie and Körner, 1995) and 1.1 to 4.6 for root *R* (Higgins and Spomer, 1976; Boone et al., 1998; Tjoelker et al., 1999a,b). This suggests that there is little difference in the overall range of Q_{10} values exhibited by leaves and roots. However, the mean Q_{10} values exhibited by mature leaves generally are higher than those of whole root systems/root segments. For example, Loveys et al. (2003) found that the mean Q_{10} of leaf *R* for several contrasting species (2.03–2.39, depending on the growth temperature) was substantially higher than that of root *R* (1.58–1.61), when determined over the 15–25 °C measurement temperature range.

One factor that needs to be considered when comparing published leaf and root Q_{10} values are the different growth conditions and range of measurement temperatures used in the different studies. Moreover, the nature of the tissue used in different studies needs to be considered. In most studies, rates of leaf *R* are measured using fully expanded individual leaves. In contrast, measurements of *R* in roots are made using whole root systems (e.g., Smakman and Hofstra, 1982; Bouma et al., 1997; Covey-Crump et al., 2002; Loveys et al., 2003) or root segments of differing age/function (e.g., Higgins and Spomer, 1976; Crawford and Palin, 1981; Sowell and Spomer, 1986; Weger and Guy, 1991; Zogg et al., 1996; Pregitzer et al., 1997, 1998; Burton et al., 2002). In whole root or shoot systems, the estimates of Q_{10} will depend on the proportion of the root or shoot system represented by immature and mature roots or leaves and the Q_{10} of each developmental stage. Although few data are available on the Q_{10} of roots of differing branching order, or size, recent studies with leaves have shown that the Q_{10} of fully expanded mature leaves is higher than that of immature leaves (Armstrong and Atkin, unpublished data). Q_{10} values of 1.5 for coarse woody roots and 2.0 for fine roots (< 2mm diameter) were reported in a *Pinus radiata* stand (Ryan et al., 1996). If the same pattern holds for roots then comparisons of leaf and root Q_{10} values should ideally be made on tissues of defined developmental age.

6. Other Factors

a. Seasonal Variation

Under field conditions, the Q_{10} of leaf *R* is higher in winter/autumn than in summer in evergreen species [(*Chamaecyparis obtusa*, (Paembonan et al., 1991); *Picea abies*, (Stockfors and Linder, 1998); *Eucalyptus pauciflora* (Atkin et al., 2000b) and *Pinus banksiana* (M. G. Tjoelker, J. Oleksyn and P. B. Reich, unpublished)], even when compared at the same measurement temperatures. However, *Eucalyptus pauciflora* showed little seasonal variation in Q_{10} over much of

Table 2. Q_{10} values for fine root respiration of forest tree species of North American biomes. Data are from Burton et al. (2002). LTER refers to Long-term Ecological Research site (US National Science Foundation).

Biome	Location	Species/stand type	Mean Annual Temperature (°C)	Q_{10} (6–24°C)
Boreal	Bonanza Creek LTER, Alaska, USA	*Populus balsamifera*	–3.3	2.4
		Picea glauca		2.9
Cold-temperate	Michigan, USA	*Acer saccharum*	3.8	2.7
		Pinus resinosa plantation		3.0
Montane cool temperate	Coweeta LTER, North Carolina, USA	Mixed hardwoods	9.4	2.4
		Quercus-Carya	11.1	3.1
		Liriodendron tulipifera	12.7	2.6
Semi-arid	Sevilleta LTER, New Mexico, USA	*Pinus edulis*	12.7	2.6
		Juniperus monosperma	12.7	2.4
Warm-temperate	Georgia, USA	Mixed *Quercus*	16.5	2.4
	Florida, USA	*Pinus elliottii* plantation	20.0	2.5

the year (Atkin et al., 2000b). Q_{10} values were only greater on days when daily average and minimum air temperatures were below 6 °C and –1 °C, respectively. In contrast, Q_{10} of tree branch and bole respiration in a *Fagus sylvatica* stand was found to be relatively constant (around 1.7) throughout the course of the year (Damesin et al., 2002).

b. Irradiance

There is growing evidence that the Q_{10} of leaf *R* is lower in the light than it is in darkness. For example, in a study by Atkin et al. (2000c) using *Eucalyptus pauciflora*, the average Q_{10} of leaf *R* (over the 6 to 25 °C range) in darkness was 2.21. Q_{10} decreased, however, when leaves were exposed to irradiances greater than 12 μmol photons m^{-2} s^{-1}, with average Q_{10} values over the same temperature range being 1.61 and 1.57 at 800 and, 2000 μmol photons m^{-2} s^{-1}, respectively. Recently, it was found that light also reduced the Q_{10} of leaf *R* in three *Plantago* species (*P. lanceolata, P. major* and *P. euryphylla*) (Atkin OK, Scheurwater I and Pons TL, in preparation). For example, in *P. major* the average Q_{10} values of leaf *R* in darkness over the 6 to 34 °C range were 2.16, 1.79 and 2.08 for plants grown at 13, 20 and 27 °C, respectively. The corresponding Q_{10} values under saturating irradiance were 1.45, 0.91 and 1.15. In *Fagus sylvatica,* Bruhn (2002) also found Q_{10} to be between 1.08 and 1.34 between 50 and 1400 μmol photons m^{-2} s^{-1} without any irradiance dependency of Q_{10} compared with a Q_{10} of 2.36 in darkness in the temperature range of 15–25 °C.

c. Water and Nutrient Availability

In the short term, water stress results in a reduction in *R* of leaves and roots. Root *R* declines during drought (Bryla et al., 1997, 2001; Burton et al., 1998). Under field conditions the relationship between soil drying and root *R* is often further complicated with occurrence of increased soil temperatures during drought. However, in a greenhouse study in which roots of citrus trees were maintained at constant temperatures, root *R* declined with decreasing soil water content over a 10-day drying period (Bryla et al., 2001). In addition, drought-induced reductions in root *R* were greater in warmer soils (25 and 35 °C) than in a cooler soil (15 °C) as soil water contents fell below 6%. Comparing the proportional differences in *R* at the three soil temperatures suggests that Q_{10} declined concurrently with soil drying. Moreover, this study demonstrated that root *R* acclimated to both soil moisture and soil temperatures (> 23 °C).

Root *R* is correlated with root N concentration (Pregitzer et al., 1998; Reich et al., 1998a,b; Eissenstat et al., 2000). In a study of *Acer saccharum* stands across sites differing in soil temperature and nitrogen availability, root *R* increased with higher net N mineralization rates and root N concentrations; whereas the Q_{10} of root *R* (5 to, 20 °C) did not differ among sites across the climate gradient (Zogg et al., 1996). Consequently, root *R* in the field has the potential to change as a complex function of soil temperature, moisture, and nitrogen availability. In addition, root *R* is also likely to be coupled to canopy carbon assimilation and internal source-sink relationships, which are also temperature-dependent processes (Pregitzer et al., 2000). There is evidence that Q_{10} is linked to carbohydrate status, especially at higher temperatures (see Section III) where substrate-limitation may begin to control respiration. We suggest that environmental factors that result in a depletion of non-structural carbohydrates in roots (e.g., drought, high temperatures) will likely be associated with a reduction in Q_{10}. However, this hypothesis remains to be tested in the field.

Although leaf *R* declines in response to short-term water deficits, less is known concerning the effects of long-term water deficits on respiratory function and Q_{10}. In a study of three deciduous tree species growing in two sites of contrasting water availability, Q_{10} differed among species (1.5 to 2.1) and was lower at the drier than the wetter site (Turnbull et al., 2001). In addition, both area- and mass-based leaf *R* were higher and light-saturated rates of photosynthesis were lower at the drier than at the wetter site, suggesting that leaf-level carbon gain was reduced at the dry site. These findings suggest that longer-term adjustments in leaf structure under low soil water availability, particularly increased leaf N, may result in increased *R*, and a lower temperature-sensitivity (Q_{10}) of dark respiration.

d. Leaf position Within a Canopy

Recently, studies have shown that there is considerable within-canopy variability in the Q_{10} of leaf *R* in some trees (Bolstad et al., 1999; Griffin et al., 2002a), with leaves in the lower part of the canopy exhibiting higher Q_{10} values than leaves in the upper canopy (Griffin et al., 2002a).

III. Mechanisms Responsible for Variation in the Q_{10} of Respiration

Section II reviewed to which extent the Q_{10} of *R* varies among and within plant tissues. Here, we discuss the factors that could contribute to this variability. We begin by first considering the impact that changes in temperature have on the mechanisms that control respiratory flux (Fig. 5). Then this information is used to assess what mechanisms are responsible for the variability in the Q_{10} of *R* and the underlying factors responsible for thermal acclimation of *R*. Much of the variability in Q_{10} values can be explained by the extent to which respiratory flux at any standard temperature is limited by enzymatic capacity, substrate supply and/or adenylates (Atkin and Tjoelker, 2003). Critically, *R* appears to become limited by respiratory capacity in the cold, whereas at higher temperatures respiratory flux is regulated by the availability of substrates and/or demand for respiratory energy. An increase in the control exerted by enzyme capacity in the cold results in the shape of *R* versus substrate concentration plots changing with measurement temperature (Fig. 6). This explains why Q_{10} is dependent on measuring temperature (Figs 2 and 4, Table 1 and Eqs. 8 and 9), and why changes in the concentration of key respiratory intermediates (e.g., reduced ubiquinone) could affect Q_{10}.

A. Impact of Temperature on the Mechanisms that Control Respiratory Flux

It is often assumed that, in short-term experiments, the temperature dependence of *R* reflects the impact of temperature on the maximum activity of respiratory enzymes. However, respiratory flux is not determined simply by potential enzyme activity. Rather, the control of respiratory flux can be shared between the availability of substrates, the demand for respiratory products (e.g., adenylates) and the capacity of the respiratory enzymes. The fact that *R* is temperature sensitive could, therefore, reflect temperature-mediated changes in control of enzymes, substrate and/or adenylates (in particular the ratio of ATP to ADP and the concentration of ADP per se; (Hoefnagel et al., 1998) over respiratory flux (Fig. 5). In turn, adenylate control over *R* will depend on the energy requirements of plant growth, maintenance and/or ion uptake (Veen, 1980; Van der Werf et al., 1988).

At moderate temperatures (e.g., 25 °C) it is unlikely that the capacity of respiratory enzymes alone limits CO_2 release and O_2 uptake (Day and Lambers, 1983; Wiskich and Dry, 1985). This is because rates of *R* measured in isolated mitochondria in vitro with saturating substrates are greater than those measured in vivo (Day and Lambers, 1983; Wiskich and Dry, 1985). Moreover, O_2 uptake in isolated mitochondria in the presence of ADP is rarely substrate-saturated

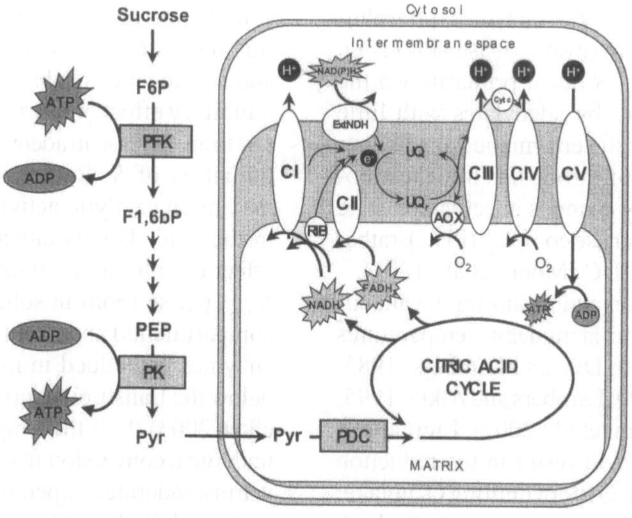

Fig. 5. Schematic representation of the plant respiratory system, including glycolysis (cytosol and plastids), the citric acid cycle (matrix of mitochondria) and the electron-transport chain (inner mitochondrial membrane). Sites of adenylate control in glycolysis are shown. F6P: fructose 6 phosphate; F1,6bP: fructose 1,6 bisphosphate; PFK, phosphofructokinase; PK: pyruvate kinase; PDC: pyruvate dehydrogenase complex; PEP: phosphoenolpyruvate; Pyr: pyruvate; CI: complex I; CII: complex II (succinate dehydrogenase); CIII: complex III (cytochrome bc_1 complex); Cyt *c*: cytochrome *c* oxidase; CIV: complex IV (cytochrome *c* oxidase); CV: ATP synthase; ExtNDH: external NAD(P)H dehydrogenase(s); RIB: internal, rotenone-insensitive NADH dehydrogenase bypass; AOX: alternative oxidase; UQ and UQ_r: reduced and oxidised ubiquinone, respectively.

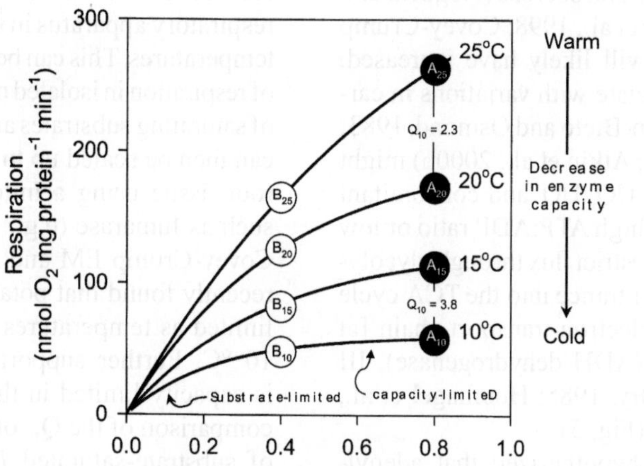

Fig. 6. Effect of measuring temperature on substrate dependence of respiration. In this example, substrate availability is represented by the redox state of the ubiquinone pool (UQ_r/UQ_t). At low UQ_r/UQ_t values, rates of O_2 uptake are limited by substrate availability (i.e. reduced ubiquinone), regardless of the measuring temperature. In measurements made in the cold (e.g., 10 °C), *R* becomes substrate-saturated at high UQ_r/UQ_t values (i.e. *R* becomes limited by enzymatic capacity at high UQ_r/UQ_t in the cold). In contrast, *R* is not limited by enzymatic capacity per se at high temperatures (e.g., 25 °C). Symbols: A_{25}, A_{20}, A_{15} and A_{10} represent rates of *R* at a UQ_r/UQ_t value of 0.8 for measurements made at 25, 20, 15 and 10 °C. Similarly, B_{25}, B_{20}, B_{15} and B_{10} represent rates of *R* at a UQ_r/UQ_t value of 0.4 for measurements made at 25, 20, 15 and 10 °C. Q_{10} values (between 10–15 and, 20–25 °C) were calculated for UQ_r/UQ_t values of 0.4 and 0.8 using Eq. 1 (Section II.A.3). At $UQ_r/UQ_t = 0.4$, Q_{10} values between 10–15 and, 20–25 °C were 2.86 and 1.70, respectively. Data are rates of respiration in the presence of ADP for soybean cotyledon mitochondria oxidizing succinate (Atkin et al., 2002).

when measured at moderate temperatures (Fig. 6; Atkin et al., 2002). Although glycolytic flux is regulated by the activity of two key enzymes, phosphofructokinase (PFK) and pyruvate kinase (PK) (Fig. 5), changes in the rate of *R* occur primarily via the control of these enzymes by adenylates with little control being exerted by the enzymatic capacity per se. Similarly, regulation of carbon flux into the mitochondria is controlled by changes in activation state of the pyruvate dehydrogenase complex (PDC), rather than by the capacity of PDC (Moore et al., 1993).

In most cases, adenylates and/or substrate availability regulate respiratory flux at moderate temperatures (Saglio and Pradet, 1980; Day and Lambers, 1983; Williams and Farrar, 1990; Lambers and Atkin, 1995; Geiger et al., 1998; Atkin, et al., 2002). Limitations in substrate availability will result in the reduction state of the ubiquinone (UQ) pool limiting O_2 uptake; reduced UQ (UQ_r) is the common substrate for both the cytochrome pathway [Cyt path; i.e. Complex III, cytochrome *c* and cytochrome *c* oxidase (Complex IV)] and alternative oxidase (AOX) in plant mitochondria (Fig. 5). Changes in the proportion of total UQ that is in the reduced form (i.e. UQ_r/UQ_t) can result in concomitant changes in the rate of mitochondrial O_2 uptake (Moore et al., 1988; Day et al., 1996). In tissues where O_2 uptake is stimulated by the addition of substrates such as glucose and sucrose (Noguchi and Terashima, 1997; Geiger et al., 1998; Covey-Crump et al., 2002), UQ_r/UQ_t will likely have increased. Variations in *R* that correlate with variations in carbohydrate content (Azcón-Bieto and Osmond, 1983; Azcón-Bieto et al., 1983; Atkin et al., 2000b) might also reflect variations in UQ_r/UQ_t and concomitant changes in O_2 uptake. A high ATP:ADP ratio or low ADP concentration will restrict flux through glycolysis (particularly at PK), entrance into the TCA cycle (via the PDC) and the electron-transport chain [at Complexes I (internal NADH dehydrogenase), III and IV] (Wiskich and Dry, 1985; Hoefnagel et al., 1998; Loef et al., 2001) (Fig. 5).

Atkin et al. (2000a) hypothesized that adenylate limitation of respiratory flux would increase as temperatures decreased, because processes such as ion uptake, growth and maintenance (which require respiratory ATP) are likely to slow down in the cold. Moreover, membranes become less fluid at low temperatures, which should result in respiration being more coupled to adenylate limitation. However, Covey-Crump et al. (2002) recently found that neither uncoupler (which removes adenylate restriction to respiratory flux) nor exogenous substrates increased O_2 uptake by roots below 10 °C, whereas substantial stimulation was observed at higher temperatures (Q_{10} was therefore higher in the presence of uncoupler and/or exogenous substrate). This suggests that the inhibitory effect of low temperature on plant *R* is not due to an increase in adenylate control and/or substrate limitation of *R*. Rather, *R* is likely limited by the maximum catalytic activity of respiratory enzymes in the cold. This is either because of the inhibitor effect of cold on the potential enzyme activity (i.e. V_{max}) per se (both in soluble and membrane-bound compartments) and/or limitations on the function of enzymes embedded in membranes at temperatures below the transition temperature (T_m; Atkin and Tjoelker, 2003). T_m is the temperature where membranes undergo a conversion from a fluid state (that exists at warm-moderate temperatures) to a gel-like state (that exits in chilled membranes). T_m as high as 15–20 °C have been reported for some species, with increases in the unsaturated fatty acid content of membranes decreasing T_m (Berry and Raison 1981). Below T_m, the function of membrane-bound respiratory enzymes is likely to be reduced owing to a gel-like state of the membrane.

To determine whether respiratory enzyme capacity limits respiratory flux at low temperatures, data are needed on the maximum potential flux of the respiratory apparatus in intact leaves and roots at low temperatures. This can be obtained via measurements of respiration in isolated mitochondria, in the presence of saturating substrates and ADP. Mitochondrial rates can then be scaled up to represent rates per gram of root tissue using a mitochondrial marker enzyme such as fumarase (e.g., Van Emmerik et al., 1992). Covey-Crump EM and Atkin OK, (in preparation) recently found that potato leaf *R* becomes capacity limited as temperatures decrease to between 5 and 10 °C. Further support for the hypothesis that *R* is capacity limited in the cold can be obtained via comparison of the Q_{10} of *R* of in vivo rates with that of substrate-saturated *R* in isolated mitochondria (which can be scaled up to the intact organ level via knowledge of the concentration of mitochondrial protein per unit mass) (Millar et al., 1998; Atkin et al., 2002). The average Q_{10} between 10 and 25 °C for in vivo respiration (1.9) is lower than the Q_{10} for substrate-saturated isolated mitochondria (2.4) in soybean cotyledons (Atkin et al., 2002). As a result, the respiratory temperature-response curves for in vivo rates of maximum potential respiratory flux and actual respiratory flux converge as the temperature decreases (i.e. capacity limits flux in the cold).

There is also evidence that enzymatic capacity limits R in isolated mitochondria per se. When assessing the effect of measuring temperature on substrate dependence of mitochondrial O_2 uptake, Atkin et al. (2002) found that O_2 uptake (in the presence of ADP) became substrate-saturated at high substrate availability (as shown by the high UQ_r/UQ_t ratios; (i.e., R becomes limited by enzymatic capacity at high UQ_r/UQ_t in the cold; Fig. 6). In contrast, R is not limited by enzymatic capacity per se at high temperatures (e.g., 25 °C; Fig. 6). Although all enzymes are likely to be slowed in the cold, recent evidence suggests that the NADH dehydrogenases (internal and external) (Dufour et al. 1996; Atkin et al., 2002) as well as the Complex V (ATP synthase) (Svensson et al. 2002) are particularly affected (Fig. 5).

Taken together, it seems likely that at moderate temperatures, the respiratory flux is regulated by the supply of substrates and/or adenylates (i.e. R is not limited by enzyme capacity), whereas in the cold R is limited by maximum activity of respiratory enzymes. A short-term decrease in temperature therefore results in control over respiratory flux shifting from regulation by adenylates/substrates *to* enzyme capacity (e.g., as shown in Fig. 6). Below, we discuss how this shift in control can be used to explain why the Q_{10} of R varies with measurement temperature.

B. Why does the Q_{10} Vary?

Variations in the Q_{10} of R could reflect one or more of the following: variations in (1) the temperature sensitivity of respiratory enzymes; (2) the extent to which R is limited by adenylates and/or substrate availability; and (3) the extent to which increases in temperature result in a transition from limitations in maximum enzyme activity to regulation by adenylates/substrate supply. Most reactions catalysed by enzymes exhibit a higher Q_{10} when substrates are saturating (Berry and Raison, 1981); thus, the Q_{10} of R is greater under conditions where respiratory flux is limited by enzymatic capacity than when R is limited by substrate supply (Atkin et al., 2002; Atkin and Tjoelker 2003). Conversely, transition from enzymatic control to limitations imposed by substrate supply (or adenylates) should be associated with a decline in the Q_{10}.

1. Measurement Temperature and Substrate Dependence of the Q_{10}

Is there experimental evidence to support the hypothesis that Q_{10} values vary depending on the extent to which enzymatic capacity limits respiratory flux? In Fig. 6, the UQ_r/UQ_t value required to saturate O_2 uptake increased with rising temperature, indicating that R was less limited at high measuring temperatures by the capacities of the Cyt pathway and AOX (Atkin et al., 2002; Atkin and Tjoelker, 2003). As a result, for a standard UQ_r/UQ_t value the relative difference between R at 10 and 15 °C was greater than that between, 20 and 25 °C (i.e. Q_{10} was greater in the low measurement-temperature range; Fig. 6). Underpinning this increase in Q_{10} in the low measurement temperature range was the fact that R was limited by enzymatic capacity in the cold, but not at higher temperatures.

Q_{10} values of root R are greater in the presence of exogenous substrates and/or uncoupler than in their absence (Covey-Crump et al., 2002), and the Q_{10} of R is often greater in leaves and roots containing higher concentrations of soluble carbohydrates (Azcón-Bieto et al., 1983; Covey-Crump et al., 2002). Moreover, the Q_{10} of O_2 uptake in isolated mitochondria increases with increasing reduction level of the ubiquinone (UQ) pool (Fig. 6; Atkin et al. 2002). At low substrate availability (e.g., a UQ_r/UQ_t value of 0.4), the Q_{10} was 2.85 between 10 and 15 °C; this increased to 3.80 at a UQ_r/UQ_t value of 0.80 (Fig. 6). These findings can be explained as follows. Typically, the V_{max} increases with measurement temperature (up to a maximum rate); however, to take advantage of this increased potential rate, sufficient substrate and/or ADP supply must be available. If substrate availability and/or adenylates limit R, then increases in measurement temperature will have less stimulatory effect on flux (Fig. 6). A further factor contributing the substrate dependence of the Q_{10} of R is the effect of temperature on the affinity constant (K_m) of the enzymes catalyzing the reaction in question [see Berry and Raison (1981) for further details]; this is shown by the fact that the rate of $R = (V_{max} * C)/(K_m + C)$ where C is concentration of available substrate (i.e. reduced UQ when considering O_2 uptake, as in Fig. 6). At very low substrate availability, the temperature dependence of R will reflect the quotient of the separate temperature dependencies of the V_{max} and the K_m (Berry and Raison, 1981). Thus, if the Q_{10} of the V_{max} and the K_m of the terminal oxidases are similar [as is the case for the CO_2 fixation by Rubisco in the chloroplast (Berry and Raison, 1981)] then changes in temperature will have little effect on O_2 uptake at low substrate concentrations but will substantially increase *R* at higher substrate concentrations. As a result, Q_{10} values of O_2 uptake (calculated over a given measurement temperature range) increase as substrate availability increases, largely as a result of temperature-dependent changes in the K_m of the terminal oxidases and *R* being less limited by enzymatic capacity at high measuring temperatures than in the cold (Fig. 6). The substrate concentration required to saturate *R* therefore rises with increasing temperature; consequently, the relative difference in rates of *R* at low and high measuring temperatures (i.e. Q_{10}) increases with increasing substrate availability. An additional factor that accentuates the effect of substrate availability on the Q_{10} of total O_2 uptake (i.e. O_2 uptake by the cytochrome and alternative pathways) is the stimulatory effect of high UQ-pool reduction levels on the activation state of the alternative oxidase (Hoefnagel and Wiskich 1998), particularly at high measuring temperatures (Atkin et al., 2002). Taken together, the Q_{10} of plant *R* is indeed higher whenever respiratory flux is limited by enzymatic capacity than when *R* is limited by substrate supply. Conversely, transition from enzymatic control to limitations imposed by substrate supply (or adenylates) is associated with a decline in Q_{10}.

Intuitively, one might expect variations in UQ_r/UQ_t to be coupled to variations in the availability of NADH and/or succinate, which in turn could reflect variations in the concentration of soluble substrates. If correct, variations in Q_{10} would be expect to be coupled to variations in the concentration of soluble substrates. As stated above, several studies have indeed reported a positive relationship between the concentration of soluble sugars and the Q_{10} of plant *R* (Berry and Raison, 1981; Azcón-Bieto et al., 1983; Covey-Crump et al., 2002). Moreover, addition of exogenous substrate increased O_2 uptake (presumably via increases in UQ_r/UQ_t) and the Q_{10} of *R* in *Plantago lanceolata* roots (Covey-Crump et al., 2002). However, no relationship was found between the concentration of soluble carbohydrates and the Q_{10} of *R* in three other studies (Breeze and Elston, 1978; Atkin et al., 2000b; Griffin et al., 2002a). Also, Millenaar et al. (2000) found no link between changes in the concentration of soluble sugars and changes in UQ_r/UQ_t in *Poa annua* roots. Thus, there is currently insufficient evidence for us to state that variations in Q_{10} are always linked to sugar-dependent variations in UQ_r/UQ_t (see Section 3.B.2 for a discussion of the other factors that may contribute to variations in the substrate dependence of the Q_{10}).

One factor that could contribute to the variation in temperature dependence of Q_{10} (see above) is the effect of measuring temperature on UQ_r/UQ_t per se. Recently, Covey-Crump EM, Bykova N, Gardeström P and Atkin OK (in preparation) found that in vivo UQ_r/UQ_t values (in potato leaf protoplasts) were relatively constant between 22 and 32 °C (Fig. 7). However, in vivo UQ_r/UQ_t increased substantially at measuring temperatures below 22 °C, reaching their maximum at 10–15 °C, before declining again at 5 °C (Fig. 7). Given that Q_{10} depends both on measurement temperature and on UQ_r/UQ_t (see above) the following is suggested: between 22–32 °C, the combination of high measurement temperatures and a low in vivo UQ_r/UQ_t value would contribute to a lower Q_{10} in intact tissues over this temperature range. Conversely, the combination of relatively low measurement temperatures and a high in vivo UQ_r/UQ_t value between 10–15 °C would contribute to a higher Q_{10} in intact tissues over this temperature range. At temperatures below 10 °C, the decline in UQ_r/UQ_t values in vivo would partially offset the stimulatory effect of low measurement temperatures on the observed Q_{10}. Thus, variations in UQ_r/UQ_t in vivo and temperature-dependent variations in the shape of *R* versus UQ_r/UQ_t plots can contribute to

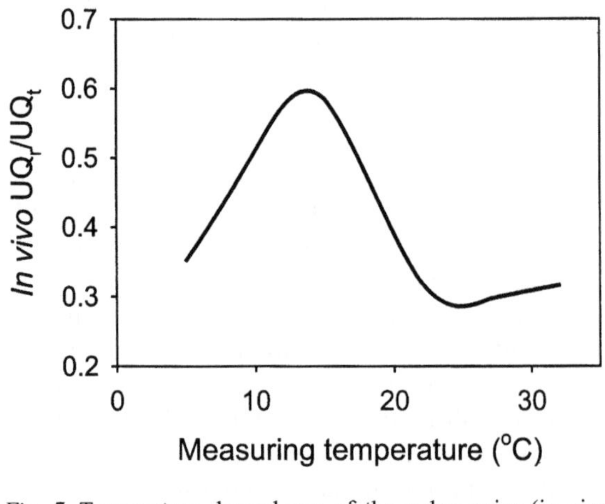

Fig. 7. Temperature dependence of the redox poise (in vivo UQ_r/UQ_t) of extracted UQ from potato protoplasts (Covey-Crump et al., in preparation). Protoplasts were fractionated to remove chloroplasts prior to determination of in vivo UQ_r/UQ_t using HPLC separation chromatographs of organic extracts.

the observed variations in measuring temperature dependence of Q_{10}.

2. Adenylates and Variable Q_{10} Values: Direct and Indirect Effects

Variations in adenylates can directly affect Q_{10}. This is because the Q_{10} of the cytochrome pathway in the presence of saturating ADP (in isolated mitochondria) is lower than that in the absence of ADP (2.55 versus 2.40; Atkin et al., 2002). In the absence of ADP, control over flux through the cytochrome pathway is mediated by the leak of protons across the inner mitochondrial membrane (Diolez et al., 1993). In contrast, control over flux through the cytochrome pathway is distributed among several steps in the presence of ADP, including the dehydrogenases, proton leak, the ATPase, CIII, Cyt *c*, and CIV (Douce and Neuburger, 1989). The Q_{10} of the proton leak is likely to be different from temperature dependence as electron flux via the cytochrome pathway (Atkin et al., 2002).

In addition to the direct effects of adenylate restriction on the Q_{10}, adenylates may indirectly affect Q_{10} via changes in UQ_r/UQ_t. In the study by Covey-Crump et al. (in preparation) cited above, concurrent measurements of UQ_r/UQ_t and adenylates showed that temperature-dependent changes in UQ_r/UQ_t (Fig. 7) were negatively correlated with changes in the ratio of ATP to ADP. This suggests that temperature-mediated increases in adenylate restriction (of glycolysis and Complex I) decreased the rate of substrate input into the UQ pool. Although adenylate restriction would also be exerted on the cytochrome pathway (Complexes III and IV) (which should increase UQ_r/UQ_t), adenylates would have no effect on the AOX per se (although adenylate-mediated changes in UQ_r/UQ_t may result in an increase in AOX activity); taken together, we suggest that adenylate restriction would have less effect on the rate of UQ oxidation than on UQ reduction (with the result that UQ_r/UQ_t was lowest under high ATP:ADP conditions). The temperature-dependent changes in ATP:ADP reflect changes in the balance between mitochondrial ATP synthesis and cellular turnover of ATP to ADP; presumably, these two groups of reactions have different temperature coefficients over the 5–22 °C temperature range in the above-mentioned study.

3. Alternative Oxidase and Variations in Q_{10}

During the 1980s and early 1990s, several authors concluded that the Q_{10} of the AOX was lower than that of the Cyt pathway (Kiener and Bramlage, 1981; McNulty and Cummins, 1987; Stewart et al., 1990). As a result, switching between the AOX and Cyt pathway could affect the Q_{10} of the overall O_2 consumption in intact tissues. However, more recent studies have not supported this conclusion. For example, Weger and Guy (1991) reported that the AOX of *Picea glauca* roots is highly sensitive to changes in temperature. Moreover, in a study using the ^{18}O-fractionation method, Gonzàlez-Meler et al. (1999) found that the Q_{10} of the AOX is similar to the Q_{10} of the Cyt pathway in mung bean leaves and hypocotyls or soybean cotyledons. Similarly, the Q_{10} of the AOX was not necessarily lower than that of the Cyt pathway in soybean cotyledon mitochondria (Atkin et al., 2002). In the presence of pyruvate (which activates the AOX; (Millar et al., 1993; Umbach et al., 1994) the Q_{10} of the AOX (2.61) is similar to/greater than the Cyt pathway; in the absence and presence of ADP, the Q_{10} of the Cyt pathway is 2.55 and 2.40, respectively (Atkin et al., 2002). Therefore, it seems unlikely that the engagement of the AOX can reduce the Q_{10} of plant *R* per se.

Engagement of the non-phosphorylating AOX might, however, affect the Q_{10} of plant *R* via its effect on adenylate restriction of mitochondrial electron transport. As discussed above, adenylate restriction can have direct and indirect effects on the Q_{10} of plant *R*. Thus, variations in engagement of the AOX might indirectly affect Q_{10} via its effects on adenylate restriction. Given that engagement of the AOX is highly variable in plant tissues [e.g., varies with age (Millar et al., 1998), growth temperature (Gonzàlez-Meler et al., 1999), phosphate supply (Gonzàlez-Meler et al., 2001), inherent relative growth rate (Millenaar et al., 2001), irradiance (Ribas-Carbó et al., 2000b; Millenaar et al., 2000; Noguchi et al., 2001), infection by *Pseodomonas syringae* (Simons et al., 1999)], AOX-mediated changes in adenylate control may, in part, explain why Q_{10} values are so variable.

4. Growth-Temperature-Dependent Variations in the Q_{10}

As discussed in Section II.B.3, measurement-temperature-corrected Q_{10} values are often higher in cold-grown plants than in their warm-grown counterparts, particularly at low measurement temperatures (e.g., Fig. 2B). One possible explanation is that growth temperature affects the level of reduction of the UQ pool (e.g., UQ_r/UQ_t might be higher in cold-grown plants), which in turn affects the Q_{10} of *R*, particularly at low measurement temperatures (Fig. 6). To our knowledge, no study has yet investigated the effect of growth temperature on in vivo UQ_r/UQ_t values. Nevertheless, cold-grown tissues often exhibit substantially higher concentrations of soluble carbohydrates than their warm-grown counterparts (Warren Wilson, 1966; Mooney and Billings, 1965; Hurry et al., 1994); such differences might be linked in some cases to differences in steady state UQ_r/UQ_t values (although not in all; see Section II.B.1).

Growth-temperature-mediated changes in Q_{10} also result from changes in the optimum temperature of *R* (see Section II.B.3; Fig. 2). This is likely to result from changes in the amounts and/or isoforms of individual enzymes, and changes in the structure and composition of membranes [e.g., degree of unsaturation of fatty acids; (Pearcy, 1978; Osmond et al., 1982)]. For example, Marie and Cummins (1984) found that the arctic herb, *Saxifraga cernua*, grown at 10 °C, had more linolenic acid and less linoeic, oleic and palmitic acids than leaves from plants grown at 20 °C.

5. Seasonal Variations in Q_{10} and Photoinhibition

Atkin et al. (2000b) speculated that the higher Q_{10} values of leaf *R* observed in *Eucalyptus pauciflora* in winter might be associated with the onset of photoinhibition [the ratio of variable to maximal fluorescence (F_v/F_m) decreases in winter, indicating a reduction in the quantum yield of photosynthesis (Blennow et al., 1998; Lambers et al., 1998)]. This hypothesis was tested in a recent study that followed changes in F_v/F_m and Q_{10} of *E. pauciflora* during recovery from cold-induced photoinhibition (Atkin OK, McDowall W, Egerton J and Ball MC, unpublished data). Severely photoinhibited plants (grown in pots that were sunken into the surrounding soil) were transferred from a field site in winter to a warm glasshouse. On the day of the transfer, the winter-sampled, field-grown plants had a mean F_v/F_m value of 0.54 ± 0.02 and Q_{10} of 2.48 ± 0.15. Within 5 days of recovery in the glasshouse, F_v/F_m values had recovered to 0.80 ± 0.02; however, the Q_{10} of *R* remained high (2.71 ± 0.07). Q_{10} values of the field-grown plants did not decrease to values similar to those exhibited by plants grown and maintained in the glasshouse (1.7–1.9) until after 30 days recovery. Thus, while Q_{10} values in winter-grown *Eucalyptus pauciflora* plants correspond with low F_v/F_m values, changes in F_v/F_m and Q_{10} can occur independently of each other. Moreover, induction of cold-induced photoinhibition under laboratory conditions was not coupled to an increase in the Q_{10} of leaf *R*. Thus, there is no evidence of a direct functional relationship between degree of photoinhibition and the Q_{10} of leaf *R*.

6. Irradiance Dependence of the Q_{10} of Leaf R

Why is the Q_{10} of leaf *R* lower in the light than darkness (Atkin et al., 2000c)? In recent years, several authors have proposed that light inhibition of *R* is due to inactivation of key enzymes that control substrate input into the mitochondria, such as the pyruvate dehydrogenase complex (PDC) and NAD^+-malic enzyme (ME) (Atkin et al., 1998a,b, 2000d; Padmasree and Raghavendra, 1998). PDC and ME are both rapidly inactivated by light (Hill and Bryce, 1992; Budde and Randall, 1990), with the timing of inactivation of ME (Hill and Bryce, 1992) and PDC (Budde and Randall, 1987) closely mirroring the time taken for light to inhibit *R* (Atkin et al., 1998a,b). One possibility is that exposure to low temperatures might accentuate the inhibitory effect of light on PDC and ME activity.

C. Mechanisms Underlying Other Changes in the Q_{10}

Our analysis shows that any treatment that alters the availability of respiratory substrate (e.g., UQ_r/UQ_t) could affect Q_{10}. For example, the tendency towards greater Q_{10} in leaves than in roots might reflect higher UQ_r/UQ_t in leaves (due to greater availability of respiratory substrates than in roots). Similarly, the higher Q_{10} observed in mature leaves, compared to immature leaves, could reflect the fact that the photosynthetic apparatus is more functional in former, with the result that the availability of respiratory substrates and UQ_r/UQ_t values are likely to be higher in mature leaves. Finally, the higher Q_{10} values in winter might reflect the effect of cold on enzymatic capacity, with *R* only becoming capacity limited in the coldest time of the year. Presumably, the lack of change in Q_{10} through the rest of the year (Atkin et al., 2000b), despite seasonal changes in temperature, would reflect long-term adjustments in the shape of the R versus UQ_r/UQ_t curves. Clearly, further work is needed to test these hypotheses.

IV. Long-term Changes in Temperature: Acclimation

This section begins with the discussion of what is meant by the term acclimation; with the demonstration that acclimation involve changes in flux that are mediated by changes in the regulation of existing respiratory enzymes and/or mediated by changes in gene expression/enzyme capacity. Then, methods of quantifying the degree of thermal acclimation of R are compared; A new method for quantifying acclimation is proposed, using information on Q_{10} and long-term temperature sensitivity of R (Eq. 12) in Section IV.B.2. Variations in the degree of acclimation among and within plant species are then discussed, as is the component of the daily temperature regime (e.g., daily maximum, minimum or mean) to which R acclimates. The coupling between the degree of acclimation and development is highlighted; in most cases, maximal acclimation requires that new leaves and/or roots be developed following a change in growth temperature. Section V presents a distinction between two types of acclimation, which differ in the extent to which the intercept of short-term temperature response functions is affected by growth temperature.

A. Defining Acclimation: Is It Simply a Change in Overall Respiratory Flux?

Temperature acclimation is commonly defined as a change in the shape of the short-term temperature-response curve of respiratory CO_2 release and/or O_2 uptake, in response to a long-term change in growth temperature (e.g., Fig. 1). The majority of examples used in this review fit this definition. However, acclimation can also be used to describe other effects of long-term changes in temperature, such as on membrane properties and freezing tolerance (Prasad et al., 1994; Uemura et al., 1995), biomass allocation/anatomy (Stefanowska et al., 1999) and/or targeted increases in specific proteins (e.g., Ribas-Carbó et al., 2000a). It also applies to cases where changes in growth temperature alter partitioning between different respiratory pathways, without changes in overall respiratory flux (Fig. 8). For example, while growth of maize at low temperature results in an increase in AOX

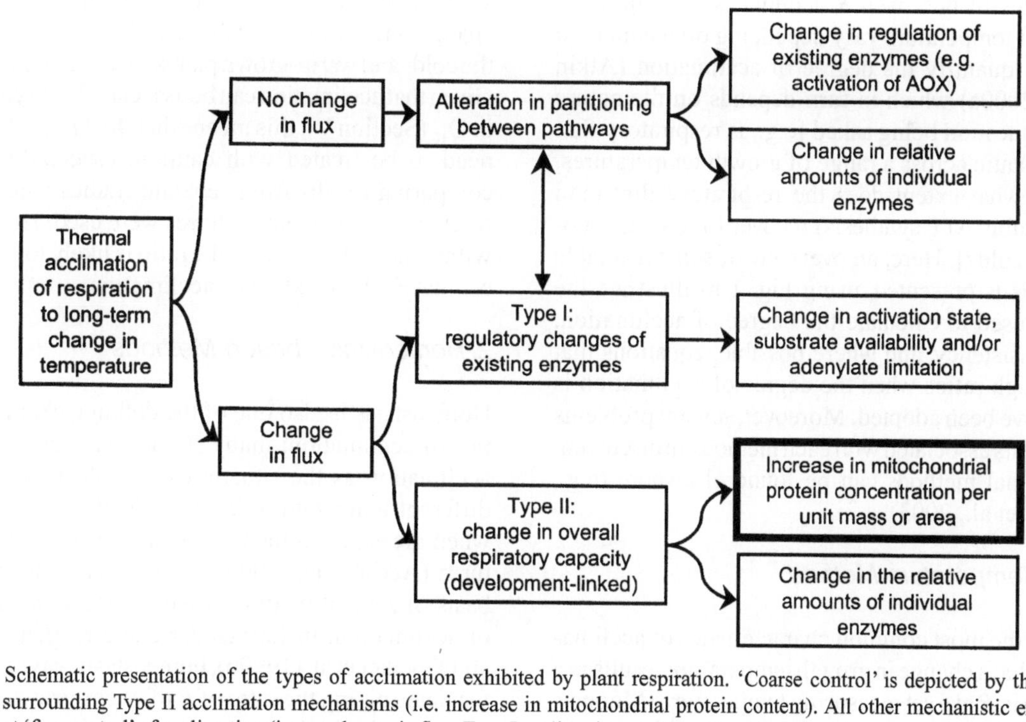

Fig. 8. Schematic presentation of the types of acclimation exhibited by plant respiration. 'Coarse control' is depicted by the heavy border surrounding Type II acclimation mechanisms (i.e. increase in mitochondrial protein content). All other mechanistic examples represent 'fine control' of acclimation (i.e. no change in flux, Type I acclimation and Type II acclimation via changes in relative amounts of individual enzymes). AOX: alternative oxidase.

protein and O_2 uptake (and a decrease in O_2 uptake via the cytochrome pathway), cold acclimation does not result in a change in *total* O_2 uptake (Ribas-Carbó et al., 2000a). Finally, growth-temperature-mediated changes in the shape of the temperature response curve of *R* do not have to be underpinned by changes in gene expression, protein content and maximum enzyme capacity (this is despite the fact that changes in growth temperature probably affect gene expression). Rather, growth-temperature-mediated changes in the regulation of existing enzymes may be sufficient to alter the shape of the temperature response curve of *R*. Examples of this are provided in Section V.A. Clearly therefore, acclimation is not simply changes in total respiratory flux that result from growth-temperature-mediated changes in enzymatic capacity (Fig. 8). However, this chapter does not attempt to deal with all changes that are associated with acclimation to a new growth temperature. Rather, it focuses on the effect of growth temperature on respiratory flux and underlying changes responsible for changes in the temperature-response curve of *R*.

B. Methodology for Quantifying the Degree of Acclimation

Estimates of how well *R* acclimates to changes in ambient temperature vary depending on the method used to quantify the degree of acclimation (Atkin et al., 2000a) which in turn depends on the nature of the question being asked [e.g., is respiratory flux homeostatic across a range of growth temperatures, and to what extent does the respiratory flux (and presumably ATP synthesis) recover following exposure to cold?]. Here, an overview of some available methods is presented, using Fig. 1 to illustrate the values used to calculate the degree of acclimation. For consistency, and where possible, equations that show high ratios when the degree of acclimation is high have been adopted. Moreover, several problems and merits associated with each method are discussed. Additional methods can be found elsewhere (e.g., Loveys et al., 2003).

1. Set Temperature Method

One of the most common characteristics of acclimation is that a change in growth temperature results in a change in *R* at a set measuring temperature (Mooney, 1963; Billings and Mooney, 1968; Chabot and Billings, 1972; Pearcy, 1977; Körner and Larcher, 1988; Collier and Cummins, 1990; Semikhatova et al., 1992; Collier, 1996; Goldstein et al., 1996; Arnone and Körner, 1997). Thus, comparison of rates of *R* at a set measuring temperature can provide an indication of the degree of acclimation. Using the *set temperature method*, the degree of acclimation can be quantified as the rate exhibited at a particular measurement temperature by a cold-grown plant divided by the rate exhibited by a warm-grown plant (Loveys et al., 2003). For example, in Fig. 1:

$$Acclim_{SetTemp} = \frac{C_x}{A} \tag{10}$$

High ratios indicate high degrees of acclimation [(with a ratio of 1.0 indicating no difference between cold- and warm-grown plants (i.e. no acclimation)]. $Acclim_{SetTemp}$ values greater than 1.0 are common, suggesting that most plant species are capable of acclimation via up-regulation of *R* in cold compared to warm temperatures, at least when defined by the differences *R* at a set temperature.

In cases where Q_{10} is constant after acclimation, this approach yields constant ratios of rates of *R* of cold-grown to warm-grown plants, regardless of the measuring temperature. However, changes in measuring temperature result in temperature-dependent changes in $Acclim_{SetTemp}$ if the mean Q_{10} differs between the cold- and warm-grown plants (Atkin et al., 2000a). Given that acclimation can be associated with changes in Q_{10} (Section V), this means that $Acclim_{SetTemp}$ ratios need to be treated with caution, especially when comparing results from separate studies where different measuring temperatures were used. However, within individual studies the ratio is likely to provide estimates of the extent of acclimation.

2. Homeostasis-based Methods

Homeostasis is also one of the defining characteristics of acclimation; many studies have defined *full* acclimation as the situation when plants grown at different temperatures exhibit identical rates of *R*, when measured at their respective growth temperatures (Atkin et al., 2000a). Homeostasis forms the basis of several methods used to assess the degree of acclimation. In Larigauderie and Körner (1995) and Tjoelker et al. (1999b), homeostasis was assessed via determining the ratio of *R* of warm-grown plants to that of *R* of cold-grown plants, each measured at their respective growth temperature. Larigauderie and

Körner (1995) defined this ratio as the *Long-Term Acclimation Ratio* (LTR_{10}), the proportional change in R of plants grown and measured at one temperature compared with a temperature 10 °C lower (LTR_{10} is analogous to the Q_{10} of the short-term temperature response function). In Fig. 1:

$$LTR_{10} = \frac{A}{D_x} \quad (11)$$

Full acclimation (i.e. perfect homeostasis) was assumed to have occurred when LTR_{10} is equal to 1.0. LTR_{10} values that are greater than 1.0 (e.g., A/D_1 in Fig. 1), but less than the Q_{10} (e.g., A/B in Fig. 1) indicate partial acclimation. No acclimation would occur when LTR_{10} and Q_{10} values are equal. By comparing LTR_{10} values with published values of Q_{10}, Larigauderie and Körner (1995) assigned a degree of acclimation to each plant species (not a specific value, but rather whether the degree of acclimation was near-full, high, medium to low or very low). This analysis suggested that only three of the 19 species exhibited full acclimation; three species exhibited a high degree, 12 species exhibited a medium-low degree, and one species exhibited a very low degree of acclimation (Larigauderie and Körner, 1995). In Tjoelker et al. (1999a,b) comparisons of LTR_{10} with measured Q_{10} values for each species suggested that two of the five boreal tree species exhibited near full acclimation, two medium-low acclimation, and one exhibited no acclimation.

The LTR_{10} method is useful in providing an insight into inter-specific differences in the degree of acclimation. Comparing LTR_{10} values with short-term Q_{10} values from the same species avoids potentially misleading conclusions concerning the degree of acclimation associated with low LTR_{10} values. For example, viewed in isolation, a low LTR_{10} such as 1.3 might suggest a very high degree of acclimation. However, if the Q_{10} of the same tissue were also low (e.g., 1.5), then little adjustment in rates of R would have occurred in response to the change in growth temperature. By comparing LTR_{10} and Q_{10} values, the magnitude of acclimation in R to temperature can be appropriately compared.

The method of comparing LTR_{10} with Q_{10} values does, however, have some drawbacks. Firstly, the method does not allow a definitive degree of acclimation value to be applied to any one species. Secondly, it requires LTR_{10} to be compared with Q_{10} values; however, deciding on which Q_{10} value to compare the LTR_{10} may be problematic if short-term temperature response data are missing, or if Q_{10} itself changes as a result of acclimation. If growth temperature affects Q_{10}, then the conclusions reached will depend on which Q_{10} value is used (i.e. should one use Q_{10} of the warm- or cold-grown plants?). Finally, the biological basis for stating that full acclimation is when LTR_{10} equals 1.0 has not been demonstrated; in some cases cold-grown plants exhibit faster rates of R than do warm-grown plants, when each is measured at their respective growth temperatures (Loveys et al. 2003).

Another limitation to the LTR_{10} method is that it does not assign a quantitative degree of acclimation value. Here, a method for assigning such a value using the range of LTR_{10} values that occur between no acclimation (i.e. $Acclim_{LTR10} = 0$, which is when the LTR_{10} and the short-term Q_{10} are equal), and full acclimation (i.e. $Acclim_{LTR10} = 1.0$) is described. This range is equivalent to the short-term Q_{10} minus 1.0. Therefore, the degree of acclimation is:

$$Acclim_{LTR_{10}} = 1 - \left(\frac{LTR_{10} - 1}{Q_{10} - 1}\right) \quad (12)$$

Thus, knowledge of the short-term Q_{10} and LTR_{10} means that a quantitative degree of acclimation to any plant tissue can be assigned. Use of this approach requires the use of mean values of Q_{10} and LTR_{10} for different growth temperatures.

Loveys et al. (2003) proposed a modification of the method of comparing LTR_{10} and Q_{10} values by: In the *homeostasis method* approach, the degree of acclimation is taken as the ratio of R exhibited by cold-grown plants divided by R of the warm-grown plants (each measured at their respective growth temperature) according to (see Fig. 1):

$$Acclim_{Homeo} = 1 - \left(\frac{D_x}{A}\right) \quad (13)$$

If D_x/A is greater than or equal to 1.0, then acclimation must have occurred. The degree of acclimation increases as $Acclim_{Homeo}$ increases. $Acclim_{Homeo}$ is the inverse of the LTR_{10} (Loveys et al., 2003); another difference is that the *homeostasis method* does not require that $Acclim_{Homeo}$ to be compared with Q_{10}. This has advantages and disadvantages; a definitive degree of acclimation is provided without the need for comparison with variable Q_{10} values that can be growth-temperature dependent. Moreover, it does not assume full acclimation, and allows for cases where cold-grown plants exhibit faster rates of *R* than warm-grown plants, each measured at their respective growth temperature. However, the method does not take into account tissues whose Q_{10} is low (and thus already lead to a high $Acclim_{Homeo}$ value, even without large adjustments in *R* occurring). Fortunately, few species exhibit Q_{10} values that differ substantially from the common mean values of 2.0–2.5 (see Section III.A). Thus, the $Acclim_{Homeo}$ ratio probably provides a good indication of the degree of acclimation of *R* to contrasting growth temperatures. It is probably most useful for comparisons of multiple species grown under identical conditions.

3. Quantifying Acclimation: Which Method to Use?

If generalizations are to be made about the degree of temperature acclimation of *R* in plants, greater effort will be needed to standardize the methods by different research groups. Ideally, acclimation is best quantified using Eq. 12, as this takes into account both the short- and long-term responses of *R* to temperature. However, in some studies it may not be possible to obtain estimates of both Q_{10} and LTR_{10}; in such cases the *Set Temperature method* (Eq. 10) should be used (with the knowledge that the $Acclim_{SetTemp}$ value is measurement-temperature dependent). Alternatively, measurements might be best made at the growth temperature (in cases where temperature control during measurements is limited) with analyses then being made using the *Homeostasis method* (Eq. 13). Ultimately, the choice of which method to use will depend on the nature of the comparison being made (e.g., pre-existing plants that experience a change in growth temperature, versus plants that develop under contrasting temperature regimes).

C. Variations in the Degree of Acclimation

The degree of acclimation exhibited by a plant species could affect its ability to adjust to a new growth temperature (and potentially its performance or fitness in that environment; Larigauderie and Körner, 1995; Tjoelker et al., 1999b). Numerous factors might account for a link between acclimation and fitness, such as the need for homeostatic rates of ATP synthesis (necessary for continued rates of growth/cellular maintenance) and CO_2 release (and thus avoid excessive loss of carbon at high temperatures). Whatever the potential benefits of acclimation (or lack of benefits), the degree of acclimation as dependent on species and environmental conditions needs to be established. For example, modelers need to know if all species acclimate to the same extent, and whether the degree of acclimation is affected by water or nutrient availability.

1. Inter-Specific Variation

There is growing evidence that the degree of respiratory acclimation in leaves and roots varies substantially, both within and among individual species (Larigauderie and Körner 1995, Tjoelker et al. 1999a; Loveys et al. 2003). For example, Larigauderie and Körner (1995) found that growth at low temperatures resulted in little or no acclimation of leaf respiration in several alpine (*Poa alpina*, *Leucanthemopsis alpina*, *Luzula alpino-pilosa*, *Carex foetida*, *Cirsium alpinum* and *Saxifraga biflora*) and lowland (*Luzula campestris*, *Carex caryophyllea* and *Cirsium acaule*) species (Table 3). In contrast, acclimation of leaf respiration occurred in *Ranunculus acris*, *Anthoxanthum odoratum*, *Leucanthemum alpinum*, *Poa pratensis*, *Taraxacum alpinum*, *T. officinale* (Larigauderie and Körner, 1995) and *Ranunculus glacialis* (Arnone and Körner, 1997). Similarly, Loveys et al. (2003) found that the degree of acclimation was highly variable among 16 species differing in inherent maximum relative growth rate (Table 3).

Are there systematic differences among plant taxa in the degree to which leaf *R* acclimates? Tjoelker et al. (1999a,b) found that broad-leave tree species exhibited a lower degree of acclimation of leaf *R* than selected conifer species (Fig. 9), suggesting that acclimation might be predicted using structural and/or functional traits. Differences in the ability to acclimate were also observed among six of the eight genera used by Larigauderie and Körner (1995). For example, the two *Taraxacum* species exhibited a greater degree of acclimation than did the two *Cirsium* species. However, Larigauderie and Körner (1995) found no evidence that within a given genus, alpine and lowland plant species differ in their extent of leaf *R* acclimation to contrasting growth temperatures. This contrasts with the result of Loveys et al. (2003), showing that slow-growing species exhibited a higher degree of leaf *R* acclimation than their fast-growing counterparts in 4 out of 6 genera. The degree of acclimation was not, however, related to inherent differences in RGR when all 16 species were

Table 3. Inter-specific variations in the degree of acclimation of leaf respiration for Type II acclimation scenarios (i.e. when leaves develop under contrasting temperatures) for plants grown under ambient CO_2 conditions. Values were calculated using rates of *R* reported in published studies. Values with a closing bracket designate measurement and/or growth temperatures used in the calculations. $Acclim_{SetTemp}$ was calculated using Eq. 10 at two measurement temperatures (shown in brackets) identical to the two growth temperatures reported in each study; for studies with three growth temperatures we only calculated $Acclim_{SetTemp}$ for the measurement temperature in the middle of the growth temperature range. Mean LTR_{10} is a mean LTR_{10} for the range of growth temperatures used in each study (calculated using 11). Where temperature intervals were not 10 °C apart, we used the equivalent of Eq. 4 to calculate the LTR_{10} over a 10 °C interval. $Acclim_{LTR10}$ (Eq. 12) was calculated in two ways; in A, mean Q_{10} and LTR_{10} values calculated from the data in each study were used; in B, we assumed that the Q_{10} varies with temperature (as per Eq. 9), with the mean temperature-corrected Q10 value for the growth temperature range used. Other abbreviations: a: mean of values from different days and times at night, c: mean of values from two and four weeks of growth (growth-T interpreted as Day-T), d: mean of values from different months. References: A, Burton et al. (2002); A, Arnone and Körner (1997); B, Cowling and Sage (1998); C, Ebrahim et al. (1998); D, Gonzàlez-Meler et al. (1999); E, Larigauderie and Körner (1995); F, Loveys et al. (2003); Thornton et al. (1996), H, Xiong et al. (2000). All values are either derived from tables or by enlarged figures from each of the references.

Species	Acclimation ratio calculated using different methods					
	$Acclim_{SetTemp}$ (Eq. 10)	Mean LTR_{10} (Eq. 11)	$Acclim_{homeo}$ (Eq. 13)	$Acclim_{LTR10}$ A (Eq. 12)	$Acclim_{LTR10}$ B (Eq. 12)	Ref.
Ranuculus glacialis	8) 0.42, 18) 0.63	8–18) 1.25	13) 0.8	0.83	0.84	A
Ranuculus repens	8) 0.93, 18) 0.80	8–18) 2.36	13) 0.42	0.21	0.14	A
Phaseolus vulgaris		20–29) 0.90 a			1.09	B
Saccharum officinarum		15–45) 1.53 d			0.38	C
Vigna radiata	19) 1.00, 28) 1.07	19–28) 2.59		–0.07	–0.40	D
Anthoxanthum alpinum		10–20) 1.4	15) 0.71		0.73	E
Anthoxanthum odoratum		10–20) 1.5	15) 0.67		0.66	E
Carex caryophyllea		10–20) 2.3	15) 0.43		0.13	E
Carex foetida		10–20) 2.4	15) 0.42		0.06	E
Cerastium uniflorum		10–20) 3.0	15) 0.33		–0.34	E
Cirsium acaule		10–20) 2.8	15) 0.36		–0.21	E
Cirsium alpinum		10–20) 2.7	15) 0.37		–0.14	E
Leucanthemopsis alpina		10–20) 2.4	15) 0.42		0.06	E
Lucanthemum vulgare		10–20) 1.0	15) 1.0		1.00	E
Luzula alpino-pilosa		10–20) 2.1	15) 0.48		0.26	E
Luzula campestris		10–20) 2.1	15) 0.48		0.26	E
Poa alpina		10–20) 2.4	15) 0.42		0.06	E
Poa pratensis		10–20) 1.0	15) 1.0		1.00	E
Ranunculus acris		10–20) 1.0	15) 1.0		1.00	E
Ranunculus acris		10–20) 1.7	15) 0.59		0.53	E
Saxifra muscoides		10–20) 3.6	15) 0.28		–0.74	E
Saxifraga biflora		10–20) 5.5	15) 0.18		–2.02	E
Taraxacum alpinum		10–20) 1.4	15) 0.71		0.73	E
Taraxacum officinale		10–20) 1.7	15) 0.59		0.53	E
Acacia aneura	23) 1.41	23–28) 0.50			1.48	F
Acacia melanoxylon	23) 0.71	18–28) 2.87	23) 0.35		–0.62	F
Achillea millefolium	23) 2.14	18–28) 0.50	23) 1.99		1.43	F
Achillea ptarmica	23) 2.83	18–28) 0.16	23) 6.10		1.73	F
Eucalyptus delegatensis	23) 1.29	18–28) 1.63	23) 0.61		0.45	F
Euclyptus dumosa	23) 1.05	18–28) 1.24	23) 0.81		0.79	F
Geum rivale	23) 1.29	18–28) 0.85	23) 1.17		1.13	F
Geum urbanum	23) 1.27	18–28) 0.80	23) 1.24		1.17	F
Luzula acutifolia	23) 1.38	18–28) 2.96	23) 0.34		–0.70	F
Luzula sylvatica		18–28) 0.85	23) 1.17		1.13	F

Continued on next page

Table 3. Continued from previous page

Species	Acclimation ratio calculated using different methods					
	$Acclim_{SetTemp}$ (Eq. 10)	Mean LTR_{10} (Eq. 11)	$Acclim_{homeo}$ (Eq. 13)	$Acclim_{LTR10}$ A (Eq. 12)	$Acclim_{LTR10}$ B (Eq. 12)	Ref.
Plantago euryphylla	23) 1.56	18–28) 0.83	23) 1.21		1.15	F
Plantago lanceolata	23) 1.43	18–28) 0.72	23) 1.39		1.24	F
Plantago major	23) 0.82	18–28) 2.07	23) 0.48		0.07	F
Poa costiniana		18–28) 1.13	23) 0.89		0.89	F
Poa trivialis	23) 1.22	18–28) 1.53	23) 0.65		0.54	F
Silene dioica	23) 0.87	18–28) 2.03	23) 0.49		0.11	F
Silene uniflora	23) 1.14	18–28) 2.45	23) 0.41		–0.26	F
Solanum tuberosum	30) 0.77	25–35) 2.20 c	30) 0.45	–0.03	–0.40	G
Colobanthus quitensis	12) 0.51	7–20) 0.98		1.03	1.01	H
Deschampsia antarctica	12) 0.48	7–20) 0.96		1.07	1.03	H

considered (Loveys et al. 2003). Thus, while there is some evidence that the degree of acclimation differs systematically among taxa in some studies, there are also many results that contradict this.

The degree of thermal acclimation of root *R* is also highly variable. Acclimation of root respiration occurs in *Plantago lanceolata* (Smakman and Hofstra, 1982; Loveys et al. 2002, 2003), *Zostera marina* (Zimmerman et al., 1989), *Citrus volkameriana* (Bryla et al., 1997, 2001), *Festuca ovina*, *Juncus squarrosus*, *Nardus stricta* (Fitter et al., 1998), *Bellis perennis*, *Poa annua* (Gunn and Farrar, 1999) and *Holcus lanatus* (E. Edwards and A.H. Fitter, pers. comm.). In contrast, there is little or no acclimation of root *R* in field-grown *Acer saccharum* and *Pinus resinosa* to seasonal changes in temperature (Burton and Pregitzer, 2003). Moreover, there was no obvious acclimation in roots of two *Picea* species (Weger and Guy, 1991; Sowell and Spomer, 1986) and *Abies lasiocarpa* (Sowell and Spomer, 1986). Similarly, while acclimation to changes in growth temperature results in near-perfect homeostasis in *Citrus volkameriana* in wet soils, no acclimation occurs in roots of the same species growing in dry soils (Bryla et al., 1997). Even in species where root *R* does acclimate, the degree of acclimation is variable. For example, in a comparison of root *R* of five cold-grown (18/12 °C) and warm-grown (30/24 °C) boreal tree species at a set measuring temperature (18 °C), warm-grown plants exhibited root *R* rates that were 50–74% of that exhibited by the cold-grown plants (Tjoelker et al. 1999a). Similarly, the degree of acclimation of root *R* was highly variable among species in the study by Loveys et al. (2003).

2. Variations with Water and Nitrogen Availability

Does the degree of acclimation vary with water and/or nutrient supply? As stated above, the degree of acclimation exhibited by *Citrus volkameriana* roots is greater in wet soils than roots growing in dry soils (Bryla et al., 1997). In contrast, low N availability does not appear to reduce the degree of acclimation exhibited by plants transferred from one growth temperature to another for several days. Atkinson and Atkin (in preparation) grew several herbaceous plant species at 25/20 °C, at both high and low N availability (2000 and 25 μM, respectively); these plants were then shifted to 15/10 °C for 7 days, and the degree of acclimation of root *R* assessed (using Eqs. 11 and 12). Although homeostasis was not observed in any of the species, plants grown at high and low N availability exhibited significant and similar degrees of acclimation. Low N supply did, however, result in a slower specific rate of *R* in all species; consequently, the absolute change in *R* following extended exposure to low temperature was lower in the low-N plants.

3. Variations That Are Development-Dependent

There is evidence that the degree of acclimation is lower in pre-existing plant tissues shifted from one temperature to another ('Type I' acclimation, as discussed in Section V.A) than in leaves/roots that develop at the growth temperature ('Type II' acclimation as discussed in Section V.B). In a comparison of

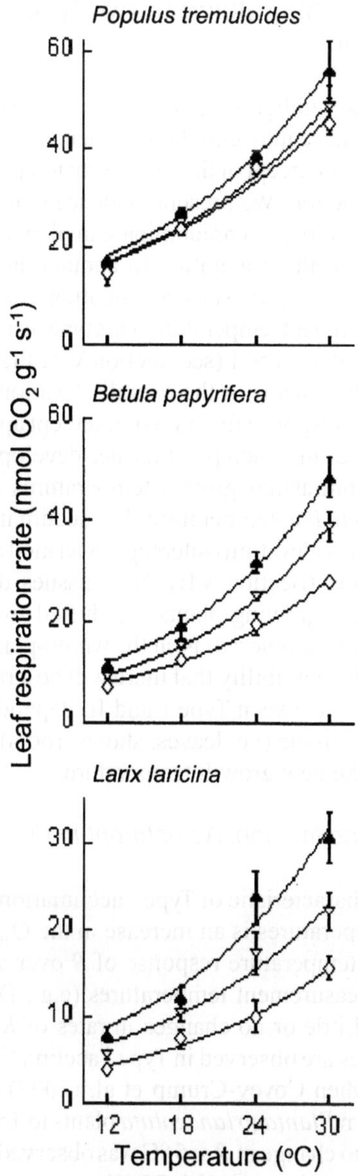

Fig. 9. Example of inter-specific variation in acclimation of R to contrasting growth temperatures. Acclimation of the short-term temperature response function of shoot dark respiration to growth temperature (▲ 18/12, ▽, 24/18, and ◇ 30/24 °C day/night) in three boreal tree species (adapted from Tjoelker et al. (1999a)).

nine species, Loveys et al. (2003) found that the degree of respiratory acclimation was greater in leaves and roots that had developed under contrasting temperatures (18, 23 and 28 °C) than in 25 °C-grown plants shifted to 15 °C for 7 days. Moreover, acclimation of R to 5 °C was substantially greater in *Arabidopsis thaliana* leaves that developed at 5 °C than that of warm-grown leaves shifted to 5 °C for several days (Fig. 10; Armstrong and Atkin, in preparation); when measured at 25 °C, the $Acclim_{SetTemp}$ ratio (Eq. 9) values were 2.37 and 1.52, and $Acclim_{LTR10}$ values (Eq. 11) were 1.52 and 0.78, for the cold-developed and cold-shifted tissues, respectively.

There are two other areas where the stage of development needs to be considered when dealing with acclimation of plant R to temperature. First, the effect of growth temperature on rates of R at a set temperature may depend on the age of plant leaves or roots. Recently, A. Armstrong and O. K. Atkin (unpublished) found that immature leaves of *Arabidopsis thaliana* exhibited near-identical rates of R at any given temperature (no acclimation), regardless of whether the tissue developed under 25 °C or 5 °C. In contrast, rates of leaf R at any given temperature were faster in mature cold-acclimated leaves (see above). These findings suggest that although thermal environment during development likely leads to long-term effects on R response to temperature (i.e. acclimation), the full extent of the acclimation response is only evident in fully developed leaves or roots. Second, the question of whether roots and leaves of the same plant differ in their ability to acclimate R to contrasting temperatures also depends on development. For example, Loveys et al. (2003) found no evidence that leaves and roots differ in their magnitude of acclimation when both tissues develop at the prevailing growth temperature. Similarly, Tjoelker et al. (1999a) also reported no systematic difference in the degree of acclimation of roots and leaves in tree seedlings that develop under contrasting temperatures. However, Loveys et al. (2003) found that roots exhibited a higher degree of acclimation than leaves when plants were shifted from 25 °C to 15 °C for 7 days. The importance of development for acclimation is further highlighted by the fact that fast-growing species exhibit higher acclimation values than their slow-growing counterparts, when plants grown at 25 °C are exposed to 15 °C for 7 days (Loveys et al., 2003); fast-growing species had probably developed more new roots following the temperature shift than their slow-growing counterparts. As Loveys et al. (2003) measured whole root systems (i.e. young and old roots); development of new roots at the new growth temperature would slowly increase the acclimation ratio of the whole root system.

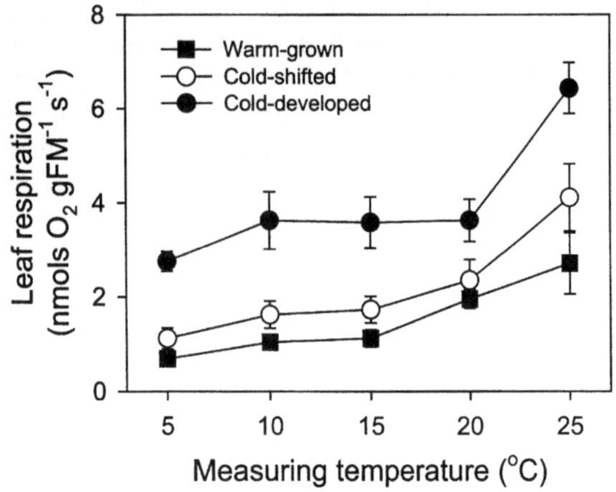

Fig. 10. Role of development for acclimation of leaf *R* to low growth temperatures. The graph shows the measurement-temperature dependence of *R* for warm-grown (25/20 °C), cold-shifted (warm-grown leaves exposed to 5 °C for 7 days) and cold-developed (5 °C) *Arabidopsis thaliana* plants (Armstrong and Atkin, unpublished data).

D. Component of the Daily Temperature Regime to Which R Acclimates

Thermal acclimation of *R* has to be taken into account when modeling responses of *R* to a warmer climate (see Sections VI and VII). The most straightforward approach may be to assume that *R* acclimates to the daily mean temperature, and then use forecasted daily mean temperatures to model, e.g., an annual release of CO_2 for a specific type of species or vegetation. So far, only a few studies have addressed the question of which component of the daily temperature regime *R* acclimates to.

For the sake of simplicity, the diurnal fluctuations in temperature can be divided into a daily minimum, mean, and maximum temperature. The general conclusion is that *R* does not acclimate to the daily mean temperature in all species and/or tissues (Fitter et al., 1998; Atkin et al., 2000b). Will (2000) and Covey-Crump et al. (2000) examined the response of *R* in *Pinus taeda* leaves and *Plantago lanceolata* roots, respectively. Bruhn et al. (in preparation) assessed whether generalizations could be made among and/or within 10 genotypes differing in photosynthetic pathway (C_3 or C_4), functional groups (woody or herbaceous) or maximum growth rate. The temperature regime that *R* acclimated to was highly variable among and within species after transfer to different thermal environments and direct generalizations are not yet possible.

V. Distinguishing Between Two Types of Acclimation

This section highlights the fact that acclimation of *R* to temperature may entail adjustments in Q_{10} (i.e. slope) and/or intercept of the short-term temperature-response function. We present evidence that adjustments in Q_{10} appear to predominate in plants shifted from one growth temperature to another, as would occur in plants (or its tissues) acclimating to seasonal changes in ambient temperature. This type of acclimation is termed as Type I (see Section V.A; Fig. 8). In contrast, adjustments in the overall elevation of the temperature-response function or intercept appear to be the manner in which plant tissues developed and grown in contrasting growth temperatures exhibit acclimation in *R* to temperature. The acclimation via adjustments in elevation (intercept) is termed as Type II acclimation (Section V.B). New tissues develop when actively growing plants are shifted from one temperature to another; thus, at the whole-plant level this raises the possibility that there will be a range of intermediates between Type I and II, depending on the extent of tissue (i.e. leaves, shoots, roots) development at the new growth temperature.

A. Type I Acclimation: Adjustment in Q_{10}

The main characteristic of Type I acclimation to low growth temperatures is an increase in the Q_{10} of the short-term temperature response of *R* over a broad range of measurement temperatures (e.g., D_1 to C_2 in Fig. 1). Little or no changes in rates of *R* at low temperatures are observed in Type I acclimation. For example, when Covey-Crump et al. (2002) shifted warm-grown *Plantago lanceolata* plants to 15/10 °C for 7 days, no change in *R* at 5 °C was observed. However, *R* measured over the 15–25 °C range was higher in the cold-acclimated plants; as a result, the overall Q_{10} of *R* was greater in the cold-acclimated plants. Similarly, acclimation of fully developed *Eucalyptus pauciflora* leaves to very low temperatures in winter was not associated with an increase in *R* measured at 0 °C (winter: 1.5 ± 0.2 nmol CO_2 gDM^{-1} s^{-1}; summer: 1.5 ± 0.4 nmol CO_2 gDM^{-1} s^{-1}), but associated with a large increase in the overall Q_{10} (winter: 3.4 ± 0.6; summer: 2.3 ± 0.2) (Atkin et al. 2000b). Among conifers, needle *R* of *Picea abies*, also exhibits higher Q_{10} values in winter than in summer (Stockfors and Linder 1998) as does needle *R* of *Pinus banksiana* (M. G. Tjoelker, J. Oleksyn and P. B. Reich unpublished).

Thus, Type I acclimation (Q_{10} adjustment) occurs in a broad range of plant taxa, and appears to be a key way in the temperature response of *R* to changes in ambient temperature of the growing environment (see Section II.B.3).

B. Type II Acclimation: Adjustment in Intercept

Type II acclimation is observed in plant leaves or roots that have fully developed in contrasting growth temperatures. Type II acclimation is associated with a change in rates of *R* at both high and low measuring temperatures, with the rate of *R* at 0 °C (y-intercept) increasing in plants acclimated to cold temperatures, and decreasing in plants acclimated to warm temperatures (Fig. 1). Type II acclimation is evidenced by shifts in the elevation or intercept of the entire short-term temperature-response function of *R*. Consequently when *R* is compared at common measurement temperatures, cold-acclimated plants have higher *R* than warm-acclimated plants. This distinction between Type I and Type II categories should not be construed to represent exclusive groupings, but rather represent two modes of temperature acclimation of *R*. Both types may occur together in plants acclimating to a changing thermal environment. For example, repeated measures of fully developed needle *R* in *Pinus banksiana* over the course of a year revealed increases in the intercept of the temperature-response function (Type II) throughout a fall to winter cooling period, with an increase in Q_{10} (Type I acclimation) evident only in midwinter when minimum air temperatures were far below 0 °C (M. G. Tjoelker, J. Oleksyn and P. B. Reich unpublished). Adjustments in the temperature-response function of *R* probably also involve changes in the optimum temperature of *R*, and changes in Q_{10}, particularly at low and high measuring temperatures, as shown in Fig. 2 (calculated using data from Semikhatova et al. (1992); see Section II.B.3).

C. Mechanisms Responsible for Changes Associated with Type I and Type II Acclimation

What mechanism(s) is responsible for the increased Q_{10} associated with Type I acclimation? As discussed in Section III.B, the Q_{10} of *R* is UQ_r/UQ_t dependent (Fig. 6). Thus, growth-temperature-mediated changes in the reduction state of the UQ pool could be responsible for some changes in Q_{10} associated with Type I acclimation; this seems the most likely explanation for the results reported by Covey-Crump et al. (2002). In addition to cold-acclimated roots exhibiting a higher Q_{10} than their warm-maintained counterparts, Covey-Crump et al. (2002) also found that root *R* was less limited by substrate availability following cold acclimation. A removal of adenylate restriction of glycolysis would increase carbon flux into the mitochondria, and presumably increase UQ_r/UQ_t. Further work is needed to assess whether UQ_r/UQ_t increases during Type I acclimation. Alternatively, Type I acclimation might be associated with a change in the extent to which *R* was limited by adenylates; for example, a transition from State 4 (-ADP) to State 3 (+ADP) respiration results in large increases in O_2 uptake at high temperatures, but little change in *R* at low temperatures.

Could the changes associated with Type II acclimation to low temperatures (increased *R* at low and high temperatures, increased Q_{10} at low temperatures and reduced optimum temperature) reflect, in part, the effects of increased substrate availability? Theoretically, increased substrate availability could result in an increase in mitochondrial O_2 uptake at both low and high temperatures, if O_2 uptake was substrate limited (both in the cold and warmer temperatures). However, substrates often only limit O_2 uptake at warmer temperatures (i.e. they do not limit flux in the cold; Atkin et al. 2002, Covey-Crump et al. 2002). Rather, changes associated with Type II acclimation are likely to be related with changes in the overall capacity and optimum temperature of the respiratory system. Previous studies have suggested that respiratory capacity differs between cold- and warm-grown plants [either via differences in capacity per mitochondrion (Klikoff, 1966, 1968) or differences in the number of mitochondria per unit area (Miroslavov and Kravkina 1991)]. Other changes associated with plants growth in cold compared with warm environments include increases in leaf and whole-plant nitrogen concentration (Tjoelker et al. 1999b; Weih and Karlsson, 2001) and increases in concentrations of non-structural carbohydrates (Farrar and Williams, 1991; Oleksyn et al., 2000). Leaf and root *R* increases with increasing tissue nitrogen concentration (Ryan, 1995; Reich et al., 1998; Pregitzer et al., 2000; Burton et al., 2002). Development at low temperature is also associated with an increase in AOX protein levels (Vanlerberghe and Mcintosh, 1992; Gonzàlez-Meler et al., 1999) and cyanide-resistant respiration in some lowland crop species (Stewart et al., 1990; Gonzàlez-Meler et al., 1999).

Growth-temperature-mediated changes in respiratory capacity could be the result of differences in the steady-state concentrations of soluble carbohydrates exhibited by warm- and cold-grown tissues (Warren Wilson, 1966; Mooney and Billings, 1965; Hurry et al., 1994), which in turn can affect respiratory gene expression (McDonnell and Farrar, 1992). In addition, growth-temperature dependent changes in plant uncoupling protein [PUMP; Maia et al. 1998; Nantes (1999]) might contribute to changes in rates of *R* at low temperatures by decreasing the degree of adenylate restriction of *R* at low temperatures. The extent to which maintenance processes (e.g. protein turnover and maintenance of cellular ion gradients) acclimate to new growth temperatures could also contribute to changes in adenylate restriction of *R* at low temperatures (via changes in the demand for ATP). Growth-temperature dependent changes in the optimum temperature of *R* are likely to reflect changes in the amounts and/or isoforms of individual enzymes, and changes in the structure and composition of membranes (e.g., degree of unsaturation of fatty acids; Lambers et al., 1998). Increases in the membrane fluidity might increase proton leakage across the inner mitochondrial membrane and substrate transport across the membrane.

D. Degree of Acclimation in Type I and Type II Scenarios

To what extent does the degree of acclimation (Section IV.B) differ depending on whether acclimation is Type I or Type II? To assess this question, $Acclim_{LTR10}$ ratios were calculated (Eq. 12) for Type I and Type II acclimation of leaf *R* using data on nine species grown at 25 °C and shifted to 15 °C for 7 days (Type I) or grown and measured at 18, 23 and 28 °C (Type II) (Loveys et al., 2003). The average $Acclim_{LTR10}$ of the nine species was 0.24 and 0.75 for the Type I and Type II scenarios, respectively; clearly, Type II acclimation (increase in intercept) results in a higher degree of acclimation than Type I acclimation (adjustment in Q_{10}).

The present knowledge of photosynthetic acclimation suggests that the development of new tissues is required for greater degree of acclimation associated with the Type II scenario. When winter rye (*Secale cereale*) and *Arabidopsis thaliana* are shifted from a warm growth temperature to 5 °C, net photosynthesis rates of pre-existing leaves only partly acclimates; to fully acclimate, both species need to develop new leaves at 5 °C (Hurry et al., 1995; Strand et al., 1997). Newly developed leaves possess the chemical composition and enzymatic machinery to allow acclimation of photosynthesis rates to a new growth temperature more fully than that of leaves that had already matured (Hurry et al., 1995; Strand et al., 1997). The results of Loveys et al. (2003) and A Armstrong and OK Atkin (unpublished; see Section IV.C.3 and Fig. 10) also support the hypothesis that development of new tissues is required for the higher degree of acclimation associated with Type II acclimation.

E. 'Fine' and 'Coarse' Control of Acclimation

Taken together, it seems likely that several levels of acclimation of *R* to temperature can be distinguished (Fig. 8). Firstly, exposure of an existing leaf/root to a new growth temperature can result in Type I acclimation (adjustment in Q_{10}), which results largely from changes in substrate availability and/or adenylate control. Changes in gene expression may also occur, but are not essential for the overall change in respiratory flux. Secondly, development of new leaves/roots following a change in growth temperature can result in Type II acclimation (adjustment in intercept of the temperature-response curve); here, two levels of control might be observed. A *'coarse'* control, where total respiratory capacity might be altered as a result of changes in the density of mitochondria and/or amount of total protein invested in the respiratory chain (Fig. 8), and a *'fine'* control, where Type II acclimation could be associated with changes in the relative amounts of particular enzymes (e.g., AOX versus Complex IV; Ribas-Carbo et al., 2000a). The changes associated with Type I acclimation would also come under *'fine'* control.

VI. Impacts of Variations in the Q_{10} and Acclimation

A. Variations in Q_{10} and Long-Term CO_2 Release

Variations in Q_{10} can have substantial impacts on the total amount of CO_2 released into the atmosphere by root and leaf *R*. For example, in a modeling exercise using temperatures recorded at a field site in S.E. Australia, Atkin *et al.* (2000a) found that annual CO_2 release was up by 47% in tissues with a Q_{10} of 3.0 compared with that of tissues with a Q_{10} of

1.5, whereas rapid acclimation reduced annual CO_2 release by 40%. Clearly, the extent of annual respiratory CO_2 release will be not be accurately predicted whenever an incorrect Q_{10} value is assumed.

Failure to take into account the temperature dependence of the Q_{10} can also result in erroneous estimates of long-term respiratory CO_2 release under field conditions. To illustrate this, seasonal variations in root *R* (relative units) were modeled using data on seasonal variation in soil temperature (measured 9 cm below the soil surface every 5 min) in June and December 1997 at the Danish National Arboretum in Hørsholm, Denmark (55° 52' N, 12° 30' E) (Leverenz et al. 1999). Predicted rates of *R* were calculated using two approaches. In the first approach, it was assumed that Q_{10} varies with soil temperature (using the slope and intercept values for root Q_{10} versus temperature plots described above), with R_1, R_2 etc values being predicted using Eq. 7 (see Section II.B.3.c). In the second approach, Q_{10} was assumed to be temperature insensitive; Q_{10} values at the mean soil temperatures experienced in June and December were estimated using Eq. 4, while Eq. 7 was used to predict R_T. In June the soil temperature range was 28.8 °C (from 5.8 to 34.6 °C) with a mean temperature of 17.2 °C, whereas in December soil temperatures ranged from –4.3 to +8.5 °C (i.e. 12.8 °C range) with a mean of 1.9 °C. For both ways of calculating R_T, *R* was set to 1 (relative units) at 0 °C. This modeling exercise showed that in June, monthly R_T would have been 24.5% greater when Q_{10} varied with measurement temperature than when assuming a constant Q_{10}. However, in December the modeled respiration was not significantly different with the use of a temperature-dependent Q_{10} resulting in only 0.4% less total respiration. Nevertheless, failure to allow for temperature-dependent variations in the Q_{10} can result in large errors in the predicted rates of *R*, particularly under conditions where plants experience large daily variations in temperature, and where the temperatures are relatively high.

B. Acclimation to Different Components of the Daily Temperature Regime on CO_2 Release

Section IV.C highlighted the fact that *R* does not always acclimate to the daily mean temperature. Here, the potential impact on the calculation of annual root CO_2 respiratory release by acclimation to the daily minimum, mean, or maximum temperature under field conditions is discussed (Bruhn 2002). For the sake of simplicity, it was assumed that root *R* acclimates to either one of the three components experienced by the roots the preceding seven days, and that homeostasis was accomplished (but see Section IV.C). Furthermore, a constant Q_{10} throughout the year of either 1.5 or 2.0 was assumed which was not allowed to vary with short-term changes in temperature (but see Section II.B). The calculated annual CO_2 release by the roots is shown in Table 4. This simple model exercise demonstrates the necessity of applying the *right* component of the daily temperature regime for predicting *R*. For example, if one assumes that root *R* acclimates to the daily mean temperature, but in fact it acclimates to the daily minimum temperature (or night-time temperature), then annual CO_2 release at 20 cm depth would be underestimated by 8% and 13% for Q_{10} equal to 1.5 and 2.0, respectively. Furthermore, if surface temperatures were used as a substitute for soil temperature, then the same underestimation would have been 41% and 61%. It is important, therefore, that further work be conducted to assess the extent to which contrasting species differ in the temperature to which respiration acclimates.

C. Impact of Acclimation on the Balance Between Respiration and Photosynthesis

1. Individual Leaves

The short-term temperature sensitivity of light-saturated photosynthesis (P_{sat}) typically differs from that of leaf *R* (in darkness). For example, a decline in temperature from 25 °C to 15 °C reduces leaf *R* and P_{sat} by 55% and 21%, respectively, in *Eucalyptus pauciflora* (Atkin et al., 2000c). As a result, the balance between dark leaf *R* and P_{sat} varies with short-term changes in temperature. However, prolonged exposure to a new growth temperature can result in photosynthetic (and respiratory) acclimation, with the result that the balance between leaf *R* and P_{sat} is re-established (Gifford, 1995; Dewar et al., 1999). Recently, a comparison of leaf *R* and P_{sat} in 16 species grown at 18, 23 and 28 °C found that the balance between leaf *R* and light-saturated photosynthesis was maintained across species and growth temperatures (Loveys et al., 2003). The maintenance of a balance probably reflects the fact that dark leaf *R* and P_{sat} are interdependent, with *R* relying on photosynthesis for substrate, whereas photosynthesis depends on *R* for a range of compounds [e.g., ATP; Krömer (1995), Hoefnagel et al. (1998), Atkin et al. (2000c), Pad-

Table 4. Theoretical impact of Q_{10} values and acclimation to different components of the diurnal temperature regime on annual root respiratory CO_2 release, for roots whose respiration rates have remained homeostatic (i.e. fully acclimated) when measured at the average maximum, minimum or mean temperatures of the preceding seven days (7dayTemp). Temperature data were taken from a field site in SE Australia (Atkin et al., 2000b), both at the soil surface and soil sub-surface (–200 mm). The rate at which *R* was assumed to remain homeostatic at the 7dayTemp (R_{homeo}) was set at 65.2 nmol CO_2 g^{-1} s^{-1}. This rate was the rate of root *R* reported by Atkin et al. (1996) for *Poa trivialis*. Respiration rates at the temperatures recorded every 30 min period (R_T) were calculated according to: $R_T = R_{homeo} * Q_{10}^{([T - 7dayTemp]/10)}$. The Q_{10} (either 1.5 or 2.0) was assumed to be constant over the year, and to be unaffected by short-term changes in temperature. The values calculated each 30 min were then used to estimate the annual rate of CO_2 release. Values are calculated annual root respiration rates (mol CO_2 g^{-1} $year^{-1}$)

Q_{10}	Soil profile/Component of diurnal temperature regime					
	Sub-surface			Surface		
	Min	Mean	Max	Min	Mean	Max
1.5	2.11	1.95	1.79	3.47	2.03	1.16
2.0	2.25	1.95	1.70	5.61	2.21	0.85

masree et al. (2002)]. What is surprising, however, is the extent to which contrasting species exhibit similar leaf *R* to P_{sat} ratios at all temperatures (Loveys et al., 2003). Leaves of plant species from contrasting habitats differ substantially in chemical composition, metabolic fluxes and physical structures; such differences might result in differences between species and growth temperature in the amount of leaf *R* needed to support photosynthesis and vice versa. Thus, even though the relationship between leaf *R* and P_{sat} is affected by environmental factors such as water availability (Turnbull et al., 2001), the available data demonstrates that temperature-mediated differences in dark leaf *R* are closely linked to concomitant differences in leaf photosynthesis.

2. Whole Plants

Does acclimation result in the proportion of daily fixed carbon respired in whole plants (i.e. *R/P*) being constant across a range of growth temperatures (when measured at the growth temperature)? Few studies have actually measured in situ rates of whole plant *R* and photosynthesis. Gifford (1995) found that *R/P* was constant for wheat (*Triticum aestivum)* grown at constant temperatures ranging from 15 to 30 °C (when measured at the respective growth temperatures). Likewise, soybean (*Glycine max*) grown at a range of growth temperatures between, 20 and 35 °C showed no differences in *R/P* ratios, owing to acclimation of *R* to temperature (Ziska and Bunce 1998). However, in that study, growth under elevated CO_2 concentration (700 µl l^{-1}) did result in reduced *R/P*, compared with growth at ambient CO_2 (350 µl l^{-1}). A study with seedlings of five boreal tree species (Tjoelker et al., 1999a) showed small increases in proportional utilization of daily fixed carbon in *R* in plants grown in warmer compared with colder growth environments (Fig. 11). For example, total *R* uses ranged from 24 to 55% of total daily net CO_2 uptake among species in plants grown at 18/12 °C, and increased to 38 to 74% in plants grown at 30/24 °C (Fig. 11); however, these increases were lower than would have occurred without acclimation of both respiration and photosynthesis to growth temperature. Moreover, in that study species differed in overall *R* losses as a proportion of daily net CO_2 uptake. Compared with the faster-growing broad-leaved species *Populus tremuloides* and *Betula papyrifera*, slower-growing conifers *Larix laricina*, *Pinus banksiana*, and *Picea mariana* used a larger proportion of net daily CO_2 uptake in *R*, especially in roots.

An example of near-perfect homeostasis of *R/P* in whole plants is shown in Fig. 12 for *Plantago major*, grown at 13, 20 and 27 °C (I. Scheurwater, T. L. Pons and O. K. Atkin, unpublished data). At any given measuring temperature, *R/P* was greater in cold-grown plants than in their warm-grown counterparts; *R/P* increased with increasing measuring temperature in all cases. Importantly, however, acclimation of whole plant *R* and whole shoot *P* resulted in a constant *R/P* across growth temperatures, as long as *R* and *P* were measured at the respective growth temperature of each treatment (Fig. 12). Similarly, Loveys et al. (2003) found no difference in the balance between daily whole plant *R* and *P* among *Silene uniflora* grown at 18, 23 and 28 °C. However, four of the six species did exhibit higher daily whole plant *R/P* values at 28 °C than at 18 and 23 °C (Loveys et al. 2003) with the percentage increase in *R/P* from 18 to 28 °C differing among species. Thus, while *R/P* is often homeostatic at moderate growth tempera-

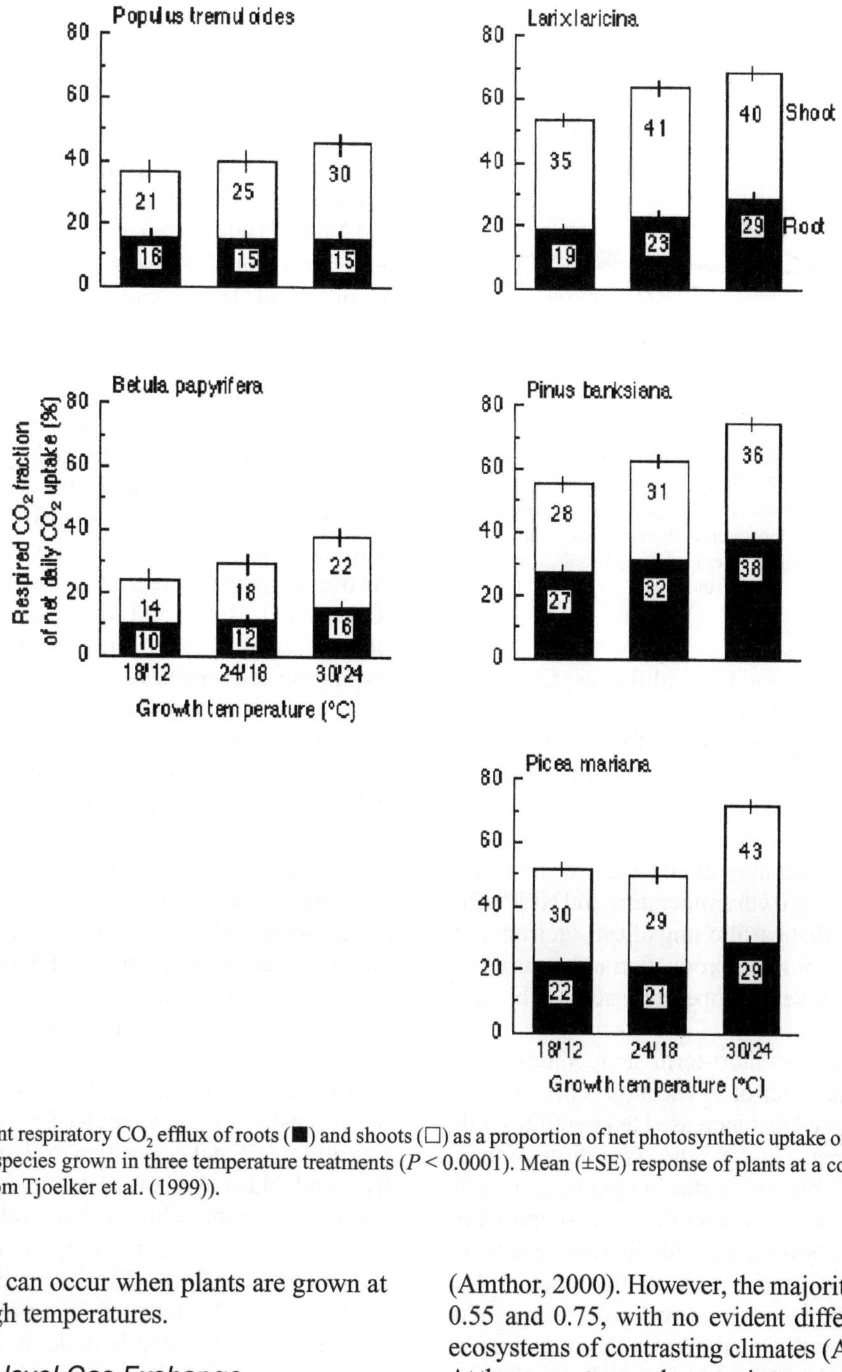

Fig. 11. Whole-plant respiratory CO_2 efflux of roots (■) and shoots (□) as a proportion of net photosynthetic uptake of CO_2 of seedlings of five boreal tree species grown in three temperature treatments ($P < 0.0001$). Mean (±SE) response of plants at a common mass (230 mg) are shown (from Tjoelker et al. (1999)).

tures, increases can occur when plants are grown at unfavorably high temperatures.

D. Ecosystem-level Gas Exchange

Is the relationship between *R* and *P* constrained at the ecosystem scale by growth temperature? In a literature review of estimates of annual (or seasonal) *R*/*P* ratios of diverse ecosystems ranging from cropping system, prairie grassland, and tropical to boreal forests, *R*/*P* values ranged between 0.32 and 0.88 (Amthor, 2000). However, the majority fell between 0.55 and 0.75, with no evident differences among ecosystems of contrasting climates (Amthor, 2000). At the ecosystem scale, net primary production (NPP) is the balance between gross primary production (GPP) and autotrophic respiration (R_a). An expression of the partitioning of carbon from net photosynthesis to biomass production is the NPP/GPP ratio or carbon-use efficiency (CUE, see Dewar et al., 1998; Cannell and Thornley, 2000). In a comparison of the NPP/GPP ratios of above-ground production

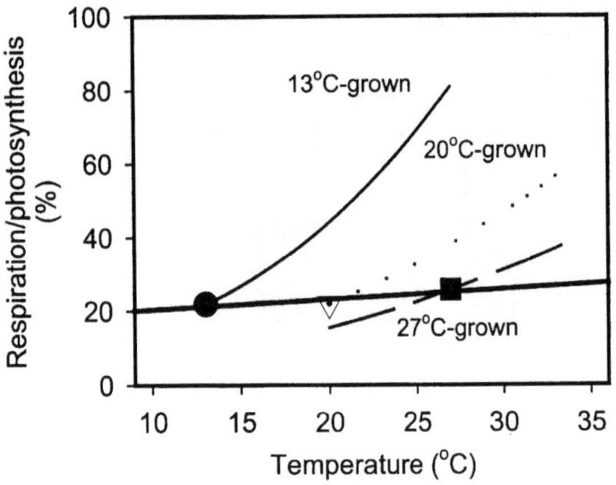

Fig. 12. Effect of growth and measuring temperature for *Plantago major* grown at 13, 20 and 27 °C on the percentage of daily fixed carbon respired (R/P) in whole plants. Symbols represent R/P at each respective growth temperature, whereas lines represent the effect of short-term changes in temperature (I. Scheurwater, T. L. Pons and O. K. Atkin., unpublished data)

in boreal, temperate and tropical forests, CUE was unrelated to mean annual temperature (range –5 to +25 °C), and averaged about 0.5 (Ryan et al., 1997; Saxe et al., 2001). Although there is little evidence that the NPP/GPP ratio is a constant, values may generally fall within a limited range, between 0.4 to 0.6 (Cannell and Thornley, 2000). The lack of an association between growth temperature and NPP/GPP or *R/P* suggests that partitioning of carbon from net photosynthesis to biomass production and respiration is largely insensitive to temperature across climatic zones.

Given the greater short-term temperature sensitivity of *R* than that of *P* (Section VI.A.1), it is frequently assumed that increased temperatures will result in an increase in *R/P* ratios (see discussion in Dewar et al., 1999). Yet studies on plants grown in controlled environments generally fail to support this assumption (Section VI.A.1). An alternative view is that since photosynthesis provides the substrate for *R*, a coupling between these two processes would be expected over the long term, with consequently little change in the *R/P* ratio of altered growth temperatures (Dewar et al., 1998, 1999). Thus, while short-term increases in *R* with temperature may be influenced by the availability of labile carbon (especially at high temperature), in the long term *R* may in part be constrained by substrate supply from *P*. Consequently, the temperature acclimation of *R* and constraints imposed by linkages between *R* and *P* suggest that the use of short-term Q_{10} values to describe long-term responses of *R* to temperature will likely greatly overestimate respiratory carbon fluxes at the ecosystem scale.

Soil CO_2 efflux is a major source of respiratory carbon efflux at the ecosystem scale, and often modeled as an exponential function of temperature (Lloyd and Taylor, 1994). Median Q_{10} values of 2.4 are reported in a global review, although values are higher in colder soils (Raich and Schlesinger, 1992). There is evidence that Q_{10} may be higher in winter than in summer (Kutsch and Kappen, 1997), indicating the potential for dynamic acclimation of soil respiration to temperature. Soil respiration includes heterotrophic respiration (R_h) (dominated by decomposers) in addition to autotrophic *R* of roots. Q_{10} may differ among the various subcomponents of soil respiration; however, roots will likely exert a strong control on the overall temperature sensitivity of soil respiration (Boone et al., 1998) and total respiratory flux (e.g., Bowden et al., 1993). There is evidence that soil respiration acclimates to a change in temperature. A study of experimental warming in a tall grass prairie (Luo et al., 2001) revealed decreased Q_{10} values (5 to 30 °C) of soil *R* with warming of 1.5 to 1.9 °C (50 mm depth). Q_{10} values declined from 2.7 to 2.4 in unclipped plots, and from 2.3 to 2.1 in clipped plots, whereas intercepts of the temperature-response functions were unchanged. The mechanistic basis for acclimation of soil respiration to temperature remains to be determined. Acclimation of soil *R* to warming has the potential to mitigate a presumed positive feedback between ecosystem *R* and climate warming.

Total ecosystem respiration (R_e) is a key component of net ecosystem exchange (NEE = GPP – R_e). In European forests, net ecosystem exchange determined from eddy flux covariance estimates (Valentini et al., 2000) revealed that while GPP was relatively constant across a broad latitudinal range (41 to 64 °N), R_e (and the ratio of R_e to NEE) increased with increasing latitude, and accounted for an observed decline in NEE with increasing latitude. In the same study, R_e was unrelated to mean annual temperature, suggesting temperature would unlikely be an important contributing factor to differences in R_e across latitude. Other recent studies indicate that the Q_{10} of R_e may be highly sensitive to seasonal variation in soil water content. In a study of three Mediterranean woodland ecosystems, the Q_{10} of R_e increased from 1.0 to more than 2.5 with increasing soil moisture (Reichstein et al., 2002). In general, the Q_{10} of ecosystem respiration and its component processes appear to be sensitive to changes in both temperature and soil moisture, and these factors may interact in their effect on the temperature sensitivity of respiration rates in the field. At the same time, an apparent independence of R_e across broad latitudinal climate gradients and other processes such as decomposition rates in mineral soil (Giardina and Ryan, 2000) suggests that factors other than climate are important in governing rates of respiration. In this regard, the role of temperature acclimation is largely untested.

E. Global Atmospheric CO_2 Concentrations

Net ecosystem production (NEP) is the balance between photosynthetic carbon gain (GPP) and total ecosystem respiration (R_e). Seasonal variations in NEP are large enough to account for observed intra-annual variation in global atmospheric CO_2 concentrations in the northern hemisphere (Keeling et al., 1996). In addition there is considerable evidence that the NPP of northern forests exceeds R_h, resulting in a positive NEP and their contribution to a terrestrial carbon sink.

To the present, none of the carbon flux models incorporate acclimation-related changes in the temperature sensitivity of R. Rather, most simulation models such as Biome-BGC, Century (Schimel et al., 1997), PnET (Aber and Federer, 1992) and several dynamic vegetation models (White et al., 2000; Cramer et al., 2001) assume that R responds to short- and long-term changes in temperature in a fixed, exponential manner (Q_{10} = 2.0). Although temperature-based changes in Q_{10} are incorporated into the Terrestrial Ecosystem Model (McGuire et al., 1992), no allowance is made for long-term adjustments associated with acclimation. The absence of acclimation in such models is not, however, surprising, as environmental physiologists have yet to provide modelers with means of predicting the degree of acclimation in contrasting plants and/or plants grown in contrasting environments. Some models predict a loss of carbon with climate warming, owing to a greater response of R_h to warming than that of NPP. Other simulations indicate increases in NEP in northern forest ecosystems as a result of the effect of increased temperatures on nitrogen mineralization in enhancing NPP (VEMAP, 1995). However, the consequences of temperature acclimation in above- and below-ground respiration have not been examined in these modeling frameworks.

Recently, coupled global circulation models (GCMs) that incorporate a positive feedback effect by the carbon cycle (e.g., Cox et al., 2000) have predicted that global warming will result in increased rates of respiratory CO_2 efflux into the atmosphere, which in turn will compound the greenhouse effect. In a comparison of non-coupled and coupled GCMs, Cox et al. (2000) found that the coupled model (which included a positive feed-back) predicted that mean annual land temperatures would be 2.5 °C higher by the year 2100 than that predicted by the non-coupled model. Critical to such coupled models is the extent to which R is temperature sensitive, both in the short- and long term. Coupled GCMs assume that the Q_{10} of R is 2.0, that Q_{10} will be constant in the future, and that respiration does not acclimate (adjustment in rates of R following a change in temperature) to long-term changes in temperature.

VII. Concluding Statements

We have shown that the sensitivity of plant R to short- and long-term changes in temperature is highly variable, which in turn reflects the effect of temperature on the regulation of the respiratory apparatus. Moreover, the response of R to long-term changes in temperature is highly dependent on the effect of temperature on plant development. Thus, much of the variability in Q_{10} values and the degrees of acclimation can be explained by considering the effect of temperature on substrate availability, enzyme capacity, demand for respiratory energy and plant development. However, there is still relatively little known about the combined effects of variability in Q_{10} and the degree of acclimation on future rates of plant R. To successfully model future rates of R, factors such as the temperature dependence of Q_{10} and the degree and speed of acclimation need to be taken into account.

References

Aber JD and Federer CA (1992) A generalized, lumped-parameter model of photosynthesis, evapotranspiration and net primary production in temperate and boreal forest ecosystems. Oecologia 92: 463–474

Amthor JS (1997). Plant respiratory response to elevated CO_2 partial pressure. In: Allen Jr LH, Kirkham MB, Olszyk DM and Whitman CE (eds) Advances in carbon dioxide effects research, pp 35–77. American Society of Agronomy, Wisconsin

Amthor JS (2000) The McCree-de Wit-Penning de Vries-Thornley respiration paradigms: 30 years later. Ann Bot 86: 1–20

Arnone JA and Körner C (1997) Temperature adaptation and acclimation potential of leaf dark respiration in two species of *Ranunculus* from warm and cold habitats. Arctic Alpine Res 29: 122–125

Atkin OK and Day DA (1990) A comparison of the respiratory processes and growth rates of selected Australian alpine and related lowland plant species. Aust J Plant Physiol 17: 517–526

Atkin OK and Tjoelker MG (2003) Thermal acclimation and the dynamic response of plant respiration to temperature. Trends Plant Sci. 8: 343–351

Atkin OK, Botman B and Lambers H (1996) The causes of inherently slow growth in alpine plants: An analysis based on the underlying carbon economies of alpine and lowland *Poa* species. Funct Ecol 10: 698–707

Atkin OK, Evans JR and Siebke K (1998a) Relationship between the inhibition of leaf respiration by light and enhancement of leaf dark respiration following light treatment. Aust J Plant Physiol 25: 437–443

Atkin OK, Evans JR, Ball MC, Siebke K, Pons TL and Lambers H (1998b). Light inhibition of leaf respiration: The role of irradiance and temperature. In: Moller IM, Gardestrom P, Gliminius K and Glaser E (eds) Plant Mitochondria: From Gene to Function, pp 25–32. Bluckhuys Publishers, Leiden

Atkin OK, Edwards EJ and Loveys BR (2000a) Response of root respiration to changes in temperature and its relevance to global warming. New Phytol 147: 141–154

Atkin OK, Holly C and Ball MC (2000b) Acclimation of snow gum *(Eucalyptus pauciflora)* leaf respiration to seasonal and diurnal variations in temperature: The importance of changes in the capacity and temperature sensitivity of respiration. Plant Cell Environ 23: 15–26

Atkin OK, Evans JR, Ball MC, Lambers H and Pons TL (2000c) Leaf respiration of snow gum in the light and dark. Interactions between temperature and irradiance. Plant Physiol 122: 915–923

Atkin OK, Millar AH, Gardeström P and Day DA (2000d). Photosynthesis, carbohydrate metabolism and respiration in leaves of higher plants. In: Leegood RC, Sharkey TD and von Caemmerer S (eds) Photosynthesis, Physiology and Metabolism, pp 153–175. Kluwer Academic Publishers, Dordrecht

Atkin OK, Zhang QS and Wiskich JT (2002) Effect of temperature on rates of alternative and cytochrome pathway respiration and their relationship with the redox poise of the quinone pool. Plant Physiol 128: 212–222

Azcón-Bieto J and Osmond CB (1983) Relationship between photosynthesis and respiration. The effect of carbohydrate status on the rate of CO_2 production by respiration in darkened and illuminated wheat leaves. Plant Physiol 71: 574–581

Azcón-Bieto J, Lambers H and Day DA (1983) Effect of photosynthesis and carbohydrate status on respiratory rates and the involvement of the alternative pathway in leaf respiration. Plant Physiol 72: 598–603

Azcón-Bieto J (1992). Relationships between photosynthesis and respiration in the dark in plants. In: Barber J, Guerrero MG and Medrano H (eds) Trends in Photosynthesis Research, pp 241–253. Intercept Ltd, Andover, Hampshire

Berry JA and Raison JK (1981). Responses of macrophytes to temperature. In: Lange OL, Nobel PS, Osmond CB and Zeigler H (eds) Physiological Plant Ecology I. Responses to the Physical Environment, pp 277–338. Springer-Verlag, Berlin

Billings WD, Godfrey PJ, Chabot BF and Bourque DP (1971) Metabolic acclimation to temperature in Arctic and alpine ecotypes of *Oxyria digyna*. Arctic Alpine Res 3: 277–289

Billings WD and Mooney HA (1968) The ecology of arctic and alpine plants. Biol Rev 43: 481–529

Blennow K, Lang ARG, Dunne P and Ball MC (1998) Cold-induced photoinhibition and growth of seedling snow gum (*Eucalyptus pauciflora*) under differing temperature and radiation regimes in fragmented forests. Plant Cell Environ 21: 407–416

Bolstad PV, Mitchell K and Vose JM (1999) Foliar temperature-respiration response functions for broad-leaved tree species in the southern Appalachians. Tree Physiol 19: 871–878

Bolstad PV, Reich P and Lee T (2003) Rapid temperature acclimation of leaf respiration rates in *Quercus alba* and *Quercus rubra*. Tree Physiol 23: 969–976

Boone RD, Nadelhoffer KJ, Canary JD and Kaye JP (1998) Roots exert a strong influence on the temperature sensitivity of soil respiration. Nature 396: 570–572

Bouma TJ, Nielsen KL, Eissenstat DM and Lynch JP (1997) Estimating respiration of roots in soil: interactions with soil CO_2, soil temperature and soil water content. Plant Soil 195: 221–232

Bowden RD, Nadelhoffer KJ, Boone RD, Melillo JM and Garrison JB (1993) Contributions of aboveground litter, belowground litter, and root respiration to total soil respiration in a temperature mixed hardwood forest. Can J For Res 23: 1402–1407

Breeze V and Elston J (1978) Some effects of temperature and substrate content upon respiration and the carbon balance of field beans (*Vicia faba* L.). Ann Bot 42: 863–876

Bruhn D, Mikkelsen TN and Atkin OK (2002) Does the direct effect of atmospheric CO_2 concentration on leaf respiration vary with temperature? Responses in two species of *Plantago* that differ in relative growth rate. Physiol Plant 114: 57–64

Bryla DR, Bouma TJ and Eissenstat DM (1997) Root respiration in citrus acclimates to temperature and slows during drought. Plant Cell Environ 20: 1411–1420

Bryla DR, Bouma TJ, Hartmond U and Eissenstat DM (2001) Influence of temperature and soil drying on respiration of individual roots in citrus: Integrating greenhouse observations into a predictive model for the field. Plant Cell Environ24: 781–790

Budde RJA and Randall DD (1987) Regulation of pea mitochondrial pyruvate dehydrogenase complex activity: Inhibition of ATP-dependent inactivation. Arch Biochem Biophys 258: 600–606

Budde RJA and Randall DD (1990) Pea leaf mitochondrial pyruvate dehydrogenase complex is inactivated in vivo in a light-dependent manner. Proc Natl Acad Sci USA 87: 673–676

Burton AJ and Pregitzer KS (2002) Measurement carbon dioxide concentration does not affect root respiration of nine tree species in the field. Tree Physiol 22: 67–72

Burton AJ and Pregitzer KS (2002) Field measurements of root respiration indicate little to no seasonal temperature acclimation for sugar maple and red pine. Tree Physiology 23: 273–280

Burton AJ, Pregitzer KS, Ruess RW, Hendrik RL and Allen MF (2002) Root respiration in North American forests: Effects of nitrogen concentration and temperature across biomes. Oecologia 131: 559–568

Burton AJ, Pregitzer KS, Zogg GP and Zak DR (1996) Latitudinal variation in sugar maple fine root respiration. Can J For Res 26: 1761–1768

Burton AJ, Pregitzer KS, Zogg GP and Zak DR (1998) Drought reduces root respiration in sugar maple forests. Ecol Appl 8: 771–778

Cannell MGR and Thornley JHM (1998) Temperature and CO_2 responses of leaf and canopy photosynthesis: A clarification using the non-rectangular hyperbola model of photosynthesis. Ann Bot 82: 883–892

Cannell MGR and Thornley JHM (2000) Modeling the components of plant respiration: Some guiding principles. Ann Bot 85: 45–54

Chabot BF and Billings WD (1972) Origins and ecology of the Sierran alpine flora and vegetation. Ecol Monog 42: 163–199

Collier DE (1996) No difference in leaf respiration rates among temperate, subarctic, and arctic species grown under controlled conditions. Can J Bot 74: 317–320

Collier DE and Cummins WR (1990) The effects of low growth and measurement temperature on the respiratory properties of five temperate species. Ann Bot 65: 533–538

Covey-Crump EM, Attwood RG and Atkin OK (2002) Regulation of root respiration in two species of *Plantago* that differ in relative growth rate: the effect of short- and long-term changes in temperature. Plant Cell Environ 25: 1501–1513

Cowling SA and Sage RF (1998) Interactive effects of low atmospheric CO_2 and elevated temperature on growth, photosynthesis and respiration in *Phaseolus vulgaris*. Plant Cell Environ 21: 427–435

Cox PM, Betts RA, Jones CD, Spall SA and Totterdell IJ (2000) Acceleration of global warming due to carbon-cycle feedbacks in a coupled climate model. Nature 408: 184–187

Cramer W and et al. (2001) Global response of terrestrial ecosystem structure and function to CO_2 and climate change: Results from six dynamic global vegetation models. Global Change Biol 7: 357–373

Crawford RMM and Palin MA (1981) Root respiration and temperature limits to the north-south distribution of four perennial maritime plants. Flora 171: 338–354

Criddle RS, Hopkin MS, Mcarthur ED and Hansen LD (1994) Plant distribution and the temperature coefficient of metabolism. Plant Cell Environ 17: 233–243

Damesin C, Ceschina E, Le Goff N, Ottorini J and Dufrêne E (2002) Stem and branch respiration of beech: From tree measurements to estimations at the stand level. New Phytol 153: 159–172

Day DA, Krab K, Lambers H, Moore AL, Siedow JN, Wagner AM and Wiskich JT (1996) The cyanide-resistant oxidase: To inhibit or not to inhibit, that is the question. Plant Physiol 110: 1–2

Day DA and Lambers H (1983) The regulation of glycolysis and electron transport in roots. Physiol Plant 8: 155–160

Dewar RC, Medlyn BE and Mcmurtrie RE (1998) A mechanistic analysis of light and carbon use efficiencies. Plant Cell Environ 21: 573–588

Dewar RC, Medlyn BE and McMurtrie RE (1999) Acclimation of the respiration/photosynthesis ratio to temperature: insights from a model. Global Change Biol 5: 615–622

Diolez P, Kesseler A, Haraux F, Varerio M, Brinkmann K and Brand MD (1993) Regulation of oxidative-phosphorylation in plant mitochondria. Biochem Soc Trans 21: 769–773

Douce R and Neuburger M (1989) The uniqueness of plant mitochondria. Annu Rev Plant Physiol Mol Biol 40: 371–414

Dufour S, Rousse N, Canioni P, Diolez P 1996) Top-down control analysis of temperature effect on oxidative phosphorylation. Biochem J 314: 743–751

Ebrahim MK, Zingsheim O, ElShourbagy MN, Moore PH and Komor E (1998) Growth and sugar storage in sugarcane grown at temperatures below and above optimum. J Plant Physiol 153: 593–602

Eissenstat DM, Wells CE, Yanai RD and Whitbeck JL (2000) Building roots in a changing environment: Implications for root longevity. New Phytol 147: 33–42

Farrar JF and Williams ML (1991) The effects of increased atmospheric carbon dioxide and temperature on carbon partitioning, source-sink relations and respiration. Plant Cell Environ 4: 819–830

Field CB (2001) Plant physiology of the 'missing' carbon sink. Plant Physiol 125: 25–28

Fitter AH, Graves JD, Self GK, Brown TK, Bogie DS and Taylor K (1998) Root production, turnover and respiration under two grassland types along an altitudinal gradient—influence of temperature and solar radiation. Oecologia 114: 20–30

Forward DF (1960) Effect of temperature on respiration. In: Ruhland W (ed) Encyclopedia of Plant Physiology, Vol 12, pp 234–258. Springer-Verlag, Berlin

Geiger M, Stitt M and Geigenberger P (1998) Metabolism in slices from growing potato tubers responds differently to addition of sucrose and glucose. Planta 206: 234–244

Giardina CP and Ryan MG (2000) Evidence that decomposition rates of organic carbon in mineral soil do not vary with temperature. Nature 404: 858–861

Gifford RM (1995) Whole plant respiration and photosynthesis of wheat under increased CO_2 concentration and temperature—long-term versus short-term distinctions for modeling. Global Change Biol 1: 385–396

Goldstein G, Drake DR, Melcher P, Giambelluca TW and Heraux J (1996) Photosynthetic gas exchange and temperature-induced damage in seedlings of the tropical alpine species *Argyroxiphium sandwicense*. Oecologia 106: 298–307

Gonzàlez-Meler MA, Ribas-Carbó M, Giles L and Siedow JN (1999) The effect of growth and measurement temperature on the activity of the alternative respiratory pathway. Plant Physiol 120: 765–772

Gonzalez-Meler MA, Giles L, Thomas RB and Siedow JN (2001) Metabolic regulation of leaf respiration and alternative pathway activity in response to phosphate supply. Plant Cell Environ 24: 205–215

Griffin KL, Turnbull M and Murthy R (2002a) Canopy position affects the temperature response of leaf respiration in *Populus deltoides*. New Phytol 154: 609–619

Griffin KL, Turnbull M, Murthy R, Lin GH, Adams J, Farnsworth B, Mahato T, Bazin G, Potasnak M and Berry JA (2002b) Leaf respiration is differentially affected by leaf vs. stand-level nighttime warming. Global Change Biol 8: 479–485

Gunn S and Farrar JF (1999) Effects of a 4 °C increase in temperature on partitioning of leaf area and dry mass, root respiration and carbohydrates. Funct Ecol 13: 12–20

Higgins PD and Spomer GG (1976) Soil temperature effects on root respiration and the ecology of alpine and subalpine plants. Bot Gaz 137: 110–120

Hill SA and Bryce JH (1992). Malate metabolism and light-enhanced dark respiration in barley mesophyll protoplasts. In:

Lambers H and van der Plas LHW (eds) Molecular, Biochemical and Physiological Aspects of Plant Respiration, pp 221–230. SPB Academic Publishing bv, The Hague

Hoefnagel MHN and Wiskich JT (1998) Activation of the plant alternative oxidase by high reduction levels of the Q-pool and pyruvate. Arch. Biochem. Biophys. 355: 262–270.

Hoefnagel MHN, Atkin OK and Wiskich JT (1998) Interdependence between chloroplasts and mitochondria in the light and the dark. Biochim Biophys Acta 1366: 235–255

Hubbard RM, Ryan MG and Lukens DL (1995) A simple, battery-operated, temperature-controlled cuvette for respiration measurements. Tree Physiol 15: 175–179

Hurry VM, Malmberg G, Gardestrom P and Oquist G (1994) Effects of a short-term shift to low temperature and of long- term cold hardening on photosynthesis and ribulose-1,5-bisphosphate carboxylase oxygenase and sucrose phosphate synthase activity in leaves of winter rye (*Secale cereale* l). Plant Physiol 106: 983–990

Hurry VM, Tobiæson M, Krömer S, Gardeström P and Öquist G (1995) Mitochondria contribute to increased photosynthetic capacity of leaves of winter rye (*Secale cereale* L) following cold-hardening. Plant Cell Environ 18: 69–76

Ivanova TL, Semikhatova OA, Judina OS and Leina GD (1989). The effect of temperature on the respiration of plants from different plant-geographic zones. In: Semikhatova OA (ed) Ecophysiological investigations of photosynthesis and respiration in plants, pp 140–166. Nauka Publishing, Nauka

James WO (1953) Plant Respiration. Clarendon Press, Oxford

Keeling RF, Piper SC and Heimann M (1996) Global and hemispheric sinks deduced from changes in atmospheric O_2 concentrations. Nature 381: 218–221

Kiener CM and Bramlage WJ (1981) Temperature effects on the activity of the alternative respiratory pathway in chill-sensitive *Cucumis sativus*. Plant Physiol 68: 1474–1478

Klikoff LG (1966) Temperature dependence of the oxidative rates of mitochondria in *Danthonia intermedia*, *Penstemon davidsonii* and *Sitanion hystrix*. Nature 212: 529–530

Klikoff LG (1968) Temperature dependence of mitochondrial oxidative rates of several plant species of the Sierra Nevada. Bot Gaz 129: 227–230

Körner C (1999) Alpine Plant Life. Functional Plant Ecology of High Mountain Ecosystems. Springer-Verlag, Berlin

Körner C and Larcher W (1988). Plant Life in Cold Environments. In: Long SF and Woodward FI (eds) Plants and Temperature. Symposium of the Society of Experimental Biologists. Volume 42, pp 25–57. The Company of Biologists Limited, Cambridge

Krömer S (1995) Respiration during photosynthesis. Annu Rev Plant Physiol Plant Mol Biol 46: 45–70

Kutsch WL and Kappen L (1997) Aspects of carbon and nitrogen cycling in soils of the bornhoved lake district. II. Modeling the influence of temperature increase on soil respiration and organic carbon content in arable soils under different managements. Biogeochem 39: 207–224

Lambers H and Atkin OK (1995). Regulation of carbon metabolism in roots. In: Madore MA and Lucas WJ (eds) Carbon Partitioning and Source-Sink Interactions in Plants. pp 226–238. The American Society of Plant Physiologists, Riverside

Lambers H, Chapin FS and Pons TL (1998) Plant Physiological Ecology. Springer-Verlag, New York

Lambers H, Van Der Werf A and Bergkotte M (1993). Respiration: the alternative pathway. In: Hendry GAF and Grime JP (eds) Methods in comparative plant ecology: A laboratory manual, pp 140–144. Chapman and Hall, London

Larigauderie A and Körner C (1995) Acclimation of leaf dark respiration to temperature in alpine and lowland plant species. Ann Bot 76: 245–252

Lawrence C and Holaday AS (2000) Effects of mild night chilling on respiration of expanding cotton leaves. Plant Sci 157: 233–244

Leverenz JW, Bruhn D and Saxe H (1999) Responses of two provenances of *Fagus sylvatica* seedlings to a combination of four temperature and two CO_2 treatments during their first growing season: Gas exchange of leaves and roots. New Phytol 144: 437–454

Lloyd J and Taylor JA (1994) On the temperature dependence of soil respiration. Funct Ecol 8: 315–323

Loef I, Stitt M and Geigenberger P (2001) Increased levels of adenine nucleotides modify the interaction between starch synthesis and respiration when adenine is supplied to discs from growing potato tubers. Planta 212: 782–791

Loveys BR, Scheurwater I, Pons TL, Fitter AH and Atkin OK (2002) Growth temperature influences the underling components of relative growth rate: An investigation using inherently fast- and slow-growing plant species. Plant Cell Environ 25: 975–987

Loveys BR, Atkinson LJ, Sherlock DJ, Roberts RL, Fitter AH and Atkin OK (2003) Thermal acclimation of leaf and root respiration: An investigation comparing inherently fast- and slow-growing plant species. Global Change Biol in press

Luo YQ, Wan SQ, Hui DF and Wallace LL (2001) Acclimatization of soil respiration to warming in a tall grass prairie. Nature 413: 622—625

Maia IG, Benedetti CE, Leite A, Turcinelli SR, Vercesi AE and Arruda P (1998) AtPUMP: An *Arabidopsis* gene encoding a plant uncoupling mitochondrial protein. FEBS Letts 429: 403–406

Marie BAH and Cummins WR (1984) Age and temperature effects on saturation of leaf fatty acids of *Saxifraga cernua,* an arctic herb. Can J Bot 62: 1018–1021

McDonnell E and Farrar JF (1992). Substrate supply and its effect on mitochondrial and whole tissue respiration in barley roots. In: Lambers H and van der Plas LHW (eds) Molecular, Biochemical and Physiological Aspects of Plant Respiration, pp 455–162. SPB Academic Publishing bv, The Hague

McGuire AD, Melillo JM, Joyce LA, Kicklighter DW, Grace AL, Moore B and Vorosmarty CJ (1992) Interactions between carbon and nitrogen dynamics in estimating net primary productivity for potential vegetaion in North America. Global Biochem Cycles 6: 101–124

McNulty AM and Cummins WR (1987) The relationship between respiration and temperature in leaves of the arctic plant *Saxifraga cernua*. Plant Cell Environ 10: 319–325

Millar AH, Atkin OK, Menz RI, Henry B, Farquhar G and Day DA (1998) Analysis of respiratory chain regulation in roots of soybean seedlings. Plant Physiol 117: 1083–1093

Millar AH, Wiskich JT, Whelan J and Day DA (1993) Organic acid activation of the alternative oxidase of plant mitochondria. FEBS Lett 329: 259–262

Millenaar FF, Gonzalez-Meler MA, Fiorani F, Welschen R, Ribas-Carbo M, Siedow JN, Wagner AM and Lambers H (2001) Regulation of alternative oxidase activity in six wild monocotyledonous species. An in vivo study at the whole root level. Plant Physiol 126: 376–387

Millenaar FF, Roelofs R, Gonzalez-Meler MA, Siedow JN, Wagner AM and Lambers H (2000) The alternative oxidase in roots of *Poa annua* after transfer from high-light to low-light conditions. Plant J 23: 623–632

Miroslavov EA and Kravkina IM (1991) Comparative analysis of chloroplasts and mitochondria in leaf chlorenchyma from mountain plants grown at different altitudes. Ann Bot 68: 195–200

Mooney HA (1963) Physiological ecology of coastal, subalpine, and alpine populations of *Polygonum bistortoides*. Ecology 44: 812–816

Mooney HA and Billings WD (1965) Effects of altitude on carbohydrate content of mountain plants. Ecology 46: 750–751

Moore AL, Dry IB and Wiskich JT (1988) Measurement of the redox state of the ubiquinone pool in plant-mitochondria. FEBS Lett 235: 76–80

Moore AL, Gemel J and Randall DD (1993) The regulation of pyruvate dehydrogenase activity in pea leaf mitochondria—the effect of respiration and oxidative phosphorylation. Plant Physiol 103: 1431—1435

Nantes IL, Fagian MM, Catisti R, Arruda P, Maia IG and Vercesi AE (1999) Low temperature and aging-promoted expression of PUMP in potato tuber mitochondria. FEBS Letts 457: 103–106

Noguchi K, Go CS, Terashima I, Ueda S and Yoshinari T (2001) Activities of the cyanide-resistant respiratory pathway in leaves of sun and shade species. Aust J Plant Physiol 28: 27–35

Noguchi K and Terashima I (1997) Different regulation of leaf respiration between *Spinacia oleracea*, a sun species, and *Alocasia odora*, a shade species. Physiol Plant 101: 1–7

Oleksyn J, Zytkowiak R, Reich PB, Tjoelker MG and Karolewski P (2000) Ontogenetic patterns of leaf CO_2 exchange, morphology and chemistry in *Betula pendula* trees. Trees 14: 271–281

Osmond DL, Wilson RF and Raper CD (1982) Fatty acid composition and nitrate uptake of soybean roots during acclimation to low temperature. Plant Physiol 70: 1639–1643

Padmasree K, Padmavathi L and Raghavendra AS (2002) Essentiality of mitochondrial oxidative metabolism for photosynthesis: Optimization of carbon assimilation and protection against photoinhibition. Crit Rev Biochem Mol Biol 37: 71–119

Padmasree K and Raghavendra AS (1998). Interaction with respiration and nitrogen metabolism. In: Raghavendra AS (ed) Photosynthesis. A Comprehensive Treatise, pp 197–211. Cambridge University Press, Cambridge

Paembonan SA, Hagihara A and Hozumi K (1991) Long-term measurement of CO_2 release from the above-ground parts of a hinoki forest tree in relation to air temperature. Tree Physiol 8: 399–405

Palta JA and Nobel PS (1989) Root respiration for agave-deserti — influence of temperature, water status and root age on daily patterns. J Exp Bot 40: 181–186

Pearcy RW (1977) Acclimation of photosynthetic and respiratory carbon dioxide exchange to growth temperature in *Atriplex lentiformis* (Torr.) Wats. Plant Physiol 59: 795–799

Pearcy RW (1978) Effect of growth temperature on the fatty acid composition of the leaf lipids in *Atriplex lentiformis* (Torr.) Wats. Plant Physiol 69: 484–486

Pisek A, Larcher W, Vegis A and Napp-Zinn K (1973). The normal temperature range. In: Precht H, Christophersen J, Hensel H and Larcher W (eds) Temperature and life, pp 102–194. Springer-Verlag, Berlin

Poorter H, Remkes C and Lambers H (1990) Carbon and nitrogen economy of 24 wild species differing in relative growth rate. Plant Physiol 94: 621–627

Prasad TK, Anderson MD and Stewart CR (1994) Acclimation, hydrogen peroxide, and abscisic acid protect mitochondria against irreversible chilling injury in maize seedlings. Plant Physiol 105: 619–627

Pregitzer KS, King JA, Burton AJ and Brown SE (2000) Responses of tree fine roots to temperature. New Phytol 147: 105–115

Pregitzer KS, Kubiske ME, Yu CK and Hendrick RL (1997) Root architecture, carbon and nitrogen in four temperate forest species. Oecologia 111: 302–308

Pregitzer KS, Laskowski MJ, Burton AJ, Lessard VC and Zak DR (1998) Variation in sugar maple root respiration with root diameter and soil depth. Tree Physiol 18: 665–670

Raich JW and Schlesinger WH (1992) The global carbon-dioxide flux in soil respiration and its relationship to vegetation and climate. Tellus Series B - Chem Phys Meteor 44: 81–99

Rakonczay Z, Seiler JR and Samuelson LJ (1997) A method for the in situ measurement of fine root gas exchange of forest trees. Env Exp Bot 37: 107–113

Reich PB, Walters MB, Ellsworth DS, Vose JM, Volin JC, Gresham C and Bowman WD (1998) Relationships of leaf dark respiration to leaf nitrogen, specific leaf area and leaf life-span: A test across biomes and functional groups. Oecologia 114: 471–482

Reich PB, Walters MB, Tjoelker MG, Vanderklein D and Buschena C (1998) Photosynthesis and respiration rates depend on leaf and root morphology and nitrogen concentration in nine boreal tree species differing in relative growth rate. Funct Ecol 12: 395–405

Reichstein M, Tenhunen JD, Roupsard O, Ourcival J-M, Rambal S, Dore S and Valentini R (2002) Ecosystem respiration in two Mediterranean evergreen Holm Oak forests: Drought effects and decomposition dynamics. Funct Ecol 16: 27–39

Ribas Carbó M, Aroca R, Gonzàlez Meler MA, Irigoyen JJ and Sanchezdiaz M (2000) The electron partitioning between the cytochrome and alternative respiratory pathways during chilling recovery in two cultivars of maize differing in chilling sensitivity. Plant Physiol 122: 199–204

Ribas-Carbo M, Robinson SA, Gonzalez-Meler MA, Lennon AM, Giles L, Siedow JN and Berry JA (2000) Effects of light on respiration and oxygen isotope fractionation in soybean cotyledons. Plant Cell Environ 23: 983–989

Rook DA (1969) The influence of growing temperature on photosynthesis and respiration of *Pinus radiata* seedlings. NZ J Bot 7: 43–55

Ryan MG (1991) Effects of climate change on plant respiration. Ecol App 1: 157–167

Ryan MG (1995) Foliar maintenance respiration of sub-alpine and boreal trees and shrubs in relation to nitrogen content. Plant Cell Environ 18: 765–772

Ryan MG, Hubbard RM, Pongracic S, Raison RJ and McMurtrie RE (1996) Foliage, fine-root, woody-tissue and stand respiration in *Pinus radiata* in relation to nitrogen status. Tree Physiol 16: 333–343

Ryan MG, Lavigne MB and Gower ST (1997) Annual carbon

cost of autotrophic respiration in boreal forest ecosystems in relation to species and climate. J Geophys Res Atm 102: 28871–28883

Saglio PH and Pradet A (1980) Soluble sugars, respiration, and energy charge during aging of excised maize root tips. Plant Physiol 66: 516–519

Saxe H, Cannell MGR, Johnsen B, Ryan MG and Vourlitis G (2001) Tree and forest functioning in response to global warming. New Phytol 149: 369–399

Schimel DS (1995) Terrestrial ecosystems and the carbon cycle. Global Change Biol 1: 77–91

Schimel DS and 26 others (1997) Continental scale variability in ecosystem processes: Models, data and the role of disturbance. Ecol Monographs 67: 251–271

Semikhatova OA, Gerasimenko TV and Ivanova TI (1992). Photosynthesis, respiration, and growth of plants in the Soviet Arctic. In: Chapin FS, Jefferies RL, Reynolds JF, Shaver GR and Svoboda J (eds) Arctic Ecosystems in a Changing Climate, pp 169–192. Academic Press, San Diego

Simons BH, Millenaar FF, Mulder L, VanLoon LC and Lambers H (1999) Enhanced expression and activation of the alternative oxidase during infection of *Arabidopsis* with pseudomonas syringae pv tomato. Plant Physiol 120: 529–538

Smakman G and Hofstra RJJ (1982) Energy metabolism of *Plantago lanceolata*, as affected by change in root temperature. Physiologia Plant 56: 33–37

Sowell JB and Spomer GG (1986) Ecotypic variation in root respiration rate among elevational populations of *Abies lasiocarpa* and *Picea engelmannii*. Oecologia 68: 375–379

Stefanowska M, Kuras M, KubackaZebalska M and Kacperska A (1999) Low temperature affects pattern of leaf growth and structure of cell walls in winter oilseed rape (*Brassica napus* L., var. *oliefera* L.). Ann Bot 84: 313–319

Stewart CR, Martin BA, Reding L and Cerwick S (1990) Respiration and alternative oxidase in corn seedlings during germination at different temperatures. Plant Physiol 92: 755–760

Stockfors J and Linder S (1998) The effect of nutrition on the seasonal course of needle respiration in Norway spruce stands. Trees 12: 130–138

Strand A, Hurry V, Gustafsson P and Gardestrom P (1997) Development of *Arabidopsis thaliana* leaves at low temperatures releases the suppression of photosynthesis and photosynthetic gene expression despite the accumulation of soluble carbohydrates. Plant J 12: 605—614

Svensson AS Johansson FI, Moller IM and Rasmusson AG (2002) Cold stress decreases the capacity for respiratory NADH oxidation inn potato leaves. FEBS Lett 517: 79–82

Tjoelker MG, Oleksyn J and Reich PB (1999) Acclimation of respiration to temperature and CO_2 in seedlings of boreal tree species in relation to plant size and relative growth rate. Global Change Biol 5: 679–691

Tjoelker MG, Oleksyn J and Reich PB (2001) Modelling respiration of vegetation: Evidence for a general temperature-dependent Q_{10}. Global Change Biol 7: 223–230

Tjoelker MG, Reich PB and Oleksyn J (1999) Changes in leaf nitrogen and carbohydrates underlie temperature and CO_2 acclimation of dark respiration in five boreal tree species. Plant Cell Environ 22: 767–778

Turnbull MH, Whitehead D, Tissue DT, Schuster WSF, Brown KJ and Griffin KL (2001) Responses of leaf respiration to temperature and leaf characteristics in three deciduous tree species vary with site water availability. Tree Physiol 21: 571–578

Uemura M, Joseph RA and Steponkus PL (1995) Cold acclimation of *Arabidopsis thaliana* — effect on plasma membrane lipid composition and freeze-induced lesions. Plant Physiol 109: 15–30

Umbach AL, Wiskich JT and Siedow JN (1994) Regulation of alternative oxidase kinetics by pyruvate and intermolecular disulfide bond redox status in soybean seedling mitochondria. FEBS Letters 348: 181–184

Valentini R, Matteucci G, Dolman AJ, Schulze ED, Rebmann C, Moors EJ, Granier A, Gross P, Jensen NO, Pilegaard K, Lindroth A, Grelle A, Bernhofer C, Grunwald T, Aubinet M, Ceulemans R, Kowalski AS, Vesala T, Rannik U, Berbigier P, Loustau D, Guomundsson J, Thorgeirsson H, Ibrom A, Morgenstern K, Clement R, Moncrieff J, Montagnani L, Minerbi S and Jarvis PG (2000) Respiration as the main determinant of carbon balance in European forests. Nature 404: 861–865

Van Der Werf A, Kooijman A, Welschen R and Lambers H (1988) Respiratory energy costs for the maintenance of biomass, for growth and for ion uptake in roots of *Carex diandra* and *Carex acutiformis*. Physiol Plant 72: 483–491

Van Der Werf A, Poorter H and Lambers H (1994). Respiration as dependent on a species' inherent growth rate and on the nitrogen supply to the plant. In: Roy J and Garnier E (eds) A Whole Plant Perspective on Carbon-Nitrogen Interactions, pp 83–103. SPB Academic Publishing bv, The Hague

Van Emmerik WAM, Wagner AM and Van der Plas LHW (1992) A quantitative comparison of respiration in cells and isolated-mitochondria from *Petunia-hybrida* suspension cultures a high yield isolation procedure. J Plant Physiol 139: 390–396

Vanlerberghe GC and Mcintosh L (1992) Lower growth temperature increases alternative pathway capacity and alternative oxidase protein in tobacco. Plant Physiol 100: 115–119

Veen BW (1980). Energy costs of ion transport. In: Rains DW, Valentine RC and Holaender C (eds) Genetic Engineering of Osmoregulation. Impact on Plant Productivity for Food, Chemicals and Energy, pp 187–195. Plenum Press, New York

Wager HG (1941) On the respiration and carbon assimilation rates of some arctic plants as related to temperature. New Phytol 40: 1–19

Warren Wilson J (1966) An analysis of plant growth and its control in Arctic environments. Ann Bot 30: 383–402

Weger HG and Guy RD (1991) Cytochrome and alternative pathway respiration in white spruce (*Picea glauca*) roots. Effects of growth and measurement temperature. Physiol Plant 83: 675–681

Weih M and Karlsson (2001) Growth response of mountain birch to air and soil temperature: is increasing leaf-nitrogen content an acclimation to lower air temperature? New Phytol 150: 147–155

White A, Cannell MGR and Friend AD (2000) The high-latitude terrestrial carbon sink: A model analysis. Global Change Biol 6: 227–245

Will R (2000) Effect of different daytime and night-time temperature regimes on the foliar respiration of *Pinus taeda*: Predicting the effect of variable temperature on acclimation. Exp Bot 51: 1733–1739

Williams JHH and Farrar JF (1990) Control of barley root respiration. Phys Plant 79: 259–266

Wiskich JT and Dry IB (1985) The tricarboxylic acid cycle in plant mitochondria: Its operation and regulation. In: Douce R and Day

DA (eds) Encyclopaedia of Plant Physiology. Higher Plant Cell Respiration, pp. 281–313, Springer Verlag, The Hague

Xiong FS, Mueller EC and Day TA (2000) Photosynthetic and respiratory acclimation and growth response of Antarctic vascular plants to contrasting temperature regimes. Amer J Bot 87: 700–710

Xu M, Debiase TA, Qi Y, Goldstein A and Liu Z (2001) Ecosystem respiration in a young ponderosa pine plantation in the Sierra Nevada Mountains, California. Tree Physiol 21: 309–318

Zimmerman RC, Smith RD and Alberte RS (1989) Thermal acclimation and whole-plant carbon balance in *Zostera marina* L. (eelgrass). J Exp Marine Biol Ecol 130: 93–109

Ziska LH and Bunce JA (1998) The influence of increasing growth temperature and CO_2 concentration on the ratio of respiration to photosynthesis in soybean seedlings. Global Change Biol 4: 637–643

Zogg GP, Zak DR, Burton AJ and Pregitzer KS (1996) Fine root respiration in northern hardwood forests in relation to temperature and nitrogen availability. Tree Physiol 16: 719–725

Chapter 8

Oxygen Transport, Respiration, and Anaerobic Carbohydrate Catabolism in Roots in Flooded Soils

Timothy D. Colmer* and Hank Greenway
School of Plant Biology, Faculty of Natural and Agricultural Sciences, The University of Western Australia, 35 Stirling Highway, Crawley, 6009, WA, Australia

Summary

Soil flooding is a severe abiotic stress for many plant species (e.g., most crops), whereas well adapted species (e.g., rice (*Oryza sativa*) and other wetland species) usually thrive. Flooded soils are usually anaerobic, so that internal O_2 transport from shoot to roots is crucial for sustaining respiration in submerged organs. Formation of aerenchyma, together with a barrier impermeable to radial O_2 loss in basal zones of roots, act synergistically to enhance O_2 diffusion to the apex of the main axis of roots of many wetland species. In some situations the soil is not anoxic, but provides a restricted O_2 supply to the roots. For roots with aerenchyma, O_2 is usually much more readily available from the intercellular gas-filled pathway, than exogenously from the soil. In both types of situations some cells/tissues may become anoxic, especially the apical regions and the stele since these are at the ends of the longitudinal (gas phase) and radial (predominately liquid phase) diffusion paths, respectively. Anoxia in root tissues becomes even more likely when shoots of plants are submerged by floodwaters. The inhibition of oxidative phosphorylation due to anoxia causes a severe energy crisis. Tolerance of anoxia requires a carbohydrate supply to fuel anaerobic catabolism, and apportionment of the scarce available energy to processes essential to survival, whereas several energy-consuming processes typical for aerobic cells

*Author for correspondence, email: tdcolmer@cyllene.uwa.edu.au

H. Lambers and M. Ribas-Carbo (eds.), Plant Respiration, 137–158.

may be reduced. In addition to O_2-deficiency, roots in waterlogged soils must also tolerate high CO_2 partial pressures (e.g., up to 43 kPa); however, information on this topic is scant.

I. Introduction

Flooded soils are usually anaerobic (Ponnamperuma, 1984); plants then depend on internal transport of O_2 into the roots, from the shoots still in contact with the atmosphere (Section III.A). When shoots are completely submerged, and in submerged aquatic plants, O_2 derived from photosynthesis in the shoots moves to the roots growing in anaerobic substrates. There are also situations where soils become deficient in O_2 (i.e. hypoxic) rather than anoxic (Section II), so that roots obtain some O_2 from the external medium as well as O_2 from the shoots via diffusion in intercellular gas-filled spaces within the roots. In both types of situations, the apical regions and the stele of roots may receive a sub-optimal O_2 supply, since these are at the ends of the longitudinal (gas phase) and radial (predominately liquid phase) diffusion paths, and both the stele and apex usually have fast rates of respiration (Sections III.A and III.C). The cessation of oxidative phosphorylation during anoxia results in a severe energy deficit, so that anoxia tolerance is presumably required in at least some tissues/organs of plants inhabiting flood-prone environments (Section IV). Unless the roots can access O_2 from the shoots, cells of many species begin to die within a few hours or days (Gibbs and Greenway, 2003).

Soil O_2 deficiency results from the consumption of O_2 by plant roots and soil organisms without appreciable replacement, since the flux of O_2 into soils is ~320,000 times less when flooding occurs, due to the 10^4-fold slower diffusivity and the 32-fold lower solubility of O_2 in water than in air (Armstrong and Drew, 2002). The anaerobiosis occurs in the bulk soil, and not necessarily in the rhizosphere that, in some species, receives substantial O_2 due to radial O_2 loss (ROL) from the interconnected gas-space system in plants (Section III.B). In addition to O_2 deficiency, flooding also changes other factors that may influence the physiology of plants; CO_2 and ethylene (C_2H_4) both accumulate. The P_{CO_2} increased from 1.8–8 kPa in aerobic soil to 18–43 kPa within two weeks of flooding pots containing several rice paddy soils (Ponnamperuma, 1984). In rhizomes of *Phragmites australis* growing in flooded soil, P_{CO_2} was up to 7.4 kPa (Brix, 1988). Furthermore, anaerobic metabolism by soil micro-organisms may produce reduced compounds, such as Mn^{2+}, Fe^{2+}, S^{2-} and carboxylic acids (Ponnamperuma, 1984; McKee and McKevlin, 1993) that are toxic to plants.

The conditions in flooded soils have profound effects on metabolism in roots; the responses of plants being dependent upon species, tissue, and other environmental variables (e.g., temperature). Possible adaptations, acclimations and responses of plant roots to soil flooding are summarized in Fig. 1. There are dramatic differences among species in tolerance to soil flooding. Intolerant species, like pea (*Pisum sativum*), are severely injured within one to four days following the onset of waterlogging (Jackson, 1979); by contrast, wetland species survive, and in many cases grow vigorously, in waterlogged soils (Justin and Armstrong, 1987). Paddy rice (*Oryza sativa*), with water depths maintained between 50 and 100 mm above the soil till seven days before maturity, yielded 8.4 and 13.6 tonnes ha^{-1} in the tropics and subtropics, respectively (Ying et al., 1998). Such yields are impressive, since rooting depths in paddy fields typically do not exceed 300 mm (Section III. A.2). Thus, although soil flooding is a severe abiotic stress for many plant species, well adapted species can thrive in such conditions.

For plants with shoots in air, O_2 will diffuse at least some distance into the roots along intercellular gas-filled spaces and aerenchyma (Section III. A). Aerenchyma refers to large interconnected gas channels that greatly enhance O_2 movement into submerged portions of plants. The P_{O2} within aerenchymatous roots decreases in a curvilinear gradient with distance from the root-shoot junction, since O_2 is consumed along the diffusion pathway (Armstrong, 1979; Armstrong et al., 2000). Furthermore, gradients in $[O_2]$ also occur in the radial direction, due to uptake of O_2 by the cells along the diffusion path (Fig. 2); the steepness of the gradients will depend upon the porosity and O_2-consumption rates in the

Abbreviations: ABA – abscisic acid; AEC – adenylate energy charge = [ATP + 0.5 ADP] / [ATP + ADP + AMP]; COP – critical O_2 pressure; P_{CO_2} – CO_2 partial pressure; pH_{cyt} – cytosolic pH; pH_{vac} – vacuolar pH; P_{O_2} – O_2 partial pressure; Q_{10} – (rate of reaction at temperature $T + 10\,°C$) / (rate of reaction at temperature T); RGR – relative growth rate; ROL – radial O_2 loss

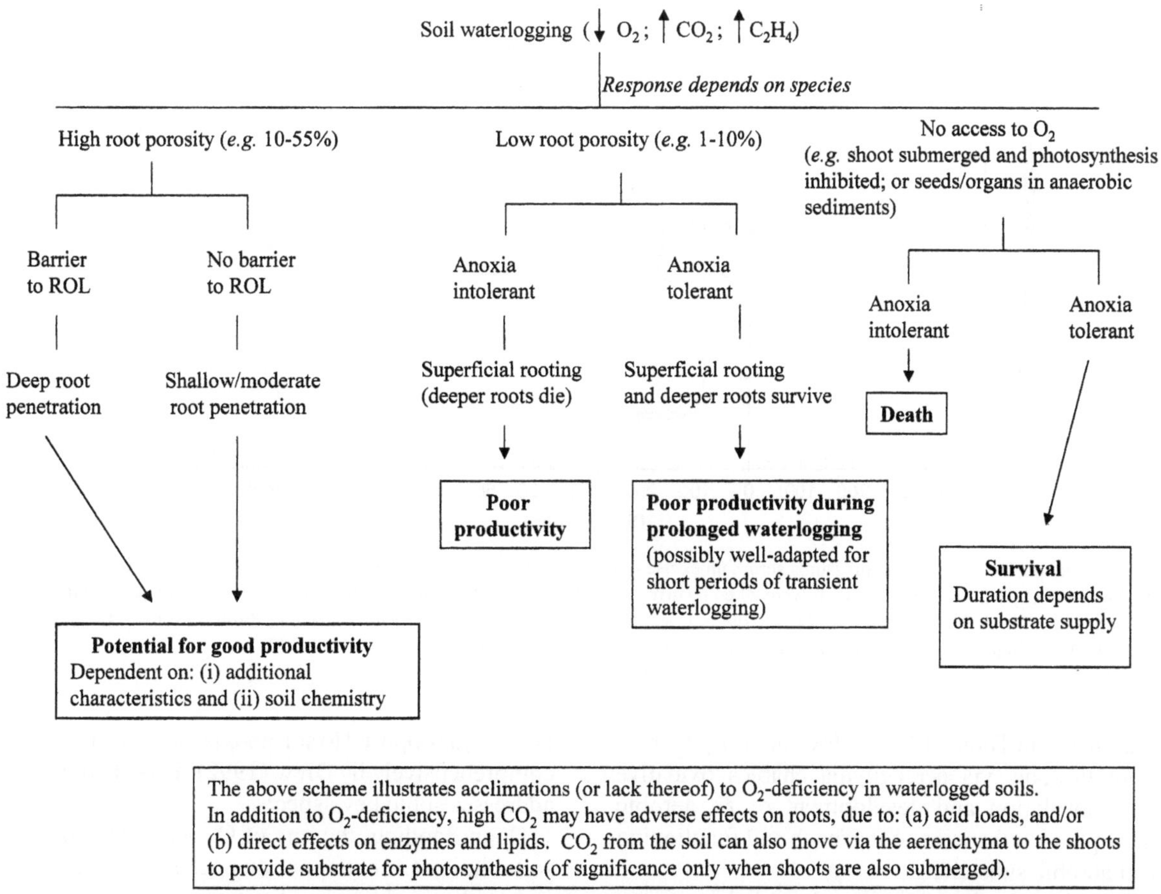

Fig. 1. Scheme showing possible responses, adaptations and acclimations of plants in response to soil flooding.

various tissues. Thus, parts of the root systems and even some tissues within individual roots of plants in flooded soils may suffer anoxia, others will experience sub-optimal $[O_2]$, while yet others may remain fully aerobic, depending on the soil O_2 status, capacity for internal O_2 movement, and root morphology; e.g., surface roots may have access to O_2 while deeper roots suffer anoxia.

Anoxia tolerance might be an important adaptation contributing to the ability of some plants to persist in flood-prone environments (Section IV). Plant species, and organs/tissues, show large differences in tolerance of anoxia (Gibbs and Greenway, 2003). For example, tips of roots of *P. sativum* died within 7 hours, and even with exogenous glucose these tips died after 24 to 36 hours of anoxia (Webb and Armstrong, 1983). On the other hand, rhizomes of the marsh species *Acorus calamus* survived for at least 90 days (Crawford and Braendle, 1996). Glycolysis linked to ethanolic fermentation is the predominant pathway for anaerobic carbohydrate catabolism during anoxia in most plant species (ap Rees et al., 1987), and provides at least some ATP during anoxia (Section IV). Anaerobic carbohydrate catabolism, together with other essential traits (Section IV.A), enable survival during anoxia; the period of survival depending on a supply of carbohydrates, the species, and tissue or organ. Growth during anoxia has only been found in a few cases; shoot extension in anoxia relies on fast rates of anaerobic carbohydrate catabolism fuelled by remobilization of carbohydrate reserves (e.g., coleoptiles of *O. sativa*, Atwell and Greenway, 1987; Perata et al., 1998; stems of *Potamogeton* spp., Summers et al., 2000; Harada and Ishizawa, 2003). Shoot extension enables submerged plants to reach flowing water containing O_2 or adequate light for photosynthesis, or enables the shoot to emerge and establish contact with the atmosphere (Ridge, 1987;

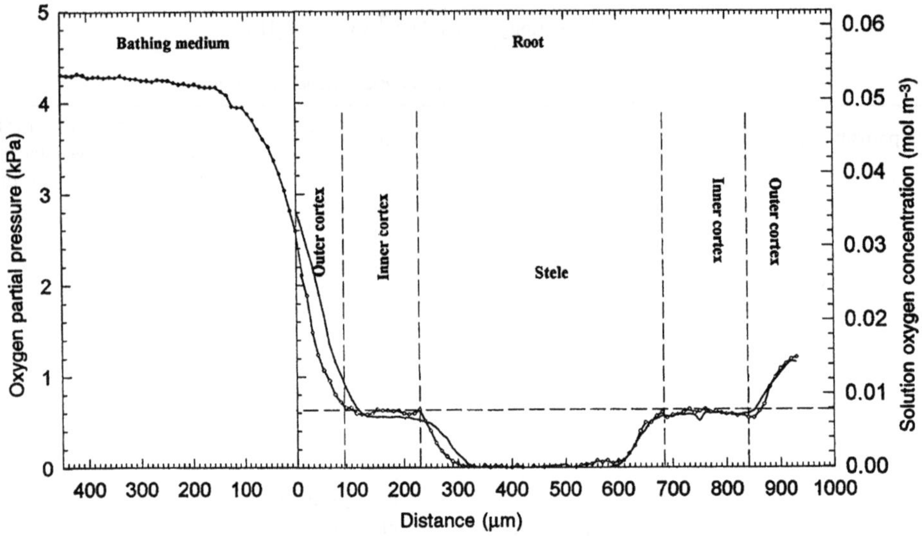

Fig. 2. Radial profile of O_2 partial pressures across an excised primary root of *Zea mays* when in a flowing nutrient solution containing 0.054 mol m^{-3} O_2, at 25°C. Measurements were taken by driving an O_2 microelectrode in steps through the root (75 mm behind the apex, total length 135 mm). The line with symbols is the in-track, and that without symbols is the out-track. Reproduced from Gibbs et al. (1998) in Australian Journal of Plant Physiology with permission of CSIRO Publishing (© CSIRO, 2002).

Voesenek and Blom, 1989). Movement of O_2 from shoot to roots via aerenchyma enables oxidative phosphorylation and development of an aerobic rhizosphere, enhancing root growth and functioning in anaerobic substrates.

This chapter: (i) summarizes knowledge on root aeration, emphasising the importance of both aerenchyma and a barrier to radial O_2 loss (ROL) for internal O_2 movement to the apex of roots in flooded soils; (ii) considers the evidence for, and consequences of, some tissues/cells in roots (e.g., stele and apical regions) being hypoxic/anoxic, while adjacent regions receive adequate O_2 for oxidative phosphorylation, depending on positions relative to the longitudinal (gas phase) and radial (predominately liquid phase) diffusion paths; (iii) evaluates possible limitations of carbohydrate supply to fuel catabolism in roots, as related to anoxia tolerance, and (iv) provides some speculation on the possible effects on respiration in roots of high P_{CO2} typical for flooded soils.

II. Soils with Low, But Not Zero, O_2

In many situations flooded soils become anoxic, and then roots depend completely on internal transport of O_2 from the shoots (Section III.A). However, soils may become hypoxic, rather than anoxic. Soil O_2 deficiency, as opposed to soil anoxia, has been reviewed comprehensively by Drew (1992). This chapter only addresses some key aspects.

Upon flooding, the rate of O_2 depletion from the soil solution and entrapped air pockets will depend on temperature and activity of micro-organisms and plant roots, so depletion will be slow at low temperatures and low content of organic matter (Drew, 1992). For soils in England, waterlogged in different seasons, O_2 almost reached zero at 200 mm depth after 10 to 15 days in January (soil temperature was 4 °C) and after five to six days in May (soil temperature was 11 to 12 °C) (Cannell et al., 1980). In a clay soil during spring in Australia (soil temperature was 11 to 13 °C), O_2 at 125 and 250 mm was depleted within two to four days after flood irrigation (Meyer et al., 1985). The depletion of O_2 over time may allow roots to acclimate prior to the onset of anoxia (Gibbs and Greenway, 2003), and/or be important for survival of roots during the time required to form aerenchyma (Thomson et al., 1990). Other situations of moderate O_2 deficiency in soils were documented for a compacted soil with low air-filled porosity (Blackwell et al., 1985), in a wetland soil with a significant flow of water through the profile (Armstrong and Boatman, 1967), and even in soils under daily trickle irrigation (Meek et al., 1983). Oxygen deficiency would not only develop during transient waterlogging, but also persist for some time following drainage, particularly in fine-textured soils in which the O_2 flux can remain low for two to four days after drainage (Blackwell, 1983).

In addition to the situations summarized above, Setter and Belford (1990) hypothesized that in some cases O_2 may be available to roots in flooded soils if substantial ROL occurs from roots of neighbors. These authors suggested that O_2 movement through the aerenchyma (being up to 70% of root cross-sections) in winter-dormant kikuyu grass (*Pennisetum cladestinum*) might provide O_2 to roots of subterranean clover (*Trifolium subterraneum*) growing during the wet winter; this species has a relatively low root porosity (e.g., 10 to 14% in roots in hypoxic solution; Gibberd et al., 2001). Unfortunately, the evidence is only anecdotal. Investigation of this possibility is warranted, since this could be a hitherto unrecognized mechanism by which plants with little aerenchyma could grow in flooded soils.

The best method to evaluate the capacity of a soil to supply O_2 to roots is by measurements of O_2 fluxes using bare Pt electrodes, as developed by Blackwell (1983). Using intact plants, the extension rate of oat (*Avena sativa*) roots in flooded soil at 10°C, was decreased by 80% once O_2-flux rates dropped below 0.25 $\mu mol\ m^{-2}\ s^{-1}$, a value close to the O_2 flux required to achieve maximum respiration rates in excised root tips at 10 °C (Blackwell and Wells, 1983). To compare with units of P_{O2}, and the commonly used terminology 'critical O_2 pressure' (COP), soil O_2 was 15 kPa when the minimum threshold of O_2 flux to sustain root extension was reached (Blackwell and Wells, 1983). The threshold rate of O_2 flux for maximum root extension was determined by measurements taken while O_2 was depleted following soil flooding (Blackwell and Wells, 1983). That is, the system was not in a steady-state, and so the roots might not have fully acclimated to the lower $[O_2]$ (see Drew, 1997; Gibbs and Greenway, 2003). A steady-state was achieved for seminal roots of *Z. mays* in a chamber perfused with nutrient solution at selected $[O_2]$ (Gibbs et al., 1998). At 4 to 5 kPa O_2, O_2 fluxes and root-extension rates were reduced to about half of those at 21 kPa O_2; and the COP for root extension would have been ~10 kPa. These two estimates of COP for extension of roots with exogenously supplied O_2 are one order of magnitude higher than that determined for roots of *O. sativa* with O_2 supplied endogenously via aerenchyma (Armstrong and Webb, 1985). The higher COP for exogenous O_2 is a consequence of the large apparent resistance to O_2 diffusion between a bulk external solution and the interior of roots (Fig. 2). The apparent resistance to radial O_2 diffusion into roots results approximately equally from resistance to diffusion across liquid-phase boundary layers adjacent to roots and 'resistance' across the epidermis/hypodermis; the latter includes a component due to O_2 consumption by these cell layers as well as a physical resistance. Thus, internal O_2 diffusion from the shoots to the root tips, through gas-filled spaces, can be significant for preventing O_2 deficiency, even when some O_2 is present in the external medium.

Importantly, substantial functioning (e.g., nutrient uptake) of seminal roots of *Triticum aestivum* occurred at $[O_2]$ that severely inhibited growth (Kuiper et al., 1994). These seminal roots would have consumed 10 to 35% of the amount of O_2 from the rooting medium when compared to that in aerated controls, as deduced from respiration rates of excised root segments when measured at a range of external $[O_2]$. The roots would also have received some O_2 from the shoots, but this would presumably have been restricted to the upper parts, since seminal roots of *T. aestivum*, at the growth stage tested by Kuiper et al. (1994) did not form aerenchyma when plants were transferred from aerated to O_2-deficient solution (Thomson et al., 1990). Relative growth rates (RGRs) of these O_2-deficient roots were only 33% of those of aerated controls (Kuiper et al., 1994), and much of the dry mass increment in the O_2-deficient roots was probably due to carbohydrate accumulation (Section IV.B.1). Nevertheless, at 11 days after commencement of the low-O_2 treatment the net uptake rates of K^+ and NO_3^- by the seminal roots were still appreciable, being 80% and 47%, respectively (Kuiper et al., 1994). For roots in hypoxic solution, the stele is likely the first tissue to suffer anoxia (Section III.C), a condition that can inhibit xylem loading and therefore solute transport to the shoots (Gibbs et al., 1998).

III. 'Avoidance' of Anoxia: Internal Aeration in Plants

A. Aerenchyma and O_2 Movement

1. Diffusion of O_2 within Roots

When in an anaerobic medium, the capacity for internal O_2 transport will largely determine respiratory activity and therefore energy status in roots, and, in the longer term, the survival, functioning and growth of roots. Aerenchyma, together with non-aerenchymatous intercellular gas-filled spaces, provides a pathway for movement of gases between shoots and the root tips (Armstrong, 1979; Jackson and Armstrong, 1999; Colmer, 2003a). In roots of many wetland species, aerenchyma is constitutive, and its volume is often further enhanced by soil flooding (Justin and Armstrong, 1987). Formation of aerenchyma following the onset of root-zone O_2 deficiency can take a few days (e.g., adventitious roots of *T. aestivum*, Malik et al., 2003). Roots of some species do not form aerenchyma (e.g., *Brassica napus*, Voesenek et al., 1999), or lose the capacity to form aerenchyma with age/developmental state (e.g., *T. aestivum* seminal roots longer than 100 mm; Thomson et al., 1990). Porosity in roots (i.e. gas-filled intercellular spaces plus aerenchyma) varied from below 1% in some non-wetland species to as much as 53% in a wetland species, for plants grown in flooded potting mix (Justin and Armstrong, 1987). Oxygen moves by diffusion in the aerenchyma in roots (Armstrong, 1979). Convective flows can occur along the shoots and rhizomes of several emergent and floating-leaved wetland species (Section III.A.2), yet even in these species O_2 movement into and along roots is diffusive (Beckett et al., 1988).

The importance of internal O_2 diffusion for maintaining respiratory metabolism in roots when in an anaerobic substrate was demonstrated by measurements of adenylate energy charge (AEC) in root tips of *Z. mays* seedlings transferred into an O_2-free medium after different pre-treatments resulting in roots with, or without, aerenchyma (Drew et al., 1985). Seventy five min after transfer of the intact roots into an O_2-free medium, the AEC in 5 mm root tips was ~0.7 for roots with aerenchyma (porosity 12.7%), compared with ~0.4 in those without aerenchyma (porosity 3.7%). AEC was 0.9 in excised root tips when aerobic (Drew et al., 1985).

Mathematical models predict the dependence of root penetration into anaerobic substrates upon internal O_2 diffusion (Armstrong, 1979; Armstrong and Beckett, 1987). The maximum length (ι) of a root dependent upon internal O_2 diffusion is determined by (Armstrong, 1979):

$$\iota = \sqrt{\left(\left(2D_o \tau \varepsilon \left(C_o - C_1\right)\right) / M\right)} \qquad (1)$$

where D_o = O_2 diffusivity in air, τ = tortuosity factor, ε = the fractional root porosity, $C_o = P_{O_2}$ in air at the root-shoot junction, $C_1 = P_{O2}$ in air spaces just behind the apex at which growth would cease, M = rate of O_2 consumption by the root tissue. Armstrong (1979) cautioned that the assumptions (radially homogeneous tissue, uniform O_2-consumption rates and porosity along the root, absence of ROL) would rarely hold; nevertheless, the equation provides a useful 'benchmark' against which experimental data can be interpreted (Thomson et al., 1992; Gibberd et al., 2001; McDonald et al., 2001).

Equation (1) highlights the importance of fractional root porosity and rates of O_2 consumption on the potential length of the O_2-diffusion path. The importance of high porosity for root growth was shown in a study of 91 species grown in flooded potting mix; species with roots of ≤5% porosity penetrated 30 to 95 mm, whereas those with ≥35% porosity reached 150 to 345 mm (Justin and Armstrong, 1987). Soil temperature is predicted to have a marked effect on the distance O_2 reaches within roots (Armstrong, 1979), since cooler conditions would decrease rates of O_2 consumption along the diffusion path; Q_{10} for O_2 diffusion being 1.1 while for respiration short-term Q_{10} values are 2 to 3 (Chapter 7, Atkin et al.). However, respiration in some species or in some conditions acclimates to changes in temperature (Atkin et al., 2000; Chapter 7, Atkin et al.), so that in these cases the longer-term effect of temperature on the maximum lengths of roots in anaerobic substrates may be less than expected based on the Q_{10} values given above. The possibility that roots might acclimate to hypoxia by decreasing O_2-consumption rates is considered in the next paragraph.

Lambers and Steingrover (1978) hypothesized that roots of waterlogging-tolerant species might decrease rates of O_2 consumption in response to hypoxia, by reduced activity of the alternative oxidase. In leaves of *O. sativa* seedlings submerged in darkness for 24 hours, declines in mRNA levels of nucleus-encoded respiratory genes (coding for alternative oxidase and cytochrome c oxidase) were observed, whereas levels of transcripts of mitochondria-encoded cytochrome c oxidase genes were maintained (Tsuji et al., 2000). For roots of *Hordeum vulgare* seedlings in N_2-flushed solution (with shoots in air), alternative oxidase protein and capacity (assayed for isolated mitochondria) decreased to very low levels, whereas mRNA levels (assayed for whole roots) were relatively constant, indicating possible translational regulation of alternative oxidase protein levels (Szal et al., 2003). Capacity of the cytochrome pathway was the same for mitochondria isolated from aerated or N_2-flushed roots (Szal et al., 2003). By contrast, anoxia increased alternative oxidase protein level in cultured cells of *Glycine max* (Amor et al., 2000). The adaptive benefit of reduced activity of the alternative oxidase in roots dependent on O_2 supply via aerenchyma (i.e. in 'hypoxic' roots), would be that reduced consumption of O_2 would enhance its diffusion further into the root system before being exhausted. On the other hand, increased expression of the alternative oxidase in anoxia has been suggested to assist in protection against oxidative stress during re-aeration (Amor et al., 2000). In roots of *H. vulgare*, alternative oxidase protein increased again within 24 hours after returning seedlings from N_2-flushed to aerated solutions (Szal et al., 2003). In order to reach more definitive conclusions regarding the adaptive significance of these possible changes in alternative oxidase activity for roots of plants in flooded soils, the responses described above, and in vivo activities of the alternative and cytochrome pathways measured using the ^{18}O-fractionation technique (Chapter 3, Ribas-Carbo et al.), should be evaluated in roots of a wider range of flooding-tolerant and -intolerant species when exposed to a range of $[O_2]$. Rates should be assessed on a tissue fresh mass or on a protein basis, since formation of aerenchyma will reduce rates on a tissue volume basis, while accumulation of non-structural carbohydrates (Section IV.B.1) can be large enough to give a misleading impression when rates are expressed on a dry mass basis.

In addition to higher porosity and lower rates of O_2 consumption, other morphological and anatomical features also enhance the capacity for longitudinal O_2 diffusion in roots. (i) Roots of larger diameter have a lower diffusive resistance per unit length, compared with roots of smaller diameter (Armstrong, 1979; Armstrong et al., 1982). (ii) Lateral roots consume O_2 from the aerenchyma in the parent root, thus lowering P_{O_2} in the aerenchyma, and decreasing diffusion to the apex of the main axis (Armstrong et al., 1983). However, if laterals emerge near the base of roots of high porosity, O_2 consumed by the laterals might have little impact on the P_{O_2} in the root base, and thus on O_2 diffusion to the tip of the main root axis (Armstrong et al., 1990). (iii) Roots with a narrow stele would consume less O_2 per unit length of the diffusion path in the aerenchyma that leads to the tip of the main axis, than roots with a relatively thick stele (Armstrong and Beckett, 1987). Respiration in the stele can be relatively fast; e.g., ~3-fold faster on a volume basis than in the cortex plus outer cell layers in roots of *Z. mays* (Armstrong et al., 1991b), and several wetland species have a relatively narrow stele (McDonald et al., 2002). (iv) A barrier to ROL in the cell layers exterior to the aerenchyma, as shown in the basal zones in roots of numerous wetland species (Armstrong, 1979; Colmer, 2003a), would diminish O_2 losses to the rhizosphere, and thus enhance longitudinal diffusion toward the tip of the main axis (Section III.B). Plant species show much variation in the expression of these morphological and anatomical traits, with a combination of these being expressed in many wetland species, and not, or to a lesser extent, in dryland species (Justin and Armstrong, 1987; McDonald et al., 2002; Colmer, 2003a; Garthwaite et al., 2003).

2. Through-flows of Gases along Rhizomes of Wetland Plants

Convective flows can transport O_2 (and other gases) along the shoots and rhizomes of several emergent and floating-leaved wetland species, but only when pressure gradients are established along the aerenchymatous pathway, and when there is a low-resistance exit to the atmosphere, enabling through-flows (Beckett et al., 1988). For example, in the yellow waterlily (*Nuphar luteum*) growing in 1.5 m of water, flow rates of gases within the aerenchymatous petioles were 40 to 56 ml min^{-1} (Dacey, 1981); these impressive rates of gas flow were aptly described by Dacey (1980) as 'internal winds.' Indeed, through-flows can result in an increase of two orders of magnitude in the effective length of aeration in culms and rhizomes above that possible via diffusion (Armstrong et al., 1991a), enabling some emergent wetland plants such as *Eleocharis sphacelata* to inhabit areas with up to 2 m water depth (Sorrell et al., 1997). Through-flows increase P_{O_2} in rhizomes above those if only diffusion occurs, and therefore enhance O_2 diffusion into roots arising from the rhizomes. For example, in *P. australis* through-flow increased P_{O_2} in the rhizome from ~9 to ~20 kPa, and increased by ~4-fold the ROL from just behind the main apex of an adventitious root arising from the rhizome (Armstrong J et al., 1992).

Rates of through-flow are determined by the pressure gradient and the resistance to flow along the aeration system. The pressure gradient can result from: (i) pressures above atmospheric generated in living shoot tissues (e.g., several wetland species; Brix et al., 1992) due to gradients in H_2O vapor concentration between the interior and exterior of an enclosed space with the surface of the enclosure containing micro-pores (Dacey, 1981; Armstrong et al., 1996a,b; Grosse, 1996), or (ii) venturi-induced suction caused by wind blowing over the open ends of broken culms, at least in *P. australis* (Armstrong J et al., 1996).

The requirements for pressurization in leaves have been evaluated using physical models. Pressurization occurs when there is humid air in an enclosed space (e.g., a leaf blade or sheath), surrounded by less humid air, and with at least some part of the surface having small enough pores to have significantly more resistance to pressure flow than to diffusion of gas molecules ('Knudsen regime'; Leuning, 1983). Since H_2O vapor occupies space, it dilutes the concentrations of other atmospheric gases in the enclosed space, resulting in concentration gradients of N_2 and O_2, so these gases diffuse in, and increase the total pressure within the enclosed space. Static pressures increase as pore size decreases; nevertheless, pressurization can occur with pores up to 3 μm in diameter, depending on other factors, such as the overall porosity and thickness (i.e. diffusion path length) of the partition between the enclosed space and atmosphere (Leuning, 1983; Armstrong et al., 1996a,b). The static pressure (ΔP_{static}) above atmospheric in a 'closed' system is described by the equation (Armstrong et al., 1996a):

$$\Delta P_{static} = (P_a + P_{wi} - P_{wa}) - P_a \qquad (2)$$

where: P_a is pressure of the atmosphere, P_{wi} is H_2O vapor pressure inside the enclosed space, and P_{wa} is H_2O vapor pressure of the atmosphere outside the enclosed space. If the enclosed space has an outlet, then pressurization will result in a flow of gases via the outlet at a rate determined by the pressure gradient and resistance to flow, and $\Delta P_{dynamic}$ will be lower than ΔP_{static}. For a given resistance in the exit pathway, maximum flow rates occur when the diameters of the pores in the surface of the enclosed space are ~0.2 μm (Armstrong et al., 1996a). If pores in the partition are smaller than ~0.2 μm, the static pressure would become even higher than at 0.2 μm, but the flow would be reduced because the resistance to flow through the partition would become a significant part of the resistance of the pathway as a whole. Leaf-to-air temperature gradients (leaf warmer) enhance pressurization, since this increases the gradient in H_2O vapor concentration across the partition, at least in the case of emergent species (Steinberg, 1996; Colmer, 2003a).

An alternative mechanism to positive pressures generated in living shoot tissues is venturi-induced suction caused by wind blowing over tall, broken culms; to date, this mechanism is only documented for *P. australis* (Armstrong J et al., 1996). Wind blowing across an open cylinder causes a localized reduction in air pressure (ΔP_{static}) below atmospheric, as described by Bernoulli's equation (Armstrong J et al., 1992):

$$\Delta P_{static} = -\tfrac{1}{2}\,\rho\, V^2 \qquad (3)$$

where: ρ is the density of air and V the wind velocity. ΔP_{static} for broken culms was ~60 % of theoretical values, since the culms are leaky and have jagged rims (Armstrong J et al., 1992). In the intact plant, gas is sucked out of taller, broken culms, and air enters via shorter culms exposed to the lower wind speeds near the water/ground surface (Armstrong J et al., 1992, 1996).

As discussed above, through-flows greatly increase aeration of rhizomes, and therefore O_2 available to diffuse into roots arising from rhizomes. Such flows enable rhizomes and therefore roots, to grow at depths deeper than accessible to roots arising from a root-shoot junction of a non-rhizomatous species, such as *O. sativa*. Maximum lengths of the roots per se in these two cases should be similar (i.e. if porosity, ROL, and O_2-consumption rates were similar), with P_{O2} at the root-rhizome or root-shoot junctions being near atmospheric. As examples, maximum root lengths for *O. sativa* and *P. australis* in a waterlogged potting mix were, respectively, 276 mm and 223 mm (Justin and Armstrong, 1987). However, since rhizomes can grow relatively deep into flooded soils, the potential soil volume explored by rhizomatous species can be larger than for non-rhizomatous species. For example, rhizomes of *Phragmites communis* growing in the field occurred at depths between 250 to 1500 mm below the soil surface (Haslam, 1970). Deep rhizomes, with roots, might give a competitive advantage for nutrient acquisition, at least in flooded soils low in nutrients. Nevertheless, the non-rhizomatous *O. sativa* is very productive in permanently flooded fields, where the plough pan is at 200 mm depth from the soil surface, and maximum lengths of root main axes is 290 mm, presumably due to horizontal growth along the plough pan (Kirk, 2003). Such data show that convective flows and the ability to access soil layers deeper than those accessible by aerenchymatous roots arising from a shoot base near the soil surface, are not required for high productivity, at least in fertile agricultural soils when these are flooded.

B. Radial O_2 Loss (ROL) from Roots to Flooded Soils

ROL may 'protect' roots against reduced toxins (e.g., Fe^{2+}) often present in flooded soils; in the oxygenated rhizosphere, Fe^{2+} would be oxidized to the insoluble Fe_2O_3 (Begg et al., 1994; Mendelssohn et al., 1995; St-Cyr and Campbell, 1996). Carboxylic acids produced by micro-organisms in anaerobic soils may also be catabolized by aerobic micro-organisms in an oxygenated rhizosphere. However, there is a cost to these putative beneficial effects; ROL along a root axis would diminish the O_2 supply to the apex of the main axis, and therefore reduce rooting depths in anaerobic substrates. For example, Armstrong (1979) estimated that even for roots of *O. sativa* with a barrier to ROL at ≥ 50 mm behind the apex (i.e. with O_2 loss only from the apical 50 mm), ~30% of the O_2 supplied via the aerenchyma might be lost to the rhizosphere.

The flux of O_2 from the root aerenchyma to a soil is determined by: (i) the concentration gradient, (ii) the physical resistance to O_2 diffusion, and (iii) the rate of O_2 consumption in the cells exterior to the aerenchyma. Roots of many, but not all, wetland species contain a 'tight' barrier to ROL in the basal zones (Armstrong, 1964, 1979; Visser et al., 2000; McDonald et al., 2002; Colmer, 2003a), as shown in Fig. 3 for an intact adventitious root of *P. australis* in an O_2-depleted medium (Armstrong et al., 2000). The $[O_2]$ at the root surface was relatively high near the tip, whereas root surface $[O_2]$ was extremely low at 30 mm and further from the apex (Fig. 3). This pattern of $[O_2]$ at the epidermis occurred despite the opposite profile for P_{O_2} in the aerenchyma; as predicted from diffusion in a tube with O_2 consumption along the path (see Armstrong, 1979), internal P_{O_2} was highest near the root base, and declined in a curvilinear gradient toward the root tip (Fig. 3). In the basal zones, the combination of high P_{O_2} in the aerenchyma, but low ROL, indicates the permeability to O_2 movement across the outer cell layers was very low.

Measurements of $[O_2]$ across an intact adventitious root of *P. australis* when in an O_2-depleted medium, showed that the greatest impedance to ROL was in the hypodermis + epidermis (Fig. 4). Oxygen consump-

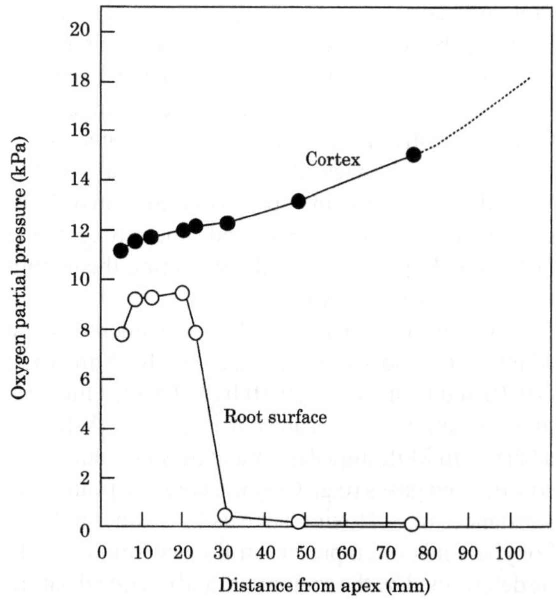

Fig. 3. Longitudinal profile of O_2-partial pressures along the surface and in the cortex of an adventitious root of *Phragmites australis* when in an O_2-depleted solution, at 23 °C. Measurements were taken using an O_2 microelectrode. The shoots were in air, and the intact 100 mm root, as well as the rhizome and other roots, were in an O_2-free solution containing 0.05% (w/v) agar to prevent convection. Reproduced from Armstrong et al. (2000) in the Annals of Botany with permission of the Annals of Botany Company.

tion by the hypodermis + epidermis, in addition to a physical barrier against O_2 diffusion, would cause the substantial drop in $[O_2]$ across the outer cell layers of the root (Armstrong et al., 2000).

A 'tight' barrier to ROL develops constitutively in the basal zones of adventitious roots of many wetland species, but was induced by growth in stagnant deoxygenated medium in several others (Colmer, 2003a), including *O. sativa* (Colmer, 2003b). By diminishing O_2 losses from the basal root zones to the soil, the barrier to ROL enhances longitudinal O_2 diffusion towards the apex of the main axis, and therefore tends to keep higher apical $[O_2]$ than otherwise would be the case; the higher $[O_2]$ will, in turn, increase the capacity for aeration of the soil surrounding the apex. Even in species with a 'tight' barrier to ROL, O_2 losses are substantial near the root tip, and also from laterals (Armstrong J and Armstrong, 1988; Conlin and Crowder, 1989; Armstrong et al., 1990; Sorrell, 1994). The quantitative significance of ROL from laterals for sediment oxygenation by *P. australis* (adventitious root main axes contain a barrier to

ROL in basal zones) was highlighted in mathematical modeling by Armstrong J et al. (1996) that predicted ~97% of O_2 loss from the root system may occur via laterals (based on P_{O_2} of 17 kPa at root-rhizome junction). The restriction of ROL to key sites (laterals and root tips) may be viewed as an 'efficient use' of the limited supply of O_2; this spatial distribution of ROL maximizes root penetration, and functioning of the laterals and tip regions, while protecting them from toxins in anaerobic soils.

Although a barrier to ROL is considered to be adaptive for roots in waterlogged soils, Armstrong (1979) and Koncalova (1990) hypothesized this feature may diminish water and nutrient uptake. Whether a barrier to ROL impedes uptake of water, nutrients, soil-derived gases (e.g., CO_2 and ethylene) and other substances (e.g., 'toxins' such as Fe^{2+}, Mn^{2+} and carboxylic acids often present in flooded soils), would be determined by the anatomical and chemical nature of the barrier, and the possibility of 'passage areas' (Armstrong et al., 2000; Beckett et al., 2001). Unfortunately, information on the nature of the barrier, and whether it impedes movement of substances from soils into roots is scant (Colmer, 2003a). Diffusive permeabilities across epidermal/hypodermal 'sleeves' isolated from roots of the wetland *Carex arenaria*, as compared with 'sleeves' from roots of the non-wetland species onion (*Allium cepa*), were 200-fold lower for water and 500-fold lower for $H_2PO_4^-$ and Ca^{2+} (Robards et al., 1979). In adventitious roots of *O. sativa*, hydraulic conductivity in the radial direction (expressed on a root surface area basis) was 0.2 to 0.5 of that in the primary seminal root of *Z. mays* (Miyamoto et al., 2001). However, experiments with 'sleeves' of the outer cell layers from adventitious roots of *O. sativa* indicated these layers do not limit the rate of water uptake (Ranathunge et al., 2003). Moreover, induction of the barrier to ROL in adventitious roots of *O. sativa* had no effect on the capacity for net NO_3^- uptake from aerobic solution, measured using ion-selective microelectrodes (Rubinigg et al., 2002). To better understand the possible impacts of a barrier to ROL on root functioning, measurements of nutrient and water uptake (in air and anoxia), as well as entry of soil-derived gases and 'toxins', are required for roots of a wider range of wetland species. Uptake rates of these various substances should be evaluated for zones of roots with a barrier to ROL, and for zones (i.e. tip of the main axis and laterals) without a barrier to ROL. A high degree of spatial resolution will be required, since at least in the case of *P. australis*, small O_2-permeable areas can occur in regions that otherwise contain a barrier to ROL (Armstrong et al., 2000).

C. Radial Diffusion of O_2 into the Stele of Roots

In roots dependent on O_2 supplied via the intercellular gas-filled spaces, such as aerenchyma, and in roots dependent on a restricted O_2 supply from the soil, anoxic zones may develop adjacent to zones with sufficient O_2 for oxidative phosphorylation. In both cases, such anoxic zones have been shown for the stele and the root apex.

For submerged tissues, $[O_2]$ gradients will develop between the epidermis and the core of the tissue. Berry and Norris (1949) hypothesized that such gradients could result in anoxia in the core, with more peripheral cells still receiving adequate O_2 for oxidative phosphorylation. Data from excised onion (*A. cepa*) root segments showed that when the $[O_2]$ was below the COP for respiration, the RQ increased, implying a switch of at least some cells to ethanolic fermentation (Berry and Norris, 1949). Severe hypoxia can also occur in the central tissues and/or most distal zones (i.e. just behind the root tip) in roots of *Z. mays* when in an O_2-deficient medium. The evidence for this consists of: (i) measurements of $[O_2]$ within roots using microelectrodes (Armstrong et al., 1994), (ii) higher pyruvate decarboxylase (PDC) activity in the stele versus the cortex, separated from the intact primary root of seedlings in hypoxic or anoxic solutions (Thomson and Greenway, 1991), and (iii) increased [alanine], a product of anaerobic catabolism, in 2 to 3 mm tips sampled from the primary root of seedlings after 24 to 31 hours in O_2-deficient (i.e. 0.06 mol m^{-3} O_2) solution (Gibbs et al., 1995). Moreover, mathematical models indicate that cells in regions of the stele may receive sub-optimal O_2, even in roots reliant on an internal O_2 supply via aerenchyma (Armstrong and Beckett, 1987). Thus, the position of any particular cell within an organ or tissue, relative to the longitudinal (gas phase) and radial (predominately liquid phase) diffusion paths, will be a major determinant of its O_2 supply.

The extent to which the stele becomes O_2 deficient depends upon several factors, such as: (i) the P_{O_2} in the cortical gas spaces (determined by the external $[O_2]$, the O_2 diffusion in the aerenchyma, and the distance from the root-shoot junction), (ii) the resistance to radial O_2 diffusion through the endodermis and stelar

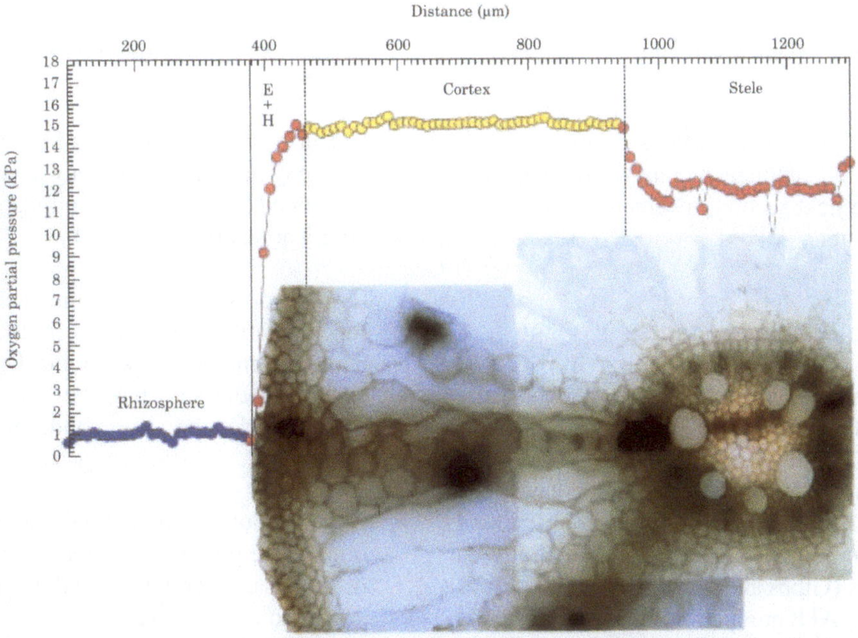

Fig. 4. Radial profile of O_2-partial pressures adjacent to, and inside, an intact adventitious root of *Phragmites australis* when in an O_2-depleted solution, at 23 °C. Measurements were taken using an O_2 microelectrode, by driving the microelectrode in steps toward and through the root. The shoots were in air, and the measurements were taken 100 mm behind the tip of an intact root 160 mm in length. Note the relatively high O_2-partial pressure in the cortex and in the relatively narrow stele, but the steep gradient of O_2-partial pressure within the epidermal/hypodermal cylinder, and very low concentrations in the solution adjacent to the root exterior, indicating that the outer layers have a very high resistance to O_2 loss from the root to the medium. Reproduced from Armstrong et al. (2000) in the Annals of Botany with permission of the Annals of Botany Company.

tissues, (iii) the radius of the stele (i.e. diffusion-path length), and (iv) the rate of O_2 consumption in the stele (Armstrong and Beckett, 1987). In contrast to the relatively high porosity and gas-phase diffusion in the root cortex, the stele usually has negligible porosity, so that radial diffusion is predominately in the liquid phase (Armstrong and Beckett, 1987; Armstrong et al., 1991b). Additionally, the rate of O_2 consumption (per unit volume) in the stele can be about three-fold faster than that in the cortex plus outer cell layers, at least in roots of *Z. mays* (Armstrong et al., 1991b). This combination of liquid-phase diffusion and relatively fast O_2-consumption rates result in steep declines in $[O_2]$ with distance into the stele (Fig. 2).

Roots of many wetland species tend to have a narrow stele; e.g., in *O. sativa* the stele occupies less than 5% of the cross-sectional area of the adventitious roots, as compared with 24 to 36% in *Sorghum bicolor* (McDonald et al., 2002). Therefore, the diffusion path from the aerenchyma to the centre of the stele is relatively short in roots of many wetland species; therefore, P_{O2} in the stele may only decline by a few kPa between the aerenchyma and the centre of the stele (Fig. 4). If O_2 deficiency did occur in the stele, or part thereof, this would have implications for root functioning, especially ion transport to the shoot (Gibbs et al., 1998); therefore, avoidance of this condition might be necessary to maintain vigorous growth in flooded soils.

Submerged plants are subject to temporal changes in O_2 status, as well as the spatial differences in O_2 supply described above. In submerged *O. sativa*, O_2 produced in photosynthesis, diffused in aerenchyma to the roots, resulting in 0.6 to 0.8 mol m^{-3} O_2 in solution at the surface of the elongating zone within a few min after starting the light period, and this increase in $[O_2]$ was followed by resumption of root extension (Waters et al., 1989). By contrast, following the onset of darkness, $[O_2]$ at the root tip declined rapidly to almost zero, and root extension slowed almost immediately and had virtually ceased within ~1 hour; furthermore, ethanol was then produced by the roots (Waters et al., 1989). Similarly, O_2 produced during photosynthesis in submerged aquatic and marine plants resulted in marked diurnal changes in $[O_2]$ in roots (Sorrell and Dromgoole, 1987; Christensen et al., 1994; Pedersen et al., 1995; Connell et al., 1999).

The occurrence of anaerobic carbohydrate catabolism in roots of intact plants during times of waterlogging and submergence makes it relevant to discuss, in the next section, mechanisms of anoxia tolerance in plants.

IV. Anoxia Tolerance in Plants

Anoxia tolerance has been reviewed comprehensively (Drew, 1997; Gibbs and Greenway, 2003; Greenway and Gibbs, 2003). The central feature of anoxia is an energy crisis; rates of ATP production have been assessed from rates of anaerobic catabolism, and range between 3% to as high as 38% of rates in aerobic tissues (Gibbs and Greenway, 2003). For *Z. mays* root tips, ATP production in anoxia was ~7% of the rate when fully aerobic (sample perfused with solution bubbled with 100 kPa O_2), measured using in vivo saturation-transfer ^{31}P-NMR spectroscopy (Roberts et al., 1985); this result was similar to rates assessed from data on anaerobic carbohydrate catabolism (Gibbs and Greenway, 2003). Anoxia tolerance has been shown for several plant organs, such as coleoptiles, stems, tubers, and rhizomes of wetland species. Although anoxia tolerance may also be of adaptive significance for roots of plants in flood-prone environments, further work is required to provide conclusive evidence.

Anoxia tolerance is crucial for rhizomes and storage organs of marsh species buried in anoxic mud over winter. Rhizomes of the marsh species *Acorus calamus* survived anoxia for more than 90 days (Crawford and Braendle, 1996). In spring, growth of shoots commences, and O_2 becomes available when the shoots start photosynthesis in illuminated water or when the shoots reach the atmosphere (Crawford and Braendle, 1996). In *Potamogeton pectinatus*, elongation of stems emerging from tubers was much faster in anoxia than in air (Summers and Jackson, 1996). Anoxia tolerance is also necessary during establishment of plants from seeds in flooded soils. The coleoptile of *O. sativa*, and of the paddy weed *Echinochloa oryzoides*, elongated during five days of anoxia, imposed three days after imbibition (Pearce and Jackson, 1991).

In the absence of comprehensive data we can only speculate if anoxia tolerance is also adaptive for roots of plants in waterlogged soils, and for plants submerged during flooding. We suggest four situations:

(i) During the initial period of waterlogging, when aerenchyma in roots of many species has not yet developed (e.g., *Z. mays*, Konings, 1982; *T. aestivum*, Thomson et al., 1990).

(ii) For roots exposed to transient waterlogging. In anoxia-intolerant species, roots below the water table decompose (Jackson and Drew, 1984), but whether roots shown to be anoxia-tolerant can survive under the same conditions has not been tested.

(iii) For roots of submerged plants during nighttime, since photosynthesis is the main O_2 source resulting in marked diurnal fluctuations in O_2 supply to the roots (e.g., *O. sativa*; Waters et al., 1989). Ethanol production in roots of submerged *O. sativa* occurred during the night (Waters et al., 1989), implying anoxia tolerance in roots is one component of submergence-tolerance.

(iv) During long-term waterlogging when parts of the stele might be anoxic, even in some species with aerenchymatous roots (Section III.C). Fast rates of anaerobic carbohydrate catabolism in the stele may then contribute to energy required for nutrient transport to the shoot (e.g., in xylem loading), and for transport of sugars to the root cells (Saglio, 1985).

A. Anaerobic Carbohydrate Catabolism: Fast and Slow Modes

There are two modes of carbohydrate catabolism as related to anoxia tolerance in plants. In certain anoxia-tolerant organs/tissues, rates of carbohydrate catabolism during anoxia can be substantially faster than when in air (Mode 1: 'Fast mode' with accelerated glycolysis; i.e. a 'Pasteur Effect'); yet in other anoxia-tolerant organs/tissues catabolism can be much slower than when in air (Mode 2: 'Slow mode,' sometimes called 'metabolic arrest') (Gibbs and Greenway, 2003).

A prominent case for the mode with fast rates of anaerobic catabolism is the stems of the aquatic plant *P. pectinatus*, in which elongation is promoted by anoxia (Summers and Jackson, 1996). The stems have an estimated rate of glycolysis six times faster in anoxia than in air (based on CO_2-evolution rates; Summers et al., 2000). Mobilization of carbohydrate reserves in the tubers (described for *P. distinctus*; Harada and Ishizawa, 2003), and translocation to the stems, almost certainly provides the required substrates. The fast rates of glycolysis linked to ethanol production, and perhaps some energy production via sucrose synthase, would provide the ATP required for extension growth (Gibbs and Greenway, 2003).

The most extreme example of the mode with slow anaerobic catabolism is germinating lettuce (*Lactuca sativa*) seeds, which can survive anoxia for at least 14 days without any measurable CO_2 evolution, indicating very slow rates of anaerobic catabolism (Pradet and Bomsel, 1978). The 'slow' mode conserves carbohydrate reserves, and thus prolongs the duration of survival; the duration also depending on the initial levels of carbohydrates. The 'slow' mode of anaerobic catabolism may not be of relevance for roots of plants with shoots in air, since levels of non-structural carbohydrates often increase upon exposure of the roots to O_2 deficiency (Section IV.B.1), but may become important during complete submergence when non-structural carbohydrates in plants are usually depleted (Section IV.B.2). The other, less extreme, examples of plant tissues/organs with a tendency toward this slower mode of carbohydrate catabolism are discussed in Gibbs and Greenway (2003); in these cases glycolysis appears to be down-regulated following acclimation to anoxia (e.g., in the coleoptile of *O. sativa*; Colmer et al., 2001).

Anoxia tolerance in tissues with either mode of carbohydrate catabolism will also require other adaptive characteristics (Drew, 1997; Chang et al., 2000; Gibbs and Greenway, 2003); even the 'fast' mode produces considerably less ATP than oxidative phosphorylation, and cells would still suffer an energy shortage. These other adaptive characteristics include:

(i) A reduction in energy required for maintenance which, for anoxia-tolerant tissues, have been assessed to be two- to nine-fold lower in anoxic than in aerated tissues (Greenway and Gibbs, 2003). Such reductions may be achieved by decreases in protein turnover, ion fluxes, and other processes (Greenway and Gibbs, 2003).

(ii) Direction of the restricted supply of energy to crucial processes for survival; e.g., regulation of pH_{cyt} (Greenway and Gibbs, 2003).

Greenway and Gibbs (2003) suggested that without these acclimations, membrane selectivity declines, culminating after some time in cell death due to the loss of integrity of the tonoplast (Zhang et al., 1992), causing release of hydrolytic enzymes from the vacuole.

Both respiration and anaerobic catabolism require adequate substrates; these might become limiting because of reductions in photosynthesis in plants in waterlogged soil and when submerged. The next section examines levels of non-structural carbohydrates in plants in flooded soils.

B. Substrate Availability: Soluble Sugars and Starch

Photosynthesis declines in many species soon after waterlogging occurs. In the case of submergence, photosynthesis is also usually reduced, with the degree of inhibition depending on the characteristics of the flood-waters. In this section we discuss whether these declines in photosynthesis result in a deficiency of carbohydrates for respiration (and ethanolic fermentation), or whether in the case of plants in waterlogged soil the decreased photosynthesis results from feedback due to increased levels of carbohydrates that accumulate due to reduced consumption resulting from decreased growth. In that case, carbohydrates should not be deficient, provided that transport of sugars continues even to the root tips. The available data indicate that carbohydrates are usually ample in plants with shoots in air and roots in O_2-deficient media (Section IV.B.1), while the opposite applies to plants when shoots are submerged also (Section IV.B.2). The inference that carbohydrates may limit survival and growth of submerged plants requires testing by adding exogenously supplied sugars.

1. During Waterlogging

Plants exposed to a low $[O_2]$ in the root medium usually have higher levels of non-structural carbohydrates in both leaves and roots, than plants in aerated conditions (Setter et al., 1987). For plants in nutrient solutions flushed with N_2, non-structural carbohydrates increased in four species from either flooding-prone or well-drained habitats (Albrecht et al., 1997), and in *T. aestivum* (Barrett-Lennard et al., 1988; Albrecht et al., 1993). For plants in waterlogged soils, non-structural carbohydrates increased 5-fold in leaves of *T. aestivum* (Malik et al., 2002), while in roots of *Medicago sativa* sucrose increased from 2–6% to 11–16% of the dry mass (Barta 1988; Castonguay et al., 1993); in leaves starch increased 2.5- to 6-fold reaching over 20% of the dry mass (Castonguay et al., 1993). The accumulation of non-structural carbohydrates occurs, despite reductions in rates of photosynthesis in plants with roots in waterlogged soils or in hypoxic nutrient solution (Castonguay et al., 1993; Albrecht et al., 1997; Malik et al., 2001, 2002). Setter et al. (1987b) argued that for plants with roots in O_2-deficient media, the evidence favored the scenario as shown in Fig. 5; that is, non-structural carbohydrates would accumulate primarily due to reduced utilization because growth is inhibited, as also concluded by others (Barta, 1987; Castonguay et al., 1993; Albrecht et al., 1997). Reductions in photosynthesis in plants in waterlogged soil may then result from feedback regulation by the high concentrations of non-structural carbohydrates in leaves (Fig. 5).

In contrast to the preceding, Albrecht et al. (1997) suggested that in *Senecio aquaticus* (a species that inhabits flooding-prone soils), reduced utilization of carbohydrates in growth and catabolism did not account for high accumulation of fructans, because after 9 days in N_2-flushed solution culture there was only 'slightly altered biomass production.' We disagree with this view. Firstly, plant dry masses were reduced by ~20%; assessed from fresh mass data presented by Albrecht et al. (1997), but since the fresh:dry mass ratios were not given, we used values of 9.5 for shoots (Lambers, 1976) and 9.7 for roots (calculated from Crawford, 1966). Furthermore, growth of *S. aquaticus*, assessed as dry mass minus non-structural carbohydrates, was reduced by 51% (assessed from Albrecht et al., 1997). The large reduction in growth (i.e. dry mass excluding non-structural carbohydrates) of *S. aquaticus* in a N_2-flushed solution is somewhat unusual for a wetland species. The responses described above for *S. aquaticus* differ from an earlier study in which root dry mass was not reduced for plants in N_2-flushed solutions for 23 days (Lambers and Steingrover, 1978). Furthermore, soluble sugars did accumulate (up to two-fold) in roots of *S. aquaticus* during the first few days; however, subsequently soluble sugar concentrations were lower (~60% of values in aerated controls) for the remainder of the experiment (Lambers et al., 1978). The reason for these differences in results between the two studies is not known. We

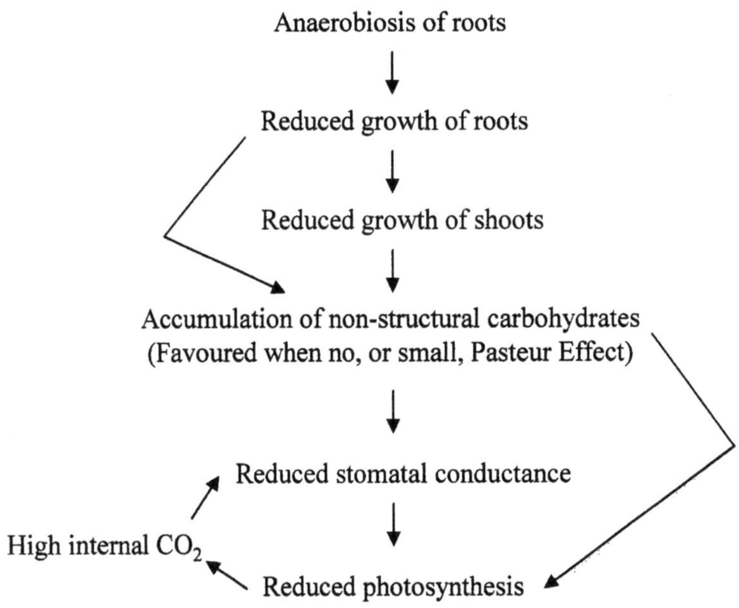

Fig. 5. Hypothesis to account for accumulation of non-structural carbohydrates in shoots and roots of plants with roots in an O_2-deficient medium. Based on Setter et al. (1987). 'Pasteur Effect' is accelerated glycolysis.

conclude that the data on *Senecio* species published by Albrecht et al. (1997) are consistent with the hypothesis that carbohydrate accumulation in plants exposed to root-zone O_2-deficiency is due to reduced growth decreasing utilization of carbohydrates. An additional contribution to carbohydrate accumulation might be reduced catabolism in hypoxic roots, if the activity of the alternative oxidase declines (discussed in Section III.A.1).

Increases in non-structural carbohydrates in plants grown with a low [O_2] in the root-zone are not universal; levels of carbohydrates in plants tend to be variable and depend on environment and time of day. As one example, in *O. sativa* with roots in N_2-flushed solutions, the sum of hexoses and sucrose increased only during summer, and not during winter, presumably due to differences in rates of photosynthesis associated with light intensity and day length (Limpinuntana and Greenway, 1979). Furthermore, 2-mm tips of seminal roots of *T. aestivum* had lower [total sugar] when in N_2-flushed, as compared with aerated solution, even though all other tissues had higher [total sugar] (Barrett-Lennard et al., 1988). The main axes of these seminal roots had lost the capacity to elongate (established when returned to aerated solution), so the failure to accumulate sugars in the apical 2 mm of roots might have resulted from severe injury; anoxic roots of *T. aestivum* suffered substantial losses of solutes, including sugars, to the medium (Greenway et al., 1992).

Carbohydrates accumulated in plants during waterlogging may be important for recovery upon soil drainage (Albrecht et al., 1993, 1997); e.g., in both seminal and adventitious roots of *T. aestivum* the RGRs for ethanol-insoluble dry mass over the first four days after return of plants from N_2-flushed to aerated solutions were 37% faster than those in continuously aerated plants (Barrett-Lennard et al., 1988). Furthermore, after 14 days of waterlogging the total accumulated non-structural carbohydrates in *T. aestivum* were the same or higher than the amount required for adventitious root growth during the first 7 days after drainage (Malik et al., 2002). Whether these reserves are of adaptive value is not sure. During the first few days after drainage, stomatal opening and photosynthesis remained at less than 35 to 50% of values in plants in continuously drained pots (Malik et al., 2001). Photosynthesis during the first few days of recovery may have been down-regulated due to feedback from high tissue carbohydrates; then there would be little benefit from accumulated carbohydrates. Alternatively, the accumulated carbohydrates may be crucial for rapid re-growth of roots if there is an inherently low capacity for photosynthesis during post-waterlogging. Leaf N, P and K concentrations were low in *T. aestivum* after 15 days of waterlogging (Trought and Drew, 1980), and these low levels may decrease photosynthetic capacity, so that the carbohydrates accumulated during waterlogging would be an important source of substrates during the time taken for photosynthetic capacity to recover following drainage.

Summing up, the hypothesis that intact plants with roots in O_2-deficient media are usually not limited by carbohydrate supply needs to be confirmed for wetland species that do not suffer substantial reductions in growth when in anaerobic rooting media. Future work should focus on the carbohydrate status of tips of roots in their most rapid phase of elongation, avoiding samples taken from roots that have reached their maximum lengths as set by the internal O_2 diffusion from the shoots (Section III.A.1). Furthermore, whether exogenous sugars can stimulate growth should also be tested. Such tests are required, because high carbohydrate levels do not rule out a possible response to exogenous sugars, as shown for anoxic *T. aestivum* roots (Waters et al., 1991) and slices of tissue from the tap root of *Beta vulgaris* (Zhang and Greenway, 1994).

2. During Submergence

In contrast to the accumulation of non-structural carbohydrates in plants during waterlogging (Section IV.B.1), carbohydrate depletion typically occurs during complete submergence, at least for *O. sativa* (Setter et al., 1987; 1997). The adverse consequence of depletion of carbohydrates for submergence tolerance in *O. sativa* has been demonstrated by manipulation of [non-structural carbohydrates] in tissues before submergence using a variety of techniques (reviewed by Jackson and Ram, 2003). It was concluded that decreases in [non-structural carbohydrates] during submergence of *O. sativa* were due to continued catabolism (and growth), while photosynthesis was reduced. Photosynthesis being limited by: (i) low irradiance, and (ii) inadequate CO_2 supplies (Jackson and Ram, 2003). It is therefore not surprising that the carbohydrate status would differ between plants completely submerged, and plants with roots in waterlogged soil and shoots in air.

Here we only discuss further one aspect of the

response of tissue [carbohydrate] to complete submergence, which is relevant to catabolism. In *O. sativa*, submergent-tolerant cultivars retained a higher concentration (total dry mass basis) of non-structural carbohydrates at the end of submergence than intolerant cultivars (Mallik et al., 1995; Singh et al., 2001; Ram et al., 2002; Jackson and Ram, 2003). So, the decreased survival in the intolerant cultivars might have been associated with a deficiency of substrate for respiration and/or anaerobic catabolism. The higher non-structural carbohydrates in the tolerant cultivars at the end of submergence has been attributed to a combination of a higher initial [carbohydrate] and slower consumption during submergence (Mallik et al., 1995; Chaturvedi et al., 1996; Singh et al., 2001; Ram et al., 2002). There is persuasive evidence for slower decreases in concentration of non-structural carbohydrates in the tolerant cultivars (Singh et al., 2001); after 15 days submergence, decreases in total non-structural carbohydrates were 1.6 times greater in shoots of submergent-intolerant compared with tolerant cultivars. A similar difference between cultivars after 7 days submergence was reported by Mazaredo and Vergara (1982). There are three possible mechanisms that could cause these differences in the rate of decreases in concentrations of non-structural carbohydrates during submergence:

i) Catabolism might be faster in the intolerant than in the tolerant cultivars. A slowing down of catabolism associated with submergence tolerance would be consistent with down-regulation of ethanol formation in anoxia-tolerant tissues during anoxia (Greenway and Gibbs, 2003).

ii) Faster decreases of [non-structural carbohydrates] in submergent-intolerant cultivars might be due to their faster shoot elongation (Mallik et al., 1995; Chaturvedi et al., 1996; Singh et al., 2001; Ram et al., 2002; Jackson and Ram, 2003). However, it is unknown whether this faster elongation is associated with increases in structural dry mass, such as proteins, cell walls and/or organic solutes for an increased volume; such changes would themselves decrease the non-structural carbohydrate concentration, when expressed on a total dry mass basis.

iii) Photosynthesis during submergence might be faster in the tolerant cultivars; this needs to be evaluated in experiments using a range of exogenous CO_2 and irradiances, but such data are presently not available.

Resolution of these possibilities is worthwhile, especially considering the clear cultivar differences in response to submergence in this important crop.

Another possible factor interfering with respiration in plants in flooded soils are high P_{CO_2}, and this will be briefly considered in the next section.

V. Effects of High Partial Pressures of CO_2 (P_{CO_2}) in Flooded Soils on Respiration

Root environments usually have substantially higher P_{CO_2} than the atmosphere (Cramer, 2002). For example, the P_{CO_2} in the gas-phase 0.3 m below the surface of a well-drained soil ranged between 0.08 kPa in winter to 1.6 kPa in summer (Johnson et al., 1994). Upon flooding, P_{CO_2} in soils (and roots) becomes even higher, since escape of CO_2 evolved from soil organisms and subterranean parts of plants is drastically curtailed because of the slow diffusion of gases in water. As an example of the high P_{CO_2} that can occur in flooded soils, P_{CO_2} reached 18 and 43 kPa in pots containing flooded soils from rice fields with pH 5.7 and 6.3, respectively (Ponnamperuma, 1984). Yet, surprisingly little is known about the responses of roots to these high P_{CO_2}. In this section we have only reviewed the aspects directly related to effects of high (i.e. several kPa) P_{CO_2} on respiration; the most relevant data available are for callus cells derived from shoots, but with P_{CO_2} suddenly raised.

Partial pressures of CO_2 in roots will be determined by rates of CO_2 evolution by the cells, CO_2 exchanges with the soil, and any movement of CO_2 to the shoots, either by longitudinal diffusion in aerenchyma or by mass flow in the xylem. Permeability of membranes to CO_2 is very high (Espie and Colman, 1982), so exchange of CO_2 between roots and soil would be fast, unless the impermeable layer to ROL which occurs in roots of many wetland species (Section III.B), is also impermeable to CO_2. Even then, the root apex would still be exposed to high P_{CO_2}. Reliable data on P_{CO2} within roots in flooded soils are not available. However, in rhizomes of *P. australis* in soil flooded for two months, P_{CO_2} was 1.2 to 3.2 kPa at 0 to 20 cm depth and 6.7 to 7.4 kPa at 50 to 80 cm depth (Brix, 1988). By inference, P_{CO_2} would have been at least as high in the roots arising from these rhizomes.

In the very different context of global warming,

the effects of increased atmospheric P_{CO_2} (i.e. ~70 Pa, as compared with current atmospheric P_{CO_2} of 36 Pa) on respiration, and other processes, have been reviewed (Drake et al., 1997, 1999). Such data do not seem relevant for predicting responses of roots to the much higher P_{CO2} in flooded soils; these can be three orders of magnitude higher than current atmospheric P_{CO2}. Metabolism could be affected by very high (i.e. several kPa) P_{CO_2} due to an acid load (i.e. a tendency to acidify the cytosol) (see Bown, 1985), or direct effects of CO_2 and/or HCO_3^- on enzymes and/or lipids (Mitz, 1979; Lorimer, 1983). The [HCO_3^-] in the various compartments will depend on the P_{CO_2} and pH, as described by the Henderson-Hasselbalch equation (Segel, 1976). To emphasize the differences between cells exposed to elevated P_{CO_2} in the atmosphere (e.g., 70 Pa; Drake et al., 1997) and cells exposed to P_{CO_2} in flooded soils (e.g., ~7 kPa in rhizomes; Brix, 1988), we compare the degree of acid loading under these two situations. Assuming pH_{cyt} of 7.5, [HCO_3^-] in the cytosol of cells exposed to 7 kPa P_{CO_2} could have risen to 63 mol m^{-3}, whereas at P_{CO_2} of 70 Pa (i.e. approximately double present atmospheric P_{CO_2}), the [HCO_3^-] in the cytosol would be only 0.63 mol m^{-3} (calculated using the Henderson-Hasselbalch equation). Thus, even if P_{CO2} in the atmosphere doubled, the acid loading would be negligible, whereas for cells of organs in flooded soils the acid loading is of the same order of magnitude as the buffering capacity of 20 to 100 mol m^{-3} (pH unit)$^{-1}$ in the cytosol of plant cells (Kurkdjian and Guern, 1989).

The inhibitory effect of high (i.e. several kPa) P_{CO_2} on respiration has been studied using callus cells of *Dianthus caryophyllus*. Uncoupled respiration in the presence of SHAM (i.e. an estimate of the maximum capacity of the cytochrome pathway) was inhibited by 50% when 20 mol m^{-3} CO_2/HCO_3^- was added (Palet et al., 1991). Given that the pH in solution increased from 5.7 to 6.0 following addition of the HCO_3^- (see Table 1 in Palet et al., 1991), [CO_2] would have been 12 mol m^{-3}, and P_{CO_2} 35 kPa; calculated using pK = 6.1 for $HCO_3^- \leftrightarrow CO_2$ (Segel, 1976) and conversion of CO_2 to P_{CO_2} at 25°C according to Armstrong (1979). That only the cytochrome pathway remained operational during this experiment was verified by addition of 1 mol m^{-3} cyanide at the end of the experiment which abolished O_2 uptake (Palet et al., 1991). The limiting reactions responsible for the reduced O_2 uptake at high [CO_2/HCO_3^-] are not known. Cytochrome oxidase may have been directly inhibited (Palet et al., 1991), a feasible suggestion since in isolated mitochondria in state 3 the flux control coefficient reached 0.47 to 0.66 (Padovan et al., 1989). Alternatively, glycolysis could have been inhibited. Furthermore, a large inhibition of succinate oxidation at high [HCO_3^-] has been found in isolated mitochondria from *Z. mays* (Cerwick et al., 1995). These hypotheses need to be explored for fibrous roots exposed to a range of P_{CO_2}, and include pre-treatments during which P_{CO_2} is raised gradually to assess possible acclimation to high P_{CO_2}.

VI. Conclusions

Waterlogging and flooding restrict gas exchange between the soil and atmosphere, so O_2 becomes deficient and usually absent altogether, while CO_2 increases. Cessation of oxidative phosphorylation during anoxia causes a severe energy crisis. In wetland species, tubers and rhizomes are anoxia-tolerant, while coleoptiles and stems even grow during anoxia. Anoxia tolerance requires a carbohydrate supply to fuel ethanolic fermentation, and apportionment of the scarce energy to processes essential to survival. In some cases, rates of glycolysis in anoxia are faster than in aerobic tissues (a 'Pasteur Effect'); e.g., in shoots that elongate to escape anoxia ('escape' is achieved if the shoot reaches an O_2 source).

In the majority of cases the cessation of oxidative phosphorylation stops growth and often causes death of cells within a few hours or days. Adaptation to flooded environments is centered on obtaining O_2, and complemented by a capacity to cope with a restricted O_2 supply to part of the roots. Some soils become hypoxic, rather than anoxic, and thus supply measurable, though insufficient O_2, to roots. Even then, an adequate supply of O_2 via aerenchyma is required for plants to thrive. Maximum root lengths are determined by the capacity for internal O_2 diffusion to the apex of the main axis; in turn, influenced by the volume of aerenchyma, respiration rates, and rates of ROL to the soil. In some wetland species, through-flows of gases greatly increase aeration of rhizomes, and the supply of O_2 available to diffuse into roots. This allows a substantial increase in rhizome depths, and therefore of the volume of soil explored by roots arising from these rhizomes.

In roots of some species, including those with aerenchyma, anaerobic carbohydrate catabolism may occur in the stele and in the terminal tip (i.e. tissues at the end of the O_2-diffusion paths), while oxidative phosphorylation continues in the cells with access to O_2, derived either from the external solution or via the cortical aerenchyma. Thus, anoxia tolerance is presumably a component of tolerance of roots to soil waterlogging and flooding, at least in some species. Currently, nearly all the data on this aspect are for roots of *Z. mays*. Future work needs to establish the degree to which these patterns of O_2 supply, and aerobic versus anaerobic catabolism, occur in tissues of roots of other species. The degree of O_2 deficiency in roots of wetland species might be much less than in aerenchymatous roots of non-wetland species; roots of wetland species typically have more aerenchyma and a much narrower stele, and these features can mitigate development of anoxic zones (e.g., see Fig. 4 for *P. australis*).

An important issue for plants in waterlogged soils, or when submerged, is whether substrates are sufficient for catabolism, be it respiration or ethanol formation. Non-structural carbohydrates usually increase in roots of plants in waterlogged soils with shoots in air, while the opposite applies when plants are submerged. The accumulation of carbohydrates in plants in waterlogged soils can be attributed to reduced consumption due to growth being inhibited. Conversely, in submerged plants photosynthesis is inhibited, so carbohydrates are depleted. The inference that carbohydrate deficiency might limit survival of submerged wetland species requires testing; e.g., by feeding exogenous sugars. In addition, for roots of plants in waterlogged soils it remains feasible that carbohydrate supply to apices might become limiting.

Another important feature of a flooded environments is a high P_{CO_2}, which is therefore bound to become high in roots. However, surprisingly little is known about the response of plants to P_{CO_2} in the range of 5–35 kPa. In this review we have only dealt with the aspects directly related to respiration, and even on this aspect there is little known. Research should focus on long-term effects, rather than use sudden applications of high P_{CO_2}, since the latter do not occur in nature.

Acknowledgments

We thank Professor W Armstrong, and the editors of this volume, for constructive comments on drafts of this chapter.

References

Albrecht G, Biemelt S and Baumgartner S (1997) Accumulation of fructans following oxygen deficiency stress in related plant species with differing flooding tolerance. New Phytol 136: 137–144

Albrecht G, Kammerer S, Praznik W and Wiedenroth EM (1993) Fructan content of wheat seedlings (*Triticum aestivum* L.) under hypoxia and following re-aeration. New Phytol 123: 471–476

Amor Y, Chevion M and Levine A (2000) Anoxia pretreatment protects soybean cells against H_2O_2-induced cell death: Possible involvement of peroxidases and of alternative oxidase. FEBS Lett 477: 175–180

ap Rees T, Jenkin LET, Smith AM and Wilson PM (1987) The metabolism of flood-tolerant plants. In: Crawford RMM (ed) Plant Life in Aquatic and Amphibious Habitats, pp 227–238. Blackwell Scientific, Oxford

Armstrong J and Armstrong W (1988) *Phragmites australis*—A preliminary study of soil-oxidizing sites and internal gas transport pathways. New Phytol 108: 373–382

Armstrong J, Armstrong W and Beckett PM (1992) *Phragmites australis*: Venturi- and humidity-induced pressure flows enhance rhizome aeration and rhizosphere oxidation. New Phytol 120: 197–207

Armstrong J, Armstrong W, Beckett PM, Halder JE, Lythe S, Holt R and Sinclair A (1996) Pathways of aeration and the mechanisms and beneficial effects of humidity- and venturi-induced convections in *Phragmites australis* (Cav.) Trin. Ex Steud. Aquat Bot 54: 177–197

Armstrong W (1964) Oxygen diffusion from the roots of some British bog plants. Nature 204: 801–802

Armstrong W (1979) Aeration in higher plants. Adv Bot Res 7: 225–332

Armstrong W and Beckett PM (1987) Internal aeration and the development of stelar anoxia in submerged roots. A multishelled mathematical model combining axial diffusion of oxygen in the cortex with radial losses to the stele, the wall layers and the rhizosphere. New Phytol 105: 221–245

Armstrong W and Boatman DJ (1967) Some field observations relating the growth of bog plants to conditions of soil aeration. J Ecol 55: 101–110

Armstrong W and Drew MC (2002) Root growth and metabolism under oxygen deficiency. In: Waisel Y, Eshel A and Kafkafi U (eds) Plant Roots: The Hidden Half, 3rd Edition, pp 729–761. Marcel Dekker, New York

Armstrong W and Webb T (1985) A critical oxygen pressure for root extension in rice. J Exp Bot 36: 1573–1582

Armstrong W, Healy MT and Webb T (1982) Oxygen diffusion in pea. I. Pore space resistance in the primary root. New Phytol 91: 647–659

Armstrong W Healy MT and Lythe S (1983) Oxygen diffusion in pea. II. Oxygen concentration in the primary root apex as affected by growth, the production of laterals and radial oxygen loss. New Phytol 94: 549–559

Armstrong W, Armstrong J and Beckett PM (1990) Measurement and modelling of oxygen release from roots of *Phragmites australis*. In: Cooper F and Findlater BC (eds) The Use of Constructed Wetlands in Water Pollution Control, pp 41–52. Pergamon Press, Oxford

Armstrong W, Armstrong J, Beckett PM and Justin SHFW

(1991a) Convective gas-flows in wetland plant aeration. In: Jackson MB, Davies DD and Lambers H (eds) Plant Life Under Oxygen Deprivation, pp 283–302. SPB Academic Publishing, The Hague

Armstrong W, Beckett PM, Justin SHFW and Lythe S (1991b) Modelling, and other aspects of root aeration by diffusion. In: Jackson MB, Davies DD and Lambers H (eds) Plant Life Under Oxygen Deprivation, pp 267–282. SPB Academic Publishing, The Hague

Armstrong W, Strange ME, Cringle S and Beckett PM (1994) Microelectrode and modelling study of oxygen distribution in roots. Ann Bot 74: 287–299

Armstrong W, Armstrong J and Beckett PM (1996a) Pressurised ventilation in emergent macrophytes: The mechanism and mathematical modelling of humidity-induced convection. Aquat Bot 54: 121–135

Armstrong W, Armstrong J and Beckett PM (1996b) Pressurised aeration in wetland macrophytes: Some theoretical aspects of humidity-induced convection and thermal transpiration. Folia Geobot Phytotaxon 31: 25–36

Armstrong W, Cousins D, Armstrong J, Turner DW and Beckett PM (2000) Oxygen distribution in wetland plant roots and permeability barriers to gas-exchange with the rhizosphere: A microelectrode and modelling study with *Phragmites australis*. Ann Bot 86: 687–703

Atkin OK, Edwards EJ and Loveys BR (2000) Response of root respiration to changes in temperature and its relevance to global warming. New Phytol 147: 141–154

Atwell GJ and Greenway H (1987) Carbohydrate metabolism of rice seedlings grown in oxygen deficient solution. J Exp Bot 38: 466–478

Barrett-Lennard EG, Leighton PD, Buwalda F, Gibbs J, Armstrong W, Thomson CJ and Greenway H (1988) Effects of growing wheat in hypoxic solutions and of subsequent transfer to aerated solutions. I. Growth and carbohydrate status of shoots and roots. Aust J Plant Physiol 15: 585–598

Barta AL (1987) Supply and partitioning of assimilates to roots of *Medicago sativa* L. and *Lotus corniculatus* L. under anoxia. Plant Cell Environ 10: 151–156

Barta AL (1988) Response of field grown alfalfa to root waterlogging and shoot removal. I. Plant injury and root content of carbohydrate and minerals. Agron J 80: 889–892

Beckett PM, Armstrong W, Justin SHFW and Armstrong J (1988) On the relative importance of convective and diffusive gas-flows in plant aeration. New Phytol 110: 463–468

Beckett PM, Armstrong W and Armstrong J (2001) Mathematical modelling of methane transport by *Phragmites*: The potential for diffusion within the roots and rhizosphere. Aquat Bot 69: 293–312

Begg CBM, Kirk GJD, MacKenzie AF and Neue HU (1994) Root-induced iron oxidation and pH changes in the lowland rice rhizosphere. New Phytol 128: 469–477

Berry LJ and Norris WE (1949) Studies on onion root respiration. I. Velocity of oxygen consumption in different segments of roots at different temperatures as a function of partial pressure of oxygen. Biochim Biophys Acta 3: 593–606

Blackwell PS (1983) Measurement of aeration in waterlogged soils: Some improvements of techniques and their application to experiments using lysimeters. J Soil Sci 34: 271–285

Blackwell PS and Wells EA (1983) Limiting oxygen flux densities for oat root extension. Plant Soil 73: 129–139

Blackwell PS, Ward MA, Lefevre RN and Cowan DJ (1985) Compaction of a swelling clay soil by agricultural traffic; effects upon conditions for growth of winter cereals and evidence for some recovery of structure. J Soil Sci 36: 633–650

Bown AW (1985) CO_2 and intracellular pH. Plant Cell Environ 8: 459–465

Brix H (1988) Light-dependent variations in the composition of the internal atmosphere of *Phragmites australis* (cav.) Trin. Ex Steudel. Aquat Bot 30: 319–329

Brix H, Sorrell BK and Orr PT (1992) Internal pressurization and convective gas flow in some emergent freshwater macrophytes. Limnol Oceanogr 37: 1420–1433

Cannell RQ, Belford RK, Gales K, Dennis CW and Prew RD (1980) Effects of waterlogging at different stages of development on the growth and yield of winter wheat. J Sci Food Agric 31: 117–132

Castonguay Y, Nadeau N and Simard RR (1993) Effects of flooding on carbohydrate and ABA levels in roots and shoots of alfalfa. Plant Cell Environ 16: 695–702

Cerwick SF, Martin BA and Reding LD (1995) The effect of carbon dioxide on maize seed recovery after flooding. Crop Sci 35: 1116–1121

Chang WWP, Huang L, Shen M, Webster C, Burlingame AL and Roberts JKM (2000) Patterns of protein synthesis and tolerance of anoxia in root tips of maize seedlings acclimated to a low-oxygen environment, and identification of proteins by mass spectrometry. Plant Physiol 122: 295–317

Chaturvedi GS, Ram PC, Singh AK, Ram P, Ingram KT, Singh BB, Singh RK and Singh VP (1996) Carbohydrate status of rainfed lowland rice in relation to submergence, drought, and shade tolerance. In: Singh VP, Singh BB and Zeigler BB(eds) Physiology of stress tolerance in plants, pp 103–122. International Rice Research Institute, Los Banos

Christensen PB, Revsbech NP and Sand-Jensen K (1994) Microsensor analysis of oxygen in the rhizosphere of the aquatic macrophyte *Littorella uniflora* (L.) Ascherson. Plant Physiol 105: 847–852

Colmer TD (2003a) Long-distance transport of gases in plants: a perspective on internal aeration and radial oxygen loss from roots. Plant Cell Environ 26: 17–36

Colmer TD (2003b) Aerenchyma and an inducible barrier to radial oxygen loss facilitate root aeration in upland, paddy and deepwater rice (*Oryza sativa* L.). Ann Bot 91: 301–309

Colmer TD, Huang S and Greenway H (2001) Evidence for down regulation of ethanolic fermentation and K^+ effluxes in the coleoptile of rice seedlings during prolonged anoxia. J Exp Bot 52: 1507–1517.

Conlin TSS and Crowder AA (1989) Location of radial oxygen loss and zones of potential iron uptake in a grass and two nongrass emergent species. Can J Bot 67: 717–722

Connell EL, Colmer TD and Walker DI (1999) Radial oxygen loss from intact roots of *Halophila ovalis* as a function of distance behind the root tip and shoot illumination. Aquat Bot 63: 219–228

Cramer MD (2002) Inorganic carbon utilization by root systems. In: Waisel Y, Eshel A and Kafkafi U (eds) Plant Roots: The Hidden Half, 3rd Edition, pp 699–715. Marcel Dekker, New York

Crawford RMM (1966) The control of anaerobic respiration as a determining factor in the distribution of the genus *Senecio*. J Ecol 54: 403–413

Crawford RMM and Braendle R (1996) Oxygen deprivation stress

in a changing environment. J Exp Bot 47: 145–159
Dacey JWH (1980) Internal winds in water lilies: An adaptation for life in anaerobic sediments. Science 210: 1017–1019
Dacey JWH (1981) Pressurized ventilation in the yellow waterlily. Ecology 62: 1137–1147
Drake BG, Gonzalez-Meler MA and Long SP (1997) More efficient plants: A consequence of rising atmospheric CO_2? Annu Rev Plant Physiol Plant Mol Biol 48: 609–639
Drake BG, Azcon-Bieto J, Berry J, Bunce J, Dijkstra P, Farrar J, Gifford RM, Gonzalez-Meler MA, Koch G, Lambers H, Siedow J, Wullschleger S (1999) Does elevated atmospheric CO_2 concentration inhibit mitochondrial respiration in green plants? Plant Cell Environ 22: 649–657
Drew MC (1992) Soil aeration and plant root metabolism. Soil Sci 154: 259–268
Drew MC (1997) Oxygen deficiency and root metabolism: Injury and acclimation under hypoxia and anoxia. Annu Rev Plant Physiol Plant Mol Biol 48: 223–250
Drew MC, Saglio PH and Pradet A (1985) Larger adenylate charge and ATP/ADP ratios in aerenchymatous roots of *Zea mays* in anaerobic media as a consequence of improved oxygen transport. Planta 165: 51–58
Espie GS and Colman B (1982) Photosynthesis and inorganic carbon transport in isolated *Asparagus* cells. Plant Physiol 70: 649–654
Garthwaite AJ, von Bothmer R and Colmer TD (2003) Diversity in root aeration traits associated with waterlogging tolerance in the genus *Hordeum*. Funct Plant Biol 30: 875–889
Gibberd MR, Gray JD, Cocks PS and Colmer TD (2001) Waterlogging tolerance among a diverse range of *Trifolium* accessions is related to root porosity, lateral root formation and 'aerotropic rooting.' Ann Bot 88: 579–589
Gibbs J, de Bruxelles G, Armstrong W and Greenway H (1995) Evidence for anoxic zones in 2–3 mm tips of aerenchymatous maize roots under low O_2 supply. Aust J Plant Physiol 22: 723–730
Gibbs J and Greenway H (2003) Mechanisms of anoxia tolerance in plants. I. Growth, survival and anaerobic catabolism. Funct Plant Biol 30: 1–47
Gibbs J, Turner DW, Armstrong W, Darwent MJ and Greenway H (1998) Response to oxygen deficiency in primary maize roots. I. Development of oxygen deficiency in the stele reduces radial solute transport to the xylem. Aust J Plant Physiol 25: 745–758
Greenway H and Gibbs J (2003) Mechanisms of anoxia tolerance in plants. II. Energy requirements for maintenance and energy distribution to essential processes. Funct Plant Biol 30: 999–1036
Greenway H, Waters I and Newsome J (1992) Effects of anoxia on uptake and loss of solutes in roots of wheat. Aust J Plant Physiol 19: 233–247
Grosse W (1996) The mechanism of thermal transpiration (= thermal osmosis). Aquat Bot 54: 101–110
Harada T and Ishizawa K (2003) Starch degradation and sucrose metabolism during anaerobic growth of pondweed (*Potamogeton distinctus* A. Benn.) turions. Plant Soil, 253: 125–135
Haslam SM (1970) The performance of *Pragmites communis* Trin. in relation to water supply. Ann Bot 34: 867–877
Jackson MB (1979) Rapid injury of peas by soil waterlogging. J Sci Food Agric 30: 143–152
Jackson MB and Armstrong W (1999) Formation of aerenchyma and the processes of plant ventilation in relation to soil flooding and submergence. Plant Biol 1: 274–287
Jackson MB and Drew MC (1984) Effects of flooding on growth and metabolism of herbaceous plants. In: Kozlowski TT (ed) Flooding and Plant Growth, pp 47–128. Academic Press, London
Jackson MB and Ram PC (2003) Physiological and molecular basis of susceptibility and tolerance of rice plants to complete submergence. Ann Bot 91: 227–241
Johnson D, Geisinger D, Walker R, Newman J, Vose, J, Elliot K and Ball T (1994) Soil pCO_2, soil respiration, and root activity in CO_2-fumigated and nitrogen-fertilised pondesora pine. Plant Soil 165: 129–138
Justin SHFW and Armstrong W (1987) The anatomical characteristics of roots and plant response to soil flooding. New Phytol 106: 465–495
Kirk GJD (2003) Rice root properties for internal aeration and efficient nutrient acquisition in submerged soil. New Phytol 159: 185–194
Koncalova H (1990) Anatomical adaptations to waterlogging in roots of wetland graminoids: Limitations and drawbacks. Aquat Bot 38: 127–134
Konings H (1982) Ethylene-promoted formation of aerenchyma in seedling roots of *Zea mays* L. under aerated and non-aerated conditions. Physiol Plant 54: 119–124
Kuiper PJC, Walton CS and Greenway H (1994) Effects of hypoxia on ion uptake by nodal and seminal wheat roots. Plant Physiol Biochem 32: 267–276
Kurkdjian A and Guern J (1989) Intracellular pH: Measurement and importance in cellular activity. Annu Rev Plant Physiol Plant Mol Biol 40: 271–303
Lambers H (1976) Respiration and NADH-oxidation of the roots of flood-tolerant and flood-intolerant *Senecio* species as affected by anaerobiosis. Physiol Plant 37: 117–122
Lambers H, Atkin OK and Millenaar FF (2002) Respiratory patterns in roots in relation to their functioning. In: Waisel Y, Eshel A and Kafkafi U (eds) Plant Roots: The Hidden Half, 3rd Edition, pp 521–552. Marcel Dekker, New York
Lambers H and Steingrover E (1978) Efficiency of root respiration of a flood-tolerant and flood-intolerant *Senecio* species as affected by low oxygen tension. Physiol Plant 42: 179–184
Lambers H, Steingrover E and Smakman G (1978) The significance of oxygen transport and of metabolic adaptation in flood-tolerance of *Senecio* species. Physiol Plant 43: 277–281
Leuning R (1983) Transport of gases into leaves. Plant Cell Environ 6: 181–194
Limpinuntana V and Greenway H (1979) Sugar accumulation in barley and rice grown in solutions with low concentrations of oxygen. Ann Bot 43: 373–381
Lorimer GH (1983) Carbon dioxide and carbamate formation: The makings of a biochemical control system. Trends Biochem Sci 8: 65–68
Malik AI, Colmer TD, Lambers H and Schortemeyer M (2001) Changes in physiological and morphological traits of roots and shoots of wheat in response to different depths of waterlogging. Aust J Plant Physiol 28: 1121–1131
Malik AI, Colmer TD, Lambers H, Setter TL and Schortemeyer M (2002) Short-term waterlogging has long-term effects on the growth and physiology of wheat. New Phytol 153: 225–236
Malik AI, Colmer TD, Lambers H and Schortemeyer M (2003) Aerenchyma formation and radial O_2 loss along adventitious

roots of wheat with only the apical portion exposed to O_2 deficiency. Plant Cell Environ 26: 1713–1722

Mallik S, Kundu C, Banerji C, Nayek DK, Chatterjee SD, Nanda PK, Ingram KT and Setter TL (1995) Rice germplasm evaluation and improvement for stagnant flooding. In: Ingram KT (ed) Rainfed Lowland Rice: Agricultural Research for High Risk Environments, pp 97–109. International Rice Research Institute, Manila

Mazaredo AM and Vergara BS (1982) Physiological differences in rice varieties tolerant and susceptible to complete submergence. In: Proceedings of the 1981 International Deepwater Rice Workshop, pp. 327–341. International Rice Research Institute, Manila

McDonald MP, Galwey NW and Colmer TD (2001) Evaluation of *Lophopyrum elongatum* as a source of genetic diversity to increase the waterlogging tolerance of hexaploid wheat (*Triticum aestivum*). New Phytol 151: 369–380

McDonald MP, Galwey NW and Colmer TD (2002) Similarity and diversity in adventitious root anatomy as related to root aeration among a range of wetland and dryland grass species. Plant Cell Environ 25: 441–451

McKee WH and McKevlin MR (1993) Geochemical processes and nutrient uptake by plants in hydric soils. Environ Toxicol Chem 12: 2197–2207

Meek BD, Ehlig CF, Stolzy LH and Graham LE (1983) Furrow and trickle irrigation effects on soil oxygen and ethylene and tomato yield. J Soil Sci Soc Am 47: 631–635

Meyer WS, Barrs HD, Smith RCG, White NS, Heritage AD and Short DL (1985) Effects of irrigation on soil oxygen status and root and shoot growth of wheat in a clay soil. Aust J Agric Res 36: 171–185

Mendelssohn IA, Keiss BA and Wakeley JS (1995) Factors controlling the formation of oxidised root channels: a review. Wetlands 15: 37–46

Mitz MA (1979) CO_2 biodynamics: a new concept of cellular control. J Theor Biol 80: 537–551

Miyamoto N, Steudle E, Hirasawa T and Lafitte R (2001) Hydraulic conductivity of rice roots. J Exp Bot 52: 1835–1846

Padovan AC, Dry IB and Wiskisch JT (1989) An analysis of the control of phosphorylation-coupled respiration in isolated plant mitochondria. Plant Physiol 90: 928–933

Palet A, Ribas-Carbo M, Argiles JM and Azcon-Bieto J (1991) Short term effects of carbon dioxide on carnation callus cell respiration. Plant Physiol 96: 467–472.

Pearce DME and Jackson MB (1991) Comparison of growth responses of barnyard grass (*Echinochloa oryzoides*) and rice (*Oryza sativa*) to submergence, ethylene, carbon dioxide and oxygen shortage. Ann Bot 68: 201–209

Pedersen O, Sand-Jensen K and Revsbech NP (1995) Diel pulses of O_2 and CO_2 in sandy lake sediments inhabited by *Lobelia dortmanna*. Ecology 76: 1536–1545

Perata P, Loreti E, Guglielminetti L and Alpi A (1998) Carbohydrate metabolism and anoxia tolerance in cereal grains. Acta Bot Neerl 47: 269–283

Ponnamperuma FN (1984) Effects of flooding on soils. In: Kozlowski TT (ed) Flooding and Plant Growth, pp 9–45. Academic Press, New York

Pradet A and Bomsel JL (1978) Energy metabolism in plants under hypoxia and anoxia. In: Hook DD and Crawford RMM (eds) Plant Life in Anaerobic Environments, pp 89–118. Ann Arbor Science, Michigan

Ram PC, Singh BB, Singh AK, Ram P, Singh PN, Singh HP, Boamfa I, Harren F, Santosa E, Jackson MB, Setter TL, Reuss J, Wade LJ, Singh VP and Singh RK (2002) Submergence tolerance in rainfed lowland rice: Physiological basis and prospects for cultivar improvement through marker-aided breeding. Field Crop Res 76: 131–152

Ranathunge K, Steudle E and Lafitte R (2003) Control of water uptake by rice (*Oryza sativa* L.): Role of the outer part of the root. Planta 217: 193–205

Ridge I (1987) Ethylene and growth control in amphibious plants. In: Crawford RMM (ed) Plant Life in Aquatic and Amphibious Habitats, pp 53–76. Blackwell Scientific, Oxford

Robards AW, Clarkson DT and Sanderson J (1979) Structure and permeability of the epidermal/hypodermal layers of the sand sedge (*Carex arenaria*, L.). Protoplasma 101: 331–347

Roberts JKM, Lane AN, Clark RA and Nieman RH (1985) Relationships between the rate of synthesis of ATP and the concentrations of reactants and products of ATP hydrolysis in maize root tips, determined by ^{31}P nuclear magnetic resonance. Arch Biochem Biophys 240: 712–722.

Rubinigg M, Stulen I, Elzenga JTM and Colmer TD (2002) Spatial patterns of radial oxygen loss and nitrate net flux along adventitious roots of rice raised in aerated or stagnant solution. Funct Plant Biol 29: 1475–1481

Saglio PH (1985) Effect of path or sink anoxia on sugar translocation in roots of maize seedlings. Plant Physiol 77: 285–290

Segel IH (1976) Biochemical Calculations. 2nd Edition, John Wiley and Sons, New York

Setter TL and Belford RK (1990) Waterlogging: How it reduces plant growth and how plants can overcome its effects. West Aust J Agric 31: 51–55

Setter TL, Ellis M, Laureles EV, Ella ES, Senadhira D, Mishra SB, Sarkarung S and Datta S (1997) Physiology and genetics of submergence tolerance in rice. Ann Bot 79 (Supplement), 67–77

Setter TL, Waters I, Greenway H, Atwell BJ and Kupkanchanakul T (1987) Carbohydrate status of terrestrial plants during flooding. In: Crawford RMM (ed) Plant Life in Aquatic and Amphibious Habitats, Special Publication No. 5 British Ecological Society, pp 411–433. Blackwell Scientific Publications, Oxford

Singh HP, Singh BB, Ram PC (2001) Submergence tolerance of rainfed lowland rice, search for physiological marker traits. J Plant Physiol 158: 883–889

Sorrell BK (1994) Airspace structure and mathematical modelling of oxygen diffusion, aeration and anoxia in *Eleocharis sphacelata* R. Br. roots. Aust J Mar Freshw Res 45: 1529–1541

Sorrell BK and Dromgoole FI (1987) Oxygen transport in the submerged freshwater macrophyte *Egeria densa* Planch. I. Oxygen production, storage and release. Aquat Bot 28: 63–80

Sorrell BK, Brix H and Orr PT (1997) *Eleocharis sphacelata*: internal gas transport pathways and modelling of aeration by pressurized flow and diffusion. New Phytol 136: 433–442

St-Cyr L and Campbell PGC (1996) Metals (Fe, Mn, Zn) in the root plaque of submerged aquatic plants collected in situ: relations with metal concentrations in the adjacent sediments and in the root tissue. Biogeochemistry 33: 45–76

Steinberg SL (1996) Mass and energy exchange between the atmosphere and leaf influence gas pressurization in aquatic plants. New Phytol 134: 587–599

Summers JE and Jackson MB (1996) Anaerobic promotion of stem extension in *Potamogeton pectinatus*. Roles for carbon dioxide,

acidification and hormones. Physiol Plant 96: 615–622
Summers JE, Ratcliffe RG and Jackson MB (2000) Anoxia tolerance in the aquatic monocot *Potamogeton pectinatus*: absence of oxygen stimulates elongation in association with an unusually large Pasteur effect. J Exp Bot 51: 1413–1422
Szal B, Jolivet Y, Hasenfratz-Sauder M-P, Dizengremel P and Rychter AM (2003) Oxygen concentration regulates alternative oxidase expression in barley roots during hypoxia and post-hypoxia. Physiol Plant 119: 494–502
Thomson CJ and Greenway H (1991) Metabolic evidence for stelar anoxia in maize roots exposed to low O_2 concentrations. Plant Physiol 96: 1294–1301
Thomson CJ, Armstrong W, Waters I and Greenway H (1990) Aerenchyma formation and associated oxygen movement in seminal and nodal roots of wheat. Plant Cell Environ 13: 395–403
Thomson CJ, Colmer TD, Watkins ELJ and Greenway H (1992) Tolerance of wheat (*Triticum aestivum* cvs. Gamenya and Kite) and triticale (*Triticosecale* cv. Muir) to waterlogging. New Phytol 120: 335–344
Trought MCT and Drew MC (1980) The development of waterlogging damage in wheat seedlings (*Triticum aestivum* L.) II. Accumulation and redistribution of nutrients by the shoot. Plant Soil 56: 187–199
Tsuji H, Nakazono M, Saisho D, Tsutsumi N and Hirai A (2000) Transcript levels of the nuclear-encoded respiratory genes in rice decrease by oxygen deprivation: Evidence for involvement of calcium in expression of the alternative oxidase 1a gene. FEBS Lett 471: 201–204
Visser EJW, Colmer TD, Blom CWPM and Voesenek LACJ (2000) Changes in growth, porosity, and radial oxygen loss from adventitious roots of selected mono- and dicotyledonous wetland species with contrasting types of aerenchyma. Plant Cell Environ 23: 1237–1245
Voesenek LACJ, Armstrong W, Bogemann GM, McDonald MP and Colmer TD (1999) A lack of aerenchyma and high rates of radial oxygen loss from the root base contribute to the waterlogging intolerance of *Brassica napus*. Aust J Plant Physiol 26: 87–93
Voesenek LACJ and Blom CWPM (1989) Growth responses of *Rumex* species in relation to submergence and ethylene. Plant Cell Environ 12: 433–439
Waters I, Armstrong W, Thompson CJ, Setter TL, Adkins S, Gibbs J and Greenway H (1989) Diurnal changes in radial oxygen loss and ethanol metabolism in roots of submerged and non-submerged rice seedlings. New Phytol 113: 439–451
Waters I, Kuiper PJC, Watkin E, Greenway H (1991) Effects of anoxia on wheat seedlings. I. Interaction between anoxia and other environmental factors. J Exp Bot 42: 1427–1435
Webb T and Armstrong W (1983) The effects of anoxia and carbohydrates on the growth and viability of rice, pea and pumpkin roots. J Exp Bot 34: 579–603
Ying J, Peng S, He Q, Yang H, Yang C, Visperas RM and Cassman KG (1998) Comparison of high-yield rice in tropical and subtropical environments. I. Determinants of grain and dry matter yields. Field Crop Res 57: 71–84
Zhang Q and Greenway H (1994) Anoxia tolerance and anaerobic catabolism of aged beetroot storage tissue. J Exp Bot 45: 567–575
Zhang Q, Lauchli A and Greenway H (1992) Effects of anoxia on solute loss from beetroot storage tissue. J Exp Bot 43: 897–905

Chapter 9

Effects of Soil pH and Aluminum on Plant Respiration[1]

Rakesh Minocha*
USDA Forest Service, NERS, PO Box 640, 271 Mast Road, Durham, NH 03824, U.S.A.

Subhash C. Minocha
Department of Plant Biology, University of New Hampshire, Durham, NH 03824, U.S.A.

Summary

Interactions among external (soil) pH, cellular pH, and their effects on respiratory metabolism are complex. While the effects of changes in the apoplastic pH on the cytosolic pH are not clearly understood, pH directly affects enzymatic reactions in the cell, and pH-regulated ion uptake has profound indirect effects on cellular respiratory metabolism. A major consequence of soil acidification is the release of aluminum in solubilized forms from its insoluble forms, which, in turn, adversely affects the uptake of cations, causes organic acid secretion, and inhibits cell division and growth in the roots. Consequently, the respiratory metabolism is redirected to meet the needs of organic acid efflux from the roots. The effects of changes in external pH on cellular pH and consequent effects of this change on respiratory metabolism, particularly through effects on soil aluminum are summarized.

*Author for correspondence, email: rminocha@hopper.unh.edu
[1] Scientific contribution number 2168 from the New Hampshire Agricultural Experiment Station.

H. Lambers and M. Ribas-Carbo (eds.), Plant Respiration, 159–176.

I. Introduction

Soil pH is affected by the biogeochemistry of the soil and by changes in ionic strength attributable to inputs from the environment in the form of acidic deposition and/or fertilizer addition. Acidic deposition from the environment (NO_X and SO_X) contributes to a lowering of the soil pH (Van Breemen, 1985), which causes leaching of nutrients as well as solubilization of aluminum (Al[2]) (Marschner, 1995). Solubilized Al competes with cation uptake, further compounding the effects of low pH on nutrient uptake (Lawerence et al., 1995). Combined actions of changes in pH, perturbation of nutrient uptake, and solubilization of Al have profound effects on growth, development, and metabolism, including respiratory metabolism in plants.

In this chapter we review the complex relationships among pH, Al solubilization, and respiratory metabolism in plants. Since in nature most pH effects occur in soil and soil solution, we focus primarily on interactions in the root; where relevant, studies with in vitro cell cultures and effects on the above-ground plant parts are also discussed. A critical limiting factor in this discussion is that we lack an understanding of the regulation of cellular pH and its response to changes in external or apoplastic pH. And because soil pH has both direct and indirect effects on plant metabolism, it is not possible to study these effects independently of one another. Finally, solubilization of Al, which is one of the primary effects of lowering of the soil pH, has numerous direct and indirect effects on respiratory metabolism.

II. Relationship Between External (Soil and Apoplast) and Internal (Symplast) pH

The regulation of intracellular pH is a fundamental physiological process of immense importance to nearly every metabolic activity in the plant cell, particularly the transport of nutrients and plant hormones, as well as to numerous enzymatic reactions (Raven and Smith, 1974; Smith and Raven, 1976; Minocha, 1987; Kurkdjian and Guern, 1989). Thus, it is not surprising that cells allocate a significant amount of energy to the regulation of cytosolic and organellar pH. The metabolic production and consumption of large quantities of H^+ and OH^- ions within the cytosol, together with transport of H^+ across cellular membrane systems, constantly challenge the homeostatic mechanisms of pH regulation in the cell (Sanders and Bethke, 2000). Cytosolic pH is tightly regulated within a relatively narrow range of 7.0–7.5, even under a wide range of pH (5.0–8.0) of the bathing solution, though local variations in the pH of cellular compartments and organelles are common (Minocha, 1987; Marschner, 1995; Sanders and Bethke, 2000). However, Smith (1984) and Gehl and Colman (1985) argued that changes in the cytosolic pH are much more pronounced in response to external pH than commonly believed, particularly when cells are challenged by an external pH outside the range of 4.5 to 8.0. Other physical and chemical factors that influence cytosolic pH include light, temperature, hypoxia, and metabolic poisons (Smith and Raven, 1979; Kurkdjian and Guern, 1989; Roberts et al., 1992). The effects of hypoxia on cytoplasmic pH, however, vary with the time of treatment (Roberts et al., 1992). Vacuolar pH generally is much lower (5.0–6.5, but as low as 2.0–4.0), and is not regulated as tightly as the cytosolic pH (Torimitsu et al., 1984; Sanders and Bethke, 2000). Chloroplast pH ranges from 6.5 to 8.0, and is affected by the rate of photosynthesis (Heldt et al., 1973; Davis, 1974).

Most animal cells are constantly bathed in a pH at or near the cytosolic pH. For animal cell culture media also, the pH is controlled experimentally by the incorporation of buffers or by rapid replacement of the medium. By contrast, most plant cell culture media are poorly buffered, and traditionally adjusted to a pH of 5.5 + 0.3 initially; the operating pH of the medium can fluctuate, and reach values that are much lower than the starting pH, especially when ammonium is the source of N (Minocha, 1987; Goodchild and Givan, 1990). In nature, except for plants growing in calcareous soils, plant roots often are surrounded by an acidic pH (Marschner, 1995). Since both pH gradients and differences in potential across the plasmalemma and organellar membranes play essential roles in transport processes and energy relationships in cells (Malkin and Niyogi, 2000), external (apoplastic and ambient environment) pH could have significant effects on respiratory metabolism and consequently on the growth of plant cells

Abbreivations: NO_X – nitrous oxides; OAA – oxalo acetic acid; PEP – phosphoenolpyruvate; ROS – reactive oxygen species; SO_X – sulfur oxides; TCA – tricarboxyloic acid

[2] All ionic forms of aluminum are represented by Al in this chapter; although the toxicity of different forms varies considerably; the most toxic form Al^{3+} occurs only at pH ≤4.5.

in vitro and that of plants in nature.

A major difficulty in interpreting experimental data on the effects of external pH on metabolic reactions in the cell is measuring cytosolic and organellar pH accurately. All direct (microelectrodes) and indirect (partitioning of weakly-acidic dyes or shift in ^{31}P NMR spectrum) methods used to date have provided, at best, a crude estimate of the steady-state pH of a given cellular compartment (Heldt et al., 1973; Moon and Richards, 1973; Walker and Smith, 1975; Boron and Roos, 1976; Burt et al., 1979; Roberts et al., 1992; Bligny and Douce, 2001; Hesse et al., 2002).

Several mechanisms of cytosolic pH regulation were discussed by Smith and Raven (1979) and later by Marschner (1995). While metabolic control of pH through enzymatic reactions was proposed by Davies (1973a,b), Raven and Smith (1974, 1978) and Smith (1984) argued that cytosolic pH might be regulated more by a biophysical rather than by a biochemical pH-stat. They suggested that H^+ transport across the plasmalemma, and probably the tonoplast, plays a major role in the regulation of cytosolic pH. This argument is supported by the existence of proton pumps in both the plasma membrane and the tonoplast. It is now believed that both biophysical and biochemical mechanisms contribute to the regulation of cytosolic pH (Marschner, 1995; Edwards et al., 1998). Smith (1984) suggested that the fine control of cytosolic pH begins to break down when cells are challenged with an external pH below 4.5, and that significant net influx of H^+ causes a downward shift in the cytosolic pH (Rent et al., 1972; Lane and Burris, 1981; Torimitsu et al., 1984; Gehl and Colman, 1985). Similarly, a decrease in the cytosolic pH occurs when anions are taken up in excess.

Finally, plant roots not only respond to pH changes in the soil, but also contribute to the modulation of ambient soil pH by a variety of mechanisms including: differential uptake of NO_3^- and NH_4^+, release of H^+ and/or OH^-, changes in the uptake and release of CO_2, and release of various organic anions and cations (Minocha, 1987; Sorensen et al., 1989; Hoffland et al., 1992; Yan et al., 1992). The modulation of apoplastic and rhizosphere pH by the roots is species specific and heavily dependent on the buffering capacity of the soil. This feature of roots further complicates the interpretation of results on the effects of changes in soil pH on physiological processes in plants. A similar modulation of the nutrient solution's pH also occurs in cell culture systems (Minocha, 1987).

III. Soil pH and Respiration

Our current understanding of the effects of external pH on the overall metabolism of cells, particularly respiratory metabolism, is rather limited and highly speculative (Minocha, 1987; Marschner, 1995; Lambers et al., 1998). In terrestrial environments, increased acidification is more pronounced than alkalinization. The effects of pH changes on plant roots may be short term (rapid) or long term, direct and acute or indirect and slow. It is generally believed that short-term effects of low pH contribute more to root physiology than long-term effects (Kochian, 1995). Lowering the pH of the bathing solution (hence the apoplastic pH) can influence plant cell metabolism in several ways. Whereas some of these effects are due to direct interactions of increased [H^+] with the cell wall and plasmalemma, others are mediated through effects on nutrient availability, uptake and assimilation, and the release and influx of cytotoxic cations. It is difficult to estimate the extent of various forms of respiration (e.g., dark respiration, photo-respiration, including non-phosphorylating respiration and alternative respiration, etc.) precisely, or to understand their metabolic significance and regulation in the plant (Lambers et al., 1998; Siedow and Day, 2000). The dozens of individual reactions of the respiratory metabolism, that are localized in various cellular compartments, are affected directly by cytosolic and/or organelle pH. The effects of pH on individual enzymes often studied in vitro are subject to misinterpretation because of the complexity of these reactions and their sites of localization in the cells. In the following section we discuss the effects of pH on respiratory metabolism in plant cells; examples are drawn from the literature on cell cultures, crop plants, and forest trees.

A. Direct Effects of Soil pH on Respiration

Root respiration constitutes a major proportion of the total respiratory carbon metabolism in plants; under certain conditions, it can account for half or more of the carbon fixed in a day (Van der Werf et al., 1992; Lambers et al., 1998). Root respiration depends on the rate of growth and nutrient supply to the root, both of which are affected directly by changes in soil pH (Van der Werf et al., 1992). The soil pH varies with soil depth resulting in strong pH gradients (up to 2 pH units) within the rhizosphere and the bulk of the soil (Marschner et al., 1991; Godbold and Jentschke,

1998). Over long periods at low pH, a reduced growth rate due to reduced ion absorption and transport, will adversely affect root respiration.

Rapid direct effects of low pH on root elongation presumably do not involve synthesis of new proteins or hormones, nor do they affect root dry weight (Peiter et al., 2001). Perhaps the primary cause is via interactions with auxin-induced H^+-driven cell elongation, which is accompanied by increased respiration (Rayle and Cleland, 1992). This pH effect would be largely apoplastic via buffering of the H^+ ions that are secreted (high pH effect), or by preventing H^+ secretion from the membrane (low pH effect) or by inhibiting cellulose microfibril sliding or interfering with cellulosic bonds by stabilizing them (freezing the cell wall) and preventing elongation of the cell wall.

Soil CO_2 concentration can increase quickly in response to flooding and acidity, and high CO_2 could adversely affect root respiration (Thomas et al., 1973; Nobel and Palta, 1989; Palta and Nobel, 1989; Qi et al., 1994; Lambers et al., 1996, 1998; Wlodarczyk et al. 2002). Suggested mechanisms of CO_2 effects on root respiration include its direct inhibition of cytochrome oxidase, malic enzymes, and other mitochondrial enzymes (Thomas et al., 1973; Gonzales-Meler et al., 1996; Lambers et al., 1998; Drake et al., 1999).

The presence and the extent of plant symbionts, pathogens, and mycorrhizal associations in the root directly affect its respiratory metabolism, and these associations also are subject to adverse effects of low pH (Marschner, 1995; Qian, 1998). Nitrogen-fixing microbes generally promote respiration due to a combination of increased nutrient uptake and increased growth rate. Mycorrhizae have a similar positive effect on root respiration and ion uptake.

The overall extent and the rates of respiration (and its regulation) in the above-ground parts of the plant have been studied extensively (Lambers et al., 1998; Siedow and Day, 2000), yet little is known about the direct effects of leaf exposure to acid mist/clouds on leaf respiration. Leaf respiration depends on three factors: maintenance of biomass, growth, and transport of nutrients (Lambers et al., 1998). It is the last one that is affected the most by changes in soil pH. Major factors that govern respiration in green tissues are the process of photosynthetic carbon fixation and transport of photosynthetic products from the leaf (Siedow and Day, 2000). In C_3 plants, the extent of photorespiration may be variable but it responds more to CO_2 levels than to pH. Also, indirect effects of low pH on leaf respiration via its effects on stomata closure can be important under acid mist conditions (Sanders and Bethke, 2000). The major effects of ambient cloud pH on respiration would occur at night when there is increased vapor condensation due to lower temperature, though there is little direct evidence for such an effect.

B. Soil pH, Ion Uptake and Respiration

The most significant indirect effects of changes in soil pH on respiratory metabolism in plants probably result from alterations in nutrient supply and uptake. A large proportion of cellular respiratory metabolism is devoted to N absorption and assimilation in plants (Lambers et al., 1996). While a part of this is perhaps related to the energy dependence of N uptake and assimilation, equally important is the needed carbon skeleton for N assimilation that is provided by the TCA cycle (Forde and Clarkson, 1999, and references therein). The interactions among NO_3^-, NH_4^+ and soil pH play a critical role in N uptake by the roots (Britto et al., 2001b; Kronzucker et al., 2001 and references therein).

Both the relative amounts of NH_4^+ and NO_3^- in the soil and their absorption by roots are affected by pH in complex ways. For example: (a) at relatively high soil pH (7.0–9.5), passive influx of free ammonia (NH_3) increases total N uptake; (b) a low vacuolar pH acts as a facilitator of NH_3 transport and trapping (in the form of NH_4^+); and (c) high-affinity transport systems for NH_4^+ apparently are unaffected by pH changes over a wide range (4.5–9.0), though this transport is presumably H^+-coupled (Wang et al., 1993a,b; Britto et al., 2001a). It should be noted that preferential uptake of NH_4^+ by roots (and also by cells in culture) results in lowering the soil (or medium) pH, while the opposite is true when NO_3^- is taken up (Forde and Clarkson, 1999; Britto and Kronzucker, 2002).

Vanhala (2002) summarized the seasonal variation in soil respiration in three forest sites that differed in N, water availability, and pH. Although seasonal variation in soil chemical properties was small, soil respiration rates decreased from spring to summer, with minimum values at the end of August. The soil respiration rates increased again in autumn, but values were not close to those in the spring. At a constant temperature (14 °C), the respiration rate was determined by moisture and pH, while at a constant moisture content, it depended heavily on the amount of organic matter (N and C-org) and pH. When the respiration rate was calculated on the basis of the N concentration, the variation was explained primarily by pH.

Inorganic phosphate (Pi) is a key nutrient that is crucial for the biosynthesis of macromolecules and for the whole gene-expression machinery. Phosphate also plays a predominant role in plant respiratory metabolism because of its involvement in energy transfer reactions (Siedow and Day, 2000). Cellular Pi levels are tightly associated with its metabolic demand, and regulated by its supply, which is significantly affected by pH of the nutrient solution. In plants, several high-affinity Pi transporters have been cloned that regulate the balance of cellular Pi and its organellar compartmentation in various tissues. Several of these are PO_4^{3-}/H^+ transporters whose activity responds to external pH. A reduction in nutritional availability of PO_4^{3-} results in a decrease in cellular Pi, ATP, and ADP, which affects numerous metabolic pathways that use these phosphorylated moieties. Apart from direct effects of Pi deficiency on the respiratory pathway (Gauthier and Turpin, 1994), significant indirect effects include altered N metabolism (Johnson et al., 1996b), an increase in amino acids and NADH (Johnson et al., 1996a; Juszczuk and Rychter, 1997), and an increase in alcohol dehydrogenase activity (Neumann et al., 2000). Some of these effects are related to the increased production of organic acids (see discussion below) that involves changes in the activities of certain respiratory enzymes, e.g., PEP carboxylase, malate dehydrogenase, PEP phosphatase, citrate synthase, phosphofructophosphatase, and NADP-dependent glyceraldehyde-3-phosphate dehydrogenase (Duff et al., 1989b; Theodorou et al., 1992; Delhaize et al., 1993a,b; Johnson et al., 1994, 1996a,b; Watt and Evans, 1999b).

An example of how pH and Al-induced reduction in Ca^{2+} uptake, Al-induced efflux of citrate (see discussion below), and availability of cellular Pi could modulate at least one key step in respiration is their effect on the properties of the vacuolar PEP phosphatase, which presumably converts PEP into pyruvate (reviewed in Givan, 1999). K^+, Mg^+, and ADP are positive effectors of this enzyme, while ATP, citrate, and Ca^{2+} are negative effectors (Duff et al., 1989a, 1991). As described later, Al causes a reduction in cellular [Ca] (which also is a direct effect of low pH), and also induces a significant efflux of citrate; both of which, combined with reduced ATP and Pi, tend to promote the conversion of PEP to pyruvate, which could then serve as a substrate for synthesis of citrate and other organic acids (Fig. 1). A reduction in the concentrations of citrate and ATP also enhances the activity of key glycolytic reactions that involve phosphofructokinase. Levi and Gibbs (1976) and Steup et al. (1976) demonstrated that high amounts (2–5 mM) of PPi promoted a breakdown of starch in the chloroplast by plastid glycolysis.

In addition to effects on the uptake of N and P, pH profoundly affects the availability as well as the uptake of many other nutrients, including Ca, Mg, Mn, Fe, and K. In turn, these nutrients play significant but complex roles in respiration (Marschner, 1995; Siedow and Day, 2000).

IV. Interactions Among Soil pH, Aluminum and Respiration

Bound as oxides and complex aluminosilicates, Al is the most abundant metal in the Earth's crust. Until recently, surface-water concentrations of Al ions (particularly Al^{3+}) have remained minimal due to the insolubility of Al hydroxide complexes at neutral pH (Macdonald and Martin, 1988). The lowering of soil pH due to acidic deposition and excess fertilization leads to the solubilization of otherwise insoluble Al in the mineral soil horizon. On the soil particles, Al competes with Mg and Ca for ion-exchange sites, leading to leaching of these nutrients into mineral soil and eventually into surface water (Lawrence et al., 1995). Al profoundly affects plant growth and development through inhibition of cell division (via inhibition of DNA, RNA and protein synthesis), reduction of cell enlargement, and cytotoxicity by direct interactions with the plasmalemma (Delhaize and Ryan, 1995; Sparling and Lowe, 1996; Ma et al., 2001; Kochian et al., 2002). The direct and indirect effects of Al on cation uptake, membrane properties, and organic acid efflux, all are tied to cellular respiration The discussion here is limited mostly to the effects of Al on membrane properties and metabolic processes that are related (Fig. 1) and presumably are responsible for effects on plant respiration.

The degree to which Al effects are seen in plants depends on environmental factors that include the amount and species of Al ions, amounts of other ions present, amount and form of organic matter present, and plant genotype (Karr et al., 1984; Kochian et al., 2002). While most plants exclude this metal or absorb it only in small quantities, several widely spread taxa

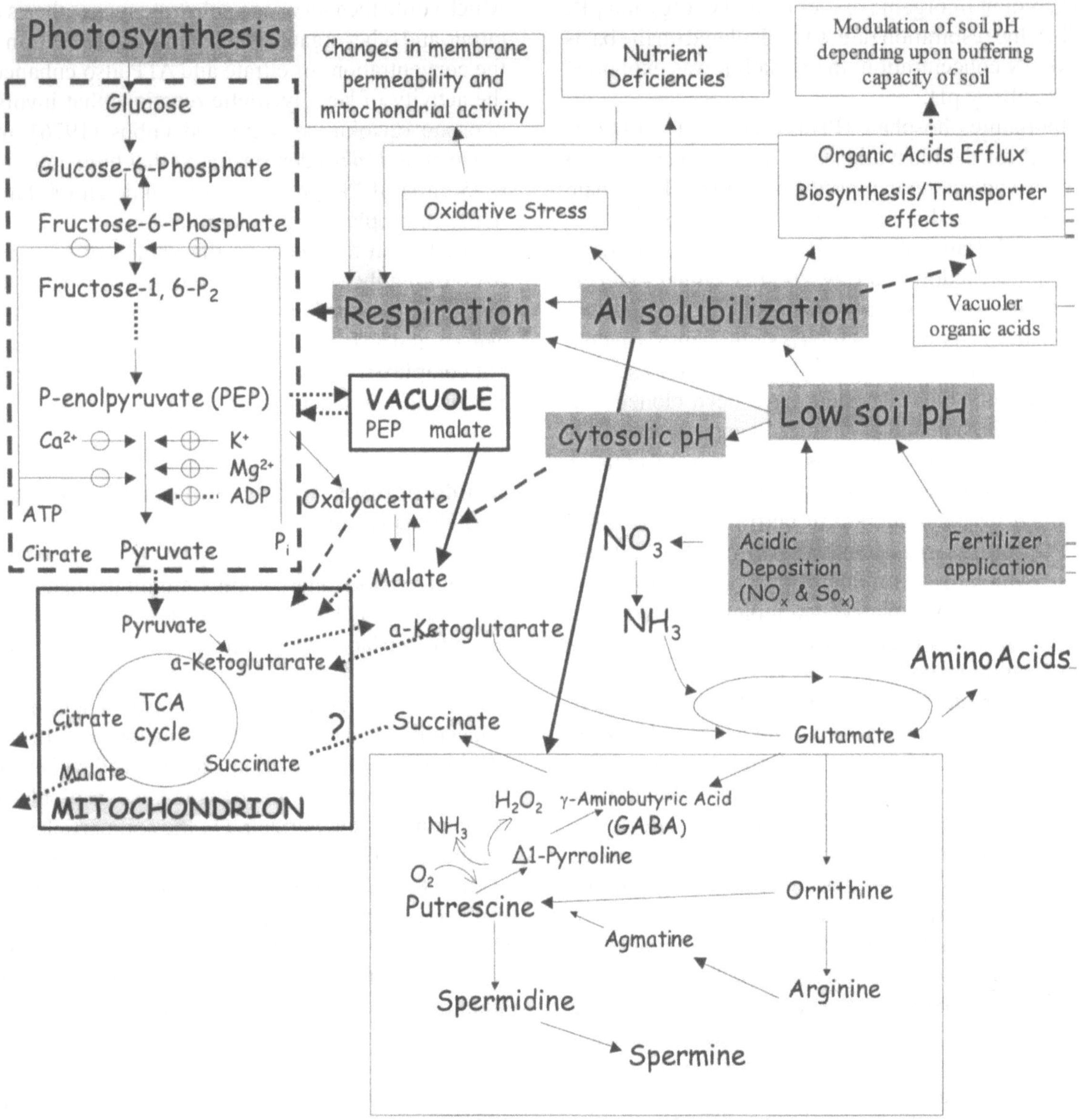

Fig. 1. Interactions among low pH, Al solubilization and the respiratory pathway in plants. Only selected parts of the pathways that are relevant to the discussion in this chapter are shown here.

are known to be hyperaccumulators of Al (Jansen et al., 2002). Soil pH dictates the total amount of Al that is solubilized, and thus available for physiological effects, and also affects the inorganic forms of Al that determine its toxicity. The Al^{3+}, $Al^{2+}(OH)$, and $Al^{+}(OH)_2$ monomers generally are regarded as the toxic forms of Al in aqueous systems; of these, Al^{3+} is many times more toxic than the hydroxyl forms, although a pH ≤4.5 is required for its formation (Delhaize and Ryan, 1995). Tolerance/resistance mechanisms have evolved in plants to cope with an acidic rhizosphere environment where Al concentrations are relatively high (Piñeros and Kochian, 2001).

The complexity of interactions among the effects of Al and pH on respiration is evident from: (a) multiple effects of soil pH on Al availability, Al speciation, and uptake, and (b) variable pH of the cellular compartments and resulting effects on Al speciation and compartmentation in the symplast (Kochian et al., 2002). The results of a study by Godbold and Jentschke (1998) who attempted to experimentally analyze the interaction between Al and pH in a nutrient solution further illustrate the complexity of this interaction. The uptake of ions by the plant is heavily dependent on the ion exchange capacity of the root apoplast, which is subject to major influence of pH (Meychik and Yermakov, 2001). Al, whose solubilization and specific forms depend on pH (the lower the soil pH, the higher the overall Al solubilization and the availability of Al^{3+}), readily displaces Ca and Mg in the apoplast (Godbold et al., 1988; Schröder et al., 1988). While the soil and the nutrient solution chemistry suggest that maximum Al toxicity should coincide with maximum Al solubility, lowering of pH to around 3.2 actually reduces Al toxicity (Grauer and Horst, 1990; Kinraide, 1993). In the root cortical cell walls of Norway spruce (*Picea abies*) seedlings, the cation-exchange capacity (for Ca, Mg, etc.) was the highest at pH 4.0, within the range of 3.2–5.0 that was tested (Godbold and Jentschke, 1998; see also, Godbold et al., 1988; Schröder et al., 1988; Keltjens, 1995). In the presence of Al, the highest cation-exchange capacity was at pH 3.2; it decreased as the pH of the solution increased, as did the amount of Ca and Mg present in the root apoplast. Higher Al in the soil or the nutrient solution was associated with reduced foliar Ca and Mg in some plants, but there was no effect on foliar Mg levels in others (Godbold et al., 1988; Schröder et al., 1988; Moustakas et al., 1995; Oleksyn et al., 1996; Lee et al., 1999; Minocha et al., 2000). The effect of Al on both shoot and root growth was maximum at pH 5.0 and minimum at pH 3.2.

A. Aluminum in the Apoplast and Symplast

A major controversy concerning the physiological effects of Al is whether its effects are an apoplastic or a symplastic phenomenon (Kochian et al., 2002). There is a large body of evidence attesting to the apoplastic effects of Al. However, under acidic conditions, Al^{+3} can enter the cell by transporters, by adsorption endocytosis, or by forming complexes with Fe^{+3}, amino acids and organic acids (Kochian, 1995), and thus show symplastic effects. Possible apoplastic interactions of Al include immobilization of cell walls, chelation with organic acids, and stimulated efflux of phosphate. Examples of the symplastic interactions are chelation by organic acids, proteins, and other organic ligands in the cytosol; compartmentation in vacuoles; and superinduction of Al-sensitive enzymes.

The plasma membrane is known to be an effective barrier to Al entry into the cytoplasm (Cuenca et al., 1991; Marienfeld and Stelzer, 1993; Marienfeld et al., 1995). Other researchers have reported that Al enters the symplast in the root cells, where it can accumulate and/or be transported to the shoots (Matsumoto et al., 1976; Asp et al., 1988; Zhang and Taylor, 1989; Lazof et al., 1994, 1997). For the primary effects of short-term exposure to Al, its entry into the symplast is not considered necessary (Olivetti and Etherton, 1991; Huang et al., 1992; Rengel, 1992). There is also evidence that Al can directly alter plasma-membrane properties by interacting with certain phospholipids and altering their peroxidation levels (Akeson and Munns, 1989; Rengel, 1996; Yamamoto et al., 1997; Zhang et al., 1997) and ion-channel proteins (Rengel, 1992). One of the few early studies of Al effects on membrane functions showed accelerated leakage of K^+ from the plasma membrane (Woolhouse, 1969). Later studies showed the activation of anion channels similar to the S-type Cl^- channel in the membrane in response to Al (Ryan et al., 1997; Zhang et al., 2001). Its ability to chelate extracellular organic acids, to affect membrane properties, and to stimulate efflux of phosphate all will have significant effects on cellular respiration rates.

Cytosolic effects of Al are generally believed to be less important, because at cytosolic pH of about 7, Al exists mostly as $Al(OH)_4^{-1}$ which has little toxicity (Martin, 1988; Kinraide and Parker, 1990; Kochian, 2000). According to Martin (1988), most of the Al in cytosol is tied up, and its available concentration is low, about 10^{-10} M. However, it could still compete for Mg binding sites even at extremely low concentrations (Kinraide and Parker, 1990; Martin, 1992). For example, binding of Al to ATP is 7 orders of magnitude greater than that of Mg, and the $Al\text{-}ATP^{4-}$ complex binds more strongly to hexokinase than Mg-ATP complex. This would result in a negative impact of Al on respiration. It has not yet been possible to make credible assertions about intracellular localization of Al because of the differences in instruments, sample preparation methods, operating conditions, and lengths of Al exposure used in various studies (Rengel, 1996).

B. Direct Effects of Aluminum on Respiration

There have been several reports on the effects of Al treatment on the rates of cellular respiration in a variety of plants. Upon exposure to Al, the respiration rates of carrot (*Daucus carota*) suspension cultures decreased significantly (Honda et al., 1997), as did the respiration rates of white spruce (*Picea glauca*) seedlings (Nosko et al., 1988). In carrot, Al exposure also caused a reduction in cellular ATP level. Al inhibited respiration in cut rose leaves (Son et al., 1994), inhibited mitochondrial respiration and oxidative phosphorylation in the roots of rice (*Oryza sativa*) seedlings (Hao and Liu, 1989), as well as red spruce (*Picea rubens*) and tobacco (*Nicotiana tabacum*) cultures (Ikegawa et al., 1998; Minocha et al., 2001; Yamamoto et al., 2002). Schaberg et al. (2000) observed reduced respiration in elongating shoots of red spruce following Al exposure. Dark respiration was inhibited in the seedlings of Japanese red pine (*Pinus densiflora*) when Al was added to the culture medium (Lee et al., 1999, 2001). Yamamoto et al. (2002) reported an increase in reactive oxygen species, inhibition of respiration, depletion of ATP, and loss of growth in tobacco and pea plants 12 h after Al exposure. These results were used to hypothesize that mitochondrial dysfunction may lead to the production of reactive oxygen species, which may be a key factor in growth inhibition by Al. Other cases where Al caused an increase in respiration are red spruce seedlings and wheat (*Triticum aestivum*) roots (McLaughlin et al., 1990, 1991; Collier et al., 1993).

In most of the above-mentioned cases, the target site for Al effects on respiration was not identified. Three kinds of effects of Al on plant respiratory metabolism can be postulated: (a) directly by interactions with the plasmalemma and organellar membranes, (b) those related to the uptake of ions, and (c) those related to Al-induced efflux of organic acids.

The effects of Al on the activities of enzymes involved in ATP generation and utilization via effects on ATP synthase and ATPases also have been reported. Increases in the activities of vacuolar ATPase and mitochondrial ATP synthase concomitant with a reduction in the plasma membrane ATPase activity were observed in an Al-resistant wheat variety (Hamilton et al., 2001). Changes in the level of transcript for vacuolar ATPase paralleled changes in its enzyme activity, indicating that this response occurs at the level of gene expression. However, no change in the transcript levels of mitochondrial ATP synthase was observed. The induction of vacuolar ATPase activity is thought to be a homeostatic mechanism required to provide energy for Na^+/H^+ antiport that delivers Na^+ to vacuoles. However, upon exposure to Al, an increase in vacuolar ATPase may be needed to maintain the cytosolic pH neutrality or to sequester Al in vacuoles via an Al^+/H^+ antiport similar to the Na^+/H^+ antiport. Currently, there is no direct evidence for such Al^+/H^+ co-transporters in plants (Hamilton et al., 2001 and references therein).

C. Aluminum, Organic Acid Metabolism, and Respiration

As discussed earlier, solubilized Al interferes with the uptake of many essential cations, particularly Ca, Mg, Fe, and Mn (Delhaize and Ryan, 1995; Ryan et al., 2001; Ma et al., 2001). The importance of these ions in respiration was also discussed in relation to the effects of pH. Al has specific interaction with the uptake of Ca and P, with the former it competes for carriers; with the latter, it forms insoluble $Al_3(PO_4)$ both in the soil and in the apoplast. Pellet et al. (1995) showed that PO_4^{3-} secretion and organic acid secretion might be regulated independently.

One of the best documented phenomena in relation to Al effects in plants is its role in inducing the efflux of low-molecular-weight organic acids from intact roots as well as from cells in culture (reviewed in Delhaize et al., 1993a,b; Delhaize and Ryan, 1995; Ma, 2000; Ma et al., 2001; Ryan et al., 2001; Kochian et al., 2002; Minocha and Long, 2004). Acids that are commonly released include: citrate, malate, succinate, and oxalate. Often, a particular acid is secreted in a species-specific manner, and more than one acid can be secreted simultaneously. Secretion of organic acids is greater in Al-resistant varieties, is Al induced, and is dependent on Al concentration. This secretion is several folds greater in the apex than in the mature parts of the root, suggesting localized detoxification and injury (Ryan et al., 1995). While the induction of organic acid secretion by Al is well established, the mechanisms of signal perception and transduction that cause the efflux, the signals that cause increased production of these acids, and the specific enzyme(s) involved in their biosynthesis are not known.

Several organic acids are present in all living organisms as intermediates of the TCA cycle, the main respiratory pathway that oxidizes malate and pyruvate to release energy (Bray et al., 2000; López-Bucio et al., 2000; Siedow and Day, 2000). Some of these are present both in the cytosol and in the mitochondria, and can accumulate in large quantities in vacuoles. Organic acids also are linked with other metabolic processes, e.g., amino-acid biosynthesis, regulation of cytosolic pH and osmotic potential, chelation of cations, and charge balance during excessive carbon uptake (Edwards et al., 1998). Questions remain with respect to Al-induced organic acid secretion; for example: since the raw materials for the synthesis of most organic acids that are secreted are in the mitochondria, what triggers their release from this organelle, or from the vacuole, into the cytosol for its eventual efflux? Is it simply a matter of equilibrium between the two compartments, or active mitochondrial and vacuolar membrane transporters are involved? If the latter is the case, then how are these transporters regulated by Al? Does Al enter mitochondria? It is believed that Al interactions could occur at any or all of these levels to cause efflux of a particular acid.

The mechanisms that regulate the secretion of organic acids as well as their biosynthesis to support increased secretion in response to Al treatment have been widely debated. Ma (2000) discussed two patterns for Al-induced organic acid secretion. In the first, there is no lag phase between Al exposure and organic acid secretion (rapid, short-term responses); activation of anion channels is implicated in this approach. The second entails a marked lag phase between Al exposure and acid secretion (slow and long-term effects); gene activation may play a part in affecting organic acid metabolism in this pattern. Furthermore, it is still not known whether Al functions from the outside or the inside of the membrane to induce rapid changes in organic acid efflux. In addition to increased organic acid efflux in response to Al, there are examples of an increase in biosynthesis and accumulation of organic acids in cells without increase in efflux (Keerthisinghe et al., 1998; Watt and Evans, 1999a; Neumann and Römheld, 2000). The lack of a positive correlation between the cellular content of an acid and its efflux may largely be due to compartmentation of the acid (Ryan et al., 2001).

The secretion of organic acids from roots in the presence of solubilized Al at low pH is an excellent example of the complex interaction among pH, Al, and respiration. In addition to solubilization of Al, a decrease in external pH causes an increase in the membrane potential (make it more negative), thus promoting the passive efflux of organic acids from the cytosol. In turn, this can enhance the release of acids from the vacuole into the cytosol. There are two additional consequences of low pH-induced organic acid efflux from the root: (a) the efflux of organic acids from the cells is accompanied by the release of K^+, and (b) the release of organic acids into the rhizosphere modulates the rhizosphere pH and soil microbe populations (Delhaize et al., 1993b; Marschner, 1995). If the efflux of both the organic acids and a cation such as K^+ continues, it will result in adverse physiological effects on osmotic potential of the root cells as well as in the carbon balance in the cells. The latter will have serious consequences for the respiratory metabolism of the root. Few experimental data are available on the prolonged effects of low pH and high Al levels on these rhizosphere interactions.

For Al-induced secretion of organic acids over extended periods, it can be argued that both the effects on anion channels and altered carbohydrate metabolism (respiration) must be involved. The extent of organic acid secretion is in the same range as the H^+/K^+ fluxes (Kochian, 2000), and the total amount of carbon secreted as organic acids can be significant; e.g., as high as 23% of the net photosynthate was secreted as organic acids by *Lupinus albus* within 13 weeks of growth (Dinkelaker et al., 1989). The role of Al in stimulating the release of organic acids from mitochondria and vacuoles is not clear. Whereas involvement of anion channels in organic acid efflux from the cell has been demonstrated (Kochian, 2000; Ryan et al., 2001), their nature and distribution in the plasmalemma, the mitochondrial membranes, and the tonoplast, and the regulation of their function by Al and other factors is poorly understood. Direct or indirect effects of Al on anion channels may include induction of more channels, and induction of second messengers to activate ion channels. It is speculated that the Al effect is not through direct interaction with the anion channels, but occurs through cascades of signal transduction steps, possibly involving Ca, kinases, phosphatases, or other regulatory proteins (Ryan et al., 2001). Al may cause transient Ca release by interfering with G protein and IP_3-mediated signal transduction pathways (Coté and Crain, 1993; Haug et al., 1994); it also may interact directly with calmodulin, a Ca-binding protein involved in signal transduction pathways. Al binds to calmodulin 10-times more strongly than Ca, though the results have been contradictory (You and Nelson, 1991). Despite recent advances in gene cloning and the availability of the entire genome sequence of several plants, little is known about the genes encoding specific anion channels involved in organic acid efflux.

The source of the organic acids that are secreted in response to Al is equally controversial. Since increased activities of enzymes of the TCA cycle are primarily responsible for their biosynthesis, increases in both enzyme activities (involving biosynthesis and post-translational modulation) and substrate availability are essential. Delhaize et al. (1993b) showed that the cellular content of malate was not affected by a higher rate of its efflux, indicating a homeostatic regulation in a constitutive manner. Basu et al. (1994) also observed increased production of malate concomitant with its increased efflux. Although in wheat no immediate effect of Al on organic acid biosynthesis via PEP carboxylase and NAD-malate dehydrogenase was observed, increased PEP carboxylase activity was associated with Al-induced organic acid secretion in maize (Gaume et al., 2001). Ryan et al. (1995) observed that both Al-sensitive and Al-resistant wheat showed similar activity of PEP carboxylase and malate dehydrogenase, key enzymes involved in malate synthesis. However, it should be noted that both of these enzymes have multiple forms (Givan, 1999; Siedow and Day, 2000). Li et al. (2000) showed that species differ in their response to Al in terms of organic acid efflux. For example, efflux of acids was rapid in wheat and slow in rye (6–10 h); isocitrate dehydrogenase, PEP carboxylase, and malate dehydrogenase were not affected; citrate synthase increased in rye but not in wheat; low temperature had no effect on malate exudation in rye but citrate exudation was lower; and inhibition of the citrate carrier (between mitochondria and cytosol) inhibited citrate secretion into the medium in both species. It was concluded that while both metabolism and carriers were affected by Al in rye, only carriers were affected in wheat.

Mugai et al. (2000) characterized citrate secretion in two bean (*Phaseolus vulgaris*) varieties that differed in Al tolerance. They showed that the resistant variety secreted more citrate than the sensitive variety. Also, the activities of citrate synthase and $NADPH^+$-isocitrate dehydrogenase were higher in the former, but those of PEP carboxylase were not significantly different under Al stress. A dual effect of Al on both organic acid production and secretion can be possible through its effects on NADP-dependent isocitrate dehydrogenase. The inhibition of this enzyme by Al could increase the cellular levels of citrate (Chiba, 1999). This might have a negative effect on the TCA cycle and respiration, causing a reduction in ATP production, followed by a passive secretion of acids due to the semipermeability of the plasma membrane being compromised.

The complexity of interactions among key metabolites involved in organic acid synthesis is illustrated by the various routes of utilization of PEP in the respiratory metabolism (Fig. 1). There are at least three known metabolic fates of PEP in the cell: (a) its conversion into pyruvate by cytosolic or plastidic PEP kinase, (b) its conversion into pyruvate by vacuolar PEP phosphatase, and (c) its carboxylation in the cytosol by PEP carboxylase to OAA (Givan, 1999; Siedow and Day, 2000). Pyruvate is then transferred into mitochondria for entry into the TCA cycle. Oxaloacetate can also enter the TCA cycle in mitochondria or it can be reduced in the cytosol to malate, which then enters the mitochondria and the joins the TCA cycle. The cytosolic malate could serve as a direct source for its efflux in response to Al. If this were to occur, its impact on TCA cycle reactions could be significant; however, little information on this is currently available.

The secretion of organic acids in response to Al has been implicated as a mechanism for Al resistance in plants (Delhaize and Ryan, 1995; Jones, 1998; Ma et al., 2001). Al resistance by organic acids secretion could operate through several mechanisms, for example, by detoxification of the extracellular Al by binding to organic acids (Delhaize and Ryan, 1995; Jones, 1998; Ma et al., 2001) or by increasing the availability of essential nutrients for uptake by roots (Jones, 1998). However, Ishikawa et al. (2000) did not observe a correlation between Al resistance and citric and malic acid exudation among various species or even between two cultivars of the same species. They suggested that there might be more effective alternative Al resistance mechanisms that operate in addition to or in place of the exudation of malic and citric acids in these species.

Several laboratories are attempting to exploit the relationship between the accumulation and/or secretion of organic acids and the resistance of plants to Al by transgenic manipulation of organic acid metabolism in order to produce Al-resistant plants (de la Fuente et al., 1997; Koyama et al., 1999; Tesfaye et al., 2001). For example, tolerance to Al was conferred to transgenic tobacco and carrot plants by overexpression of a citrate synthase gene and to alfalfa (*Medicago sativa*) by overexpression of a malate dehydrogenase gene. However, Delhaize et al. (2003) failed to see any effect of either an overexpression of a mitochondrial citrate synthase or a reduction in cytosolic isocitrate dehydrogenase on accumulation or efflux of citrate in transgenic tobacco. Nevertheless, these studies provide evidence for the potential role of Al in the respiratory metabolism of plants.

In summary, the inhibition of Ca and Mg uptake and the induction of phosphate extrusion by Al would together have a potential negative effect on respiration in plants. However, its effects on the induction of organic acid secretion should enhance certain reactions of the respiratory metabolism in plant cells.

D. Aluminum and Non-Phosphorylating Respiration

It is believed that overall flux of carbon through the TCA cycle is governed by reoxidation of NADH via the regenerative electron-transport chain and the cellular utilization of ATP (Simons and Lambers, 1998; Siedow and Day, 2000; Vanlerberghe and Ordog, 2002). The overall rate of oxygen consumption is regulated by ADP and Pi levels in the cell. Other controlling factors are availability and turnover of the substrate, with some modulation provided by the non-phosphorylating oxidation reactions. This might yield less ATP per cycle of electron flow, allowing the TCA cycle to continue. This mechanism also would allow continued production of organic acids for efflux in response to Al. The extent of non-phosphorylating or cyanide-resistant alternative oxidation reactions is subject to factors such as the amount of α-keto acids in the cell as well as a variety of environmental stresses (Simons and Lambers, 1998; Vanlerberghe and Ordog, 2002; Millenaar and Lambers, 2003). Conditions that reduce the flow of electrons through the cytochrome-based electron transport chain enhance alternative respiration. It is thought that the mechanism in part involves stimulation of the alternative oxidase by increased pyruvate in response to its reduced utilization as NADH builds up in the cells. Once it is activated, a high rate of alternative respiration may continue, allowing the sustained production of precursors and intermediates of the TCA cycle. Unfortunately, there is no direct evidence on the effects of Al on alternative oxidase. However, it can be argued that in the presence of Al, overall ATP utilization may be reduced, and enhanced alternative oxidase may serve a useful purpose in allowing a continued supply of organic acids for efflux. This situation might be similar to the situation with rotenone-insensitive, NADH dehydrogenase-driven electron flow, which bypasses Complex I (Siedow and Day, 2000). This non-phosphorylating electron flow also allows higher levels of respiration to continue under conditions of low ATP demand and utilization. The alternative oxidase also prevents the accumulation of superoxide and other hydroxyl radicals, which usually are formed under conditions of a highly reduced state of the electron transport chain. A possible role of these alternative oxidation reactions under adverse environmental conditions (chilling, drought, osmotic stress, etc.) has been discussed (Simons and Lambers, 1998; Lambers et al., 1998; Siedow and Day, 2000; Millenaar and Lambers, 2003).

E. Aluminum and Photorespiration

Photosynthesis, photorespiration, and dark respiration have strong direct and indirect metabolic interactions in regulating carbon and energy flow within the cell. Although occurring in different organelles, these three pathways share several intermediates as well as products, and affect the energy status of the cell in both competitive and complementary ways. Cytosolic and organellar pH, cytosolic content of Ca, Fe, Mg, P, and other ions, enzyme activities (all of which are affected by ambient pH), and the presence of solubilized Al in the environment have significant effects on photosynthesis, photorespiration and dark respiration (Moustakas et al., 1993; Simons et al., 1994; Marschner et al., 1995). Unfortunately, there are no data on the concurrent effects of Al on photosynthesis and photorespiration.

F. Aluminum, Oxidative Stress, and Respiration

Oxidative stress results from conditions that promote the formation of reactive oxygen species (ROS) that damage or kill cells. Among the numerous environmental factors that cause oxidative stress is Al (Bray et al., 2000). Generally, ROS are formed during certain redox reactions and during reduction of oxygen or oxidation of water by mitochondrial or chloroplast electron-transfer chains. A possible sequence of events leading to an increase in ROS during stress was discussed by Møller (2001). As much as 1–5% of the consumed oxygen could go into ROS production depending on the plant species and the overall rate of respiration (Møller, 2001). To prevent oxidative damage, plant cells are equipped with a scavenging system consisting of antioxidants of low molecular weight (e.g., ascorbate, glutathione, and vitamin E), and several protective antioxidant enzymes (Asada and Takahashi, 1987; Kuzniak, 2002). Polyamines and flavanoids also may provide some protection against free radicals (Bray et al., 2000). When these defenses are overwhelmed, as occurs during biotic and abiotic stress, the mitochondria are damaged by oxidative stress (Møller, 2001).

There is evidence that Al causes oxidative stress in plants, and there might be an overlap between protective mechanisms for Al stress and oxidative stress (Ezaki et al., 1998, 2000; Richards et al., 1998). Basu et al. (2001) showed that Al caused an induction of mitochondrial superoxide dismutase in oil seed rape (*Brassica napus*). Sakihama and Yamasaki (2002) demonstrated that Al stimulated the phenolic-dependent lipid peroxidation in barley (*Hordeum vulgare*) roots. A strong link between lipid peroxidation (oxidative stress) and cell viability under Al exposure also was reported by Ingo and Wolfgang (2000). Ogawa et al. (2000) found that Al treatment of roots activated several of the antioxidant enzymes in the needles of Hinoki cypress (*Chamaecyparis obtusa*). Lidon et al. (1999) reported that ethylene production, acyl peroxidation, and activities of antioxygenic enzymes in maize shoots were affected by Al. For example, activities of copper/zinc-superoxide dismutase, catalase and glutathione reductase were inhibited while those of ascorbate peroxidase and dehydroascorbate reductase were increased in response to Al. An Al-resistant wheat cultivar had higher levels of superoxide dismutase, and reduced glutathione and malondialdehyde, suggesting a lower level of oxidative stress compared to an Al-sensitive cultivar (Dong et al., 2002). In our studies with red spruce suspension cultures, Al caused a dose-dependent decrease in several antioxidant enzymes, e.g., glutathione reductase, monodehydroascorbate reductase, and ascorbate peroxidase (R. Minocha, S. Minocha, S. Long and L. Jahnke, unpublished). Genes for mitochondrial superoxide dismutase, glutathionine S-transferase, peroxidase, and blue copper-binding protein respond to Al and oxidative stress in various plant species. These studies suggest that there may be common protective mechanisms among several types of stress, including pathogen stress, oxidative stress, heat shock stress, and Al stress.

G. Aluminum, Polyamines, and Respiration

The aliphatic polyamines (putrescine, spermidine, and spermine) play an important role in the growth and development of all living organisms (Cohen, 1998). The polyamine metabolism interacts with the respiratory metabolism in two ways (Fig. 1): (a) their biosynthesis depends on a continuous supply of glutamate (derived from α-keto glutarate) for the production of ornithine and arginine, the primary substrates for polyamine biosynthesis; and (b) their oxidative catabolism yields succinate that could contribute to the cytoplasmic pool of this acid that can be secreted. Polyamines carry a net positive charge at cellular pH and also play a role in modulating cellular pH (Cohen, 1998). In addition, polyamines can affect several mitochondrial functions via electrostatic interactions (Votyakova et al., 1999).

Abiotic stress conditions including low pH, high SO_2, nutrient deficiency or oversupply, and high Al cause increased cellular concentrations of polyamines, particularly putrescine (Dohmen et al., 1990; Santerre et al., 1990; Flores, 1991; Minocha et al., 1992, 1996, 1997, 2000; Wargo et al., 2002). There often is an inverse relationship between the cellular concentrations of putrescine and those of Ca, Mg, Mn, and K in response to Al treatment (Minocha et al., 1992, 1996; Zhou et al., 1995); these cations also respond to apoplastic pH. A key distinction between the polyamines (organic cations) and the inorganic cations is that the cytoplasmic availability of the latter can change only in response to external stimuli by recompartmentalization, because their cellular levels are derived entirely from uptake and transport, which depend upon their availability in the soil solution. By contrast, polyamines are synthesized within the cell, allowing adjustment of their cellular concentrations to meet physiological needs in situ. For example, substitution of Ca by putrescine at certain sites in the cell could increase Ca availability for key signal transduction pathways at times of Ca deficiency, e.g., in the presence of high concentrations of Al.

On the basis of extensive work from our group using cell cultures and mature conifer and hardwood trees in the field, we have proposed that putrescine could be used as a potential biochemical indicator of Al stress (indirectly Ca deficiency) in visually healthy trees (Minocha et al., 1996, 1997, 2000; Wargo et al., 2002). The availability of early biochemical indicators can aid in assessing the current status of (Al) stress in visually healthy trees, which, in turn, can be useful in planning potential treatments and management strategies designed to alleviate the deleterious effects of stress or remediate the cause(s) of stress.

V. Conclusions and Perspectives

Profound metabolic and, consequently, growth and developmental changes occur when plants are exposed to high acidity in the soil. The factors that regulate soil pH, and the cellular signals that modulate the resulting changes in metabolism, are complex and poorly understood. Oxidative metabolism is only one of a multitude of cellular responses that are affected by changes in pH, directly, and through changes in soil nutrient levels, including a major effect through Al toxicity. The latter (Al toxicity) in itself is a complex phenomenon, an understanding of which is attracting increasing attention. A better understanding of the complexity of interactions between pH and Al will allow us to produce genetically improved plant varieties that can cope with the deleterious consequences of man made climatic changes that contribute to environmental acidity.

Potential resistance to low pH is a multigenic trait. Only a few mutant/resistant varieties of commercially important plants are available that show high yields in acid soils. As a result, breeding for low-pH resistance is a formidable task. In recent years, powerful techniques have become available to analyze global changes in gene activity in response to one or more environmental factors. Techniques such as Serial Analysis of Gene Expression (SAGE), Differential Display Reverse Transcriptase (DDRT) PCR, macroarrays, and microarrays (DNA chips) allow simultaneous analysis of the expression of thousands of genes within short periods of a specific treatment (Kuhn, 2001; Bohnert et al., 2001; Wu et al., 2001; Seki et al., 2002). These techniques have not yet been applied to studies of the effects of pH changes in the environment on molecular changes in the plant cell, particularly those related to respiratory metabolism. Currently there is no EST (Expressed Sequence Tags) database for genes responding to low pH or Al. High-throughput molecular techniques should allow us to distinguish between direct effects of pH changes on gene expression in the presence or absence of indirect consequences due to nutrient uptake and Al toxicity. Likewise, high-throughput analysis of thousands of metabolites in the cells should reveal changes in respiratory and related metabolic changes in the cell. Initial results on the effects of salinity, drought, cold, and pathogen infection on gene expression have been revealing with respect to changes that occur within hours of such treatments; many of the genes identified are related to respiratory metabolism (Bohnert et al., 2001; Seki et al., 2002). It is hoped that similar studies will lead to the identification of the genes whose expression changes in response to alterations in pH and Al, and also to the identification of signal-transduction pathways that are responsible for these changes. These studies should then lead to the cloning of pH-inducible regulatory elements (promoters, enhancers, etc), which would be useful in altering the expression of other genes of interest whose expression may impart tolerance or resistance to such pH changes.

Acknowledgments

We thank Dr. Curtis Givan and Dr. John Wallace for a critical review, and Stephanie Long and Kenneth Dudzik for an editorial review of the manuscript.

References

Akeson MA and Munns DN (1990) Lipid bilayer permeation by neutral aluminum citrate and by three α-hydroxy carboxylic acids. Biochim Biophys Acta 984: 200–206

Asada K and Takahashi M (1987) Production and scavenging of active oxygen in photosynthesis. In: Kyle DJ, Osmond, CB and Arntzen, CJ (eds) Photoinhibition, pp: 227–289. Elsevier Science Publishers, New York

Asp H, Bengtsson B and Jensén P (1988) Growth and cation uptake in spruce (*Picea abies* Karst.) grown in sand culture with various aluminum contents. Plant Soil: 111: 127–133

Basu A, Basu U and Taylor GJ (1994) Induction of microsomal membrane proteins in roots of an aluminum-resistant cultivar of *Triticum aestivum* L. under conditions of aluminum stress. Plant Physiol 104: 1007–1013

Basu U, Good AG and Taylor GJ (2001) Transgenic *Brassica napus* plants overexpressing aluminum-induced mitochondrial manganese superoxide dismutase cDNA are resistant to aluminum. Plant Cell Environ 24: 1269–1278

Bligny R and Douce R (2001) NMR and plant metabolism. Curr Opin Plant Biol 4: 191–196

Bohnert HJ, Ayoubi P, Borchert C, Bressan RA, Burnap RL, Cushman JC, Deyholos M, Fischer R, Galbraith D, Hasegawa PM, Jenks M, Kawasak S, Koiwa H, Kore-Eda S, Lee BH, Michalwoski CB, Misawa E, Nomura M, Ozturk N, Postier B, Prade R, Song CP, Tanaka Y, Wang H and Zhy JK (2001) A genomics approach towards salt stress tolerance. Plant Physiol Biochem 39: 295–311

Boron WF and Roos A (1976) Comparison of microelectrode, DMO and methylamine method for measuring intracellular

pH. Am J Physiol 231: 799–809
Bray EA, Bailey-Serres J and Weretilnyk E (2000) Response to abiotic stress. In: Buchanan BB, Gruissem W and Jones RL (eds) Biochemistry and Molecular Biology of Plants, pp 1158–1203. American Society of Plant Physiologists, Rockville, MD
Britto DT and Kronzucker HJ (2002) NH_4^+ toxicity in higher plants: A critical review. J Plant Physiol 159: 56–584
Britto DT, Glass ADM, Kronzucker HJ, Siddiqi MY (2001a) Cytosolic concentrations and transmembrane fluxes of NH_4^+ /NH_3. An evaluation of recent proposals. Plant Physiol 125: 523–526
Britto DT, Siddiqi MY, Glass ADM and Kronzucker HJ (2001b) Futile transmembrane NH_4^+ cycling: A cellular hypothesis to explain ammonium toxicity in plants. Proc Natl Acad Sci USA 98: 4255–4258
Burt CT, Cohen SM and Barany M (1979) Analysis of intact tissue with ^{31}P NMR. Annu Rev Biophys Bioeng 8: 1–25
Chiba H (1999) Masters Thesis (Japanese). Okayama University, Okayama, Japan
Cohen SS (1998) A Guide to the Polyamines. Oxford University Press, New York
Collier DE, Ackermann F, Somers DJ, Cummins RW and Atkin OK (1993) The effect of aluminum exposure on root respiration in an aluminum-sensitive and an aluminum-tolerant cultivar of *Triticum aestivum*. Physiol Plant 87: 447–452
Coté GG and Crain RC (1993) Biochemistry of phosphoinositides. Annu Rev Plant Physiol Plant Mol Biol 44: 333–356
Cuenca G, Herrera R and Mérida T (1991) Distribution of aluminum in accumulator plants by X-ray microanalysis in *Richeria grandis* Vahl leaves from a cloud forest in Venezuela. Plant Cell Environ 14: 437–441
Davies DD (1973a) Control of and by pH. Symp Soc Biol 27: 513–529
Davies DD (1973b) Metabolic control in higher plants. In: Milborrow BV (ed) Biosynthesis and its Control in Plants, pp 1–20. Academic Press, London
Davis RF (1974) Photoinduced changes in electrical potentials and H^+ activities of the chloroplast, cytoplasm, and vacuole of *Phaeoceros laevis*. In: Zimmerman U, Dainty J (eds) Membrane Transport in Plants, pp: 197–201. Springer-Verlag, New York
de la Fuente JM, Ramirez-Rodriguez V, Cabrera-Ponce JL and Herrera-Estrella L (1997) Aluminum tolerance in transgenic plants by alteration of citrate synthesis. Science 276: 1566–1568
Delhaize E and Ryan PR (1995) Aluminum toxicity and tolerance in plants. Plant Physiol 107: 315–321
Delhaize E, Craig S, Beaton CD, Bennet RJ, Jagadish VC and Randall PJ (1993a) Aluminum tolerance in wheat (*Triticum aestivum* L.) I. Uptake and distribution of aluminum in root apices. Plant Physiol 103: 685–693
Delhaize E, Ryan PR, Hocking PJ and Richardson AE (2003) effects of altered citrate synthase and isocitrate dehydrogenase expression on internal citrate concentrations and citrate efflux from tobacco (*Nicotiana tabacum* L.) roots. Plant Soil 248: 137–144
Delhaize E, Ryan PR and Randall PJ (1993b) Aluminum tolerance in wheat (*Triticum aestivum* L.) II. Aluminum-stimulated excretion of malic acid from root apices. Plant Physiol 103: 695–702
Dinkelaker B, Romheld V and Marschner H (1989) Citric-acid excretion and precipitation of calcium citrate in the rhizosphere of white lupin (*Lupinus albus* L.). Plant Cell Environ 12: 285–292
Dohmen GP, Koppers A and Langebartels C (1990) Biochemical response of Norway spruce (*Picea abies* (L.) Karst.) towards 14-month exposure to ozone and acid mist: Effects on amino acid, glutathione and polyamine titers. Environ Pollut 64: 375–383
Dong B, Sang WL, Jiang X, Zhou JM, Kong FX, Hu W and Wang LS (2002) Effects of aluminum on physiological metabolism and antioxidant system of wheat (*Triticum aestivum L.*). Chemosphere 47: 87–92
Drake BG, Azcon BJ, Berry J, Bunce J, Dijkstra P, Farrar J, Gifford RM, Gonzales-Meler MA, Koch G, Lambers H, Siedow J and Wullschleger S (1999) Does elevated atmospheric CO_2 concentration inhibit mitochondrial respiration in green plants? Plant Cell Environ 22: 649–657
Duff SMG, Lefebvre DD and Plaxton WC (1989a) Purification and characterization of a phosphoenolpyruvate phosphatase from *Brassica-nigra* suspension cells. Plant Physiol 90: 734–741
Duff SMG, Moorhead GB, Lefebvre DD, and Plaxton WC (1989b) Phosphate starvation inducible 'bypass' adenylate and phosphate dependent glycolytic enzymes in *Brassica nigra* suspension cells. Plant Physiol 90: 1275–1278
Duff SMG, Plaxton WC and Lefebvre DD (1991) Phosphate-starvation response in plant cells — De novo synthesis and degradation of acid phosphatases. Proc Natl Acad Sci USA 88: 9538–9542
Edwards S, Nguyen BT, Do B and Roberts JKM (1998) Contribution of malic enzyme, pyruvate kinase, phosphoenolpyruvate carboxylase, and the Krebs cycle to respiration and biosynthesis and to intracellular pH regulation during hypoxia in maize root tips observed by nuclear magnetic resonance imaging. Plant Physiol 116: 1073–1081
Ezaki B, Gardner RC, Ezaki Y, Kondo H and Matsumoto H (1998) Protective roles of two aluminum (Al)-induced genes, HSP150 and SED1 of *Saccharomyces cerevisiae*, in Al and oxidative stresses. FEMS Microbiol Lett 159: 99–105
Ezaki B, Gardner RC, Ezaki Y and Matsumoto H (2000) Expression of aluminum-induced genes in transgenic *Arabidopsis* plants can ameliorate aluminum stress and/or oxidative stress. Plant Physiol 122: 657–665
Flores HE (1991) Changes in polyamine metabolism in response to abiotic stress. In: Slocum R and Flores HE (eds) The Biochemistry and Physiology of Polyamines in Plants, pp 214–225. CRC Press, Boca Raton
Forde BG and Clarkson DT (1999) Nitrate and ammonium nutrition of plants: Physiological and molecular perspectives. Adv Bot Res 30: 1–90
Gaume A, Mächler F and Frossard E (2001) Aluminum resistance in two cultivars of *Zea mays* L.: Root exudation of organic acids and influence of phosphorus nutrition. Plant Soil 234: 73–81
Gauthier DA and Turpin DH (1994) Inorganic phosphate (Pi) enhancement of dark respiration in the Pi-limited green alga *Selenastrum minutum* — Interactions between H^+/Pi co-transport, the plasmalemma H^+-ATPase, and dark respiratory carbon flow. Plant Physiol 104: 629–637
Gehl KA and Colman B (1985) Effect of external pH on the internal pH of *Chlorella saccharophila*. Plant Physiol 77: 917–921
Givan CV (1999) Evolving concepts in plant glycolysis: Two centuries of progress. Biol Rev 74: 277–309
Godbold DL, Fritz E and Holtermann A (1988) Aluminum toxicity

and forest decline. Proc Natl Acad Sci USA 85: 3888–3892
Godbold DL and Jentschke G (1998) Aluminium accumulation in root cell walls coincides with inhibition of root growth but not with inhibition of magnesium uptake in Norway spruce. Physiol Plant 102: 553–560
Gonzalez-Meler MA, Ribas Carbo M, Siedow JN and Drake BG (1996) Direct inhibition of plant mitochondrial respiration by elevated CO_2. Plant Physiol 112: 1349–1355
Goodchild JA and Givan CV (1991) Stimulation of bicarbonate incorporation by ammonium in nonphotosynthetic cell-suspension cultures of *Acer-pseudoplatanus*. Physiol Plant 82: 537–542
Grauer UE and Horst WJ (1990) Effect of pH and nitrogen source on aluminium tolerance of rye (*Secale cereale* L.) and yellow lupin (*Lupinus luteus* L.). Plant Soil 127:13–21
Hamilton CA, Good AG and Taylor GJ (2001) Induction of vacuolar ATPase and mitochondrial ATP synthase by aluminum in an aluminum-resistant cultivar of wheat. Plant Physiol 125: 2068–2077
Hao LN and Liu HT (1989) Effects of aluminum on physiological functions of rice seedlings. Acta Bot Sin 31: 847–853
Haug A, Shi B and Vitorello V (1994) Aluminum interaction with phosphoinositide-associated signal transduction. Arch Toxicol 68: 1–7
Heldt HW, Werdan K, Milovancev M and Geller G (1973) Alkalization of the chloroplast stroma caused by light-dependent proton flux into the thylakoid space. Biochim Biophys Acta 314: 224–241
Hesse SJA, Ruijter GJG and Dijkema C (2002) Intracellular pH homeostasis in the filamentous fungus *Aspergillus niger*. Eur J Biochem 269: 3485–3494
Hoffland E, Vandenboogaard R, Nelemans J and Findenegg G (1992) Biosynthesis and root exudation of citric and malic-acids in phosphate-starved rape plants. New Phytol 12: 675–680
Honda M, Ito K and Hara T (1997) Effect of physiological activities on aluminum uptake in carrot (*Daucus carota* L.) cells in suspension culture. Soil Sci Plant Nutr 43: 361–368
Huang JW, Shaff JE, Grunes DL and Kochian LV (1992) Aluminum effects on calcium fluxes at the root apex of aluminum-tolerant and aluminum-sensitive wheat cultivars. Plant Physiol 98: 230–237
Ikegawa H, Yamamoto Y and Matsumoto H (1998) Cell death caused by a combination of aluminum and iron in cultured tobacco cells. Physiol Plant 104: 474–478
Ingo R and Wolfgang B (2000) The role of lipid peroxidation in aluminum toxicity in soybean cell suspension cultures. J Biol Sci 55: 957–964
Ishikawa S, Wagatsuma T, Sasaki R and Ofei-Manu P (2000) Comparison of the amount of citric and malic acids in Al media of seven plant species and two cultivars each in five plant species. Soil Sci Plant Nutr 46: 751–758
Jansen S, Broadley MR and Robbrecht E (2002) Aluminum hyperaccumulation in angiosperms: A review of its phylogenetic significance. Bot Rev 68: 235–269
Johnson JF, Allan DL and Vance CP (1994) Phosphorus stress-induced proteoid roots show altered metabolism in *Lupinus albus*. Plant Physiol 104: 657–665
Johnson JF, Allan DL, Vance CP and Weiblen G (1996a) Root carbon dioxide fixation by phosphorus-deficient *Lupinus albus* — Contribution to organic acid exudation by proteoid roots. Plant Physiol 112: 19–30
Johnson JF, Vance CP and Allan DL (1996b) Phosphorus deficiency in *Lupinus albus* — Altered lateral root development and enhanced expression of phosphoenolpyruvate carboxylase. Plant Physiol 112: 31–41
Jones DL (1998) Organic acids in the rhizosphere — A critical review. Plant Soil 205: 25–44
Juszczuk IM and Rychter AM (1997) Changes in pyridine nucleotide levels in leaves and roots of bean plants (*Phaseolus vulgaris* L.) during phosphate deficiency. J Plant Physiol 151: 399–404
Karr MC, Coutinho J and Ahlrichs JL (1984) Determination of aluminum toxicity in Indiana soils by petri dish bioassays. Proc Ind Acad Sci 93: 85–88
Keerthisinghe G, Hocking PJ and Ryan PR (1998) Effect of phosphorus supply on the formation and function of proteoid roots of white lupin (*Lupinus albus* L.) Plant Cell Environ 21: 467–478
Keltjens WG (1995) Magnesium uptake by Al-stressed maize plants with special emphasis on cation interactions at root exchange sites. Plant Soil 171: 141–146
Kinraide TB (1993) Aluminum enhancement of plant-growth in acid rooting media — A case of reciprocal alleviation of toxicity by 2 toxic cations. Physiol Plant 88: 619–625
Kinraide TB and Parker DR (1990) Apparent phytotoxicity of mononuclear hydroxy-aluminum to 4 dicotyledonous species. Physiol Plant 79: 283–288
Kochian LV (1995) Cellular mechanisms of aluminum toxicity and resistance in plants. Annu Rev Plant Physiol Plant Mol Biol 46: 237–260
Kochian LV (2000) Molecular physiology of mineral nutrient acquisition, transport, and utilization. In: Buchanan BB, Gruissem W and Jones RL (eds) Biochemistry and Molecular Biology of Plants, pp 1158–1203. American Society of Plant Physiologists, Rockville
Kochian LV, Pence NS, Letham DLD, Piñeros MA, Magalhaes JV, Hoekenga OA and Garvin DF (2002) Mechanisms of metal resistance in plants: Aluminum and heavy metals. Plant Soil 247: 109–119
Koyama H, Takita E, Kawamura A, Hara T and Shibata D (1999) Over expression of mitochondrial citrate synthase gene improves the growth of carrot cells in Al-phosphate medium. Plant Cell Physiol 40: 482–488
Kronzucker HJ, Britto DT, Davenport RJ and Tester M (2001) Ammonium toxicity and the real cost of transport. Trends Plant Sci 6: 335–337
Kuhn E (2001) From library screening to microarray technology: Strategies to determine gene expression profiles and to identify differentially regulated genes in plants. Ann Bot 87: 139–155
Kurkdjian A and Guern J (1989) Intracellular pH: Measurement and importance in cell activity. Annu Rev Plant Physiol Plant Mol Biol 40: 271–303
Kuzniak E (2002) Transgenic plants: An insight into oxidative stress tolerance mechanisms. Acta Physiol Plant 24: 97–113
Lambers H, Atkin OK and Scheurwater I (1996) Respiration patterns in roots in relation to their functioning. In: Waisel Y, Eshel A, Kafkaki U (eds) Plant Roots: The Hidden Half, pp 323–362. Marcel Dekker, New York
Lambers H, Chapin III FS and Pons TL (1998) Plant Physiological Ecology. Springer-Verlag, New York
Lane A and Burris JE (1981) Effect of environmental pH on the

internal pH of *Chlorella pyrenoidosa*, *Scenedesmus quadricauda*, and *Euglena mutabilis*. Plant Physiol 68: 43–442

Lawrence GB, David MB and Shortle WC (1995) A new mechanism for calcium loss in forest-floor soils. Nature 378: 162–165

Lazof DB, Goldsmith JG, Rufty TW and Linton RW (1994) Rapid uptake of aluminum into cells of intact soybean root tips. A microanalytical study using secondary ion mass spectroscopy. Plant Physiol 106: 1107–1114

Lazof DB, Goldsmith JG and Linton RW (1997) The in situ analysis of intracellular aluminum in plants. Prog Bot 58: 112–159

Lee CH, Jin HO and Izuta T (1999) Growth, nutrient status and net photosynthetic rate of *Pinus densiflora* seedlings in various levels of aluminum concentrations. J Korean For Soc 88: 249–254

Lee CH, Jin HO and Kim YK (2001) Effects of Al and Mn on the growth, nutrient status and gas exchange rates of *Pinus densiflora* seeds. J Korean For Soc 90: 74–82

Levi C and Gibbs M (1976) Starch degradation in isolated spinach chloroplasts. Plant Physiol 57: 933–935

Li XF, Ma JF and Matsumoto H (2000) Pattern of aluminum-induced secretion of organic acids differs between rye and wheat. Plant Physiol 123: 1537–1543

Lidon FC, Barreiro MG, Ramalho JC and Lauriano JA (1999) Effects of aluminum on nutrient accumulation in maize shoots: Implications on photosynthesis. J Plant Nutr 22: 397–416

López-Bucio J, Nieta-Jacobo MF, Ramírez-Rodríguez V and Herrera-Estrella L (2000) Organic acid metabolism in plants: From adaptive physiology to transgenic varieties for cultivation in extreme soils. Plant Sci 160: 1–13

Ma JF (2000) Role of organic acids in detoxification of aluminum in higher plants. Plant Cell Physiol 41: 383–390

Ma JF, Ryan PR and Delhaize E (2001) Aluminium tolerance in plants and the complexing role of organic acids. Trends Plant Sci 6: 273–278

Macdonald TL and Martin RB (1988) Aluminum ion in biological systems. TIBS 13: 15–19

Malkin R and Niyogi K (2000) In: Buchanan BB, Gruissem W, and Jones RL (eds) Biochemistry and Molecular Biology of Plants, pp 568–628. American Society of Plant Physiologists, Rockville

Marienfeld S and Stelzer R (1993) X-ray microanalyses in roots of Al-treated *Avena sativa* plants. J Plant Physiol 141: 569–573

Marienfeld S, Lehmann H and Stelzer R (1995) Ultrasound investigations and EDX analyses of Al-treated oat (*Avena sativa*) roots. Plant Soil 171: 167–173

Marschner H (ed) (1995) Mineral Nutrition of Higher Plants. 2nd ed, Academic Press, London

Marschner H, Häussling M and George E (1991) Ammonium and nitrate uptake rates and rhizosphere-pH in non-mycorrhizal roots of Norway spruce (*Picea abies* (L.) Karst.) Trees-Struct Funct 5:14–21

Martin RB (1988) Bioinorganic chemistry of aluminum. In Sigel H (Ed) Metal Ions in Biological Systems, Vol 24. Aluminum and its Role in Biology, pp 1–57. Marcel Dekker, New York

Martin RB (1992) Aluminum speciation in biology. In: Chadwick DJ and Whelan J (eds) Aluminum in Biology and Medicine, pp 5–25. Wiley, New York

Matsumoto H, Hirasawa E, Torikai H and Takahashi E (1976) Localization of absorbed aluminum in pea root and its binding to nucleic acids. Plant Cell Physiol 17: 127–137

McLaughlin SB, Andersen CP, Edwards NT, Roy WK and Layton PA (1990) Seasonal patterns of photosynthesis and respiration of red spruce saplings from two elevations in declining southern Appalachian Stands. Can J For Res 20: 485–495

McLaughlin SB, Andersen CP, Hanson PJ, Tjoelker MG and Roy WK (1991) Increased dark respiration and calcium deficiency of red spruce in relation to acidic deposition at high-elevation southern Appalachian Mountain sites. Can J For Res 21: 1234–1244

Meychik NR and Yermakov LP (2001) Ion exchange properties of plant root cell walls. Plant Soil 234: 181–193

Millenaar FF and Lambers H (2003) The alternative oxidase: In vivo regulation and function. Plant Biol 5: 2–15

Minocha R and Long S (2004) Effects of aluminum on organic acid metabolism and secretion into the culture medium and the reversal of Al effects by exogenous addition of organic acids in cell suspension cultures of red spruce (*Picea rubens* Sarg.). Tree Physiol 24: 55–64

Minocha R, Minocha SC, Long S and Shortle WC (1992) Effects of aluminum on DNA synthesis, cellular polyamines, polyamine biosynthetic enzymes, and inorganic ions in cell suspension cultures of a woody plant, *Catharanthus roseus*. Physiol Plant 85: 417–424

Minocha R, Shortle WC, Coughlin DJ and Minocha SC (1996) Effects of Al on growth, polyamine metabolism, and inorganic ions in suspension cultures of red spruce (*Picea rubens*). Can J For Res 26: 550–559

Minocha R, Shortle WC, Lawrence GB, David MB and Minocha SC (1997) A relationship among foliar chemistry, foliar polyamines, and soil chemistry in red spruce trees growing across the northeastern United States. Plant Soil 191: 109–122

Minocha R, Aber JD, Long S, Magill AH and McDowell W (2000) Foliar polyamine and inorganic ion content in relation to soil and soil solution chemistry in two fertilized forest stands at the Harvard Forest, Massachusetts. Plant Soil 222: 119–137

Minocha R, McQuattie C, Fagerberg W, Long S and Noh EW (2001) Effects of aluminum in red spruce (*Picea rubens*) cell cultures: Cell growth and viability, mitochondrial activity, ultrastructure, and potential sites of intracellular aluminum accumulation. Physiol Plant 113: 486–498

Minocha SC (1987) pH of the medium and the growth and metabolism of cells in culture. In: Bonga JW and Durzan DJ (eds) Cell and Tissue Culture in Forestry, pp 125–141. Martinus Nijhoff Publishers, Dordrecht, The Netherlands

Møller IM (2001) Plant mitochondria and oxidative stress: Electron transport, NADH turnover, and metabolism of reactive oxygen species. Annu Rev Plant Physiol Plant Mol Biol 52: 561–591

Moon RB and Richards JH (1973) Determination of intracellular pH by ^{31}P magnetic resonance. J Biol Chem 248: 7276–7278

Moustakas M, Ouzounidou G and Lannoye R (1993) Rapid screening for aluminum tolerance in cereals by use of the chlorophyll fluorescence test. Plant Breed 111: 343–346

Moustakas M, Ouzounidou G and Lannoye R (1995) Aluminum effects on photosynthesis and elemental uptake in an aluminum-tolerant and non-tolerant wheat cultivar. J Plant Nutr 18: 669–683

Mugai EN, Agong SG and Matsumoto H (2000) Aluminium tolerance mechanisms in *Phaseolus vulgaris* L.: Citrate synthase activity and TTC reduction are well correlated with citrate secretion. Soil Sci Plant Nutr 46:939–950

Neumann G and Römheld V (2000) The release of root exudates as affected by the plant's physiological status. In: Pinton R, Varanini Z, Nannipieri Z (eds) In the Rhizosphere: Biochemistry and Organic Substances in Soil-Plant Interface. pp 41–93, Marcel Dekker, New York

Neumann G, Massonneau A, Langlade N, Dinkelaker B, Hengeler C, Romheld V and Martinoia E (2000) Physiological aspects of cluster root function and development in phosphorus-deficient white lupin (*Lupinus albus* L.) Ann Bot 85: 909–919

Nobel PS and Palta JA (1989) Soil O_2 and CO_2 effects on root respiration of cacti. Plant Soil 120: 263–271

Nosko P, Brasnard P, Kramer JR and Kershaw KA (1988) The effect of aluminum on seed germination and early seedling establishment growth and respiration of white spruce *Picea glauca*. Can J Bot 66: 2305–2310

Ogawa T, Matsumoto C, Takenaka C and Tezuka T (2000) Effect of Ca on Al-induced activation of antioxidant enzymes in the needles of Hinoki cypress (*Chamaecyparis obtuse*). J For Res 5: 81–85

Oleksyn J, Karolewski P, Giertych MJ, Werner A, Tjoelker MG and Reich PB (1996) Altered root growth and plant chemistry of *Pinus sylvestris* seedlings subjected to aluminum in nutrient solution. Tree 10: 135–144

Olivetti GP and Etherton B (1991) Aluminum interactions with corn root plasma membrane. Plant Physiol 96 (suppl): 142

Palta JA and Nobel PS (1989) Influence of soil O_2 and CO_2 on root respiration of *Agave deserti*. Physiol Plant 76: 187–192

Peiter E, Yan F and Schubert S (2001) Lime-induced growth depression in *Lupinus* species: Are soil pH and bicarbonate involved? J Plant Nutr Soil Sci 164: 165–172

Pellet DM, Grunes DL and Kochian LV (1995) Organic acid exudation as an aluminum tolerance mechanism in maize (*Zea mays* L.). Planta 196: 788–95

Piñeros MA and Kochian LV (2001) A patch-clamp study on the physiology of aluminum toxicity and aluminum tolerance in maize. Identification and characterization of Al^{3+}-induced anion channels. Plant Physiol 125: 292–305

Qi JE, Marshall JD and Mattson KG (1994) High soil carbon-dioxide concentrations inhibit root respiration of Douglas-fir. New Phytol 128: 435-442

Qian XM (1998) Influence of liming and acidification on the activity of the mycorrhizal communities in a *Picea abies* (L.) Karst. stand. Plant Soil 199: 99–109

Raven JA and Smith FA (1974) Significance of hydrogen ion transport in plant cells. Can J Bot 52: 1035–1048

Raven JA and Smith FA (1978) Effect of temperature and external pH on the cytoplasmic pH of *Chara corallina*. J Exp Bot 29: 853–856

Rayle DL and Cleland RE (1992) The acid growth theory of auxin-induced cell elongation is alive and well. Plant Physiol 99: 1271–1274

Rengel Z (1992) Role of Ca in Al toxicity. New Phytol 121: 499–513

Rengel Z (1996) Uptake of aluminum by plant cells. New Phytol 134: 389–406

Rent RK, Johnson RA and Barr CE (1972) Net H^+ influx in *Nitella clavata*. J Membrane Biol 7: 231–244

Richards KD, Schott EJ, Sharma YK, Davis KR and Gardner RC (1998) Aluminum induces oxidative stress genes in *Arabidopsis thaliana*. Plant Physiol 116: 409–418

Roberts JKM, Hooks MA, Miaullis AP, Edwards S and Webster C (1992) Contribution of malate and amino acid metabolism to cytoplasmic pH regulation in hypoxic maize root tips studied using Nuclear Magnetic Resonance spectroscopy. Plant Physiol 98: 480–487

Ryan PR, Delhaize E and Jones DL (2001) Function and mechanism of organic anion exudation from plant roots. Annu Rev Plant Physiol Plant Mol Biol 52: 527–560

Ryan PR, Delhaize E and Randall PJ (1995) Characterization of Al stimulated efflux of malate from the apices of Al-tolerant wheat roots. Planta 196: 103–110

Ryan PR, Reid RJ and Smith FA (1997) Direct evaluation of the Ca^{2+}-displacement hypothesis for Al toxicity. Plant Physiol 113: 1351–1357

Sakihama Y and Yamasaki H (2002) Lipid peroxidation induced by phenolics in conjunction with aluminum ions. Biol Plant 45: 249–254

Sanders D and Bethke P (2000) Membrane Transport. In Buchanan BB, Gruissem W, and Jones RL (eds) (2000) Biochemistry and Molecular Biology of Plants, pp 110–158. American Society of Plant Physiologists, Rockville, MD

Santerre A, Markiewicz M and Villanueva VR (1990) Effect of acid rain on polyamines in *Picea*. Phytochemistry 29: 1767–1769

Schaberg PG, Dehayes DH, Hawley GJ, Strimbeck GR, Cumming JR, Murakami PF and Borer CH (2000) Acid mist and soil Ca and Al after the mineral nutrition and physiology of Red Spruce. Tree Physiol 20: 73–85

Schröder WH, Bauch J and Endeward R (1988) Microbeam analysis of Ca exchange and uptake in the fine roots of spruce: influence of pH and aluminum. Trees 2: 96–103

Seki M, Narusaka M, Abe H, Kasuga M, Yamaguchi-Shinozaki K, Carninci P, Hayashizaki Y and Shinozaki K (2001) Monitoring the expression pattern of 1300 *Arabidopsis* genes under drought and cold stresses by using a full-length cDNA microarray. Plant Cell 13: 113–123

Siedow JN and Day DA (2000) Respiration and photorespiration. In: Buchanan BB, Gruissem W, and Jones RL eds. Biochemistry and Molecular Biology of Plants, pp 672–726. American Society of Plant Physiologists, Rockville, MD.

Simon L, Kieger M, Sung SS and Smalley TJ (1994) Aluminum toxicity in tomato: Part 2. Leaf gas exchange, chlorophyll content, and invertase activity. J Plant Nutr 17: 307–317

Simons BH and Lambers H (1998) The alternative oxidase: Is it a respiratory pathway allowing a plant to cope with stress? In: Lerner HR (ed) Plant Responses to Environmental Stress: From Phytohormones to Genome Reorganization. pp. 265-286. Plenum Press, New York

Smith FA (1984) Regulation of the cytoplasmic pH of *Chara corallina*: Response to changes in external pH. J Exp Bot 35: 43–50

Smith FA and Raven JA (1976) H^+ transport and regulation of cell pH. In: Lüttge U and Pitman MG (eds) Encyclopedia of Plant Physiology, New Ser, 2A, pp: 317–346. Springer-Verlag, Berlin

Smith FA and Raven JA (1979) Intracellular pH and its regulation. Annu Rev Plant Physiol 30: 28–311

Son KC, Gu EG, Byoun HJ and Lim JH (1994) Effects of sucrose, BA, or aluminum sulfate in the preservative solutions on photosynthesis, respiration, and transpiration of cut rose leaf. J Korean Soc Hort Sci 35: 480–486

Sorensen KU, Terry RE, Jolley VD and Brown JC (1989) Iron-stress response of inoculated and non-inoculated roots of an iron inefficient soybean cultivar in a split-root system. J Plant Nutr 12: 437–447

Sparling DW and Lowe TP (1996) Environmental hazards of aluminum to plants, invertebrates, fish, and wildlife. Rev Environ Contam 145: 1–127

Steup M, Peavey DG and Gibbs M (1976) The regulation of starch metabolism by inorganic phosphate. Biochem Biophys Res Commun 72: 1554–1561

Tesfaye M, Temple SJ, Allan DL, Vance CP and Samac DA (2001) Overexpression of malate dehydrogenase in transgenic alfalfa enhances organic acid synthesis and confers tolerance to aluminum. Plant Physiol 127: 1836–1844

Theodorou ME, Cornel FA, Duff SMG and Plaxton WC (1992) α-subunit of pyrophosphate-dependent phosphofructokinase in black mustard suspension cells. J Biol Chem 267: 21901–21905

Thomas M, Richardson JA and Ranson SL (eds) (1973) Plant Physiology. Longman, London

Torimitsu K, Yazaki Y, Nagasuka K, Ohta E and Sakata M (1984) Effect of external pH on the cytoplasmic and vacuolar pH's in mung bean root-tip cells: A ^{31}P nuclear magnetic resonance study. Plant Cell Physiol 25: 1403–1409

Van Breemen N (1985) Acidification and decline of Central European forests. Nature 315: 16

Van der Werf A, Welschen R and Lambers H (1992) Respiratory losses increase with decreasing inherent growth rate of a species and with decreasing nitrogen supply. A search for explanations for these observations. In: Lambers H, van der Plas LHW (eds) Molecular, Biochemical and Physiological Aspects of Plant Respiration, pp 421–432. SPB Academic Publishing, The Hague

Vanhala P (2002) Seasonal variation in the soil respiration rate in coniferous forest soils. Soil Biol Biochem 34: 1375–1379

Vanlerberghe GC and Ordog SH (2002) Alternative oxidase: Integrating carbon metabolism and electron transport in plant respiration. In: Foyer CH and Noctor G (eds) Photosynthetic Assimilation and Associated Carbon Metabolism. Kluwer Academic Publishers, Dordrecht

Votyakova TV, Wallace HM, Dunbar B and Wilson SB (1999) The covalent attachment of polyamines to proteins in plant mitochondria. Eur. J. Biochem 260: 250–257

Walker NA and Smith FA (1975) Intracellular pH in *Chara corallina* by DMO distribution. Plant Sci Lett 4: 125–132

Wang MY, Siddiqi MY, Ruth TJ and Glass ADM (1993a) Ammonium uptake by rice roots I. Fluxes and subcellular-distribution of NH^{4+}-N^{13}. Plant Physiol 103: 1249–1258

Wang MY, Siddiqi MY, Ruth TJ and Glass ADM (1993b) Ammonium uptake by rice roots. II. Kinetics of NH^{4+}-N^{-13} influx across the plasmalemma. Plant Physiol 103: 1259–1267

Wargo PM, Minocha R, Wong B, Long RP, Horsley SB and Hall TJ (2002) Measuring stress and recovery in lime fertilized sugar maple in the Allegheny Plateau area of northwestern Pennsylvania. Can J For Res 32: 629–641

Watt M and Evans JR (1999a) Linking development and determinancy with organic acid efflux from proteoid roots of white lupin grown with low phosphorus and ambient or elevated atmospheric CO_2 concentration. Plant Physiol 120: 705–716

Watt M and Evans JR (1999b) Proteoid roots: Physiology and development. Plant Physiol 121: 317–323

Wlodarczyk T, Stepniewski W and Brzezinska M (2002) Dehydrogenase activity, redox potential, and emissions of carbon dioxide and nitrous oxide from Cambisols under flooding conditions. Biol Fert Soils 36: 200–206

Woolhouse HW (1969) Differences in the properties of the acid phosphatase of plant roots and their significance in the evolution of edaphic ecotypes. In: Rorison, IH (ed) Aspects of the Mineral Nutrition of Plants, pp: 357–380. Blackwell Scientific Publications, Oxford

Wu SH, Ramonell K., Gollub J and Somerville S (2001) Plant gene expression profiling with DNA microarrays. Plant Physiol Biochem 39: 917–926

Yamamoto Y, Hachiya A and Matusumoto H (1997) Oxidative damage to membranes by a combination of aluminum and iron in suspension-cultured tobacco cells. Plant Cell Physiol 38: 1333–1339

Yamamoto Y, Kobayashi Y, Devi SR, Rikiishi S and Matsumoto H (2002) Aluminum toxicity is associated with mitochondrial dysfunction and the production of reactive oxygen species in plant cells. Plant Physiol 128: 63–72

Yan F, Schubert S, and Mengel K (1992) Effect of low root medium pH on net proton release, root respiration, and root-growth of corn (*Zea mays* L) and broad bean (*Vicia faba* L.). Plant Physiol 99: 415–421

You G and Nelson DJ (1991) Al^{3+} versus Ca^{2+} ion binding to methionine and tyrosine spin-labeled bovine brain calmodulin. J Inorg Biochem 41: 283–291

Zhang G and Taylor G (1989) Kinetics of aluminum uptake by excised roots of aluminum-tolerant and aluminum-sensitive cultivars of *Triticum aestivum* L. Plant Physiol 91: 1094–1099

Zhang G, Slaski JJ, Archambault DJ and Taylor GJ (1997) Alteration of plasma membrane lipids in aluminum-resistant and aluminum-sensitive wheat genotypes in response to aluminum stress. Physiol Plant 99: 302–308

Zhang WH, Ryan PR and Tyerman SD (2001) Malate-permeable channels and cation channels activated by aluminum in the apical cells of wheat roots. Plant Physiol 125: 1459–1472

Zhou X-H, Minocha R and Minocha SC (1995) Physiological responses of suspension cultures of *Catharanthus roseus* to aluminum: Changes in polyamines and inorganic ions. J Plant Physiol 145: 277–284

Chapter 10

Understanding Plant Respiration: Separating Respiratory Components versus a Process-Based Approach

Tjeerd J. Bouma*
Netherlands Institute of Ecology, P.O. Box 140, 4400 AC Yerseke, The Netherlands

Summary

A revolution in plant modeling in the 1970s highlighted the need for better respiration algorithms. Subsequent research enhanced conceptual insights into processes underlying growth and maintenance respiration. This chapter offers an overview of the most important basic concepts used to partition respiration into energy-utilizing components for growth, maintenance and ion uptake, both with respect to modeling and for experimental measurements.

Conceptual models can offer a simplified representation of the mechanisms of respiratory energy partitioning in plant. Comparing different conceptual models demonstrates that plant growth can be simulated with

*email: t.bouma@nioo.knaw.nl

H. Lambers and M. Ribas-Carbo (eds.), Plant Respiration, 177–194.

different equations, each without containing detailed information on the underlying respiratory processes. For that reason, process-based models are more useful to quantify the relative importance of energy-consuming processes and identifying quantitatively important gaps in our knowledge. Allometric modeling is a rapidly developing research area, and seems to offer promising perspectives for the future.

Experimental methods to relate respiration to underlying energy-utilizing components for growth, maintenance and ion uptake may be divided in three distinct approaches: i) correlative or regression approaches, ii) black-box approaches, and iii) process-based approaches. The first method especially has enhanced our insight into the differences between fast- and slow-growing species. Although useful in the past, the second approach is outdated, and should no longer be used. More studies should use process-based approaches to further progress our understanding. Especially combining experimental and theoretical process-based approaches seems to offer interesting perspectives. Our understanding of the relation between the energy-utilizing components with environmental conditions is still limited, but gradually increasing by including environmental conditions in process-based research. In addition to down-scaling by studying energy-utilizing processes in depth, up-scaling as done in allometric modeling may offer valuable insights into our understanding of the use of respiratory energy.

I. Introduction

The concept of dividing respiration into energy-utilizing components for growth and maintenance was first introduced in the early 1900s by microbiologists interested in the efficiency of fermentation processes. Wohl and James (1942) later presented this concept to plant science. It was not until the 1960s, however, that scientists began to explore causal relationships between these respiratory components (Audus, 1960; Olson, 1964), and included the concept in early plant carbon balance calculations (Hiroi and Monsi, 1964; Monsi, 1968). During that time, a revolution in plant modeling made it clear that better respiration algorithms were needed which led to the development of conceptual insights into processes underlying growth and maintenance respiration (McCree, 1970, 1974; Thornley, 1970; Penning de Vries, 1972, 1975; Penning de Vries et al., 1974). Veen (1980) later distinguished a third component respiration related to ion uptake. Overall, a great deal of research has focused on quantifying the relative importance of each respiratory component, without modifying the basic concept of dividing respiration into energy-utilizing components for growth, maintenance and ion uptake. For an extensive historical review, see Amthor (2000a), from which this brief summary is an excerpt.

This chapter presents an overview of the most important basic concepts used to divide respiration into energy-utilizing components, and propose new directions for future research [see also reviews by Amthor (2000a), Cannell and Thornley (2000), and Thornley and Cannell (2000)]. Basic equations, definitions and modeling exercises presented here will provide the background that is required to understand various

Abbreviations: $[H/I_j]$ – the stoichiometry between protons and ion_j for a given membrane passage (mol H^+ $[mol\ ion_j]^{-1}$); [H/P] – the stoichiometry of the H^+-ATPase in a given membrane (mol H^+ $[mol\ ATP]^{-1}$); 'X' – may represent C, CH_2O, O_2, CO_2 or ATP, depending on how respiration is assessed. In case that 'X' represents C, CH_2O, 'X' is generally expressed in grams. However, if 'X' represents O_2, CO_2 or ATP, 'X' is generally expressed as moles.; 'Z' – an attribute other than the dry weight of living biomass (M) such as e.g., tissue N concentration; CUE – carbon-use efficiency (= G / P); e_i – specific cost of a given individual processes, out a total of i (subscript) processes; G – plant growth rate ($g_{growth}\ s^{-1}$ or g $C_{growth}\ s^{-1}$); g_R ($\Leftrightarrow$ 1/Y) – specific costs for tissue construction $\Leftrightarrow$ growth coefficient (g 'X' g_{growth}^{-1}); g_U ($\Leftrightarrow$ 1/U) – specific costs for ion uptake $\Leftrightarrow$ uptake coefficient (g 'X' mol_{ion}^{-1}); k_D – the turnover rate of degradable mass [g $C_{senescence}$ (g $C_{total})^{-1}\ s^{-1}$]; K_d-value – degradation constant for proteins; k_S – senescence rate [g $C_{senescence}$ (g $C_{total})^{-1}\ s^{-1}$]; M – living biomass (g); M_D – degradable biomass (g); M_j – the number of active membrane passages for ion_j (-); M_{ND} – non-degradable biomass (g); m_R ($\Leftrightarrow$ m) – specific costs for maintenance $\Leftrightarrow$ maintenance coefficient (g 'X' $g^{-1}\ s^{-1}$); m_Z – specific maintenance costs per unit 'Z'; NNUR $\Leftrightarrow$ NU_NR – net nitrate uptake rate ($mol_{nitrate}\ g^{-1}\ s^{-1}$).; NUR – net nutrient uptake rate ($mol_{ion}\ g^{-1}\ s^{-1}$).; P – photosynthetic rate (g C s^{-1}); R – overall respiration rate (g 'X' s^{-1}); r – overall respiration rate per unit dry mass (g 'X' $g^{-1}\ s^{-1}$; r = R/M). In process based calculations, r is the sum of the respiratory costs for i (subscript) individually, but simultaneously ongoing processes ($r = \Sigma\ [v_i * e_i]$); R_G – growth respiration rate (g 'X' s^{-1}); r_G – growth respiration rate per unit dry mass (g 'X' $g^{-1}\ s^{-1}$); RGR – relative growth rate ($g_{growth}\ g^{-1}\ s^{-1}$).; R_M – maintenance respiration rate (g 'X' s^{-1}); r_M – maintenance respiration rate per unit dry mass (g 'X' $g^{-1}\ s^{-1}$); r_{Mgrad} – respiration rate required to maintain a gradient for ion_j (mol ATP $g^{-1}\ s^{-1}$); r_U – ion uptake respiration rate per unit dry mass (g 'X' $g^{-1}\ s^{-1}$); v_i – rate of a given individual processes, out a total of i (subscript) processes; Y_G – growth yield $\Leftrightarrow$ unit C growth per unit of C utilized [g C_{growth} (g $C_{utilized})^{-1}$]; Φ_j – the efflux rate of ion j per unit dry (root) biomass (mol $ion_j\ g^{-1}\ s^{-1}$)

experimental approaches to quantify individual energy-utilizing components, and the consequences for interpretation of such measurements (e.g., black-box versus process-based estimates of maintenance costs). I also present recent insights into factors affecting respiration rates and the division into energy-utilizing components, as well as some speculative ideas for future research.

II. Basic Equations Used to Define Respiratory Components

Mathematically, respiration was first partitioned into growth and maintenance components as follows (De Wit et al., 1970; McCree, 1970; Thornley, 1970):

$$R = R_G + R_M = g_R G + m_R M \quad (1)$$

where R is overall respiration (g 'X' s^{-1}), R_G and R_M are growth and maintenance respiration (g 'X' s^{-1}), g_R is specific costs for tissue construction (also referred to as the growth coefficient; g 'X' g_{growth}^{-1}), G is growth rate (g_{growth} s^{-1}), m_R is the maintenance coefficient (g 'X' g^{-1} s^{-1}), and M is biomass (g). Note that the specific costs, g_R and m_R, are *not* defined as constants; their values may vary depending on growth conditions and the stage of plant development. 'X' in each unit represents C, CH_2O, O_2, CO_2 or ATP, depending on how respiration is assessed. If 'X' represents O_2, CO_2 or ATP, 'X' is generally expressed as moles rather than grams.

The equation can also be expressed on a dry mass basis which is more common in recent literature:

$$r = r_G + r_M = g_R RGR + m_R \quad (2)$$

where r is overall respiration per unit dry mass (g 'X' g^{-1} s^{-1}; r = R/M), r_G and r_M are growth and maintenance respiration (g 'X' g^{-1} s^{-1}; $r_G = R_G/M$ and $r_M = R_M/M$), and RGR is relative growth rate (g_{growth} g^{-1} s^{-1}).

Veen (1980) later modified the basic equation for root respiration to include a term for ion uptake (i.e. transport processes):

$$r = r_G + r_U + r_M = g_R\, RGR + g_U\, NUR + m_R \quad (3)$$

where r_U is ion uptake respiration (g 'X' g^{-1} s^{-1}), g_U is specific costs for ion uptake (also referred to as a uptake coefficient; g 'X' mol_{ion}^{-1}), and NUR is net nutrient uptake rate (mol_{ion} g^{-1} s^{-1}). Specific costs of g_U is *not* a constant. Many studies simplify equation 3 by replacing nutrient uptake with nitrogen uptake, which requires considerably more respiratory energy for uptake than all other nutrients combined (Clarkson, 1998). In the simplified version, NUR is often indicated as NNUR (Van der Werf et al., 1988) or NU_NR (e.g., Bouma et al., 1996).

Equations 1 to 3 have been represented in different forms using alternative symbols (e.g., 1/Y instead of g_R, 1/U instead of g_U, and m instead of m_R (notations De Visser and Lambers, 1983; Lambers et al., 1983).

Based on these equations, growth respiration is defined as the respiratory energy required to convert non-structural carbohydrates into new plant constituents, such as proteins, lipids, organic acids, and structural carbohydrates (Penning de Vries et al., 1974). Root ion uptake respiration is defined as the respiratory energy required for nutrient uptake needed to sustain growth (Veen, 1980). Furthermore, according to Eqs. 2 and 3, maintenance respiration should equal all the remaining respiration not associated with growth and ion uptake. More formally, maintenance respiration is defined as the sum of all energy-consuming processes that maintain cellular structure, including any acclimation to environmental changes (cf. Penning de Vries, 1975). This is a valuable conceptual definition, although there may be some discussion on the processes that should be included or excluded in the model. In general, turnover of energetically costly cellular components, such as proteins and lipids, and restoration of intracellular gradients across membranes to offset leakages along the electrochemical potential gradients, are regarded as the most important maintenance processes (Penning de Vries, 1975).

Although separating respiration into components for growth, maintenance, and ion uptake may be conceptually useful, it is rather artificial from a biochemical perspective. That is, biochemical processes involved, for example, in protein synthesis required for growth of new tissue (costs = growth respiration) are the same as those involved in protein synthesis to enable acclimation to environmental change, or to replace damaged proteins (costs = maintenance respiration). Overall, each respiratory component (growth, maintenance and ion uptake) contains numerous processes, and each of these processes has its own rate and specific costs.

III. Model Approaches Used to Define Respiratory Components

Since 1970, there has been a substantial amount of research dedicated to the study of growth and maintenance respiration in an effort to develop better respiration algorithms in plant modeling. Therefore, a brief overview of the concepts behind the most important modeling approaches is given, as an introduction to experimental methods. It is important to distinguish conceptual models from process-based models, as they illustrate different points. Process-based models can reveal the relative costs of individual processes that make up a plant's C balance, while conceptual models illustrate that similar model results can be obtained by alternative (simple) formulations, even though they are based on a different set of assumptions.

A. Conceptual Models

Thornley and Cannell (2000) presented an insightful overview of the basic concepts used to model respiration and whole-plant and ecosystem C balances. They summarized three methods to conceptually model maintenance respiration, from which I present an excerpt. In the first method, Method A, maintenance respiration has priority over growth respiration, and is defined as:

$$G = Y_G(P - m_R M) - k_S M \quad (4.1)$$

and

$$R = R_G + R_M = (1 - Y_G)(P - m_R M) + m_R M \quad (4.2)$$

where G is plant growth rate (g C_{growth} s^{-1}), Y_G is growth yield per unit of C utilized [growth yield; g C_{growth} (g $C_{utilized}$)$^{-1}$], P is photosynthesis (g C s^{-1}), and k_S is senescence rate [g $C_{senescence}$ (g C_{total})$^{-1}$ s^{-1}]. Maintenance respiration is given priority over growth by subtracting any carbon required for maintenance ($m_R M$) from P to obtain the total amount of carbon available for growth which is then multiplied by Y_G. Although this approach is widely used (Spitters et al., 1989; Agren et al., 1991; Ryan et al., 1996,), and has been used to obtain realistic growth curves (Fig. 2 in Thornley and Cannell, 2000), it has weaknesses. Since maintenance respiration is calculated as a fixed cost without feedback control, the model is sensitive to specific maintenance cost estimates (i.e. m_R). Consequently, m_R may be used as a tool to fit the model, rather than as a well described mechanistic parameter. Furthermore, it is difficult to realistically tie respiration to overall plant growth (De Wit et al., 1970; Thornley, 1971).

Many simulation models replace the $m_R M$ term in Eqs. 4.1 and 4.2 with $m_Z Z$, where Z is an attribute other than the dry weight of living biomass (M), and m_Z is specific maintenance costs per unit Z. Tissue N concentration is often used for Z (Ryan et al., 1996), since respiration and N concentration are usually well correlated in both leaf and root tissue (Reich et al., 1997, 1998a,b; Pregitzer et al., 1998). Amthor (2000a) concluded that a mechanistic rationale for using $m_Z Z$ might be found in functional relationships between tissue activity that is proportional to tissue N concentration, and maintenance processes (e.g., macromolecular turnover, ion leakage). Indeed, relationships between tissue N concentration and respiration rate via nocturnal carbohydrate export rates appear evident (Bouma et al., 1995; Noguchi et al., 2001), as photosynthesis is dependent on leaf N concentration (Reich et al., 1997, 1998a). Depending on the measuring technique used to estimate the maintenance component of respiration (see 'Experimental methods'), such costs of nocturnal carbohydrate export rates will be included in this maintenance component. Based on the definition of maintenance, one may, however, argue that such costs should be included in the growth component of respiration.

In the second method, Method B, growth has priority over maintenance respiration, and is defined as:

$$G = Y_G P - (m_R + k_S)M \quad (5.1)$$

and

$$R = R_G + R_M = (1 - Y_G)P + m_R M \quad (5.2)$$

An interesting aspect of this second method is that, irrespective of environmental conditions, it predicts a relatively fixed ratio between respiration and photosynthesis and for carbon-use efficiency (CUE = G/P) over a wide range of values for m_R — provided proportionality between maintenance and senescence rates is maintained (Thornley and Cannell, 2000). This agrees with many experimental observations in support of the paper by McCree and Troughton (1966), but is not necessarily always the case (reviewed by Cannell and Thornley, 2000). Also, if m_R

and k_S are regarded as competing rate constants, the ratio of m_R to k_S may link tissue mortality and maintenance costs. Such linkages could be useful when explaining plant tissue lifespan from an efficiency optimization perspective. For example, maintenance costs accumulated over a particular tissue's lifetime, such as the lifetime of a root, can considerably exceed the initial construction cost, and may therefore be an important factor in determining tissue senescence from a cost-benefit perspective (Eissenstat and Yanai, 1997; Bouma et al., 2001). Equations 5.1 and 5.2, however, rely on the assumption that energy supply for maintenance is solely derived at the expense of the tissues being maintained, rather than from new photosynthates which is somewhat unrealistic.

In the third method, Method C, growth is separated into degradable mass, such as proteins, and non-degradable mass, such as cell walls, and is defined as:

$$G = Y_G(P + k_D M_D) - k_D M_D - k_S M_{ND} \qquad (6.1)$$

and

$$R = R_G + R_M = (1 - Y_G)P + (1 - Y_G)\, k_D M_D \qquad (6.2)$$

where k_D is the turnover rate of degradable mass [g $C_{senescence}$ (g C_{total})$^{-1}$ s^{-1}] and M_D and M_{ND} are degradable and non-degradable mass (g). In this concept, all respiration is associated with biosynthesis. Maintenance respiration is defined as the portion of overall respiration associated with replacement of degraded materials using degraded C as the sole C source, which is rather unrealistic. All other respiration is defined as being linked to growth. This method relies on fairly arbitrary separation between growth and maintenance respiration and is not widely used, but Eqs. 6.1 and 6.2 generate realistic time curves that are very similar to those from Eqs. 4.1 and 4.2 (for M, P and the ratio of R to P), and Eqs. 5.1 and 5.2 (for M, P, and the ratios of R to P and R_G to R_M) (see Fig. 2 in Thornley and Cannell, 2000). Thus, plant growth can be modeled without much detailed information on underlying respiratory processes.

Although the results of growth models that lack detailed information on underlying respiratory processes may indicate that modelers do not necessarily need physiologists to make a model with reasonable output, physiological research can benefit significantly from modeling. This is particularly true for process-based models; however, a too literal use of conceptual models has sometimes resulted in rather crude experimental approaches (Shinano et al., 1996; Bouma et al., 2000; see 'Experimental methods').

B. Process-Based Models

Process-based respiration models are considerably more elaborate than the conceptual models described above (Thornley and Cannell, 1996; Riedo et al., 1998). These types of models distinguish various physiological plant processes, such as organ growth, phloem loading, nitrate and ammonium uptake, root and shoot nitrate reduction, N_2 fixation, and uptake of all other ions. However, similar to the conceptual models, process-based models still require a residual maintenance respiration term that cannot be fully described based on the physiologically processes. The residual term accounts for poorly defined respiratory costs, such as protein turnover, maintenance of solute gradients, damage repair, and 'futile' cycles that are not easily quantified. Despite this shortcoming, process-based models are particularly useful for quantifying the relative importance of energy-consuming processes, and help identify quantitatively important gaps in our present knowledge. Models of this type developed for grasslands and forests showed that detailed accounting of individual respiration processes reduced the relative importance of the residual maintenance term from 58–64% to 46–48%, mainly by proper accounting for phloem loading (Thornley and Cannell, 2000). Enhancing our knowledge to enable process-based modeling of the yet undefined residual maintenance respiration term is important, as maintenance costs cumulated over the lifespan of a tissue will generally greatly exceed the initial costs for tissue construction (Eissenstat and Yanai, 1997; Bouma et al., 2001).

C. Theoretical Allometric Models

The introduction of a general model for the origin of allometric scaling laws (West et al., 1997) has led to a rapid development of all kinds of plant relations, such as vascular systems (West et al., 1999), biomass distribution (Enquist and Niklas, 2002), and life history in relation to production, demography and reproduction (Enquist et al., 1999; Niklas and Enquist, 2002). Considering the progress made in modeling ontogenetic growth of animals based on allocation of metabolic energy between maintenance and growth (West et al., 2001), new insight into the use

Table 1. Methods used to define various components of respiration

Method	Parameters
Experimental	
Correlative	g_R , g_U , m_R
Black-box	m_R
Process-based	rate (v_i) and specific cost (e_i) of individual processes of g_R , g_U , and m_R
Theoretical	
Process-based	rate (v_i) and specific cost (e_i) of individual processes of g_R , g_U , and m_R
Correlative	g_R

of respiratory energy in plants may also be expected in the near future.

IV. Methods Used to Define Respiratory Components

As illustrated by the conceptual and the process-based models, plant respiration can either be separated into components for growth, maintenance and ion uptake, or defined by each individual respiratory process. The two approaches lead to different research methods, all having certain strengths, as well as particular limitations. Each approach has value depending on the specific applications. The approaches and their associated parameters are summarized in Table 1.

A. Experimental Methods

1. Correlative (or Regression) Approach

Correlative (or regression) approaches are commonly used to estimate respiratory costs of growth, maintenance and ion uptake; cf. Eqs. 2 and 3 (Lambers et al., 1983; Bouma et al., 1996). To use this method, respiration is plotted against relative growth rate (RGR). The slope of the regression line equals *specific* costs for growth (g_R). Alternatively, respiration can be plotted against both RGR and net nutrient uptake (NUR), giving a regression plane. The slopes of the regression lines for RGR and NUR equal g_R and *specific* costs for nutrient uptake (g_U), respectively. In both cases, the regression line intercepts are used to estimate *specific* costs for maintenance (m_R).

Single regression analysis (Eq. 2) requires variation in RGR, whereas multiple regression analysis (Eq. 3) requires NUR and RGR to vary independently. Such variation can be obtained by combining groups of plants that i) were grown under different conditions, ii) have manipulated shoot/root-ratios, or iii) have different developmental stages (Veen, 1980; Lambers et al., 1983; Van der Werf et al., 1988; Bouma et al., 1996 and references therein), provided values of g_R, g_U and m_R are not altered. Results produced by regression approaches are within the same range as those obtained by theoretical calculations (Table 2).

There are two fundamental problems associated with regression approaches. Firstly, it is difficult to relate specific (maintenance) processes to specific regression components, due to the correlative nature of the method. For example, in roots, maintenance of ion gradients is part of g_U when ion leakage along electrochemical potential gradients is proportional to nutrient uptake (NUR), but part of m_R when ion leakage is proportional to the living biomass (M). Secondly, the method yields single estimates for g_R, g_U and m_R, which, according to their definition, need not be constant. Within a single species, the assumption of constant *specific* costs for growth (g_R), nutrient uptake (g_U), and maintenance (m_R) is, however, fairly reasonable, because specific costs will only change when underlying processes change (e.g., changes over time in tissue composition or protein turnover rate per unit protein). Within a single species, constant values for g_R and g_U may seem more reasonable than a constant value for m_R, as many processes underlie maintenance (Bouma et al., 2000). It is emphasized that constant *specific* costs for growth (g_R), nutrient uptake (g_U), and maintenance (m_R) are different from constant respiratory costs for growth ($r_G = g_R$ RGR; Eq. 2) maintenance (r_M) and ion uptake ($r_U = g_U$ NUR; Eq. 3), because the latter is not likely to occur. Both the absolute values and relative importance of r_G, r_M and r_U can change rapidly without any changes in the *specific* costs associated with growth (g_R), nutrient uptake (g_U), and maintenance (m_R) (Fig. 1; Van der

Table 2. Comparison of results of multiple regression approach among species with the values of theoretical calculations (Bouma et al., 1996; Scheurwater et al., 1998). The values in the table represent estimates of specific costs as defined in Eq. 3: $r = r_G + r_U + r_M = g_R$ RGR + g_U NUR + m_R where r is overall respiration per unit dry mass (nmol O_2 g^{-1} s^{-1}), g_R is specific costs for tissue construction (also referred to as a growth coefficient; mmol O_2 g_{growth}^{-1}), RGR is relative growth rate (μg_{growth} g^{-1} s^{-1}), g_U is specific costs for ion uptake [also referred to as a uptake coefficient; mol O_2 (mol NO_3^-)$^{-1}$], NUR is net nutrient uptake rate [(nmol NO_3^-) g^{-1} s^{-1}], and m_R is a maintenance coefficient (nmol O_2 g^{-1} s^{-1}). Regarding the problems with the previously generally used inhibitor-titration method (Chapter 3, Ribas-Carbo), we present all data on oxygen and dry weight basis as indicated in Bouma et al. (1996). Data may be converted to ATP values, assuming that the ATP/O_2-ratio generally ranges between 4.8 and 3 (Noguchi et al., 2001; Robinson et al., 1995).

species	g_R [mmol O_2 (g DW)$^{-1}$]	g_U [mol O_2 (mol NO_3^-)$^{-1}$]	m_R [nmol O_2 (g DW)$^{-1}$ s^{-1}]	methods	refs
Dactylis glomerata	6.5	0.41 ± 0.013	25.5± 2.08	linear regression + calc.	1
Festuca ovina	6.25	1.22 ± 0.013	20 ± 2.66	linear regression + calc.	1
Solanum tuberosum	9.9 + 3.6	0.67 + 0.09	10.2 + 2.4	3-component regression	2
Solanum tuberosum	9.8 + 1.7	0.39 + 0.10	14.8 + 4.6	3-component regression	2
Solanum tuberosum		0.65 + 0.049		no growth	2
Solanum tuberosum			9.9	no growth, no uptake	2
Carex acutiformis	6.2	0.83	4.3	3-component regression	3
Carex diandra	6.4	1.16	7.0	3-component regression	3
Zea Mays	10.9	1.6	4.0	3-component regression	4
24 herb. species	5.5 to 8.0			chemical composition	5
	4.9			ammonium–fed plant	6
	4.3 to 5.5			chemical composition	2
		0.43	3.7 – 16.2	theoretical calculations	2

References: 1 = Scheurwater et al. (1998); 2 = Bouma et al. (1996); 3 = Van der Werf et al. (1988); 4 = Veen (1980), 5 = Poorter et al. (1991); 6 = Penning de Vries et al. (1974).

Werf et al., 1988).

In general, multiple regressions (Eq. 3) are more difficult to perform than simple linear regressions (Eq. 2), since it is often difficult to vary uptake rates (NUR) and root relative growth rates (RGR) independently. An alternative approach to multiple regression analysis (especially when many species are compared) is to estimate g_U by combining linear regression with theoretical estimates for g_R and empirical measurements for NUR (Scheurwater et al., 1998). This approach has enhanced our understanding of the respiration rates of fast- and slow-growing species (Scheurwater et al. 1998).

2. Black-Box Approach

In the absence of growth and uptake, respiration must, by definition, be associated with maintenance. Based on this definition, a number of methods were developed for estimating maintenance respiration. These methods are so-called 'black-box' approaches. Typical examples of this approach include the 'dark-decay' method, the 'zero-growth' method, and the 'mature-tissue' method.

The 'dark-decay' method assumes that only the most critical tissue maintenance processes occur when carbohydrate supply is limited. Hence, maintenance cost is estimated from reduced respiration of plants kept in darkness for a prolonged period of time. Despite criticism (Breeze and Elston, 1983; Denison and Nobel, 1988; Gary, 1989; Bouma et al., 2000), this method is occasionally still used due to a lack of alternatives. Figure 2 illustrates problems associated with the 'dark-decay method.' Respiration of inactive roots maintained in dry soil for 1 month should only consist of maintenance respiration (Espeleta and Eissenstat, 1998; Espeleta et al., 1998; Eissenstat et al., 1999). However, root respiration declined by an additional 50% when plants were placed in the dark for more than 3 days (Bouma et al., 2000). Lack of information on physiological functioning during dark-decay hampers useful interpretation of this type of measurement which is clearly illustrated by the increase in respiration during the first 2 days of the dark period (Fig. 2).

The 'zero-growth' method is a slightly improved

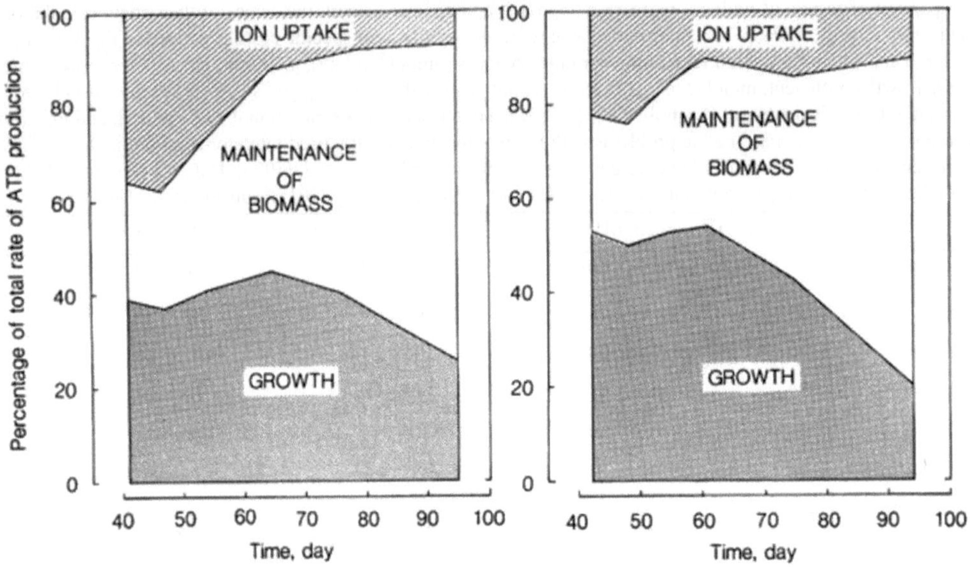

Fig. 1. Time course of the relative contribution of respiration for growth, ion uptake and maintenance in roots of *Carex diandra* (left) and *Carex acutiformis* (right) (Van der Werf et al. 1988).

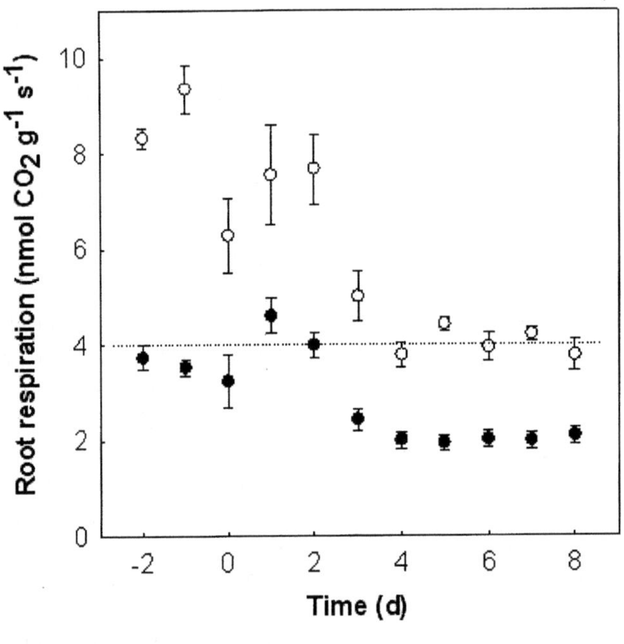

Fig. 2. Root respiration of Volkamer lemon (*Citrus volkameriana* Tan. and Pasq.) seedlings grown in wet (open symbols) or dry (closed symbols) surface soil (Bouma et al., 2000). Plants were measured for 3 days under normal cycles of day and night (i.e. up to day 0) before being placed in darkness for 24 h day^{-1} ('dark-decay' treatment) for 8 days.

derivation of the 'dark-decay' method. This method attempts to estimate maintenance cost at the 'light-compensation point' (defined here as the point at which light intensity limits photoassimilation of carbon for growth, but not maintenance respiration). As in the 'dark-decay' method, the physiological functionality of this method is unclear.

The 'mature-tissue' method estimates maintenance

Table 3. Overview of the relative importance of protein turnover and carbohydrate export for leaf respiration. Energy consumption is expressed as percentage of the total respiration.

Species	Energy consumption by protein turnover (%)	References	Energy consumption by carbohydrate export (%)	References
Alocasia odora	16 to 24	Noguchi et al. (2001)	15 to 26	Noguchi et al. (2001)
Phaseolus vulgaris	41 to 61	Noguchi et al. (2001)	23 to 33	Noguchi et al. (2001)
Phaseolus vulgaris	17 to 21 full-grown leaves 17 to 37 expanding leaves	Bouma et al. (1994)	13 to 39	Bouma et al. (1995)
Solanum tuberosum			11 to 32	Bouma et al. (1995)
Range of species			12 to 52	Table 6 in Bouma et al. (1995)
Lolium perenne	27 to 36	Barneix et al. (1988)		
Triticum aestivum	41 to 61	Zagdanska (1995)		

costs by measuring respiration on fully-grown tissues. A major complication with this method is that other processes not related to maintenance may have significant respiratory energy consumption. For example, in mature leaves, costs of carbohydrate export can be significant (Table 3), or in mature, non-growing root segments, respiration changes with root age (Fig. 3), without any clear reason why. Despite these complications, however, the simplicity of the method still makes it popular (e.g., Ryan, 1995).

3. Process-Based Approach

The process-based approach estimates both rate (v_i) and specific cost (e_i) of individual processes. Total respiratory costs (r) are the sum of all ongoing processes (subscript i), defined as:

$$r = \Sigma (v_i * e_i) \quad (7)$$

The advantage of this approach is that it eliminates the 'black-box' term. Challenges associated with the method, however, include identifying all relevant respiratory processes and estimating v_i and e_i for each process. Although considerable progress has been made in accounting for many physiological processes (Amthor, 2000a; Cannell and Thornley, 2000), considerable residual maintenance respiration remains unexplained (Thornley and Cannell, 2000). The relative importance of protein turnover, maintenance of solute gradients and 'futile' cycles for the process-based estimates are discussed below.

a. Protein Turnover

Protein turnover has several important regulatory functions in plant metabolism [see Vierstra (1993) for a review on the topic]. It may also play an important role in development and plant phenotypic plasticity required to acclimate to changes in environmental conditions.

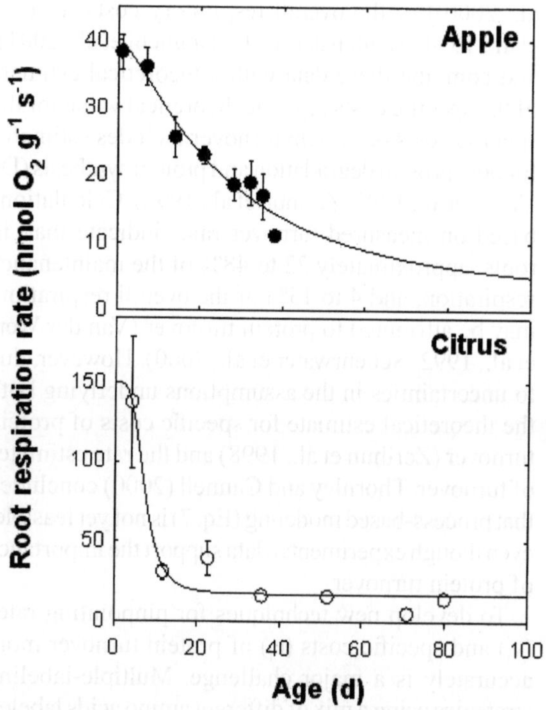

Fig. 3. Decline of root respiration as a function of root age. Respiration was measured on detached fully grown root segments collected from 20-year-old apple trees (*Malus domestica* Borkh.) (Red chief Delicious on M26 rootstock; Bouma et al., 2001) and 9-year-old bearing red grapefruit (*Citrus paradisi* Macf.) trees on sour orange rootstock (*Citrus aurantium* L.; Bouma et al., 2001). Similar patterns have been found for 25-yr-old Concord grapevines (*Vitis labruscana* Bailey; Comas et al., 2001).

Early contentions that protein turnover is an important component of maintenance respiration (Penning de Vries, 1975) was originally supported by strong relationships between respiration and tissue N concentration (Reich et al., 1997, 1998a,b; Pregitzer et al., 1998;). Recent measurements using protein synthesis inhibitors show that protein turnover may account for 15 to 60% of overall leaf respiration (Table 3), with the proportion varying depending on leaf age and plant species (Bouma et al., 1994; Noguchi et al., 2001).

It is difficult to obtain independent measurements of both the rate (v_i) and the specific costs (e_i) of protein turnover. Hence, most studies measure either protein turnover rate [e.g., degradation rate by ^{14}C leucine labeling (Van der Werf et al., 1992; Scheurwater et al., 2000)] or the overall respiratory costs (Bouma et al., 1994; Zagdanska, 1995; Noguchi et al., 2001), and combine these data with a theoretical estimate of the specific costs (e_i). The theoretical value for the specific costs of protein turnover includes estimates for both protein degradation and protein synthesis (De Visser et al., 1992; Zerihun et al., 1998). Calculations based on measured turnover rates indicate that, in roots, approximately 22 to 48% of the maintenance respiration, and 4 to 15% of the overall respiration, may be attributed to protein turnover (Van der Werf et al., 1992; Scheurwater et al., 2000). However, due to uncertainties in the assumptions underlying both the theoretical estimate for specific costs of protein turnover (Zerihun et al., 1998) and the rate estimates of turnover, Thornley and Cannell (2000) concluded that process-based modeling (Eq. 7) is not yet feasible, even though experimental data support the importance of protein turnover.

To develop new techniques for pinpointing rates (v_i) and specific costs (e_i) of protein turnover more accurately is a major challenge. Multiple-labeling strategies using a mix of different amino acids labeled with ^{13}C and ^{15}N may offer possibilities. Although many questions remain with respect to protein turnover, even less information is available on the turnover of other cellular components, such as membranes and chlorophyll. Further research in this area is urgently needed (Amthor, 2000a).

b. Maintenance of Solute Gradients

Both experimental and theoretical approaches support the hypothesis that maintaining ion gradients (i.e. nutrient uptake required to offset energetically-favorable cellular leakage) is an important and necessary energy-consuming process (Penning de Vries, 1975). Unfortunately, quantitative experimental data are currently insufficient to enable process-based modeling (Eq. 9) of the costs associated with this process (Thornley and Cannell, 2000). It is possible, however, to calculate the minimal energy required to transport (an)ions across cell membranes against an electrochemical potential gradient (see section on the 'theoretical methods: process-based approach'). Thus, quantifying the cost of maintaining ion gradients is more a question of quantifying 'leakage' rates (v_i), rather than obtaining the specific costs (e_i) of ion transport (Bouma and De Visser, 1993). From various methods used to study efflux [for examples, see references in Bouma and De Visser (1993) and Clarkson (1998)], quantifying the difference between the net uptake and the influx of labeled ions is most elegant, and allows measurements of diurnal patterns (Clarkson, 1998).

Experimental observations indicate that ion efflux is quantitatively important as an energy-requiring process, but questions remain concerning how efflux differs between plants grown in natural environments and those grown in hydroponics or intensive crop-production systems (Clarkson, 1998). A source of confusion is that many studies present specific costs of *net uptake* rather than specific costs of *influx* (i.e. specific costs per transported ion), where net uptake is defined as influx minus efflux (i.e. 'leakage'). From perturbation effects, Bouma et al. (1996) concluded that costs associated with maintaining ion gradients accounted for up to 33% of the overall costs of nitrate influx in potato. Similarly, the specific costs of net ion uptake measured in *Zea mays* (Veen, 1980) and two *Carex* species (Van de Werf et al., 1988) were two to four times greater than the theoretically expected costs of nitrate influx (Bouma et al., 1996) which indicates that maintaining ion gradients form a major part of the overall costs of nitrate uptake. Furthermore, Scheurwater et al. (1998) concluded that the specific costs of ion uptake were three times higher in a slow-growing grass species than in a fast-growing one, with the latter being close to the theoretical value. This difference among the grass species would be partially due to the energy requirements of re-uptake compensating 'leakage' (Scheurwater et al., 1999). Similar effects seem to explain respiratory differences between wild-type and slow-growing GA-deficient tomato mutants (*Solanum lycopersicum* L.; Nagel and Lambers, 2002). The question remains whether ion efflux is associated with cellular inefficiency and cannot be avoided, or if it has a particular function, such as regulating cytosolic concentrations.

c. 'Futile' Cycles and 'Inefficient' Metabolic Pathways

Some processes in the plant may seem inefficient and counterproductive. For example, to some extent protein turnover and ion efflux may be regarded as inefficient and counterproductive, when compared to using existing proteins or avoiding ion leakage. Processes like the continuous production of a simultaneously degrading product are generally referred to as 'futile' cycles. Using the term 'futile' cycles for apparently wasteful processes probably reflects more our lack of knowledge than that it tells anything about the functioning of the plant. One might speculate that 'futile' cycles have some functional role in cellular regulation, with the cost of the 'futile' cycle being smaller than the risks associated with smaller possibilities for regulation.

The alternative respiratory pathway is a clear example of an 'inefficient' metabolic pathway, whose function has been topic of discussion for a long time (Chapter 1, Lambers et al.). In comparison with the cytochrome pathway, where the ratio of ATP/O_2 is 6, the alternative pathway, which is not coupled to proton extrusion, is three times less efficient, and has an ATP/O_2 ratio of only 2. Thus, the relative contribution of both respiratory pathways is important for quantifying actual energy production (Lambers and Van der Werf, 1988; Lambers et al., 1996). The relative activity of both respiratory pathways can be determined by using the difference in discrimination for stable oxygen isotopes between the alternative and cytochrome pathways (Robinson et al., 1995; Chapter 3, Ribas-Carbo). Using this technique, recent studies indicate that the alternative respiratory pathway may play an important role during stress, such as when plants are exposed to chilling temperatures (Gonzalez-Meler et al., 1999; Ribas-Carbo et al., 2000b). Future studies should provide more insight when the alternative respiratory pathway is active.

B. Theoretical Methods

1. Process-Based Approach

A theoretical, process-based approach is similar to the experimental, process-based approach in that it is based on the rate (v_i) and the specific costs (e_i) of individual processes (cf. Eq. 9). However, with the theoretical approach, values of v_i and/or e_i are not measured, but calculated. As seen in the previous section, the experimental process-based method sometimes needs theoretical calculations for part of the data which shows the compatibility of both approaches. Theoretical, process-based calculations have been applied most intensively for estimating the specific costs of tissue construction (i.e. growth coefficient; g_R). Based on available substrates, individual chemical compounds present in plant tissue, and biochemical pathways, Penning de Vries et al. (1974) calculated the specific costs for growth (g_R; g 'X' g_{growth}^{-1}) and conversion efficiency (Y_G = unit growth per unit substrate utilized). For a detailed review of this work, see Amthor (2000a).

Limitations associated with the process-based approach include the need to determine both complete biomass composition of the plant tissue, and the exact biochemical pathways involved which are not known for all (secondary) chemical compounds. For that reason, Penning de Vries et al. (1974) simplified the method. They calculated growth costs by categorizing compounds into five groups, carbohydrates, N compounds, lipids, lignins and organic acids, and argued that differences in construction costs within these groups are negligible compared with differences among groups. This simplification allowed the method to be used with only proximate biomass composition.

In addition to estimating tissue construction costs, theoretical process-based calculations can also be used to estimate other costs, such as protein turnover (e.g., De Visser et al., 1992; Zerihun et al., 1998) and ion uptake (e.g., Bouma and De Visser, 1993; Clarkson, 1998). To calculate the energy requirements for ion transport, it is possible to use the stoichiometry of the ion carriers driven by proton-motive force, as well as the stoichiometry of any H^+-ATPase that generates the proton-motive force (i.e. electrochemical H^+ gradients across cell membranes that drive ion transport). For example, respiration required to maintain a gradient for ion_j (r_{Mgrad}; mol ATP g^{-1} s^{-1}) is calculated as:

$$r_{Mgrad} = v_i * e_i \quad (8.1)$$

$$v_i = \Phi_j \quad (8.2)$$

$$e_i = M_j\,[H/I_j]\,[H/P] \quad (8.2)$$

where v_i is the rate of the relevant process, e_i is the specific cost of the relevant process, Φ_j is the efflux of ion j (mol ion_j g^{-1} s^{-1}), M_j is the number of active membrane passages for ion_j, $[H/I_j]$ is the stoichiometry between protons and ion_j (mol H^+ [mol $ion_j]^{-1}$), and [H/P] is the stoichiometry of the H^+-ATPase (mol H^+ [mol ATP$]^{-1}$) (Bouma and De Visser, 1993). Similarly, the costs for net uptake are obtained if Φ_j is replaced by the difference between influx and efflux. Experimentally, specific costs of uptake are often overestimated when the net uptake rates are divided by the respiratory costs of the overall influx (see Section IV.A.3.b. Ion Gradient Maintenance).

2. Correlative Approach

The process-based calculation to estimate costs of tissue construction, as originally introduced by Penning de Vries et al. (1974), offered great progress understanding true growth yield. However, the method requires knowledge of both the biomass composition and the biochemical pathways, and is therefore very laborious. McDermit and Loomis (1981) provided a less labor-intensive method to determine the specific costs for growth, by using elemental composition (C, H, N and O content). Based on this analysis, a correlative method was derived that allows estimation of the costs of tissue construction based on C concentration and ash concentration only (Vertregt and Penning de Vries, 1987).

V. Relations with Environmental Conditions

Respiratory energy can be either separated into grouped components for growth, maintenance and ion uptake (Eq. 3; Table 1), or determined for defined individual processes (Eq. 7; Table 1). The question arises to what extent the results of such analyses (Table 1) depend on the species, developmental stage, growth conditions and conditions during the measurements. Differences among species are to be expected (Tables 2 and 3), and may in some cases be related to growth strategies [e.g., slow- versus fast-growing species (Scheurwater et al., 1998, 1999, 2000); sun versus shade species (Noguchi et al., 2001)]. Several well studied abiotic factors that affect total respiration include temperature (Chapter 7, Atkin et al.), water stress (Chapter 6, Flexas et al.), and CO_2 concentration (Chapter 13, Gonzalez-Meler). However, little is known about the effect of these factors on the components of respiration ($g_R + g_U + m_R$) or the rates of individual processes.

A. Respiratory Components

Atkin et al. (2000) and Atkin et al. (Chapter 7) reviewed the Q_{10} response and species-specific acclimation of respiration to short- and long-term temperature change. However, few studies have distinguished short- and long-term temperature effects on g_R, g_U, and m_R separately (Amthor, 2000a). Theoretically, one might expect specific costs for tissue construction (g_R) to be temperature independent. Similarly, specific costs for influx may also be expected to be temperature independent, whereas efflux rates, and thus the specific costs for net uptake (g_U), may show temperature dependence. The few available data suggest m_R to be temperature dependent (Mariko and Koizumi, 1993; Marcelis and Baan Hofman-Eijer, 1995), but we lack knowledge on underlying processes.

Water stress can reduce root respiration (Palta and Nobel, 1989a,b; Bryla et al., 1997, 2001) which is probably due to decreased rates of both growth (Espeleta and Eissenstat, 1998; Espeleta et al., 1998) and ion uptake (Eissenstat et al., 1999). In citrus (*Citrus volkameriana* Tan. and Pasq.), reductions in the rate of root respiration due to water stress were much less in cold soils than in warm soils, while respiration in completely dry soils was similar at any given temperature (Bryla et al., 2001). This suggests that carbon costs associated with root maintenance in dry soil are temperature independent.

Carbon dioxide concentration may alter respiration directly or indirectly, by altering related processes such as photosynthesis and plant growth. Both direct effects (Nobel and Palta, 1989; Palta and Nobel, 1989c; Qi et al., 1994; Burton et al. 1997; Clinton and Vose, 1999; Mc Dowell et al., 1999; Chapter 13, Gonzalez-Meler) and the absence of any direct effects (Bouma et al., 1997a,b; Tjoelker et al., 1999, 2001; Amthor, 2000b; Bouma and Bryla, 2000; Amthor et al., 2001; Burton and Pregitzer, 2002; Chapter 13, Gonzalez-Meler) on leaf and root respiration have been reported. McDowell et al. (1999) examined whether direct inhibitory effects of high CO_2 concentrations on root respiration was due to a reduced respiratory component for growth or maintenance. Based on analysis of covariance, they concluded that the effect was only on maintenance, and speculated that this may explain why respiration of the roots of some fast-growing species does not show CO_2 inhibition. However, Bouma et al. (1997b) found no direct effect of changing soil CO_2 concentrations on root respiration for both fast- (*Phaseolus vulgaris* L. CIAT breeding line DOR 364) and slow-growing (*Citrus volkameriana* Tan. and Pasq.) species, nor did they find any effect on growth or maintenance respiration of the fast-growing species. Thomas et al. (1993) and Thomas and Griffin (1994) found for leaves of soybean (*Glycine max* L. Merr.) and cotton (*Gossypium hirsutum* L. cv Coker 315) only an effect on the maintenance respiration, whereas the growth respiration was unaffected. However, Hamilton et al. (2001) recently found no such effects under long-term elevated [CO_2] conditions. Poorter et al. (1997) calculated that changes in tissue composition due to growth under elevated [CO_2] could account for a decrease of up to 30% in the specific costs for of tissue construction (also called growth coefficient; g_R), whereas the construction costs (glucose required to produce 1 g of leaf) remained relatively unchanged. A recent review by Amthor (2000b) revealed that in most tree species, there is no direct inhibitory effect of elevated [CO_2] on dark respiration of leaves, or twigs with leaves.

Due to the controversy in the available data, additional experiments are needed to establish whether direct responses actually occur in certain plant species, and, if so, to determine any underlying mechanisms involved (Burton et al., 1997; Amthor, 2000b; Chapter 13, Gonzalez-Meler). Part of the contrasting results may be due to CO_2-diffusion leakages when making respiration measurements at high [CO_2] concentrations (Burton and Pregitzer, 2002; Chapter 13, Gonzalez-Meler). Artifacts due to CO_2-diffusion leakages may be prevented by using gas-exchange systems with a design that makes the CO_2 readings insensitive to CO_2 leakage (e.g., see open gas exchange system in Bouma et al. 1997a) or by using designs for the respiration cuvettes in which the CO_2 concentrations in the 'analysis' air remains close to ambient air concentrations whereas the CO_2 concentrations around the plant tissues are much higher (e.g., for roots see 'surrounding' and 'headspace' measurements in figure 1 in Bouma et al. 1997a). Alternatively, one may correct the measurements for CO_2-diffusion leakages (Burton and Pregitzer, 2002; Chapter 13, Gonzalez-Meler).

B. Individual Energy-Requiring Processes

For several processes, overall respiratory costs can be modeled according to Eq. 7 [reviewed by Cannell and Thornley (2000) and Amthor (2000a)]. Despite experimental indications that the respiratory costs of protein turnover and maintenance of solute ion gradients are quantitatively important, data are still insufficient to accurately model the respiratory costs of protein turnover and maintenance of solute gradients (Thornley and Cannell, 2000). Below, a brief exploration is given of factors that may affect the rates (v_i) and specific costs (e_i) of protein turnover and maintenance of solute gradients.

With regard to protein turnover, energy costs per unit nitrogen are higher in plant species adapted to high-light conditions than in those adapted to shade conditions (Noguchi et al., 2001). It is also higher in developing young leaves than in fully-grown leaves (Bouma et al., 1994). Based on this evidence, and assuming the specific costs of protein turnover are constant [see theoretical estimates in De Visser et al. (1992) and Zerihun et al. (1998)], it appears that turnover is fastest (i.e. highest degradation constant or shortest half-life) in the most active tissues. This assumption is further supported by results of Scheurwater et al. (2000) who found a higher degradation constant (K_d-value) in a fast-growing grass species than in a slow-growing grass species.

Adaptation to environmental conditions may induce increased turnover rates. Specific examples include increased protein degradation relative to synthesis upon nutrient limitation (Davies, 1979; Hartfeld and Vierstra, 1997), and enhanced turnover for N reallocation from shaded to sun-exposed leaves in order to optimize the nitrogen-use efficiency (cf. Pons et al., 1993; Pons and Pearcy, 1994; Pons and Bergkotte, 1996). It is, however, unlikely that protein turnover is the only process causing a relationship between respiration and tissue N concentration (e.g., Reich et al., 1997, 1998a,b; Pregitzer et al., 1998). Part of this relationship will be due to the effects of tissue N concentration on processes that consume energy itself (e.g., carbohydrate export; Bouma et al., 1995; Noguchi et al., 2001).

Where present knowledge is still rather limited with respect to factors influencing protein turnover, even less is known about the factors influencing the costs for maintaining ion gradients. The reason for this is that it is difficult to obtain reliable estimates of ion efflux rates. Experimental estimates may be affected by both experimental artifacts (e.g., perturbation; Bloom and Sukrapanna, 1990), or by plant growth conditions (Clarkson, 1998). Specific costs for membrane transport can be calculated within certain limits, and therefore, may be expected to be relatively constant. For an extended discussion on factors that might affect costs of the maintenance of ion gradients, see a review by Clarkson (1998).

VI. Future Research Directions

Although the concept of separating respiration into components associated with growth, uptake and maintenance respiration remains difficult to capture using experimental approaches, this conceptual approach is still widely used. Several recent reviews (Amthor, 2000a; Thornley and Cannell, 2000) offer valuable recommendations for future research on the relationship of respiration and its underlying processes. The main message has been elegantly formulated by Thornley and Cannell (2000): 'If we are to proceed beyond description towards understanding, then it seems essential that measurements of respiratory fluxes are accompanied by measurements of other processes (fluxes), and of the status of the tissue being investigated (e.g., nutrients status, sugar concentration, N concentration and categories)'. Apart from this plea for 'down-scaling' the research, I suggest that we should also attempt to benefit from current developments in 'up-scaling,' as explained below.

A. Down-Scaling

At the present state of knowledge, research aimed at experimental estimates of growth (g_R), uptake (g_U) and maintenance (m_R; Eq. 3) respiration using correlative or black-box approaches is less useful than research toward process-based approaches, which estimate the rate (v_i) and the specific costs (e_i) of individual processes (Eq. 7). Dividing respiration into energy-utilizing components for growth, maintenance and ion uptake, and estimating g_R, g_U and m_R may be useful to identify contrasting plant growth strategies or identify effects of environmental conditions (Scheurwater et al., 1998). However, even in these types of studies, research methods should develop toward process-based approaches.

At present, the main challenges to using process-based approaches include:

- Developing new methods for measuring rates and specific costs of each process of interest, and, ideally, relating each rate measurement to simultaneous respiration measurements.
- Literature studies for obtaining new or improved theoretical estimates of the rates and specific costs of each process.

The availability of very sensitive detection methods for stabile isotopes in simple and complex biochemical compounds, combined with the commercial availability of many (multiple) labeled compounds (e.g., ^{13}C, ^{18}O, ^{15}N) may offer opportunities to pinpoint turnover rates of various cellular components, including proteins. Similarly, these methods may also offer valuable possibilities for a new impulse in research on the energetic importance of processes such as maintaining (ion) gradients or carbohydrate export.

With rapid development in various scientific disciplines, it may now also become feasible to explore the relative importance of the respiratory energy required for groups of processes that have been neglected until now. Two arbitrarily chosen examples of potentially relevant processes are DNA turnover and defense against herbivory. For DNA, it has been shown that the bulk of (e.g., UV-induced) damage is repaired by photo-reactivation, and thus does not require respiratory energy (Britt, 1997, 1999). The generality of this finding for other molecular processes still needs to be established. For (induced) plant defense compounds, data become available in terms of costs for plant fitness (Karban and Baldwin, 1997; Karban et al., 1999; Tollrian and Harvell, 1999; Van Dam and Baldwin, 2001). It would be interesting to quantify these costs in terms of respiration, as a fraction of the overall carbon budget.

B. Up-Scaling

Apart from the need for process-based research, it may also be important to do 'up-scaling'. The process-based modeling (Section III.B) showed that integration of several detailed studies into a model at a higher integration level, generates valuable insights into processes for future research. In the field of macro-ecology, major progress has been made since the introduction of a general model for the origin of allometric-scaling laws in biology (West et al., 1997). This basic model has been used to explain allometric relationships of the vascular system of plants (West et al., 1999), intraspecific- and interspecific-scaling relationships among leaf, stem, and root biomass of seed plants (Enquist and Niklas, 2002), and life-history variables to rates of production, demography and reproduction (Enquist et al., 1999; Niklas and

Enquist, 2002). It would be interesting to explore to what extent this type of analysis would lead to better (and perhaps different) understanding of the sinks for plant respiratory energy. This approach seems promising, due to the recent progress made on modeling ontogenetic growth based on allocation of metabolic energy between maintenance and growth in animals (West et al., 2001).

Acknowledgments

I am grateful to Daniela Lud for providing insight into the energy costs of DNA repair, to David Bryla, Han van Hulzen and Hans Lambers for critical comments on earlier versions of the manuscript, and to Maryann Resendes for editing this chapter.

References

Agren GI, McMurtrie RE, Parton WJ, Pastor J and Shugart HH (1991) State-of-the-art models of production-decomposition linkages in conifer and grassland ecosystems. Ecol Applic 1: 118–138

Amthor JS (1984) The role of maintenance respiration in plant growth. Plant Cell Environ 7: 561–569

Amthor JS (2000a) The McCree-de Wit-Penning de Vries-Thornley respiration paradigms: 30 years later. Ann Bot 86: 1–20

Amthor JS (2000b) Direct effect of elevated CO_2 on nocturnal in situ leaf respiration in nine temperate deciduous tree species is small. Tree Physiol 20: 139–144

Amthor JS, Koch GW, Willms JR and Layzell DB (2001) Leaf O_2-uptake in the dark is independent of coincident CO_2 partial pressure. J Exp Bot 52: 2235–2238

Atkin OK, Edwards EJ and Loveys BR (2000) Response of root respiration to changes in temperature and its relevance to global warming. New Phytol 147: 141–154

Audus LJ (1960) Effect of growth-regulating substances on respiration. In: Ruhland W (ed) Encyclopedia of Plant Physiology, Vol XII, Plant Respiration Inclusive Fermentations and Acid Metabolism, Part 2. pp 360–387. Springer-Verlag, Berlin

Barneix AJ, Cooper HD, Stulen I and Lambers H (1988) Metabolism and translocation of nitrogen in two *Lolium perenne* populations with contrasting rates of mature leaf respiration and yield. Physiol Plant 72: 631–636

Bloom AJ and Sukrapanna SS (1990) Effects of exposure to ammonium and transplant shock upon the induction of nitrate absorption. Plant Physiol 94: 85–90

Bouma TJ and De Visser R. (1993) Energy requirements for maintenance of ion concentrations in roots. Physiol Plant 89: 133–142

Bouma TJ and Bryla DR (2000) On the assessment of root and soil respiration for soils of different textures: interactions with soil moisture contents and soil CO_2 concentrations. Plant Soil 227: 215–221

Bouma TJ, De Visser R, Janssen JHJA, De Kock MJ, Van Leeuwen PH and Lambers H (1994) Respiratory energy requirements and rate of protein turnover in vivo determined by the use of an inhibitor of protein synthesis and a probe to assess its effect. Physiol Plant 92: 585–594

Bouma TJ, De Visser R, Van Leeuwen PH, De Kock MJ and Lambers H (1995) The respiratory energy requirements involved in nocturnal carbohydrate export from starch-storing mature source leaves and their contribution to leaf dark respiration. J Exp Bot 46: 1185–1194

Bouma TJ, Veen B and Broekhuysen G (1996) Analysis of root respiration of *Solanum tuberosum* as related to growth, ion uptake and maintenance of biomass. Plant Physiol Biochem 34: 795–806

Bouma TJ, Nielsen KL, Eissenstat DM and Lynch JP (1997a) Estimating respiration of roots in soil: interactions with soil CO_2, soil temperature and soil water. Plant Soil 195: 221–232

Bouma TJ, Nielsen KL, Eissenstat DM and Lynch JP (1997b) Soil CO_2 concentration does not affect root growth or root respiration in citrus or bean. Plant Cell Environ 20: 1495–1505

Bouma TJ, Bryla DR, Li Y and Eissenstat DM (2000) Is maintenance respiration in roots a constant? In: Alexia Stokes (ed) The Supporting Roots of Trees and Woody Plants: Form, Function and Physiology. Developments in Plant and Soil Sciences 87, pp 391–396. Kluwer Academic Publishers, Dordrecht

Bouma TJ, Eissenstat DM, Yanai RD, Elkin A, Hartmond U and Flores D (2001) Estimating age-dependent costs and benefits of roots with contrasting lifespan: comparing apples and oranges. New Phytol 150: 685–695

Breeze and Elston (1983) Examination of a model and data describing the effect of temperature on the respiration of crop plants. Ann Bot 51: 611–616

Britt AB (1997) Genetic analysis of DNA repair in plants. In: Lumsden (ed) Plants and UV-B: Responses to Environmental Change, pp 77–93. Cambridge University Press, Cambridge

Britt AB (1999) Molecular genetics of DNA repair in higher plants. Trend Plant Sci 4: 20–25

Bryla DR, Bouma, TJ and Eissenstat DM (1997) Root respiration in citrus acclimates to temperatures and slows during drought. Plant Cell Environ 20: 1411–1420

Bryla DR, Bouma TJ, Eissenstat DM and Hartmond U (2001) Influence of temperature and soil drying on respiration of individual roots in citrus: Integrating greenhouse observations into a predictive model for the field. Plant Cell Environ 24: 781–790

Burton AJ and Pregitzer KS (2002) Measurement carbon dioxide concentration does not affect root respiration of nine tree species in the field. Tree Physiol 22: 67–72

Burton AJ, Zogg GP, Pregitzer KS and Zak DR (1997) Effects of measurement CO_2 concentration on sugar maple respiration. Tree Physiol 17: 421-427

Cannell MGR and Thornley JHM (2000) Modelling the components of plant respiration: Some guiding principles. Ann Bot 85: 45–54

Clarkson D (1998) Mechanisms for N-uptake and their running costs; is there scope for more efficiency? In: Lambers H, Poorter H and Van Vuuren MMI (eds) Inherent Variation in Plant Growth: Physiological Mechanisms and Ecological Consequences. pp 221–235. Backhuys Publishers, Leiden

Clinton BD and Vose JM (1999) Fine root respiration in mature eastern white pine (*Pinus strobes*) in situ: The importance of CO_2 in controlled environments. Tree Physiol 19: 475–479

Comas LH, Eissenstat DM and Lakso AN (2001) Assessing root death and root system dynamics in a study of grape canopy pruning. New Phytol 147: 171–178

Davies DD (1979) Factors affecting protein turnover in plants. In: Hewitt EJ and Cutting CV (eds) Nitrogen assimilation of plants. pp 369–396. Academic Press, London

Denison RF and Nobel PS (1988) Growth of *Agave deserti* without current photosynthesis. Photosynthetica 22: 51–57

De Visser R and Lambers H (1983) Growth and the efficiency of root respiration in *Pisum sativum* as dependent on the source of nitrogen. Physiol Plant 58: 813–817

De Visser R, Spitters CJT and Bouma TJ (1992) Energy costs of protein turnover: Theoretical calculation and experimental estimation from regression of respiration on protein concentration of full-grown leaves. In: Lambers H and Van der Plas LHW (eds) Molecular, Biochemical and Physiological Aspects of Plant Respiration. pp 493–508. SPB Academic Publishing, The Hague

De Wit CT, Brouwer R. and Penning de Vries FWT (1970) The simulation of photosynthetic systems. In: Setlik I (ed) Prediction and Measurements of Photosynthetic Productivity. pp 47–70. Centre for Agricultural Publishing and Documentation, Wageningen

Eissenstat DM and Yanai RD (1997) The ecology of root lifespan. Adv Ecol Res 27: 1–60

Eissenstat DM, Whaley EL, Volder A and Wells CE (1999) Recovery of citrus surface roots following prolonged exposure to dry soil. J Exp Bot 50: 1845–1854

Enquist BJ and Niklas KJ (2002) Global allocation rules for patterns of biomass partitioning in seed plants. Science 295: 1517–1520

Enquist BJ, West GB, Charnov EL and Brown JH (1999) Allometric scaling of production and life-history variation in vascular plants. Nature 401: 907–911

Espeleta JF and Eissenstat DM (1998) Responses of citrus fine roots to localized soil drying: A comparison of seedlings with adult fruiting trees. Tree Physiol 18: 113–119

Espeleta JF, Eissenstat DM and Graham JH (1998) Citrus root responses to localized drying soil: A new approach to studying mycorrhizal effects on the roots of mature trees. Plant Soil 206: 1–10

Gary C (1989) Temperature and time-course of the carbon balance of vegetative tomato plants during prolonged darkness: Examination of a method of estimating maintenance respiration. Ann Bot 63: 449–458

Gonzalez-Meler MA, Ribas-Carbo M, Giles L and Siedow JN (1999) The effect of growth and measurement temperature on the activity of the alternative respiratory pathway. Plant Physiol 120: 765–772

Hartfeld PM and Vierstra RD (1997) Protein degradation. In: Dennis DT, Layzell DB, Lefebvre DD and Turpin DH (eds) Plant Metabolism, 2nd edition. pp 26–36. Longman Singapore publishers, Singapore

Hiroi T and Monsi M (1964) Physiological and ecological analyses of shade tolerance in plants. 4. Effect of shading on distribution of photosynthate in *Helianthus annuus*. Botanical Magazine, Tokyo 77: 1–9.

Karban R and Baldwin IT (1997) Induced Responses to Herbivory. University of Chicago Press, Chicago

Karban R, Agrawal AA, Thaler JS and Adler LS (1999) Induced plant responses and information content about risk of herbivory. Trends Ecol Evol 14: 443–447

Lambers H and Van der Werf A (1988) Vatriation in the rate of root respiration of 2 carex species. A comparison of related methods to determine the energy requirements for growth, maintenance and ion uptake. Plant Soil 111: 207–211

Lambers H, Szaniawski RK and De Visser R. (1983) Respiration for growth, maintenance and ion uptake. An evaluation of concepts, methods, values and their significance. Physiol Plant 58: 556–563

Lambers H, Atkin OK and Scheurwater I (1996) Respiratory patterns in roots in relation to their functioning. In: Waisel Y, Eshel A and Kafkaki K (eds) Plant Roots. The Hidden Half, pp 323–362. Marcel Dekker Inc., New York

Marcelis LFM and Baan Hofman-Eijer LR (1995) Growth and maintenance respiratory costs of cucumber fruits as affected by temperature, ontogeny and size of the fruits. Physiol Plant 93: 484–492

Mariko S and Koizumi H (1993) Respiration for maintenance and growth in *Reynoutria japonica* ecotypes from different altitudes on Mt Fuji. Ecol Res 8: 241–246

McCree KJ (1970) An equation for the rate of respiration of white clover plants grown under controlled conditions. In: Setlik I (ed) Prediction and Measurements of Photosynthetic Productivity. pp 221–229. Centre for Agricultural Publishing and Documentation, Wageningen

McCree KJ (1974) Equations for the rate of dark respiration of white clover and grain sorghum, as functions of dry weight, photosynthetic rate, and maintenance. Crop Sci 14: 509–514

McCree KJ and Troughton JH (1966) Non-existence of an optimum leaf area index for the production rate of white clover grown under constant conditions. Plant Physiol 41: 1615–1622

McDermit DK and Loomis RS (1981) Elemental composition of biomass and its relation to energy content, growth efficiency and growth yield. Ann Bot 48: 275–290

McDowell NG, Marshall JD, Qi JG and Mattson K (1999) Direct inhibition of maintenance respiration in western hemlock roots exposed to ambient soil carbon dioxide concentrations. Tree Physiol 19: 599–605

Monsi M. (1968) Mathematical models of plant communities. In: Eckardt FE (ed) Functioning of Terrestrial Ecosystems at the Primary Production Level, pp 131–149. UNESCO, Paris

Nagel OW and Lambers H (2002) Changes in the acquisition and partitioning of carbon and nitrogen in the gibberellin-deficient mutants *A70* and *W335* of tomato (*Solanum lycopersicum* L.) Plant Cell Environ 25: 883–891

Niklas KJ and Enquist BJ (2002) On the vegetative biomass partitioning of seed plant leaves, stems, and roots. Am Nat 159: 482–497

Nobel PS and Palta JA (1989) Soil O_2 and CO_2 effects on root respiration of cacti. Plant Soil 120: 263–271

Noguchi K, Go C-S, Miyazawa S-I, Terahima I, Ueda S and Yoshinari T (2001) Costs of protein turnover and carbohydrate export in leaves of sun and shade species. Aust J Physiol 28: 37–47

Olson JS (1964) Gross and net production of terrestrial vegetation. J Ecol 52: 99–118

Palta JA and Nobel PS (1989a) Influence of water status, temperature, and root age on daily patterns of root respiration for two cactus species. Ann Bot 63: 651–662

Palta JA and Nobel PS (1989b) Root respiration for *Agave deserti*: Influence of temperature, water status, and root age on daily

patterns. J Exp Bot 40: 181–186
Palta JA and Nobel PS (1989c) Influence of soil O_2 and CO_2 on root respiration for *Agave deserti*. Physiol Plant 76: 187–192
Penning de Vries FWT (1972) Respiration and growth. In: Rees AR, Cockshull KE, Hand, DW and Hurd RJ (eds) Crop Processes in Controlled Environments, pp 327–347. Academic Press, London
Penning de Vries FWT (1975) The costs of maintenance processes in plant cells. Ann Bot 39: 77–92
Penning de Vries FWT, Brunsting AHM and Van Laar HH (1974) Products, requirements and efficiency of biosynthesis: A quantitative approach. J Theor Biol 45: 339–377
Pons TL and Bergkotte M (1996) Nitrogen allocation in response to partial shading of a plant: Possible mechanisms. Physiol Plant 98: 571–577
Pons TL and Pearcy RW (1994) Nitrogen reallocation and photosynthesis acclimation in response to partial shading in soybean plants. Physiol Plant 92: 636–644
Pons TL, Van Rijnberk H, Scheurwater I and Van der Werf A. (1993) Importance of the gradient in photosynthetically active radiation in a vegetation stand for leaf nitrogen allocation in 2 monocotyledons. Oecologia 95: 416–424
Poorter H, VanBerkel Y, Baxter R, DenHertog J, Dijkstra P, Gifford RM, Griffin KL, Roumet C, Roy J and Wong SC (1997) The effect of elevated CO_2 on the chemical composition and construction costs of leaves of 27 C-3 species. Plant Cell Environ 20: 472–482
Pregitzer KS, Laskowski MJ, Burton AJ, Lessard VC and Zak DR (1998) Variation in sugar maple root respiration with root diameter and soil depth. Tree Physiol 18: 665–670
Qi J, Marshall JD and Mattson KG (1994) High soil carbon dioxide concentrations inhibit root respiration of Douglas fir. New Phytol 128: 435–442
Reich PB, Walters MB and Ellsworth DS (1997) From tropics to tundra: Global convergence in plant functioning. Proc Nat Acad Sci USA 94: 13730–13734
Reich PB, Walters MB, Tjoelker MG, Vanderklein D and Buschena C (1998a) Photosynthesis and respiration rates depend on leaf and root morphology and nitrogen concentration in nine boreal tree species differing in relative growth rate. Funct Ecol 12: 395–405
Reich PB, Walters MB, Ellsworth DS, Vose JM, Volin JC, Gresham C and Bowman WD (1998b) Relationships of leaf dark respiration to leaf nitrogen, specific leaf area and leaf lifespan: A test across biomes and functional groups Oecologia 114: 471–482
Ribas-Carbo M, Berry JA, Yakird D, Giles L, Robinson SA, Lennon AM and Siedow JN (1995) Electron partitioning between the cytochrome and alternative pathways in plant mitochondria. Plant Physiol 109: 829–837
RibasCarbo M, Lennon AM, Robinson SA, Giles L, Berry JA and Siedow JN (1997) The regulation of electron partitioning between the cytochrome and alternative pathways in soybean cotyledon and root mitochondria. Plant Physiol 113: 903–911
Ribas-Carbo M, Robinson SA, Gonzalez-Meler MA, Lennon AM, Giles L, Siedow JN and Berry JA (2000a) Effects of light on respiration and oxygen isotope fractionation in soybean cotyledons. Plant Cell Environ 23: 983–989
Ribas-Carbo M, Aroca R, Gonzalez-Meler MA, Irigoyen JJ and Sanchez-Diaz M (2000b) The electron partitioning between the cytochrome and alternative respiratory pathways during chilling recovery in two cultivars of maize differing in chilling sensitivity. Plant Physiol 122: 199–204
Riedo M, Grub A, Rosset M. and Fuhrer J (1998) A pasture simulation model for dry matter production and fluxes of carbon, nitrogen, water and energy. Ecol Model 105: 141–183
Robinson SA, Ribas-Carbo M, Yakird D, Giles L, Reuveni Y and Berry JA (1995) Beyond SHAM and cyanide — opportunities for studying the alternative oxidase in plant respiration using oxygen-isotope discrimination. Aust J Plant Physiol 22: 487–496
Ryan MG (1991) A simple method for estimating gross carbon budgets for vegetation in forest ecosystems. Tree Physiol 9: 255–266
Ryan MG (1995) Foliar maintenance respiration of subalpine and boreal trees and shrubs in relation to nitrogen content. Plant Cell Environ 18: 765–772
Ryan MG, Hubbard RM, Pongracic S, Raison RJ and McMurtrie RE (1996) Foliage, fine-root, woody tissue and stand respiration in *Pinus radiata* in relation to nitrogen status. Tree Physiol 16: 333–343
Scheurwater I, Cornelissen C, Dictus F, Welschen R and Lambers H (1998) Why do fast- and slow-growing grass species differ so little in their rate of root respiration, considering the large differences in rate of growth and ion uptake? Plant Cell Environ 21: 995–1005
Scheurwater I, Clarkson DT, Purves JV, Van Rijt G, Saker LR, Welschen Rand Lambers H (1999) Relatively large nitrate efflux can account for the high specific respiratory costs for nitrate transport in slow-growing grass species. Plant Soil 215: 123–134
Scheurwater I, Dunnebacke M, Eising R and Lambers H (2000) Respiratory costs and rate of protein turnover in the roots of a fast-growing (*Dactylis glomerata* L.) and a slow-growing (*Festuca ovina* L.) grass species. J Exp Bot 51: 1089–1097
Shinano T, Osaki M, Tadano T (1996) Problems in the methods of estimation of growth and maintenance respiration. Soil Sci Plant Nutr 42: 773–784
Spitters CJT, Van Kraalingen DWG and Van Keulen H (1989) A simple and universal crop simulator; SUCROS87. In: Rabbinge R, Ward SA and van Laar HH (eds) Simulation and Systems Management in Crop Protection, pp 147–181. PuDOC, Wageningen
Thomas RB and Griffin KL (1994) Direct and indirect effects of atmospheric carbon-dioxide enrichment on leaf respiration of Glycine-max (L) merr. Plant Physiol 104: 355–361
Thomas RB, Reid CD, Ybema R and Strain BR (1993) Growth and maintenance components of leaf respiration of cotton grown in elevated carbon dioxide partial pressure. Plant Cell Environ 16: 539–546
Thornley JHM (1970) Respiration, growth and maintenance in plants. Nature 227: 304–305
Thornley JHM (1971) Energy, respiration and growth in plants. Ann Bot 35: 721–728
Thornley JHM and Cannell MGR (1996) Forest response to elevated [CO_2], temperature and nitrogen supply, including water dynamics: model generated hypotheses compared with observations. Plant Cell Environ 19: 1131–1138
Thornley JHM and Cannell MGR (2000) Modelling the com-

ponents of plant respiration: representation and realism. Ann Bot 85: 55–67

Tjoelker MG, Oleksyn J and Reich PB (1999) Changes in leaf nitrogen and carbohydrates underlie temparture and CO2 acclimation of dark respiration in five boreal tree species. Plant Cell Environ 22: 767–778

Tjoelker MG, Oleksyn J, Lee TD and Reich PB (2001) Direct inhibition of leaf dark respiration by elevated CO2 is minor in 12 grassland species. New Phytol 150: 419–424

Tollrian R and Harvell CD (1999) The evolution of inducible defenses: Current ideas. In: Tollrian R and Harvell CD (eds) The Ecology and Evolution of Inducible Defenses, pp 306–321. Princeton University Press, Princeton

Van Dam NM and Baldwin IT (2001) Competition mediates costs of jasmonate-induced defences, nitrogen acquisition and transgenerational plasticity in *Nicotiana attenuata*. Funct Ecol 15: 406–415

Van der Werf A. Kooijman A, Welschen R and Lambers H (1988) Respiratory energy costs for maintenance of biomass, for growth and for ion uptake in roots of *Carex diandra* and *Carex acutiformis*. Physiol Plant 72: 483–491

Van der Werf A, Van den Berg G, Ravenstein HJL, Lambers H and Eising R (1992) Protein turnover: A significant component of maintenance in roots? In: Lambers H and Van der Plas LHW (eds) Molecular, Biochemical and Physiological Aspects of Plant Respiration, pp 483–497. SPB Academic publishing, The Hague

Veen BW (1980) Energy costs of ion transport. In: Rains DW, Valentine RC and Hollaender A (eds) Genetic Engineering of Osmoregulation. Impact on Plant Productivity for Food, Chemicals and Energy, pp 187–195. Plenum Press, New York

Vertregt N and Penning de Vries FWT (1987) A rapid method for determining the efficiency of biosynthesis of plant biomass. J Theor Biol 128: 109–119

Vierstra RD (1993) Protein degradation in plants. Ann Rev Plant Physiol Plant Mol Biol 44: 385–410

West GB, Brown JH and Enquist BJ (1997) A general model for the origin of allometric scaling laws in biology. Science 276: 122–126

West GB, Brown JH and Enquist BJ (1999) A general model for the structure and allometry of plant vascular systems. Nature 400: 664–667

West GB, Brown JH and Enquist BJ (2001) A general model for genetic growth. Nature 413: 628–631

Wohl K and James WO (1942) The energy change associated with plant respiration. New Phytol 41: 230–256

Zagdanska B (1995) Respiratory energy demand for protein turnover and ion transport in wheat leaves upon water deficit. Physiol Plant 95: 428–436

Zerihun A, McKenzie BA and Morton JD (1998) Photosynthate costs associated with the utilization of different nitrogen-forms: influence on the carbon balance of plants and shoot-root biomass partitioning. New Phytol 138: 1–11

Chapter 11

Respiratory/Carbon Costs of Symbiotic Nitrogen Fixation in Legumes

Frank R. Minchin* and John F Witty
Institute of Grassland and Environmental Research, Plas Gogerddan, Aberystwyth, Ceredigion SY23 3EB, U.K.

Summary

This chapter presents an overview of the respiratory/carbon costs of symbiotic nitrogen fixation. The various theoretical costings for nitrogen fixation suggest that respiration directly associated with nitrogenase activity will require between 1.77 and 3.01 g C g^{-1}-N (4.35–7.00 mol CO_2 mol^{-1} N_2), while respiration of the entire nitrogen–fixing nodules will require between 2.78 and 4.81 g C g^{-1}-N (6.51–11.19 mol CO_2 mol^{-1} N_2). Early attempts to measure these costs were beset by methodological problems, but some reliable approaches were developed. Measured values based on root respiration during the period of active nitrogen fixation are in the range of 5–10 g C g^{-1}-N (11.6–23.4 mol CO_2 mol^{-1} N_2), with an average value of 6.5 g C g^{-1}-N (15.1 mol CO_2 mol^{-1} N_2). On a nodule basis, values in the range of 3–5 g C g^{-1}-N (7–12 mol CO_2 mol^{-1} N_2) appear to represent the 'normal' for legume nodules, while values below about 2.5 g C g^{-1}-N are likely to be erroneous. The implications of these costings are considered in terms of the need for legumes to carefully regulate nitrogen fixation and the requirement for such regulation systems to be operational in any novel nitrogen–fixing plants.

*author for correspondence, email: frank.minchin@bbsrc.ac.uk

H. Lambers and M. Ribas-Carbo (eds.), Plant Respiration, 195–205.

I. Introduction

Symbiotic nitrogen fixation in legumes involves a remarkable mutualism between legume plants and a range of soil-borne diazotrophic bacteria, collectively known as rhizobia (see Gallon and Chaplin, 1987; Sprent and Sprent, 1990 for details). A dialogue of chemical signals between rhizobia and plant allows the bacteria to enter the roots of the host without triggering pathogenic defense mechanisms. The infected roots respond by forming nodules, the bacteria then enter into the central region of these outgrowths and rapidly multiply inside the infected cells to form numerous symbiosomes. These consist of one or more bacteria, which are separated from the host plant cytosol by means of a specialized peribacteroid membrane. At this stage the bacteria are referred to as bacteroids and, if the process has worked effectively, will have differentiated into forms that produce large quantities of nitrogenase; the enzyme system responsible for nitrogen fixation. As a result of this differentiation process the bacteroids are unable to use most of the ammonia produced by nitrogen fixation, which is then exported into the host cell cytosol. At the same time the bacteroids are completely dependent on the host cell for their supply of carbon compounds and all other nutrients. Thus, the legume nodule represents a symbiotic relationship in which large numbers of bacteria receive protection and sustenance from a host plant, which, in turn, receives fixed nitrogen.

However, there is a problem that must be overcome by the symbiosis. Before N_2 gas can be reduced, the triple bond between the two atoms must be broken; an energy-intensive reaction, requiring at least 160 kcal mol^{-1} N_2. The industrial Haber-Bosch process of nitrogen fixation requires both heat (300–400 °C) and pressure (35–100 MPa) and consumes fossil fuels (normally natural gas) to achieve this reaction. However, the symbiotic process must proceed at normal levels of pressure and temperature and utilize photosynthates produced by the host plant as the energy source.

During the 1970s and 1980s there was a great deal of interest in quantifying the costs of symbiotic nitrogen fixation in terms of respiration and carbon use. This interest was primarily driven by two questions; (a) do these costs reduce yield, in comparison with nitrogen-fertilized plants? and (b) could the carbon costs be reduced, allowing for greater yield production? The former question was also relevant to the 'holy grail' of nitrogen-fixation research; the production of cereal crops that were largely or completely dependent on biological nitrogen fixation.

These two decades also saw a plethora of reviews on the energetics and carbon costs of nitrogen fixation. Indeed, the research area came close to achieving the dubious distinction of attracting more review articles than original papers. However, since this heyday interest in this topic has waned which is somewhat surprising given the importance of energy production to biological nitrogen fixation. The reduction in interest can probably be understood in terms of the need for increasingly specialized equipment to make meaningful measurements, and also the movement of key personnel into other areas of research.

Given this background we have decided to produce a relatively short overview of the respiratory/carbon costs of symbiotic nitrogen fixation and to then consider the implications of these costs in the light of current ideas on nitrogen fixation in legumes.

II. Respiratory/Carbon Costs of Nitrogen Fixation

A. Theoretical Costs

The first problem with discussing carbon costs is the choice of units. Previous publications have used a miscellany of units; representing carbon by weight (mg or g), as mol CO_2 or as mol carbohydrate, with nitrogen being quantified by weight, as mol N, as mol N_2 or as mol NH_3. Nitrogen fixation has also been measured indirectly, and expressed in terms of acetylene reduction (mol C_2H_2) or hydrogen production (mol H_2). This situation is further complicated by comparisons between nitrogen fixation and nitrate reduction, as one mol N_2 contains twice as much nitrogen as one mol NO_3^-. To minimize this confusion we have decided to re-calculate previously published costs in terms of g C g^{-1}-N, with important values also expressed in terms of mol CO_2 mol^{-1} N_2.

The theoretical costs of nitrogen fixation by legumes have been previously reviewed by Minchin et al. (1981), Pate et al. (1981), Schubert (1982), Atkins (1984), Neves and Hungria (1987), Layzell et al. (1988), Schulze et al. (1994) and others. All these

Abbreviations: ARA – acetylene reduction assay; EAC – electron-allocation coefficient for nitrogen; HUP – uptake hydrogenase; PEP – phosphoenolpyruvate; PEPC – phosphoenolpyruvate carboxylase

authors agree that at least one mol H_2 is produced for each mol N_2 reduced (see below) so that, at its most efficient, biological nitrogen fixation by nitrogenase follows the overall equation:

$$N_2 + 8H^+ + 8e^- + 16ATP \rightarrow 2NH_3 + H_2 + 16ADP$$

giving a theoretical costing mol^{-1} N_2 of 16 mol ATP and 4 mol reductants to produce $8e^-$. Converting from mol ATP into a carbon cost will depend on the efficiency of the oxidative phosphorylation process; that is the P/O ratio. This value is normally assumed to be 3 for higher plant cells, based upon the use of the cytochrome path (giving 36 mol ATP mol^{-1} glucose), but may be as low as 2 for bacteria (giving 24 mol ATP mol^{-1} glucose). Thus, the cost of ATP production will be 0.44 to 0.67 mol glucose mol^{-1} N_2 (1.13–1.79 g C g^{-1}-N), for a P/O ratio of 3 or 2, respectively. To this must be added 0.33 mol glucose for the production of the reductants, giving an overall theoretical cost for nitrogenase activity of 1.98 to 2.57 g C g^{-1}-N.

The above equation assumes an electron-allocation co-efficient to N_2 (EAC) of 0.75 (i.e. 75% of the total electron flow to nitrogenase is used in N_2 reduction), giving 1 mol H_2 mol^{-1} N_2 fixed. Measured values of EAC for legume nodules range from 0.4 to 0.7, although more recent estimates cover the range 0.6 to 0.7 (Hunt and Layzell, 1993). An EAC of 0.6 increases H_2 production to 2 mol mol^{-1} N_2 fixed, with a cost increase of 2 mol ATP plus 1 reductant, equal to 0.36–0.44 g C g^{-1}-N (P/0 3 or 2, respectively). For an EAC of 0.7, the additional carbon costs are 0.10 to 0.12 g C g^{-1}-N. Then again, in some symbiotic systems, the H_2 produced by nitrogenase can be recycled by an uptake hydrogenase (HUP), with the potential generation of 2–3 mol ATP mol^{-1} H_2 (again, depending on the P/O ratio) giving a saving of 0.08 mol glucose mol^{-1} N_2 (0.21 g C g^{-1}-N). Thus, variations in EAC and the presence of HUP can extend the range of theoretical costs to 1.87–3.01 g C g^{-1}-N, for P/O ratios of 3 or 2, respectively.

This range of costings is, of course, only for the nitrogenase reaction, and does not take into account the costs of ammonium assimilation and export. These are also variable because some legume nodules assimilate ammonia into the amide asparagine, while others produce the ureides allantoate and allantoic acid. Theoretical costs, for amide and ureide production, respectively, have been calculated as 0.4 or 0.5 g C g^{-1}-N (Atkins, 1984), 0.1 or 0.5 g C g-1-N (Neves and Hungria, 1987) and 0.9 or 1.5 g C g-1-N (Layzell et al., 1988), with an additional 0.25 to 0.3 g C g-1-N for transport costs (Atkins, 1984; Layzell et al., 1988). Using the most recent estimates of Layzell et al. (1988) increases the theoretical nodule respiratory costs to between 2.92 and 4.76 g C g^{-1}-N.

Another complication to respiratory measurements is the anaplerotic production of oxalacetate from phosphoenolpyruvate (PEP) and CO_2 by the enzyme phosphoenolpyruvate carboxylase (PEPC). This is now recognized as a central reaction in the carbon metabolism of nodules because oxalacetate is an essential intermediate in the production of both malate (the primary carbon source for bacteroids; Urdvardi and Day, 1997) and asparagine. Therefore, when asparagine is the main product of ammonia assimilation some of the CO_2 produced from bacteroid respiration associated with nitrogenase activity will be consumed by PEPC, resulting in a reduced CO_2 output from the nodules. This reduction was calculated as 0.56 mol CO_2 mol^{-1} N_2 fixed (0.24 g C g^{-1}-N) compared to ureide-producing nodules (Layzell et al., 1988).

Taken together, these various theoretical costings suggest that respiration directly associated with nitrogenase activity (nitrogenase-linked respiration, NLR) should cost between 1.87 and 3.01 g C g-1-N (4.35–7.00 mol CO_2 mol^{-1} N_2), while respiration of nitrogen-fixing nodules should cost between 2.78 and 4.81 g C g-1-N (6.51–11.19 mol CO_2 mol^{-1} N_2), depending on P/O ratio, EAC, HUP, export product, transport costs and PEPC involvement. The situation is summarized in Table 1.

The final theoretical cost to be added for nitrogen fixation by legume nodules is that associated with growth and maintenance. This has been calculated as between 0.5 and 1.8 g C g^{-1}-N (Pate et al., 1981; Schubert, 1982; Atkins, 1984; Layzell et al., 1988), but is unlikely to make a significant contribution to nodule respiration during short-term measurements (Layzell et al., 1988).

B. Measured Costs

Given the range of theoretical costings shown in Table 1 it is not surprising that a major effort was made to obtain actual measurements of the carbon costs of nitrogen fixation. However, with the benefit of hindsight it is possible to see that the 1970s was a methodological nightmare. Thus, nitrogen fixation

Table 1. Theoretical carbon costs of nitrogen fixation (g C g^{-1}-N)

Parameter	Range	Variable
Nitrogenase activity	1.98 to 2.57	P/O ratio
Additional H_2 production	0.10 to 0.44	EAC
H_2 re-cycling	–0.21 to 0	Presence of HUP
Total for nitrogenase	1.87 to 3.01	
Ammonia assimilation	0.90 to 1.50	Assimilation product
Transport	0.25 to 0.30	
CO_2 uptake	–0.24 to 0	PEPC involvement
Total for nodule respiration	2.78 to 4.81	

was measured either directly as N increment or ^{15}N accumulation, or indirectly by the acetylene reduction assay (ARA), while respiration was almost invariably measured as CO_2 efflux, but from a variety of organs; attached roots, detached roots and detached nodules. By the early 1980s published values for carbon costs ranged from 1.1 to 19.4 g C g-1-N and several reviews began to question the validity of these measurements (Minchin et al., 1981; Pate et al., 1981, Schubert, 1982).

These concerns were justified within a few years by the recognition of errors in the ARA (Minchin et al., 1983), which resulted from the acetylene-induced closure of a variable barrier, which controls oxygen diffusion into the nodule (Witty et al., 1984). This barrier is also closed by root disturbance (Minchin et al., 1986b) and, especially, nodule detachment (Sheehy, 1987). The recognition of these errors has a profound impact on the interpretation of published measurements of carbon costs; for example, Schubert (1982) listed 30 published costings, but discarding those measurements which used ARA, disturbed roots or detached nodules leaves only 9 apparently valid costings, all of which relate to the respiration of attached nodulated roots.

1. Carbon Costings on a Root Respiration Basis

The measurement of carbon costs by comparing nodulated root respiration and whole plant N increments over an extended period of time seems to be a valid approach, which produced a consistent series of values during the 1970s and 80s (Table 2). Although large ranges were reported in some cases these could be explained as early or late periods during the plant's growth cycle when roots were active but nitrogen fixation activity was low (e.g., Ryle et al., 1978). Values obtained during the period of active nitrogen fixation are in the range of 5–10 g C g^{-1}-N (11.6–23.4 mol CO_2 mol^{-1} N_2), with an average value of 6.5 g C g^{-1}-N (15.1 mol CO_2 mol^{-1} N_2).

However, these values include growth and maintenance (G and M) respiration of both roots and nodules, and are still somewhat removed from the carbon costing on a nodule basis. Nevertheless, subtracting the theoretical costs for G and M respiration (2.2 g C g^{-1}-N, Schubert, 1982) reduces the costings to a range of c. 3–8 g C g^{-1}-N and a mean of 4.3 g C g^{-1}-N (10 mol CO_2 mol^{-1} N_2). This agrees well with values obtained on a nodule basis.

2. Nodule-Based Respiratory Costs of Nitrogen Fixation

Despite the rejection of many early nodule-based measurements, because of errors associated with closure of the oxygen diffusion barrier, 4 robust measurement techniques were developed in the 1980s (Table 3). Thus, at a discussion session held at the 5th International Symposium on Nitrogen Fixation the participants were able to agree that values in the range of 3–5 g C g^{-1}-N (7–12 mol CO_2 mol^{-1} N_2) represented the 'normal' for legume nodules, while values below about 2.5 g C g^{-1}-N were likely to be erroneous (Minchin, 1983).

The 4 early techniques listed in Table 3 all employ very different approaches. Thus, Ryle's laboratory measured the respiration of attached nodulated roots before and after the removal of a nodule sub-sample, Pate's laboratory enclosed attached nodules within micro-respirometers, and Warembourg's laboratory employed simultaneous ^{14}C and ^{15}N labeling of whole plants. All these approaches have limitations in terms

Table 2. Root respiration costs of nitrogen fixation (g C g^{-1}-N)

Species	Value	Reference
Soybean	6.0 to17.0	Ryle et al., 1978
	6.3	Ryle et al., 1979a
	5.2	Finke et al., 1982
	5.9 to18.8	Warembourg, 1983
	5.8	Patterson and LaRue, 1983
Cowpea	6.0	Herridge and Pate, 1977
	5.4	Layzell et al., 1979
	6.8	Ryle et al., 1979a
	8.9	Minchin et al., 1980
	6.1 to 7.0	Neves et al., 1981
White lupin	4.0 to 6.5	Pate, 1973
	6.9	Layzell et al., 1979
	10.2	Pate et al., 1979a
	7.4	Pate et al., 1979b
Pea	5.9	Minchin and Pate, 1973
White clover	6.6	Ryle et al., 1979a
Pigeonpea	5.0 to15.0	Rao et al., 1984
Chickpea	7.6 to14.3	Hooda et al., 1990

of either labor intensity or costs of measuring equipment and the only technique that attracted attention outside the developer's laboratory is that of Witty et al. (1983).

This approach determines the linear regression between CO_2 efflux and nitrogenase activity (measured as acetylene reduction or H_2 production under argon), which is produced during closure of the oxygen diffusion barrier, or changes in external oxygen concentration, and uses the slope of the regression as a measure of carbon costs. It has the advantage of being valid with disturbed systems, including detached nodules, but has the disadvantage of requiring that both CO_2 efflux and nitrogenase activity are measured simultaneously in a flow-through gas system. In terms of nodule-based carbon cost measurements it provides the closest approximation to the respiration associated directly with nitrogenase activity. This is because the use of acetylene or argon prevents respiration related to ammonia assimilation, and CO_2 uptake by the associated PEPC activity. However, ARA and H_2 under argon are not direct measurements of N_2 fixation and accurate conversions require either calibration of the ARA with $^{15}N_2$ or determination of EAC for conversion of the H_2 data. For comparative purposes in this review all calculations have been made using the theoretical conversion factor of 4 mol C_2H_4 or H_2 mol^{-1} N_2.

The most recent method developed in Schulze's laboratory is based on a comparison of root respiration from nodulated and nitrogen-fed roots with the assumption that growth respiration per unit assimilated N will be the same in both root types. Initial results from this approach were quite encouraging (Schulze et al., 1994), but more recent values have been lower (Schulze et al., 1999) and the method still awaits verification.

3. Variations in Carbon Costs Due to Environmental Stress and Genotypic Differences

The linear regression method has also been used in a number of laboratories to measure effects of environmental stress on carbon costs, with variable results. Water stress was found to increase carbon costs in soybean (Durand et al., 1987) and subterranean clover (Davey and Simpson, 1990), but not peas (Gonzalez et al., 1998), while nitrate application was found by some authors to increase carbon costs (Minchin et al., 1986a; Faurie and Soussana, 1993; Escuredo et al., 1996; Arrese-Igor et al., 1997; Matamoros et al., 1999), but other authors reported no effects (Schuller et al., 1988; Gordon et al., 1989; Minchin et al., 1989). The one consistent result has been a lack of effect by treatments that directly affect photosynthate supply, such as stem girdling, low root temperature, defoliation or prolonged darkness (Walsh et al., 1987; Gordon et al., 1989, 1990; Matamoros et al., 1999).

It was originally hoped that measurements of

Table 3. Nodule-based respiratory costs of nitrogen fixation (g C g^{-1}-N)

Method	Species	Value	Reference
Removal of nodule sub-sample	Cowpea	3 to 4	Ryle et al., 1979a
	Soybean	3 to 5	Schubert and Ryle, 1980
	Soybean	3.3 to 3.7	Ryle et al., 1984
Micro-respirometers	Cowpea	3.3	Pate et al., 1981
	Lupin	3.6	Pate et al., 1981
	Soybean	2.9	Rainbird et al., 1984
Simultaneous $^{14}C/^{15}N$ labeling	Soybean	2.5 to 7.6	Warembourg, 1983
	Red clover	3.2	Fernandez and Warembourg, 1987
	Soybean	4.0	Warembourg and Roumet, 1989
Linear regressions of CO_2 vs ARA/H_2	Various	3.4 to 8.4	Witty et al., 1983
	Soybean	3.6	Ryle et al., 1984
	Soybean	3.4	Rainbird et al., 1984
	Soybean	5.1	Heytler and Hardy, 1984
	Soybean	2.5	Kanamori et al., 1984
	Soybean	3.7	Walsh et al., 1987
	Various	1.9 to 3.7	Davey and Simpson, 1988
	Pea	2.9 to 3.6	Rosendahl, 1988
	Subterranean clover	3.4	Davey and Simpson, 1990
	Acacia	2.5 to 6.0	Sun et al., 1992
Calculations based on root respiration	Pea	3.5	Schulze et al., 1994
	Faba bean	4.0	
	Pea	2.0	Schulze et al., 1999
	Faba bean	2.9	

environmental stress on carbon costs would provide some insights into the physiological changes involved. However, as noted by Faurie and Soussana (1993), the linear regression method does not directly measure nitrogenase-linked respiration within the bacteroids, but measures the respiration of all the central plant and bacterial tissue within the oxygen diffusion barrier. Thus, interpretation of these measurements in terms of physiological or biochemical changes is problematic.

In the early 1980s there was hope that varying the carbon costs of nitrogen fixation could lead to increases in plant growth. However, variations in carbon costs due to pea genotype (Witty et al., 1983) were not reflected in plant growth (J. F. Witty and A. Cresswell, unpublished) while Skøt et al. (1986) found that increasing carbon costs of peas by using different rhizobial strains actually increased shoot dry weight. Similar positive or neutral correlations have been reported for supernodulating mutants of soybean and pea (Schuller et al., 1988; Rosendahl et al., 1989) and for lucerne (Twary and Heichel, 1991). Thus, Vance and Heichel (1991) were able to state that the carbon costs of nitrogen fixation did not appear to be directly related to legume productivity. This assertion has recently been challenged by Schulze et al. (2000) and Adgo and Schulze (2002) who found negative correlations between carbon costs and plant growth in peas and faba beans.

4. Carbon Costs of Nitrogen Fixation Versus Nitrate Reduction

The theoretical carbon costs of nitrate reduction to ammonia are lower than those of nitrogen fixation (0–1.5 g C g^{-1}-N; Atkins, 1984), with the zero value relating to the situation where nitrate reduction is confined to leaves and is able to directly utilize photosynthetically produced reductants (although this could incur extra costs in relation to pH balancing; Raven, 1985). Allowing for the cost of ammonia assimilation and transport increases these values to 0.8–2.4 g C g^{-1}-N, which is still lower than the range of 2.8–4.8 g C g^{-1}-N calculated for nitrogen fixation in nodules (Table 1). Thus, legume physiologists have been fascinated by the question as to whether these theoretical differences can be measured in terms of root respiration or plant growth.

For comparisons of root respiration most authors have reported higher values for nodulated roots (Ryle et al., 1978, 1979b, 1983; Pate et al., 1979a; Haystead et al., 1980; Atkins et al., 1980; Finke et al., 1982) with differences from nitrate-fed roots in the order of 10-15% on a specific respiration basis. However, there have been occasional reports of nitrate-fed roots having higher respiration rates than their nodulated counterparts (Lambers et al., 1980; de Visser and Lambers, 1983). Nevertheless, as noted by Atkins (1984), interpretation of many of these measurements is complicated by differing rates of nitrogen accumulation, uncertainty about the siting of nitrate reduction and possible changes in the energetics of other plant functions. Indeed, the measured carbon costs of nitrate reduction by legume roots (3.7, Minchin et al., 1980; 3.9, Neves et al., 1981) are higher than the theoretical value of 2.4 g C g^{-1}-N.

Comparative growth measurements of nitrogen-fixing and nitrate-reducing plants have given more diverse results, with some authors reporting no difference in growth (Gibson, 1966; Pate et al., 1979a; Atkins et al., 1980; Lambers et al., 1980; de Visser, 1985), and others reporting greater growth by nitrate-fed plants (Ryle et al., 1981; Finke et al., 1982; de Visser and Lambers, 1983; Arnott, 1984; Silsbury, 1984). Such increased growth may be related to the need for greater carbon sequestering in nodulated root development, to the detriment of leaf production, during the early vegetative stages (Mahon and Child, 1979; Minchin et al., 1980; Neves et al., 1981). However, it is also possible that all these growth measurements have a more fundamental flaw in that comparisons of nitrogen nutrition regimes have been conducted in controlled environments in which conditions were more-or-less identical for both plant types. This assumes that all other environmental parameters (i.e. water and nutrient supply, temperature, light quality and quantity) are either optimal or equally sub-optimal for the growth of both nitrogen-fixing and nitrate-fed plants; an assumption which has not been exhaustively tested.

Furthermore, the agricultural relevance of these experiments must be questioned as, under field conditions, legumes can avail of both fixed and combined nitrogen and the energy costs of nitrogen fixation do not appear to be a significant factor in terms of plant productivity (Herridge and Brockwell, 1988). Then again, with grain legumes there appears to be little correlation between nitrate supply and seed yield (Schubert and Ryle, 1980; Minchin et al., 1981).

III. Implications of High Carbon Costs

A. The Need to Regulate Nitrogen Fixation

From the above it is clear that both nitrogen fixation and nitrate reduction are energy-intensive processes with a poisonous end-product (ammonia). Thus, in both cases, there is a need for the plant to carefully regulate these processes. With nitrate reduction this can be done at the levels of both gene expression for nitrate reductase and substrate (nitrate) uptake (Stitt et al., 2002). However, with nitrogen fixation the expression of the nitrogenase gene is largely under control of the bacteroids, while it is probably impossible to prevent the egress of N_2 gas into living biological tissue under normal atmospheric conditions. Thus, the legumes have to find alternative means to regulate nitrogen fixation activity in the nodules.

One approach is to regulate carbon metabolism within nodules and there is now evidence that this occurs with both sucrose synthase (Gordon et al., 1997) and PEPC (Wadham et al., 1996; Woo and Xu, 1996). However, nitrogen fixation is a highly reductive process, which requires microaerobic conditions to be produced by restricting O_2 movement into the central infected zone of the nodules (Bergersen and Goodchild, 1973; Tjepkema and Yokum, 1973). If O_2 flow was restricted to a constant rate then any reduction in bacteroid respiration due to reduced carbon metabolism within the nodule would produce an increase in central zone oxygen concentration, leading to damage of the nitrogen-fixing system. To counteract this problem the nodule has developed a variable oxygen-diffusion barrier, which can regulate O_2 flow so as to match the requirements of carbon metabolism within the infected cells (Witty et al., 1986). The nature of this variable barrier appears to be highly complex, with components in most zones of the nodule (Minchin, 1997). Closure of the oxygen-diffusion barrier has been shown to parallel alterations in carbon metabolism under some stress conditions, such as drought or shoot removal, but under other, non-stress conditions, (e.g., nitrate application to the plant) closure of the barrier proceeds reductions in carbon metabolism (Gordon et al., 2002).

Although the mechanisms regulating nitrogen fixation reside within the nodules there is now considerable evidence that the overall control process resides within the shoot of the host plant and operates through a nitrogen feed-back mechanism similar to that involved in nitrate-uptake regulation (Stitt et al., 2002). This N feed-back hypothesis suggests that when the shoot receives more N than it can use in growth some of this excess N 'spills over' into the phloem where it is translocated to the nodulated roots and interacts with the nodule physiology so as to reduce nitrogen fixation. Much of the evidence for this feed-back regulation is circumstantial (see reviews by Hartwig, 1998; Serraj et al., 1999; 2001), but one piece of direct evidence has been provided by Neo and Layzell (1997). These authors fed ammonia gas to the shoots of soybean and lupin at a concentration that substantially increased the levels of amino compounds in the phloem sap (particularly glutamine) without inhibiting photosynthesis. Within 4 hours they observed an inhibition of nitrogen fixation activity, which operated through closure of the oxygen-diffusion barrier.

The N feed-back hypothesis of nitrogen fixation regulation is embarrassing for researchers at 2 levels. At the physiological/biochemical level, despite several decades of research on legumes we have no clear understanding as to how variations in phloem amino compounds could interact with the nodule's carbon metabolism, the oxygen-diffusion barrier, or even with N flows within the nodule. This is especially true for amide transporting legumes where it is difficult to envisage how nodular activity could be regulated by the return of amides in the phloem sap. At the agronomic level, research on nitrogen fixation has been justified on the basis of it being a limiting factor for legume growth and yield. The N feed-back hypothesis implies that nitrogen fixation is regulated by the N demand of the shoot; i.e. nitrogen-fixation activity is limited by the growth rate of the legume. However, growth under water deficit conditions may be an exceptional case as the nodules appear to be hypersensitive to this stress and fixation is reduced prior to any effects on photosynthesis (Durand et al, 1987).

B. Implications for Novel Nitrogen-fixing Plants

Given the need for high levels of nitrogen-fertilizer application to cereal crops it has long been an aspiration to transfer nitrogen-fixation activity, at levels found in the legumes, to the monocotyledonous cereals. With continuing advances in molecular biology it seems likely that this transfer of nitrogenase activity will one day be achieved, through either the generation of novel symbiotic systems or the direct transfer of the nitrogenase and associated genes, along with the appropriate microaerobic conditions. However, such an impressive feat of manipulation will still leave major problems to be overcome. In addition to the exacting O_2 requirements of nitrogenase, the high carbon costs associated with nitrogen-fixation activity will require the novel plants to divert large amounts of carbon resources to the sites of nitrogen fixation while carbon will also be needed for the utilization or storage of the potentially poisonous fixation products. Unless the plants can regulate this novel nitrogenase activity it is likely to be fatal.

Thus, it seems likely that any novel nitrogen-fixing plants will also need to develop a feed-back regulatory system similar to that of legumes. It is possible that this could be achieved by modifications of the nitrate-uptake regulation system, as the feed-back component may be similar to that of fixation regulation (Soussana et al., 2002). However, until we understand more about the regulation within legumes it will be impossible to predict what modifications will be necessary. A possibly less complex and more practical route to obtaining nitrogen-fixing cereals would be the use of molecular techniques to alter the seed quality of grain legumes.

References

Adgo E and Schulze J (2002) Nitrogen fixation and assimilation efficiency in Ethiopian and German pea varieties. Plant Soil 239: 291–299

Arnott RA (1984) An analysis of the uninterrupted growth of white clover swards receiving either biologically fixed nitrogen or nitrate in solution. Grass Forage Sci 39: 305–310.

Arrese-Igor C, Minchin FR, Gordon AJ and Nath AK (1997) Possible causes of the physiological decline in soybean nitrogen fixation in the presence of nitrate. J Exp Bot 48: 905–913

Atkins CA (1984) Efficiencies and inefficiencies in the legume/Rhizobium symbiosis—A review. Plant Soil 82: 273–284

Atkins CA, Pate JS, Griffiths GL and White ST (1980) Economy of carbon and nitrogen in nodulated and non-nodulated (NO_3-grown) cowpea [*Vigna unguiculata* (L.) Walp.]. Plant Physiol 66: 978–983

Bergersen FJ and Goodchild DJ (1973) Aeration pathways in soybean root nodules. Aust J Biol Sci 26: 729–740

Davey AG and Simpson RJ (1988) Nitrogenase activity by subterranean clover and other pasture legumes. Aust J Plant Physiol 15: 657–667

Davey AG and Simpson RJ (1990) Nitrogen fixation by subterranean clover at varying stages of nodule dehydration. III. Efficiency of nitrogenase functioning. J Exp Bot 41: 1189–1197

de Visser R (1985) Efficiency of respiration and energy requirements of N assimilation in roots of *Pisum sativum*. Physiol Plant 65: 209–218

de Visser R and Lambers H (1983) Growth and the efficiency of root respiration of *Pisum sativum* as dependent on the source of nitrogen. Physiol Plant 58: 533–543

Durand J-L, Sheehy JE and Minchin FR (1987) Nitrogenase activity, photosynthesis and nodule water potential in soyabean plants experiencing water deprivation. J Exp Bot 38: 311-321

Escuredo PR, Minchin FR, Gogorcena Y, Iturbe-Ormaetxe I, Klucas RV and Becana M (1996) Involvement of activated oxygen in nitrate-induced senescence of pea root nodules. Plant Physiol 110: 1187–1195

Faurie O and Soussana J-F (1993) Oxygen-induced recovery from short-term nitrate inhibition of N_2 fixation in white clover plants from spaced and dense stands. Physiol Plant 89: 467–475

Fernandez MP and Warembourg FR (1987) Distribution and utilization of assimilated carbon in red clover during the first year of vegetation. Plant Soil 97: 131–143

Finke RL, Harper JE and Hageman RH (1982) Efficiency of nitrogen assimilation by N_2-fixing and nitrate-grown soybean plants (*Glycine max* [L.] Merr.). Plant Physiol 70: 1178–1184

Gallon JR and Chaplin AE (1987) An Introduction to Nitrogen Fixation. Cassell, London

Gibson AH (1966) The carbohydrate requirements for symbiotic nitrogen fixation: A 'whole plant' growth analysis approach. Aust J Biol Sci 19: 499–515

Gonzalez EM, Aparicio-Tejo PM, Gordon AJ, Minchin FR, Royuela M and Arrese-Igor C (1998) Water-deficit effects on carbon and nitrogen metabolism of pea nodules. J Exp Bot 49: 1705–1714

Gordon AJ, Macduff JH, Ryle GJA and Powell CE (1989) White clover N_2-fixation in response to root temperature and nitrate. II. N_2-fixation, respiration and nitrate reductase activity. J Exp Bot 40: 527–534

Gordon AJ, Kessler W and Minchin FR (1990) Defoliation-induced stress in nodules of white clover. I. Changes in physiological parameters and protein synthesis. J Exp Bot 41: 1245–1253

Gordon AJ, Minchin FR, Skøt L and James CL (1997) Stress-induced declines in soybean N_2 fixation are related to nodule sucrose synthase activity. Plant Physiol 114: 937–946

Gordon AJ, Skøt L, James CL and Minchin FR (2002) Short-term metabolic responses of soybean root nodules to nitrate. J Exp Bot 53: 423–428

Hartwig UA (1998) The regulation of symbiotic N_2 fixation: A conceptual model of N feedback from the ecosystem to the gene expression level. Perspect Plant Ecol Evol Syst 1: 92–120

Haystead A, King J, Lamb WIC and Marriott C (1980) Growth and carbon economy of nodulated white clover in the presence and absence of combined nitrogen. Grass Forage Sci 35: 153–158

Herridge DF and Brockwell J (1988) Contributions of fixed nitrogen and soil nitrate to the nitrogen economy of irrigated soybean. Soil Biol Biochem 20: 711–717

Herridge DF and Pate JS (1977) Utilization of net photosynthate for nitrogen fixation and protein production in an annual legume. Plant Physiol 60: 759–764

Heytler PG and Hardy RWF (1984) Calorimetry of nitrogenase-mediated reductions in detached soybean nodules. Plant Physiol 75: 304–310

Hooda RS, Sheoran IS and Singh R (1990) Partitioning and utilization of carbon and nitrogen in nodulated roots and nodules of chickpea (*Cicer arietinum*) grown at two moisture levels. Ann Bot 65: 111–120

Hunt S and Layzell DB (1993) Gas exchange of legume nodules and the regulation of nitrogenase activity. Ann Rev Plant Physiol Plant Mol Biol 44: 483–511

Kanamori T, Yoneyama T and Ishizuka J (1984) relationships of dinitrogen fixation (acetylene reduction) with respiration, ATP, and magnesium in soybean nodules. Soil Sci Plant Nutr 30: 231–237

Lambers H, Layzell DB and Pate JS (1980) Efficiency and regulation of root respiration in a legume: Effects of the N source. Physiol Plant 50: 319–325

Layzell DB, Rainbird RM, Atkins CA and Pate JS (1979) Economy of photosynthate use in nitrogen-fixing legume nodules Plant Physiol 64: 888–891

Layzell DB, Gaito ST and Hunt S (1988) Model of gas exchange and diffusion in legume nodules I. Calculation of gas exchange rates and energy costs of N_2 fixation. Planta 173: 117–127

Mahon JD and Child JJ (1979) Growth response of inoculated peas (*Pisum sativum*) to combined nitrogen. Can J Bot 57: 1687–1693

Matamoros MA, Baird LM, Escuredo PR, Dalton DA, Minchin FR, Iturbe-Ormaetxe I, Rubio MC, Moran JF, Gordon AJ and Becana M (1999) Stress-induced legume root nodule senescence. Physiological, biochemical, and structural alterations. Plant Physiol 121: 97–111

Minchin FR (1983) Poster discussion 7A. Photosynthesis and hydrogen metabolism in relation to nitrogenase activity in legumes. In: Veeger C and Newton WE (eds) Advances in Nitrogen Fixation Research, pp 491–492. Nijhoff/Junk, The Hague

Minchin FR (1997) Regulation of oxygen diffusion in legume nodules. Soil Biol Biochem 29: 881–888

Minchin FR and Pate JS (1983) The carbon balance of a legume and the functional economy of its root nodules. J Exp Bot 24: 259-271

Minchin FR, Summerfield RJ and Neves MCP (1980) Carbon metabolism, nitrogen assimilation and seed yield of cowpea (*Vigna unquiculata* L. Walp) grown in an adverse temperature regime. J Exp Bot 31: 1327–1345

Minchin FR, Summerfield RJ, Hadley P, Roberts EH and Rawsthorne S (1981) Carbon and nitrogen nutrition of nodulated

roots of grain legumes. Plant Cell Environ 4: 5–26
Minchin FR, Witty, JF, Sheehy JE and Muller M (1983) A major error in the acetylene reduction assay: Decreases in nodular nitrogenase activity under assay conditions. J Exp Bot 34: 641–649
Minchin FR, Minguez MI, Sheehy JE, Witty JF and Skøt L (1986a) Relationships between nitrate and oxygen supply in symbiotic nitrogen fixation by white clover. J Exp Bot 37: 1103–1113
Minchin FR, Sheehy JE and Witty JF (1986b) Further errors in the acetylene reduction assay. Effects of plant disturbance. J Exp Bot 37: 1581–1589
Minchin FR, Becana M and Sprent JI (1989) Short-term inhibition of legume N_2 fixation by nitrate. II. Nitrate effects on nodule oxygen diffusion. Planta 180: 46–52
Neo HH and Layzell DB (1997) Phloem glutamine and the regulation of O_2 diffusion in legume nodules. Plant Physiol 113: 259–267
Neves MCP and Hungria M (1987) The physiology of nitrogen fixation in tropical grain legumes. CRC Crit Rev Plant Sci 6: 267–321
Neves MCP, Minchin FR and Summerfield RJ (1981) Carbon metabolism, nitrogen assimilation and seed yield of cowpea (*Vigna unquiculata*) plants dependent on nitrate-nitrogen or on one of two strains of *Rhizobium*. Trop Agric 58: 115–132
Pate JS (1973) Uptake, assimilation, and transport of nitrogen compounds by plants. Soil Biol Biochem 5: 109–119
Pate JS, Layzell DB and Atkins CA (1979a) Economy of carbon and nitrogen in a nodulated and nonnodulated (NO_3-grown) legume. Plant Physiol 64: 1083–1088
Pate JS, Layzell DB and McNeil DL (1979b) Modelling the transport and utilization of carbon and nitrogen in a nodulated legume. Plant Physiol 63: 730–737
Pate JS, Atkins CA and Rainbird RM (1981) Theoretical and experimental costings of nitrogen fixation and related processes in nodules of legumes. In: Gibson AH and Newton WE (eds) Current Perspectives in Nitrogen Fixation, pp105–116. Australian Academy of Science, Canberra
Patterson TG and LaRue TA (1983) Root respiration associated with nitrogenase activity (C_2H_2) of soybean, and a comparison of estimates. Plant Physiol 72: 701–705
Rainbird RM, Hitz WD and Hardy RWF (1984) Experimental determination of the respiration associated with soybean/*Rhizobium* nitrogenase function, nodule maintenance, and total nodule nitrogen fixation. Plant Physiol 75: 49–53
Rao AS, Luthra YP, Sheoran IS and Singh R (1984) Partitioning of carbon and nitrogen during growth and development of pigeonpea (*Cajunus cajan* L.). J Exp Bot 35: 774–784
Raven JA (1985) Regulation of pH and generation of osmolarity in vascular plants: A cost-benefit analysis in relation to efficiency of use of energy, nitrogen and water. New Phytol 101: 25–77
Rosendahl L (1988) Rhizobium strain effect on nitrogen accumulation in pea relates to PEP carboxylase activity in nodules and asparagine in root bleeding sap. In: O'Gara F, Manian S and Drevon JJ (eds) Physiological Limitations and the Genetic Improvement of Symbiotic Nitrogen Fixation, pp 51–55. Kluwer Academic Publishers, Dordrecht
Rosendahl L, Vance CP, Miller SS and Jacobsen E (1989) Nodule physiology of a supernodulating pea mutant. Physiol Plant 77: 606–612
Ryle GJA, Powell CE and Gordon AJ (1978) Effect of source of nitrogen on the growth of fiskeby soya bean: The carbon economy of whole plants. Ann Bot 42: 637–648
Ryle GJA, Powell CE and Gordon AJ (1979a) The respiratory costs of nitrogen fixation in soyabean, cowpea, and white clover I. Nitrogen fixation and the respiration of the nodulated root. J Exp Bot 30: 135–144
Ryle GJA, Powell CE and Gordon AJ (1979b) The respiratory costs of nitrogen fixation in soybean, cowpea and white clover. II. Comparisons of the cost of nitrogen fixation and the utilization of combined nitrogen. J Exp Bot 30: 145–153
Ryle GJA, Arnott RA and Powell CE (1981) Distribution of dry weight between root and shoot in white clover dependent on N_2 fixation or utilizing abundant nitrate nitrogen. Plant Soil 60: 29–39
Ryle GJA, Arnott RA, Powell CE and Gordon AJ (1983) Comparisons of the respiratory effluxes of nodules and roots in six temperate legumes. Ann Bot 52: 469–477
Ryle GJA, Arnott RA, Powell CE and Gordon AJ (1984) N_2 fixation and the respiratory costs of nodules, nitrogenase activity, and nodule growth and maintenance in fiskeby soyabean. J Exp Bot 35: 1156–1165
Schubert KR (1982) The Energetics of Biological Nitrogen Fixation. American Society of Plant Physiologists, Maryland
Schubert KR and Ryle GJA (1980) The energy requirements for nitrogen fixation in nodulated legumes. In: Summerfield RJ and Bunting AH (eds) Advances in Legume Science, pp 85–96. HMSO, London
Schuller KA, Minchin FR and Gresshoff PM (1988) Nitrogenase activity and oxygen diffusion in nodules of soybean cv. Bragg and a supermodulating mutant: Effects of nitrate. J Exp Bot 39: 865–877
Schulze J, Adgo E and Schilling G (1994) The influence of N_2-fixation on the carbon balance of leguminous plants. Experienta 50: 906–912
Schulze J, Adgo E and Merbach W (1999) Carbon costs associated with N_2 fixation in *Vicia faba* L. and *Pisum sativum* L. over a 14-day period. Plant Biol 1: 625–631
Schulze J, Beschow H, Adgo E and Merbach W (2000) Efficiency of N_2 fixation in *Vicia faba* L. in combination with different *Rhizobium leguminosarum* strains. J Plant Nutr Soil Sci 163: 367–373
Serraj R, Sinclair TR and Purcell LC (1999) Symbiotic N_2 fixation response to drought. J Exp Bot 50: 143–155
Serraj R, Vadez V and Sinclair TR (2001) Feedback regulation of symbiotic N_2 fixation under drought stress. Agronomie 21: 621–626
Sheehy JE (1987) Photosynthesis and nitrogen fixation in legume plants. CRC Crit Rev Plant Sci 5: 121–159
Silsbury JH (1984) Comparison of the growth rates of dinitrogen fixing subterranean clover swards with those assimilating nitrate ions. Plant Soil 80: 201–213
Skøt L, Hirsch PR and Witty JF (1986) Genetic factors in Rhizobium affecting the symbiotic carbon costs of N_2 fixation and host plant biomass production. J Appl Bacteriol 61: 239–246
Soussana J-F, Minchin FR, Macduff JH, Raistrick N, Abberton MT and Michaelson-Yeates TPT (2002) A simple model of feedback regulation for nitrate uptake and N_2 fixation in contrasting phenotypes of white clover. Ann Bot 90: 139–147
Sprent JI and Sprent P (1990) Nitrogen Fixing Organisms. Pure and Applied Aspects. Chapman and Hall, London
Stitt M, Muller C, Matt P, Gibon Y, Carillo P, Morcuende R, Scheible W-R and Krapp A (2002) Steps towards an integrated

view of nitrogen metabolism. J Exp Bot 53: 959–970

Sun JS, Simpson RJ and Sands R (1992) Nitrogenase activity and associated carbon budgets in seedlings of *Acacia mangium* measured with a flow-through system of the acetylene reduction assay. Aust J Plant Physiol 19: 97–107

Tjepkema JD and Yokum CS (1973) Respiration and oxygen transport in soybean nodules. Planta 115: 59–72

Twary SN and Heichel GH (1991) Carbon costs of dinitrogen fixation associated with dry matter accumulation in alfalfa. Crop Sci 31: 985–992

Urdvardi MK and Day DA (1997) Metabolite transport across symbiotic membranes of legume nodules. Ann Rev Plant Physiol Plant Mol Biol 48: 493–523

Vance CP and Heichel GH (1991) Carbon in N_2 fixation: Limitation or exquisite adaptation. Ann Rev Plant Physiol Plant Mol Biol 42: 373–392

Wadham C, Winter H and Schuller KA (1996) Regulation of soybean nodule phosphoenolpyruvate carboxylase in vivo. Physiol Plant 97: 531–535

Walsh KB, Vessey JK and Layzell DB (1987) Carbohydrate supply and N_2 fixation in soybean. The effect of varied daylength and stem girdling. Plant Physiol 85: 137–144

Warembourg FR (1983) Estimating the true cost of dinitrogen fixation by nodulated plants in undisturbed conditions. Can J Microbiol 29: 930–937

Warembourg FR and Roumet C (1989) Why and how to estimate the cost of symbiotic N_2 fixation? A progressive approach based on the use of ^{14}C and ^{15}N isotopes. Plant Soil 115: 167–177

Witty JF, Minchin FR and Sheehy JE (1983) Carbon costs of nitrogenase activity in legume root nodules determined using acetylene and oxygen. J Exp Bot 34: 951–963

Witty JF, Minchin FR, Sheehy JE and Minguez MI (1984) Acetylene-induced changes in the oxygen diffusion resistance and nitrogenase activity of legume root nodules. Ann Bot 53: 12–20

Witty JF, Minchin FR, Skøt L and Sheehy JE (1986) Nitrogen fixation and oxygen in legume root nodules. Oxf Surv Plant Mol Cell Biol 3: 275–314

Woo KC and Xu S (1996) Metabolite regulation of phospho-*enol*pyruvate carboxylase in legume root nodules. Aust J Plant Physiol 23: 413–419

Chapter 12

Respiratory Costs of Mycorrhizal Associations

David R. Bryla*
USDA ARS Horticultural Crops Research Laboratory, 3420 NW Orchard Ave., Corvallis, OR 93648, U.S.A.

David M. Eissenstat
Department of Horticulture, The Pennsylvania State University, University Park, PA 16802, U.S.A.

Summary

Mycorrhizal fungi form symbiotic and often mutually beneficial relationships with the roots of most terrestrial plants. In this chapter we review current literature concerned with plant respiratory requirements for supporting this important plant-fungal association, and its effect on the overall plant carbon economy. Controlled studies indicate that mycorrhizal respiratory costs are considerable, consuming between 2 to 17% of the photosynthate fixed daily, varying depending on the host and fungal species involved, the stage of colonization, and the environmental conditions. Respiratory energy is required by the mycobiont for construction of new intraradical

*Author for correspondence, email: brylad@onid.orst.edu

H. Lambers and M. Ribas-Carbo (eds.), Plant Respiration, 207–224.

and extraradical fungal tissue (including reproductive structures), for maintenance and repair of existing fungal tissue, and for cellular processes in the fungal tissue associated with the absorption, translocation and transfer of nutrients from the soil to the host. Additional respiration is also required by the host plant for stimulated root cellular processes, and potentially for increased production of root biomass. Field studies of these important processes will eventually lead us to better understand how significant mycorrhizal fungi are to the total carbon budgets of natural and managed plant communities.

I. Introduction

It is difficult to estimate the actual costs of root respiration in natural soils without considering the role of mycorrhizal fungi. These fungi are an integral part of nearly all plant communities (Van der Heijden et al., 1998). They are ubiquitous in most natural and agricultural soils, and are capable of forming close symbiotic associations with the roots of nearly 90% of the terrestrial plant species investigated thus far (Newman and Reddell, 1987; Trappe, 1987). Like many pathogenic soil fungi, mycorrhizal fungi penetrate living roots of plants to acquire energy-rich carbohydrates needed for their growth, maintenance, and function. However, unlike the pathogens, external hyphae produced by mycorrhizas absorb soil nutrients and translocate them to the fungal-plant interface where they are transferred to root epidermal and cortical cells. In effect, these fungi act as extensions of the plant's root system, increasing uptake of many soil-derived nutrients, and in some cases, even water, thereby improving the host plant's productivity (Smith and Read, 1997) and reproductive fitness (Koide and Dickie, 2002). Thus, mycorrhizal fungi incur both costs and benefits to the overall carbon economy of the host plant (Koide and Elliott, 1989; Tinker et al., 1994; Douds et al., 2000).

In this chapter, we focus on the respiratory costs of mycorrhizal associations, particularly with regard to growth, maintenance, and ion uptake by both the fungus and the host root system colonized by the fungus. While mycorrhizal fungi may be beneficial during at least some stage of a plant's development, studies indicate that below-ground carbohydrate costs are often higher when plants are associated with these fungi than when plants are not. In fact, under conditions where soil resources are non-limiting and the fungi are providing little or no benefit to the plant, carbon consumption by mycorrhizal fungi can actually reduce plant growth (Buwalda and Goh, 1982; Molina and Chamard, 1983; Koide, 1985; Ingestad et al, 1986; Modjo and Hendrix, 1986; Rousseau and Reid, 1991; Peng et al., 1993; Taylor and Harrier, 2000). The total amount of carbon required by the association is the sum of several component costs, including direct export of carbon from the host to the fungus (for growth and metabolism), as well as additional carbon use such as increased plant respiratory and non-respiratory (exudation, cell death, etc.) requirements (Finlay and Söderström, 1992). Controlled studies using whole-plant ^{14}C-labeling techniques suggest that the total cost of the association ranges from 3 to 36% of the carbon fixed daily by photosynthesis, with the largest proportion of the carbon allocated to respiration (Table 1).

The two most common and well studied types of mycorrhizal associations, categorized according to symbiont morphology and host-taxon relationships, are arbuscular (previously called vesicular-arbuscular) mycorrhizas and ectomycorrhizas. Arbuscular mycorrhizal fungi occur on many different species of herbaceous and woody plants including most crop plants and tropical trees. They are obligatorily dependent on the host plant for their source of carbohydrate energy, and form, within the root cortical cells of the host, characteristic branched haustorical structures known as arbuscules, which are likely sites for significant transfer of carbohydrate and nutrients between the host and fungus. Other distinguishing features sometimes found, depending on the fungal species, include large hyphal coils (also likely sites for carbohydrate and nutrient transfer), usually located within the epidermal cells, and terminal and intercalary swellings known as vesicles (likely storage organs containing abundant lipids) formed within and between the cortical cells. Arbuscular mycorrhizas are most noted for their ability to enhance plant uptake of diffusion-limited inorganic ions such as phosphate, copper and zinc. Ectomycorrhizal fungi, on the other hand, occur mainly on forest trees of temperate and boreal regions. They are characterized by a dense sheath or mantle of fungal tissue that encloses the colonized root, and by a plexus of

Abbreviations: PPFD – photosynthetic photon flux density; Q_{10} – temperature coefficient of respiration

Table 1. Amount of carbon required by arbuscular and ectomycorrhizal fungi associated with various host plants. Values were estimated using whole-plant ^{14}C pulse-chase labeling techniques.

Host species	Fungal species	Percentage of C fixed by photosynthesis			Reference
		Fungal Biomass	Fungal respiration	Total to fungus	
Arbuscular mycorrhizas					
Allium porrum	*Glomus mosseae*	2	5	7	Snellgrove et al. (1982)
Citrus aurantium	*G. intraradices*	n.d.	n.d.	6	Koch and Johnson (1984)
Cucumis sativus	*G. caledonium*	2	7	9	Pearson and Jakobsen (1993)
C. sativus	*G. fasciculatum*	5	15	20	Jakobsen and Rosendahl (1990)
C. sativus	*G. sp.*	9	8	17	Pearson and Jakobsen (1993)
C. sativus	*Scutellospora calospora*	2	17	19	Pearson and Jakobsen (1993)
Glycine max	*G. fasciculatum*	3	5-14	8-17	Harris et al. (1985)
Poncirus trifoliate x *C. aurantium*	*G. intraradices*	n.d.	n.d.	6-8	Douds et al. (1988)
P. trifoliate x *C. aurantium*	*G. intraradices*	n.d.	n.d.	11	Koch and Johnson (1984)
Vicia faba	*G. mosseae*	n.d.	n.d.	10	Pang and Paul (1980)
V. faba	*G. mosseae*	1	3	4	Paul and Kucey (1981); Kucey and Paul (1982)
Ectomycorrhizas					
Pinus ponderosa	*Hebeloma crustuliniforme*	2-3	5-8	7-11	Anderson and Rygiewicz (1995)
P. ponderosa	*H. crustuliniforme*	3	4	7	Rygiewicz and Anderson (1994)
Pinus taeda	*Pisolithus tinctorius*	n.d.	n.d.	6-36	Reid et al. (1983)
Salix viminalis	*Thelephora terrestris*	1-8	2-4	3-12	Durall et al. (1994)

n.d. – not determined.

hyphae known as the Hartig net, which penetrates the root intercellularly and surrounds the epidermal and cortical cells. Most ectomycorrhizal fungi are also obligate symbionts, but some may be able to act as saprotrophs (Haselwandter et al., 1990). They can hydrolyze proteins and organic phosphates, and increase plant uptake of both organic and inorganic forms of nitrogen and phosphorus (Tinker and Nye, 2000). Other distinctive groups of mycorrhizas that are less common and not as well studied include ericoid and ectendomycorrhizas, which are formed in association with ericaceous plant families, and orchid mycorrhizas (Wilcox, 1996). Thus far, only the carbohydrate requirements of arbuscular and ectomycorrhizas have been examined in detail, which restricts our discussion to these two groups of fungi.

II. Total Respiratory Costs

The total respiratory costs of mycorrhizal associations are a considerable component of the overall carbon economy of the host plant. Of those studies listed in Table 1, where respiration attributed to the fungus was separated from other fungal carbohydrate requirements, 47 to 89% of the carbon allocated to the mycorrhizal association was consumed by respiration. Respiration in these studies varied depending on the host plant and fungal species, the stage of mycorrhizal development, and the environmental conditions under which the host plant and fungus were grown.

A. Variation among Plant and Fungal Species

Root respiration varies among plant species (Lambers et al., 2002). Respiration also appears to vary among mycorrhizal fungal species associated with a particular host. In cucumber (*Cucumis sativus*), for example, the proportion of assimilated ^{14}C allocated to below-ground respiration in plants colonized by the arbuscular mycorrhizal fungus *Scutellospora calospora* was 16.5% higher than that in non-colonized plants, but only 6.5 or 7.6% higher when plants were colonized by two other arbuscular species, *Glomus caledonium* or an unclassified *Glomus* sp., respectively (Pearson and Jakobsen, 1993). Bidartondo et al. (2001) observed that of four ectomycorrhizal fungi they examined, *Paxillus involutus* produced the fewest mycorrhizal connections to its host plant and respired less carbohydrates per unit biomass than did the other fungi. Variability in respiration among species may be due to differences in 1) the amount and quality (e.g., number of arbuscules and vesicles) of fungal biomass produced and maintained by the mycobiont both within the host's root system and in the surrounding rhizosphere (Graham et al., 1982b; Giovannetti and Hepper, 1985; Estaún et al., 1987; Lioi and Giovannetti, 1987; Wong et al., 1989; 1990; Jakobsen et al., 1992; Burgess et al., 1994; Lerat et al., 2003), 2) the metabolic activity of the fungus (Lewis and Harley, 1965a,b,c; Söderström and Read, 1987; Bago et al., 2002), 3) the level of stimulation of cellular activities (e.g., cell wall and cytoplasmic invertases) and changes in carbohydrate metabolism (e.g., sucrose synthase) in the epidermal and cortex regions of the colonized roots (Wright et al., 1998, 1999), and/or 4) the extent of growth promotion (or depression) of the root system (Krishna et al., 1985; Bryla and Koide, 1990).

B. Changes with Mycorrhizal Development and Plant Age

Respiratory costs are expected to be especially high during early stages of colonization when most of the new fungal tissue is being produced, but decrease as the association matures, much in the same way that root respiratory costs decline with root age (Bouma et al., 2000, 2001). Carbohydrate substrates and respiratory energy are required during this period for construction of new intra- and extraradical fungal components, and for modifications in the cellular structures of the host (Graham and Eissenstat, 1994). Table 2 shows that although intraradical fungal biomass in soybean (*Glycine max*) roots colonized by *Glomus fasciculatum* increased from 115 mg per plant at 6 weeks to 266 mg per plant at 9 weeks, the specific rate of ^{14}C incorporation into fungal biomass was, in fact, lower at 9 weeks than at 6 weeks. The distribution of assimilated carbon to fungal respiration also decreased during this 3-week period from 18.2 mg ^{14}C plant^{-1} to 9.7 mg ^{14}C plant^{-1}, consequently lowering the plant's cost of supporting the association.

Further evidence that mycorrhizal cost decreases with age can be found in the ectomycorrhizal literature. Cairney et al. (1989) found in *Eucalyptus pilularis* roots colonized by *Pisolithus tinctorius* that young mycorrhizas accumulated more ^{14}C than older mycorrhizas, with much of the carbon transfer occurring during the first few weeks after inoculation. By 90 days after inoculation, all ^{14}C translocation to mycorrhizas had stopped. This information led them to hypothesize that in mature root systems only a small portion of the roots would require significant amounts of photosynthate to support mycorrhizal associations at any one time, which appeared to be the case when Cairney and Alexander (1992) compared allocation of ^{14}C with younger and older mycorrhizas of *Tylospora fibrillose* on *Picea sitchensis*. In this later study, they measured the ratio of activity in young to older mycorrhizas, and found that the ratio progressively increased from 2:1, when newly colonized seedlings were first transferred to a peat substrate, to 54:1 by 38 weeks after transfer. Similarly, Durall et al. (1994) examined carbon allocation in *Salix viminalis* inoculated with *Thelephora terrestris*, and found that the proportion of ^{14}C allocated to mycorrhizal respiration decreased as the plants aged from 50 to 98 days.

C. Impact of Environmental Conditions

Respiration associated with mycorrhizas is usually influenced by a combination of environmental factors that either directly affect the symbiosis by altering fungal growth and metabolism, or indirectly affect it by influencing photosynthesis and supply of carbohydrates provided by the host. Factors that have received some attention in the literature and will be discussed here include soil nutrient availability, soil temperature and moisture, light intensity, elevated atmospheric CO_2 concentrations, and ozone pollution.

1. Soil Nutrient Availability

Mycorrhizal respiration in many plant-fungal combinations is likely very dependent on soil nutrient availability, as this can affect the total amount of fungal biomass produced by the symbiosis, and also the proportion of biomass allocated to various fungal structures (some of which may have higher or lower respiratory requirements than others). Typically, mycorrhizas develop more readily under nutrient-poor conditions than under nutrient-rich or heavily fertilized conditions (Hayman, 1970; Chambers et al., 1980; Amijee et al., 1989; de Miranda et al., 1989; Jones et al., 1990; Koide and Li, 1990; Wallander and Nylund, 1991; 1992; Henry and Kosola, 1999, Nilsson and Wallander, 2003), and should therefore require proportionally more photosynthates for growth and metabolism when soil nutrients are limited. This was the case in a study by Baas and Lambers (1988) that examined the effect of increasing soil phosphorus on

Table 2. Dry weights, mycorrhizal colonization, distribution of assimilated ^{14}C, and specific rate of ^{14}C incorporation in *Glycine max – Rhizobium japonicum – Glomus fasciculatum* associations at six and nine weeks after emergence (from Harris et al., 1985).

Component	Six weeks	Nine weeks
Dry weights (g)		
Shoot	4.90	11.28
Roots	1.75	3.37
Nodules	0.14	0.41
Mycorrhiza		
Intraradicle	0.12	0.27
Extraradicle	0.16	0.24
Mycorrhizal colonization (%)		
Root length	68	76
Root mass	6.6	7.9
Distribution of assimilated ^{14}C (%)		
Biomass		
Shoot	51.0	61.2
Roots	9.7	9.4
Nodules	2.0	1.7
Mycorrhiza	2.7	2.8
Respiration		
Shoot	6.3	3.9
Roots + soil	5.2	6.5
Nodules	9.4	9.8
Mycorrhiza	13.7	4.7
Specific rate of ^{14}C incorporation (mg ^{14}C g^{-1} d. wt. day^{-1})		
Shoot	13.9	11.4
Roots	5.4	5.3
Nodules	18.8	8.7
Mycorrhiza	15.6	10.9

root respiration of *Plantago major* spp. *pleiosperma* grown with or without *G. fasciculatum*. At 38 days after transplanting, although plant growth was mostly unresponsive to colonization at any level of phosphorus, both colonized root length and root respiration of plants grown with the fungus decreased with increasing soil phosphorus availability, while root respiration of uncolonized plants remained unchanged (Fig. 1). Likewise, Lu et al. (1998) speculated that depression of soil respiration after the addition of nitrogen fertilizers to ectomycorrhizal Douglas-fir (*Pseudotsuga menziesii*) seedlings, grown in relatively fertile soil, was probably due to reduced root and mycorrhizal mycelial growth. Soil nutrient status will especially reduce mycorrhizal colonization and its corresponding respiration when conditions, such as irradiance or temperature, limit carbohydrate assimilation and transport below ground (Graham et al. 1982a; Son and Smith, 1988).

Soil nutrient status also affects the rate of nutrient acquisition by both the fungus and the host, the nutritional status of the host, and the responsiveness of the host to colonization, all of which will impact respiration associated with the symbiosis. Plants that are deficient in a particular nutrient tend to use less of the nutrient to produce a unit of biomass than plants with adequate levels of the nutrient in their tissues (Eissenstat et al., 1993). As mentioned previously, mycorrhizal fungi are most commonly found to increase plant uptake of phosphorus. However, improved plant phosphorus status tends to reduce plant allocation to roots and reduce mycorrhizal colonization (Smith and Read, 1997). Specific rates of mycorrhizal root respiration likely diminish with an increase in plant phosphorus concentration. For example, root respiration in mycorrhizal *Citrus volkameriana* seedlings grown in high-phosphorus soil was only 72% of that in mycorrhizal seedlings grown

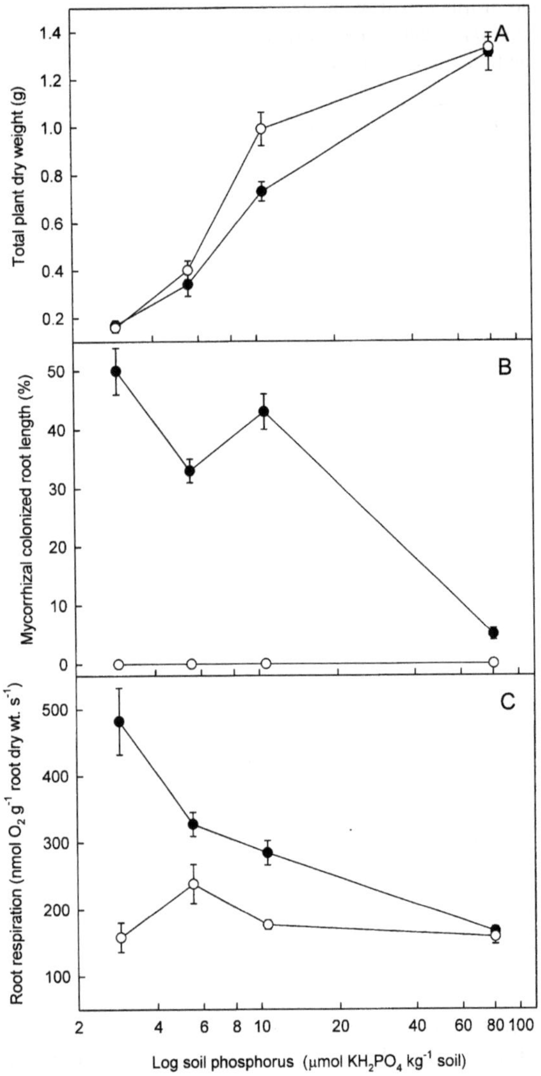

Fig. 1. (A) Total plant dry weight, (B) colonized root length, and (C) root respiration rate of mycorrhizal (●) and non-mycorrhizal (○) *Plantago major* spp. *pleiosperma* grown at four soil phosphorus levels. Vertical bars are SE. Data from Baas and Lambers (1988).

in low-phosphorus soil (Peng et al., 1993). Reduced respiration was primarily due to fewer lipid-rich vesicles of *Glomus intraradices* in high-phosphorus seedlings, and to lower maintenance requirements. It should be noted that some of the faster respiration rates in low-phosphorus mycorrhizal plants may also be a result of lower phosphorus nutrition, and not increased colonization and activity of the mycorrhizal fungus.

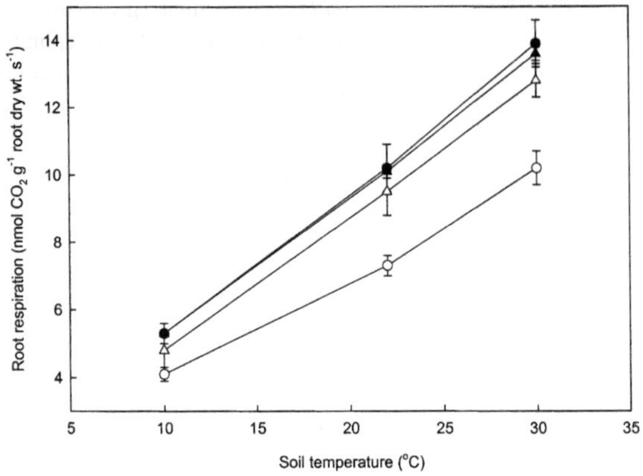

Fig. 2. Root respiration of mycorrhizal (●, ▲) and non-mycorrhizal (○, △) *Picea abies* seedlings exposed to various soil temperatures. Seedlings were grown in semi-hydroponic sand culture supplied with $(NH_4)_2SO_4$ (●, ○) or KNO_3 (▲, △) nutrient solution. Vertical bars are SE. Data from Eltrop and Marschner (1996).

2. Soil Temperature and Moisture

The respiratory response of mycorrhizas to changes in soil temperature was investigated by Eltrop and Marschner (1996). *Picea abies* seedlings were grown with or without *P. tinctorius*, and fertilized with either ammonium sulfate or potassium nitrate. They found that root respiration was significantly faster in mycorrhizal than in non-mycorrhizal plants when supplied with ammonium, but not when supplied with nitrate (Fig. 2). However, regardless of the treatment effects, the response to soil temperature was similar. Whether plants were mycorrhizal or not, or fertilized with NH_4-N or NO_3-N, the temperature coefficients of respiration, Q_{10}, from 10 to 30 °C were similar, ranging from 1.58 to 1.64. Burton et al. (2002) further demonstrated that the average Q_{10} for roots of arbuscular and ectomycorrhizal tree species were nearly identical. These few results suggest that the respiratory response of plant roots and mycorrhizal fungi to soil temperature is alike, although more research is required before any generalizations can be made.

One might also expect mycorrhizas to have little effect on the change in respiration with soil moisture; exposure to dry soil leads to a gradual decline in root respiration in many plant species (Palta and Nobel, 1989a,b; Bryla et al., 1997; Burton et al., 1998; Bouma and Bryla, 2000). In isolated fine roots of mature *C. volkameriana* trees, Espeleta and Eissenstat (1998) observed that respiration was similar between mycorrhizal and non-mycorrhizal roots exposed to 8 weeks of localized drought. After 15 weeks of drought, however, mycorrhizal roots exhibited 34% slower respiration and 21% less mortality than non-mycorrhizal roots did. *Citrus* is apparently capable of suppressing mycorrhizal root respiration after prolonged exposure to dry soil, thereby preventing excessive carbon expenditure in maintenance respiration. Thus, the ability of mycorrhizal associations to reduce root respiration under drought may delay root shedding, and serve as a way by which the fungus guarantees survival under hostile environmental conditions.

3. Light Conditions

On an absolute basis, low light availability reduces root respiration (Lambers et al., 2002) and also limits mycorrhizal respiration if photoassimilation is appreciably reduced (Bücking and Heyser, 2003). In addition, low light can reduce the percentage of root length colonized by mycorrhizal fungi (Tester et al., 1986), especially under high phosphorus conditions (Son and Smith, 1988), which would also tend to reduce mycorrhizal root respiration.

Compared with plants grown under high-light conditions, shaded plants have proportionally less root biomass which indicates that shaded plants have proportionally less total below-ground carbon expenditure than plants exposed to full light. This observation has been used to support theories of optimality in shoot and root growth, where shading leads to greater allocation of photosynthate to leaf production so that light is less limiting to overall plant growth (Brouwer, 1983; Bloom et al., 1985). However, evidence in support of this preferential allocation to shoots for shaded plants often involves comparisons between small, shaded plants and larger, non-shaded plants (Reich, 2002). Proportional allocation to roots and shoot changes continuously with plant size, making such comparisons misleading. When ontogenetic effects of plant size are taken into account using an allometric approach, most studies found no evidence of an allocation shift towards leaf production at low light (reviewed by Reich, 2002).

While biomass allocation may not be affected by light regime, there is some evidence that proportionally more photosynthate is typically allocated to maintain the metabolism of mycorrhizal root tissues when plants are grown under lower light conditions. Gansert (1994) used a PC-controlled cuvette system to measure respiration in situ of individual fine roots on 10-year-old beech saplings growing in the shaded understory and in a natural light gap of a mature beech forest. Roots were colonized by the hyphae of several ectomycorrhizas including *Xerocomus chrysenteron*, *Lactarius subdulcis* and *Russula ochroleuca*, and the rate of root respiration was correlated with colonization (expressed as percent dry weight of the total root biomass) at both sites. Although net CO_2 assimilation and mycorrhizal root respiration throughout the season was much faster in saplings growing in the light gap than in those growing in the understory, the ratio of respiration to net CO_2 assimilation, both measured on a unit dry weight basis, was substantially higher in the understory saplings (Table 3), indicating a higher relative cost of below-ground respiration when light conditions were low.

Low light may also increase the relative cost of mycorrhizal associations. At 990 μmol m^{-2} s^{-1}, mycorrhizal colonization with *P. tinctorius* increased the proportion of assimilated carbon allocated to root respiration in *Picea abies* by only 0.6 to 3.4% at soil temperatures ranging from 10 to 30°C, while at a PPFD of 290 μmol m^{-2} s^{-1}, the proportion was increased by 1.8 to 7.4% (Fig. 3). This indicates that, regardless of soil temperature, proportionally more carbon was required to support the symbiosis when plants were grown under low-light conditions than when they were grown under higher-light conditions.

4. Elevated Atmospheric CO_2 and Ozone Pollution

Rising atmospheric CO_2 concentrations and increasing levels of air pollutants, such as ozone, are expected to impact considerably the respiration associated with roots, mycorrhizas and other soil microorganisms over this century (Andrews et al., 1999; Ball and Drake, 1998; Hungate et al., 1997).

Elevated [CO_2] can increase mycorrhizal development and associated respiration by stimulating host photosynthesis, and thereby enhancing carbon allocation to the fungus (Ineichen et al., 1995; Sanders, 1996; Jifon et al., 2002). Photosynthesis of *Plantago lanceolata* was stimulated by elevated [CO_2] (600 μl l^{-1}) far more than plant growth, especially when plants were associated with *Glomus mosseae* (Staddon et al., 1999). Based on plant dry mass measurements, most of the extra carbon fixed by photosynthesis at elevated CO_2 appeared to be respired by the mycorrhizal fun-

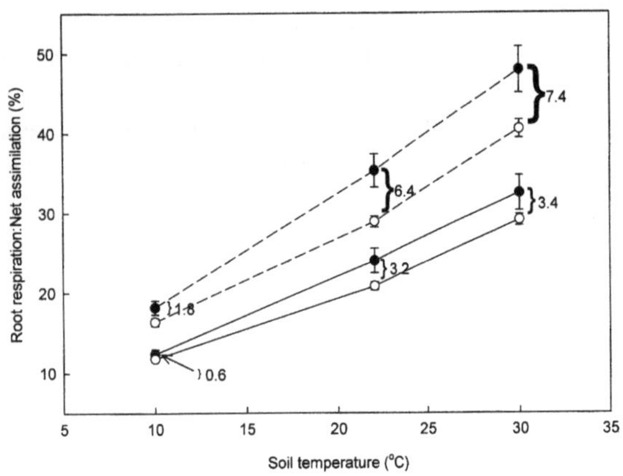

Fig. 3. Root respiration:net CO_2 assimilation ratio of mycorrhizal (●) and non-mycorrhizal (○) *Picea abies* seedlings exposed to various soil temperatures. Seedlings were grown in semi-hydroponic sand culture at a PPFD of 290 µmol m^{-2} s^{-1} (- - - -) or at 990 µmol m^{-2} s^{-1} (——). Vertical bars are SE. Data from Eltrop and Marschner (1996).

gus and the roots, which suggests that mycorrhizal fungi may increase carbon transfer through the system at elevated [CO_2] in addition to altering soil carbon pools, as shown by others (Rillig et al., 1998; Rouhier and Read, 1998; Sanders et al., 1998).

Ozone pollution, on the other hand, often reduces the formation of mycorrhizas (Ho and Trappe, 1981; Reich et al., 1986; Stroo et al., 1988; Simons and Kelly, 1989; Meier et al., 1990), and can affect the respiratory activity of the association (McCool and Menge, 1983; Gorissen et al., 1991; Scagel and Anderson, 1997; Anderson, 2003). Anderson and Rygiewicz (1995) studied the allocation and metabolism of carbon in *Pinus ponderosa* seedlings inoculated with *Hebeloma crustuliniforme* under ozone stress. Seedlings were grown in root 'mycocosms' that enabled them to measure carbon fluxes through the root system as well as through a portion of intact extramatrical hyphae, while maintaining symbiotic integrity (Rygiewicz and Anderson, 1994). Ozone reduced hyphal respiration and carbon accumulation by the fungus in mycorrhizal plants, although mycorrhizal seedlings exhibited greater biomass-weighted respiration rates than non-mycorrhizal controls (Table 4). Ozone tended to shift allocation patterns in mycorrhizal seedlings, making them more similar to non-mycorrhizal seedlings grown without ozone. Reductions in carbon allocation to mycorrhizas caused by increasing levels of air contaminants may eventually reduce vigor of the association, thus affecting any plant-derived benefits associated with forming the symbiosis.

III. Components of Mycorrhizal Respiration

As mentioned in previous chapters, root respiration depends on three major energy-requiring processes: root growth, maintenance of root biomass, and uptake of mineral nutrients. Respiration of roots associated with mycorrhizas has additional processes specifically related to the symbiosis. These include growth and maintenance of the fungal tissue, and ion uptake by the fungus (as well as transport and transfer of the ions to the host plant).

A. Construction Costs and Growth Respiration

Mycorrhizal growth respiration depends both on the amount of root and fungal tissue produced, and on the chemical composition of each tissue. Tissue containing high quantities of lipids and proteins, for example, will have higher respiratory costs of construction than tissue with more carbohydrates.

Table 3. Net CO_2 assimilation rate (C_{in}) and fine root respiration rate (C_{out}) measured on 10-year-old *Fagus sylvatica* saplings growing in the understory and in a natural gap of a mature beech forest (from Gansert, 1994).

	Understory			Gap		
	Net CO_2 assimilation (mg C plant^{-1} day^{-1})	Root respiration (mg C plant^{-1} day^{-1})	C_{in} : C_{out}	Net CO_2 assimilation (mg C plant^{-1} day^{-1})	Root respiration (mg C plant^{-1} day^{-1})	C_{in} : C_{out}
May	56 ± 15	5 ± 1	10	548 ± 95	20 ± 5	28
June	111 ± 19	7 ± 1	17	560 ± 113	30 ± 3	19
July	55 ± 21	9 ± 1	6	740 ± 141	48 ± 2	15
August	51 ± 22	7 ± 1	8	494 ± 113	54 ± 4	9
September	0 ± 17	6 ± 1	0	262 ± 100	39 ± 4	7
October	–2 ± 18	6 ± 0	0	241 ± 140	33 ± 2	7

Table 4. Biomass-weighted retention and respiratory loss of ^{14}C in mycorrhizal and non-mycorrhizal *Pinus ponderosa* seedlings exposed to two levels of ozone (from Anderson and Rygiewicz, 1995).

Ozone exposure (μmol mol^{-1} h^{-1})	Mycorrhizal status	Retained					Respired				
		Needle	Stem	Coarse roots	Fine roots	Fungus	Shoot	Root	Fungus	Root + fungus	Total
0	Mycorrhizal	39.5	27.9	20.9	25.1	18.7	45.8	15.1	52.5	20.3	31.4
	SE	4.0	2.4	2.8	2.8	2.0	11.0	2.4	15.4	2.5	3.8
0	Non-myco.	34.2	19.2	12.8	15.8	0.0	30.4	9.5	0.0	9.5	20.4
	SE	4.8	2.1	2.8	4.9	0.0	7.3	1.7	0.0	1.7	4.2
39.3	Mycorrhizal	39.8	24.6	17.2	24.2	11.2	52.9	16.8	31.2	18.6	32.6
	SE	6.7	1.4	1.5	7.9	2.3	11.3	2.7	5.1	2.1	5.7
39.3	Non-myco.	32.2	19.8	10.5	18.7	0.0	25.5	8.7	0.0	8.7	17.6
	SE	7.2	3.7	1.8	2.9	0.0	4.4	0.6	0.0	0.6	2.1

Allocation values were biomass-weighted to normalize for differences in plant component fraction size. Units are percent allocated divided by tissue component dry weight.

By using daily construction costs and subtracting the carbon retained in new root growth, Peng et al. (1993) estimated that daily growth respiration accounted for 16% of the total root and soil respiration associated with mycorrhizal colonization in *C. volkameriana* (Fig. 4; Table 5). Respiratory energy is required during colonization for growth of new internal and external fungal structures, as well as for any cellular modifications and changes in carbon allocation to the host root.

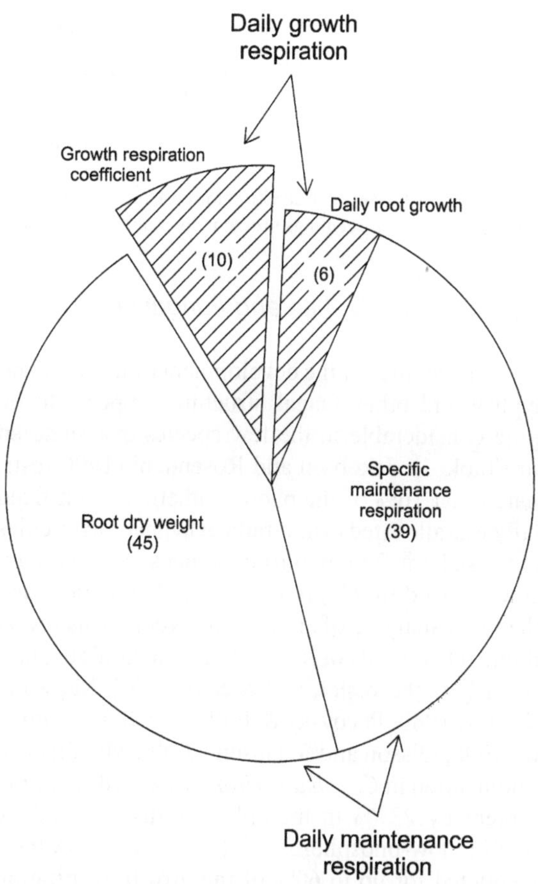

Fig. 4. Differences in total root and soil respiration between mycorrhizal and non-mycorrhizal *Citrus volkameriana* seedlings grown in high-phosphorus soil. Data from Peng et al. (1993).

1. The Host Root

Mycorrhizas tend to increase root:shoot partitioning in some species (Bryla and Koide, 1990, Eissenstat et al., 1993), but not in others (Fredeen and Terry, 1988; Thomson et al., 1986; Berta et al. 1991, 1996). Increased root biomass and root growth rate accounted for one-third of the difference in growth respiration between mycorrhizal and non-mycorrhizal *C. volkameriana* plants (Fig. 4); the other two-thirds was attributed to building more expensive roots and fungal structures (see below). Baas et al. (1989) suggested that mycorrhizal plants might have higher relative root growth rates than uncolonized plants of equal size, because of a shift in the carbon balance during the development of the symbiosis. For the most part, plants colonized by mycorrhizas do not necessarily allocate biomass to roots and shoots in the same proportion as nutritionally equivalent uncolonized plants. This altered pattern of allocation will profoundly influence plant productivity, and, consequently, any respiratory costs associated with the symbiosis. Thus, the possibility exists that faster instantaneous relative growth rates of the host root system result in faster rates of respiration for growth by mycorrhizal plants.

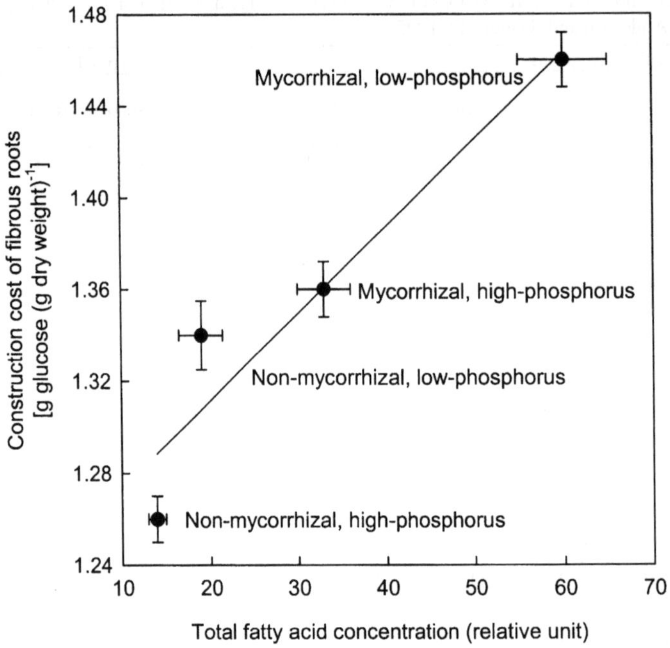

Fig. 5. Relationship between construction cost and relative total fatty acid content of fibrous roots from mycorrhizal and non-mycorrhizal *Citrus volkameriana* seedlings grown in low- and high-phosphorus soil. Data from Peng et al. (1993).

2. *Intraradical Hyphae and Fungal Organelles*

Carbon demand for the development of mycorrhizal hyphae and other fungal structures appears to be quite considerable in the few species examined so far (Table 1). Jakobsen and Rosendahl (1990) estimated that 16% of the photosynthetic carbon fixed daily was allocated to intraradical hyphae, arbuscules and vesicles in 22-day-old cucumber seedlings heavily colonized by *G. fasciculatum*. Histological and chemical analyses of arbuscular mycorrhizas reveal abundant amounts of lipids in the fungal tissue, particularly in the vesicles (Cox et al., 1975; Nagy and Nordby, 1980; Pacovsky & Fuller, 1988; Graham et al., 1995; Olsson and Johansen, 2000). Mycorrhizal colonization in *C. volkameriana* increased root lipid content by 227% in high-phosphorus soil and by 307% in low-phosphorus soil (Fig. 5), and may have accounted for up to 60% of the growth respiration (Fig. 4).

In comparison, ectomycorrhizas also contain abundant lipids (Olsson, 1999) as well as considerable amounts of sterols (Antibus and Sinsabaugh, 1993) and insoluble polysaccharides (Ling-Lee et al., 1977, Piché et al., 1981) in the fungal tissue, but produce considerably more fungal biomass than arbuscular mycorrhizas. While arbuscular mycorrhizal fungi usually represent less than 10% of the colonized root mass (but see Hepper, 1977), values representing 20 to 40% of the root mass are more common in ectomycorrhizas due to the large amount of hyphae associated with the mantle and the Hartig net (Vogt et al., 1982; 1991). Thus, construction costs and growth respiration associated with ectomycorrhizas are expected to represent a substantial host expense.

3. *Extraradical Hyphae*

Growth of the extraradical hyphae begins soon after root penetration, and can total more than 10 m of hyphae cm^{-3} of soil in arbuscular mycorrhizas (Sanders et al., 1977; Tisdall and Oades, 1979; Jakobsen and Rosendahl, 1990; Jakobsen et al., 1992), with hyphal diameters ranging from 2-30 μm (Read, 1992). In one case, Miller et al. (1995) calculated arbuscular hyphal lengths and dry weights as high as 81 m cm^{-3} and 339 μg cm^{-3}, respectively, in a pasture soil, and 111 m cm^{-3} and 457 μg cm^{-3}, respectively, in a tall grass prairie soil located in mid-western U.S. If we assume a mean hyphal radius of 5 μm, then about 0.08% of the soil volume is occupied by hyphae in soil containing 10 m cm^{-3} of hyphae. In contrast, root length density is

Table 5. Below-ground respiratory components in mycorrhizal and non-mycorrhizal *Citrus volkameriana* seedlings grown in low- or high-phosphorus soil (from Peng et al., 1993).

	Low-phosphorus		High-phosphorus	
	Mycorrhizal	Non-mycorrhizal	Mycorrhizal	Non-mycorrhizal
Root dry weight (mg)	134	78	273	230
Root growth (mg day^{-1})	9.62	3.90	14.8	13.4
Daily cost (µmol CO_2 day^{-1})				
Construction cost	468	174	671	563
Root growth (carbon)[a]	356	135	518	446
Total respiration[b]	555	234	820	600
Growth respiration[c]	114	40	153	117
Maintenance respiration[d]	330	138	558	372
Ion uptake respiration[e]	111	56	109	111
Cost per unit root dry weight [mmol CO_2 (g new root)$^{-1}$]				
Construction cost	48.7	44.7	45.3	42.0
Growth respiration coefficient	11.8	10.1	10.3	8.7
Specific respiration rates [mmol CO_2 (g whole-root system)$^{-1}$ d^{-1}]				
Total respiration	4.14	3.00	3.00	2.61
Maintenance + ion uptake respiration	3.33	2.51	2.44	2.10

[a] Calculated based on ash content, N content, and heat of combustion. [b] Measured by gas exchange. [c] Construction cost – root growth. [d] Total respiration – (growth + ion uptake respiration). [e] Estimated based on the change in whole-plant N content.

typically about 0.1 m cm^{-3} of soil near the soil surface (Marschner, 1995), although in pastures root length density can exceed 1 m cm^{-3} of soil (Newman et al., 1989). If we assume an average root radius of 0.2 mm, then roots with a root length density of 0.1 m cm^{-3} of soil would occupy about 1.3% of the soil volume, or 16 times more soil volume than external hyphae. If we further assume similar tissue densities of fungal biomass and plant biomass, then fine root biomass is about 16-fold greater than hyphal biomass outside the root. Extraradical hyphae produced by arbuscular mycorrhizas would therefore have considerably less carbon costs associated with their construction than the same length of fine roots produced by the host. Theoretical explorations have emphasized the efficiency of mycorrhizal hyphae based on their very small diameter (Yanai et al., 1995).

Mean hyphal length produced by ectomycorrhizal fungi associated with seedlings of *Pinus sylvestris*, *Pinus taeda*, and *S. viminalis* can reach 3-80 m cm^{-1} of root (Read and Boyd, 1986; Jones et al., 1990; Rousseau et al., 1994), and in *P. sylvestris*, extraradical hyphae accounted for 15% of root dry weight with *Laccaria laccata* and 123% of root dry weight with *Paxillus involutus* (Colpaert et al., 1992). Likewise, in ectomycorrhizal *Pinus pinaster* with *Hebeloma cylindrosporum*, extraradical hyphae accounted for up to 20% of root dry weight (Plassard et al., 1994). Under field conditions, the total amount of hyphal biomass produced in forest soils by ectomycorrhizal fungi (i.e. 1.25 to 2.0 kg m^{-2}) was nearly equivalent to the fine root biomass production (Högberg et al., 2001; Wallander et al., 2001; Högberg and Högberg, 2002). Thus, extraradical hyphal biomass in ectomycorrhizal plants can greatly exceed that found in arbuscular mycorrhizal plants, and should consequently have even faster rates of respiration for growth than the arbuscular fungi.

B. Maintenance Respiration

Mycorrhizal associations require respiratory energy to maintain existing fungal structures and activities, as well as any host cellular activities linked to the presence of the symbiont. Eighty-four percent of the difference in daily total root and soil respiration between mycorrhizal and non-mycorrhizal *C. volkameriana* seedlings was considered maintenance respiration (Fig. 4; Table 5). Higher maintenance respiration in the mycorrhizal seedlings was attributed to both a larger root system and to apparently greater specific rates of maintenance respiration (which, in this case, also included respiration associated with ion uptake, microbial respiration, and growth respiration of the extraradical hyphae). Increased maintenance respiration also appeared to account for most of the respiration associated with mycorrhizal colonization in common bean (*Phaselous vulgaris*) (Nielsen et al., 1998).

Assuming maintenance respiration represents most of the respiratory costs associated with mycorrhizal fungi, maintenance respiration of mycorrhizal tissue appears to be considerable. Per unit biomass, mycelial respiration by mycorrhizal fungi is several orders of magnitude faster than host root respiration (Martin et al., 1987; Rygiewicz and Anderson, 1994; Eltrop and Marschner, 1996). Per unit length, however, the fungi cost considerably less to maintain than roots. Coupled with lower construction costs, mycorrhizal associations enable the host plant to explore more soil volume per unit of carbon invested, and increase the efficiency of nutrient capture when soil resources are limited (Eissenstat and Volder, in press).

C. Ion Uptake Respiration

Mycorrhizal fungi often increase the ability of many plants to acquire soil nutrients, and therefore, may increase the energy demand required for ion uptake. Baas et al. (1989) attributed 13% of the increased respiration associated with mycorrhizal colonization in *Plantago major* to increased nutrient uptake. However, in many cases, mycorrhizas do not tend to appreciably enhance plant uptake of mobile soil ions including nitrate (Tinker and Nye, 2000), which quantitatively is the most important ion associated with uptake respiration (Veen, 1981). Thus, the importance of mycorrhizas on ion uptake respiration may be somewhat limited. Hawkins et al. (1999), for example, found no increase in ion uptake respiration due to colonization by *G. mosseae* in wheat when plants were grown hydroponically under non-limiting nutrient conditions. Colonization by *G. intraradices* also had no effect on ion uptake respiration in *C. volkameriana* seedlings grown under high-phosphorus soil conditions, but did increase ion uptake respiration when plants were grown under low-phosphorus conditions (Table 5). This was likely due to the fact that under low-phosphorus conditions, seedlings colonized by the fungi were larger and had faster rates of ion uptake by roots and hyphae than uncolonized seedlings.

Ectomycorrhizal fungi can also utilize organic forms of nitrogen by producing extracellular proteinases (Abuzinadah and Read, 1986; Zhu et al., 1990; Maijala et al., 1991), thereby providing the host plant access to nitrogen sources that would otherwise be unavailable. Respiratory costs associated with nitrogen uptake by this process are unknown.

IV. Other Respiratory Costs

There are other respiratory costs associated with mycorrhizal fungi that are potentially important to the carbon economy of the host plant, but these costs have received relatively little attention in the literature. They include respiration associated with fungal reproduction, respiration of microorganisms residing in the region of soil surrounding the extraradical hyphae, termed the mycorrhizosphere, and respiration associated with forming hyphal links with neighboring plants.

A. Fungal Reproduction

Most mycorrhizal studies have been done with young plants under laboratory or glasshouse conditions, and therefore provide no information on the carbon requirements of the mycorrhizal fruiting bodies or spores typically associated with mature vegetation in the field. The development of spores and sporocarps by mycorrhizal fungi represents significant production of fungal biomass in a relatively short time, capable of exceeding several kg m^{-2} yr^{-1} (e.g., Sieverding et al., 1989; Johnson, 1994), and depends on current assimilate from the host (Last et al., 1979; Lamhamedi et al., 1994; Högberg et al., 2001). Mycorrhizal spores are especially rich in lipids and fatty acids which comprise more than 45 to 95% of their carbon pool (Jabaji-Hare, 1988; Bago et al., 1999; Olsson and Johansen, 2000). Reproduction by the fungi thus will require large amounts of carbon for respiration and structural build-up, particularly when conditions are most favorable for sporulation such as late in the growing season (Menge, 1984; An et al., 1993; Smith and Read, 1997).

B. Microorganisms Associated with the Mycorrhizosphere

Mycorrhizas strongly influence rhizosphere microbial populations by altering the nutrient and carbon physiology of the host plant, and by changing the chemical and physical properties of the soil environment (Azcón-Aguilar and Barea, 1992; Linderman, 1992). Several microorganisms also influence the establishment of mycorrhizal fungi (Garbaye, 1994; Perotto and Bonfante, 1997), and facilitate the release of soil nutrients prior to mycorrhizal uptake (Toro et al., 1997). It is expected that most or even all of the energy required by these microorganisms, which may include both beneficial and non-beneficial species, is derived from host assimilates. However, their greatest impact on respiratory costs is probably through their direct and indirect effects on mycorrhizal associations.

C. Hyphal Links between Plants

Many mycorrhizas have low host specificity and are capable of forming inter- and intraspecific hyphal connections between neighboring plants. A number of studies have reported a net transfer of carbon between plants linked by these connections (e.g., Francis and Read, 1984; Grime et al., 1987; Simard et al., 1997), although the quantitative significance of this transfer to the carbon status of the host and recipient plants remains questionable (Fitter et al., 1998; Robinson and Fitter, 1999). Using imaging plate autoradiography, Wu et al. (2001) recorded continual movement of ^{14}C-labeled photosynthetic products between *Pinus densiflora* seedlings linked by extraradical hyphae of *Pisolithus tinctorius* or by an unidentified ectomycorrhizal fungus. Regardless of the mycorrhizal species, within 3 days of labeling, ^{14}C was detected in the extraradical hyphae and the colonized roots of the unlabeled 'receiver' seedling. Reverse-labeling demonstrated that carbon also moved from the 'receiver' seedling to the extraradical hyphae. However, carbon movement between the plants themselves by way of interlinking hyphae was not detected. Therefore, although evidence for carbon transfer between plants (for tissue growth and maintenance) was lacking in this study, the data do illustrate that hyphal links between plants may help reduce the costs associated with supporting the mycorrhiza. In fact, two other reports indicate that as much as 10% of the carbon of an arbuscular mycorrhizal root can be derived from another plant when linked by fungal hyphae (Watkins et al., 1996; Graves et al., 1997). Hyphal links might also reduce respiratory costs associated with ion uptake if nitrogen and other nutrients are transferred between plants (e.g., Frey and Schüepp, 1992).

V. Conclusions

Ever since mechanistic evidence for the dependence of mycorrhizal fungi on host carbon was first presented 30 years ago (Ho and Trappe, 1973), a fair amount of research has been devoted to elucidating the host energy demands for supporting mycorrhizal associations. We now know from laboratory and greenhouse studies that a considerable amount of photosynthate is required by mycorrhizas, and at least half of it is used for respiratory processes. However, there are still many questions about mycorrhizal respiration that remain unanswered. Respiratory costs of ericaceous and orchid mycorrhizas, for example, are entirely unknown, despite the growing realization of their importance in many ecosystems. We also have very little understanding of mycorrhizal respiratory costs on mature plants. Only recently have studies shown the significance of mycorrhizal fungi as important pathways of carbon flux from plants to soil to atmosphere under field conditions (Johnson et al., 2002a, b). New techniques need to be developed to measure respiration of mycorrhizal roots under field conditions (e.g., Espeleta et al., 1998; Bryla et al., 2001; Kutsch et al., 2001; Johnson et al., 2002a,b), and provide answers to the relevance of this symbiosis to overall carbon economy of associated plants.

Acknowledgments

The authors wish to thank Dr. Paul Schreiner and Ms. Maryann Resendes for critically reading the manuscript.

References

Abuzinadah RA and Read DJ (1986) The role of proteins in the nitrogen nutrition of ectomycorrhizal plants. III. Protein utilization by *Betula*, *Picea* and *Pinus* in ectomycorrhizal association with *Hebeloma crustuliniforme*. New Phytol 103: 506–514

Amijee F, Tinker PB and Stribley DP (1989) The development of endomycorrhizal root systems. VII. A detailed study of effects of soil phosphorus on colonization. New Phytol 111: 435–446

An ZQ, Guo BZ and Hendrix JW (1993) Populations of spores and propagules of mycorrhizal fungi in relation to the life-cycles of tall fescue and tobacco. Soil Biol Biochem 25: 813–817

Anderson CP (2003) Source-sink balance and carbon allocation below ground in plants exposed to ozone. New Phytol 157: 213–228

Andersen CP and Rygiewicz PT (1995) Allocation of carbon in mycorrhizal *Pinus ponderosa* seedlings exposed to ozone. New Phytol 131: 471–480

Andrews JA, Harrison KG, Matamala R and Schlesinger WH (1999) Separation of root respiration from soil respiration using carbon-13 labeling during Free-Air Carbon Enrichment (FACE). J Soil Sci Soc Am 63: 1429–1435

Antibus RK and Sinsabaugh RL (1993) The extraction and quantification of ergosterol from ectomycorrhizal fungi and roots. Mycorrhiza 3: 137–144

Azcón-Aguilar C and Barea JM (1992) Interactions between mycorrhizal fungi and other rhizosphere microorganisms. In: Allen MF (ed) Mycorrhizal Functioning. An Integrative Plant-Fungal Process, pp. 163–198. Chapman and Hall, New York

Baas R and Lambers H (1988) Effects of vesicular-arbuscular mycorrhizal infection and phosphate on *Plantago major* spp. *pleiosperma* in relation to the internal phosphate concentration. Physiol Plant 74: 701–707

Baas R, Werf AVD and Lambers H (1989) Root respiration and growth in *Plantago major* as affected by vesicular-arbuscular mycorrhizal infection. Plant Physiol 91: 227–232

Bago B, Pfeffer PE, Douds DD, Brouillette J, Bécard G and Shachar-Hill Y (1999) Carbon metabolism in spores of the arbuscular mycorrhizal fungus *Glomus intraradices* as revealed by nuclear magnetic resonance spectroscopy. Plant Physiol 121: 263–272

Bago B, Pfeffer PE, Zipfel W, Lammers P and Shachar-Hill Y (2002) Tracking metabolism and imaging transport in arbuscular mycorrhizal fungi. Plant Soil 244: 189–197

Ball AS and Drake BG (1998) Stimulation of soil respiration by carbon dioxide enrichment of marsh vegetation. Soil Biol Biochem 30: 1203–1205

Berta G, Tagliasacchi AM, Fusconi A, Gerlero D and Trotta A (1991) The mitotic cycle in root apical meristems of *Allium porrum* L. is controlled by the mycorrhizal fungus *Glomus* sp. Strain E. Protoplasma 161: 12–16

Berta G, Fusconi A, Lingua G, Trotta A and Sgorbati S (1996) Influence of arbuscular mycorrhizal infection on nuclear structure and activity during root morphogenesis. In: Azcón-Aguilar C and Barea JM (eds) Mycorrhizas in Integrated Systems: From Genes to Plant Development, pp 174–177. European Commission, Brussels

Bidartondo MI, Ek H, Wallander H and Söderström B (2001) Do nutrient additions alter carbon sink strength of ectomycorrhizal fungi? New Phytol 151: 543–550

Bloom AJ, Chapin III FS and Mooney HA (1985) Resource limitations in plants—an economic analogy. Annu Rev Ecol Syst 16: 363–392

Bouma TJ and Bryla DR (2000) On the assessment of root respiration for soils of different textures: Interactions with soil moisture and soil CO_2 concentrations. Plant Soil 227: 215–221

Bouma TJ, Bryla DR, Li Y and Eissenstat DM (2000) Is maintenance respiration in roots constant? In: Stokes A (ed) The Supporting Roots of Trees and Woody Plants: Form, Function and Physiology, Vol 87, pp 391–396. Kluwer Academic Publishers, Dordrecht

Bouma TJ, Yanai RD, Elkin AD, Hartmond U, Flores-Alva DE and Eissenstat DM (2001) Estimating age-dependent costs and benefits of roots with contrasting life span: Comparing apples and oranges. New Phytol 150: 685–695

Brouwer R (1983) Functional equilibrium: Sense or nonsense? Neth J Agric Sci 31: 335–348

Bryla DR and Koide RT (1990) Role of mycorrhizal infection in the growth and reproduction of wild vs. cultivated plants. II. Eight wild accessions and two cultivars of *Lycopersicon esculentum* Mill. Oecologia 84: 82–92

Bryla DR, Bouma TJ and Eissenstat DM (1997) Root respiration in citrus acclimates to temperature and slows during drought. Plant, Cell Environ 20: 1411–1420

Bryla DR, Bouma TJ, Hartmond U and Eissenstat DM (2001) Influence of temperature and soil drying on respiration of individual roots in citrus: Integrating greenhouse observations into a predictive model for the field. Plant, Cell Environ 24: 781–790

Bücking H and Heyser W (2003) Uptake and transfer of nutrients in ectomycorrhizal associations: Interactions between photosynthesis and phosphate nutrition. Mycorrhiza 13: 59–68

Burgess T, Dell B and Malajczuk N (1994) Variation in mycorrhizal development and growth stimulation of 20 isolates of *Pisolithus* inoculated onto *Eucalyptus grandis* W. Hill ex Maiden. New Phytol 127: 731–739

Burton AJ, Pregitzer KS, Zogg GP and Zak DR (1998) Drought reduces root respiration in sugar maple forest. Ecol Appl 8: 771–778

Burton AJ, Pregitzer KS, Ruess RW, Hendrik RL and Allen MF (2002) Root respiration in North American forests: Effects of nitrogen concentration and temperature across biomes. Oecologia 131: 559–568

Buwalda JG and Goh KM (1982) Host-fungus competition for carbon as a cause of growth depressions in vesicular-arbuscular mycorrhizal ryegrass. Soil Biol Biochem 14: 103–106

Cairney JWG and Alexander IJ (1992) A study of spruce (*Picea sitchensis* (Bong.) Carr.) ectomycorrhizas. II. Carbohydrate allocation to ageing *Picea sitchenis*/*Tylospora fibrillosa* (Burt.) Donk ectomycorrhizas. New Phytol 122: 153–158

Cairney JWG, Ashford AE, and Allaway WG (1989) Distribution of photosynthetically fixed carbon within root systems of *Eucalyptus pilularis* plants ectomycorrhizal with *Pisolithus tinctorius*. New Phytol 112: 495–500

Chambers CA, Smith SE and Smith FA (1980) Effects of ammonium and nitrate ions on mycorrhizal infection, nodulation and growth of *Trifolium subterraneum*. New Phytol 85: 47–62

Colpaert JV, van Assche JA and Luijtens K (1992) The growth of the extramatrical mycelium of ectomycorrhizal fungi and the growth response of *Pinus sylvestris* L. New Phytol 120: 127–135

Cox G, Sanders FE, Tinker PB and Wild JA (1975) Ultrastructural evidence relating to host-endophyte transfer in a vesicular-arbuscular mycorrhiza. In: Sanders FE, Mosse B and Tinker PB (eds) Endomycorrhizas, pp. 297–312. Academic Press, London

de Miranda JCC, Harris PJ and Wild A (1989) Effects of soil and plant phosphorus concentration on vesicular-arbuscular mycorrhiza in sorghum plants. New Phytol 112: 405–410

Douds DD Jr, Johnson CR and Koch KE (1988) Carbon cost of the fungal symbiont relative to net leaf P accumulation in a split-root VA mycorrhizal symbiosis. Plant Physiol 86: 491–496

Douds DD, Pfeffer PE and Shachar-Hill Y (2000) Carbon partitioning, cost and metabolism of arbuscular mycorrhizae. In: Kapulnick Y and Douds DD (eds) Arbuscular Mycorrhizas: Physiology and Function, pp. 107–130. Kluwer Academic Press, New York

Durall DM, Jones MD and Tinker PB (1994) Allocation of ^{14}C-carbon in ectomycorrhizal willow. New Phytol 128: 109–114

Eissenstat DM, Graham JH, Syvertsen JP and Drouillard DL

(1993) Carbon economy of sour orange in relation to mycorrhizal colonization and phosphorus status. Ann Bot 71: 1–10

Eltrop L and Marschner H (1996) Growth and mineral nutrition of non-mycorrhizal and mycorrhizal Norway spruce (*Picea abies*) seedlings grown in semi-hydroponic sand culture. II. Carbon partitioning in plants supplied with ammonium or nitrate. New Phytol 133: 479–486

Espeleta JF and Eissenstat DM (1998) Responses of citrus fine roots to localized soil drying: A comparison of seedlings with adult fruit trees. Tree Physiol 18: 113–119

Espeleta JF, Eissenstat DM and Graham JH (1998) Citrus root responses to localized drying soil: A new approach to studying mycorrhizal effects on the root of mature trees. Plant Soil 206: 1–10

Estaún V, Calvet C and Hayman DS (1987) Influence of plant genotype on mycorrhizal infection: Response of three pea cultivars. Plant Soil 103: 295–298

Finlay R and Söderström B (1992) Mycorrhiza and carbon flow to the soil. In: MF Allen (ed), Mycorrhizal Functioning. An Integrative Plant-Fungal Process, pp. 134–160. Chapman & Hall, New York

Fitter AH, Graves JD, Watkins NK, Robinson D and Scrimgeour C (1998) Carbon transfer between plants and its control in networks of arbuscular mycorrhizas. Funct Ecol 12: 406–412

Francis R and Read DJ (1984) Direct transfer of carbon between plants connected by vesicular-arbuscular mycorrhizal fungi. Nature 307: 53–56

Fredeen AL and Terry N (1988) Influence of vesicular-arbuscular mycorrhizal infection and soil phosphorus level on growth and carbon metabolism of soybean. Can J Bot 66: 2311–2316

Frey B and Schüepp H (1992) Transfer of symbiotically fixed nitrogen from berseem (*Trifolium alexandrium* L.) to maize via vesicular arbuscular mycorrhizal hyphae. New Phytol 122: 447–454

Gansert D (1994) Root respiration and its importance for the carbon balance of beech saplings (*Fagus slyvatica* L.) in a montane beech forest. Plant Soil 167: 109–119

Garbaye J (1994) Helper bacteria: A new dimension to the mycorrhizal symbiosis. New Phytol 128: 197–210

Giovannetti M and Hepper CM (1985) Vesicular-arbuscular mycorrhizal infection in *Hedysarum coronarium* and *Onobrychis viciifolia*: Host-endophyte specificity. Soil Biol Biochem 17: 899–900

Gorissen A, Joosten NN and Jansen AE (1991) Effects of ozone and ammonium-sulfate on carbon partitioning to mycorrhizal roots of juvenile Douglas fir. New Phytol 119: 243–250

Graham JH and Eissenstat DM (1994) Host genotype and the formation and function of VA mycorrhizae. Plant Soil 159: 179–185

Graham JH, Leonard RT and Menge JA (1982a) Interactions of light intensity and soil temperature with phosphorus inhibition of vesicular-arbuscular mycorrhizal formation. New Phytol 91: 683–690

Graham JH, Linderman RG and Menge JA (1982b) Development of external hyphae by different isolates of mycorrhizal *Glomus* spp. in relation to root colonization and growth of Troyer citrange. New Phytol 91: 183–189

Graham JH, Hodge NC and Morton JB (1995) Fatty acid methyl ester profiles for characterization of Glomalean fungi and their mycorrhizae. Appl Environ Microbiol 61: 58–64

Graves JD, Watkins NK, Fitter AH, Robinson D and Scrimgeour C (1997) Instraspecific transfer of carbon between plants linked by a common mycorrhizal network. Plant Soil 192: 153–159

Grime JP, Mackey JML, Hillier SH and Read DJ (1987) Floristic diversity in a model system using experimental microcosms. Nature 328: 420–422

Harris D, Packovsky RS and Paul EA (1985) Carbon economy of soybean-*Rhizobium-Glomus* associations. New Phytol 101: 427–440

Haselwandter K, Bobleter O and Read DJ (1990) Degradation of ^{14}C-labelled lignin and dehydropolymer of coniferyl alcohol by ericoid and ectomycorrhizal fungi. Arch Microbiol 153: 352–354

Hawkins HJ, Cramer MD and George E (1999) Root respiratory quotient and nitrate uptake in hydroponically grown non-mycorrhizal and mycorrhizal wheat. Mycorrhiza 9: 57–60

Hayman DS (1970) *Endogone* spore number in soil and vesicular-arbuscular mycorrhiza in wheat as influenced by season and soil treatment. Trans British Mycol Soc 54: 53–63

Henry A and Kosola K (1999) Root age and phosphorus effects on colonization of *Andropogon gerardii* by mycorrhizal fungi. Soil Biol Biochem 31: 1657–1660

Hepper CM (1977) A colorimetric method for estimating vesicular-arbuscular mycorrhizal infection in roots. Soil Biol Biochem 9: 15–18

Ho I and Trappe JM (1973) Translocation of ^{14}C from *Festuca* plants to their endomycorrhizal fungi. Nature 244: 30–31

Ho I and Trappe JM (1981) Effects of ozone exposure on mycorrhiza formation and growth of *Festuca arundinacea*. Environ Exp Bot 24: 71–74

Högberg MN and Högberg P (2002) Extramatrical ectomycorrhizal mycelium contributes one-third of microbial biomass and produces, together with associated roots, half the dissolved organic carbon in a forest. New Phytol 154: 791-795.

Högberg P, Nordgren A, Buchmann N, Taylor AFS, Ekblad A, Högberg MN, Nyberg G, Ottosson-Lofvenius M and Read DJ (2001) Large-scale forest girdling shows that current photosynthesis drives soil respiration. Nature 411: 789–792

Hungate BA, Holland EA, Jackson RB, Chapin FS III, Mooney HA and Field CB (1997) The fate of carbon in grasslands under carbon dioxide enrichment. Nature 388: 576–579

Ineichen K, Wiemken V and Wiemken A (1995) Shoots, roots and ectomycorrhiza formation of pine seedlings at elevated atmospheric carbon dioxide. Plant Cell Environ 18: 703–709

Ingestad T, Arveby A and Kähr M (1986) The influence of ectomycorrhiza on nitrogen nutrition and growth of *Pinus sylvestris* seedlings. Physiol Plant 68: 575–582

Jabaji-Hare S (1988) Lipid and fatty acid profiles of some vesicular-arbuscular mycorrhizal fungi: Contribution to taxonomy. Mycologia 80: 622–629

Jackobsen I and Rosendahl L (1990) Carbon flow into soil and external hyphae from roots of mycorrhizal cucumber plants. New Phytol 115: 77–83

Jackobsen I, Abbott LK and Robson AD (1992) External hyphae of vesicular-arbuscular mycorrhizal fungi associated with *Trifolium subterraneum* L. 1. Spread of hyphae and phosphorus inflow into roots. New Phytol 120: 371–380

Jifon JL, Graham JH, Drouillard DL and Syvertsen JP (2002) Growth depression of mycorrhizal *citrus* seedlings grown at high phosphorus supply is mitigated by elevated CO_2. New Phytol 153: 133–142

Johnson C (1994) Fruiting of hypogeous fungi in dry sclerophyll forest in Tasmania, Australia — seasonal variation and annual production. Can J Bot 98: 1173–1182

Johnson D, Leake JR and Read DJ (2002a) Transfer of recent photosynthate into mycorrhizal mycelium of an upland grassland: Short-term respiratory losses and accumulation of ^{14}C. Soil Biol Biochem 34: 1521–1524

Johnson D, Leake JR, Ostle N, Ineson P and Read DJ (2002b) In situ $^{13}CO_2$ pulse-labelling of upland grassland demonstrates a rapid pathway of carbon flux from arbuscular mycorrhizal mycelia to the soil. New Phytol 153: 327–334

Jones MD, Durall DM and Tinker PB (1990) Phosphorus relationships and production of extramatrical hyphae by two types of willow ectomycorrhizas at different soil phosphorus levels. New Phytol 115: 259–267

Koch KE and Johnson CR (1984) Photosynthesis partitioning in split-root citrus seedlings with mycorrhizal and nonmycorrhizal root systems. Plant Physiol 75: 26–30

Koide RT (1985) The nature of growth depressions in sunflower caused by vesicular-arbuscular mycorrhizal infection. New Phytol 99: 449–462

Koide R and Elliott G (1989) Cost, benefit and efficiency of the vesicular-arbuscular mycorrhizal symbiosis. Funct Ecol 3: 249–255

Koide RT and Li M (1990) On host regulation of the vesicular-arbuscular mycorrhizal symbiosis. New Phytol 114: 59–65

Koide RT and Dickie IA (2002) Effects of mycorrhizal fungi on plant populations. Plant Soil 244: 307–317

Krishna KR, Shetty KG, Dart PJ and Andrews DJ (1985) Genotype dependent variation in mycorrhizal colonization and response to inoculation of pearl millet. Plant Soil 86: 113–125

Kucy RMN and Paul EA (1982) Carbon flow, photosynthesis, and N_2 fixation in mycorrhizal and nodulated faba beans (*Vicia faba* L.). Soil Biol Biochem 14: 407–412

Kutsch WL, Staack A, Wojtzel J, Middlehoff U and Kappen L (2001) Field measurements of root respiration and total soil respiration in an alder forest. New Phytol 150: 157–168

Lambers H, Atkins OK and Millenaar FF (2002) Respiratory patterns in roots in relation to their functioning. In: Waisel Y, Eshel A and Kafkafi U (eds) Plant Roots. The Hidden Half. Third Edition, pp. 521–552. Marcel Dekker, Inc., New York

Lamhamedi MS, Godbout C and Fortin JA (1994) Dependence of *Laccaria bicolor* basidiome development on current photosynthesis of *Pinus strobus* seedlings. Can J For Res 24: 1797–1804

Last FT, Pelham J, Mason PA and Ingleby K (1979) Influence of leaves on sporophore production by fungi forming sheating mycorrhizas with *Betula* spp. Nature 180: 168–169

Lerat S, Lapointe L, Gutjahr S, Piché Y and Vierheilig H (2003) Carbon partitioning in a split-root system of arbuscular mycorrhizal plants is fungal and plant species dependent. New Phytol 157: 589–595

Lewis DH and Harley JH (1965a) Carbohydrate physiology of mycorrhizal roots of beech. I. Identity of endogenous sugars and utilization of exogenous sugars. New Phytol 64: 224–231

Lewis DH and Harley JH (1965b) Carbohydrate physiology of mycorrhizal roots of beech. II. Utilization of exogenous sugars by uninfected and mycorrhizal roots. New Phytol 64: 238–255

Lewis DH and Harley JH (1965c) Carbohydrate physiology of mycorrhizal roots of beech. III. Movement of sugars between host and fungus. New Phytol 64: 256–269

Linderman RG (1992) Vesicular-arbuscular mycorrhizae and soil microbial interactions. In: Bethlenfalvay GJ and Linderman RG (eds) Mycorrhizae in Sustainable Agriculture, pp 45–70. ASA Special Publication Number 54, ASA-CSSA-SSSA, Madison

Ling-Lee M, Ashford AE and Chilvers GA (1977) A histochemical study of polysaccharide distribution in eucalypt mycorrhizas. New Phytol 78: 329–335

Lioi L and Giovannetti M (1987) Variable effectivity of three vesicular-arbuscular mycorrhizal endophytes in *Hedysarum coronarium* and *Medicago sativa*. Biol Fertil Soils 4: 193–197

Lu SJ, Mattson KG, Zaerr JB and Marshall JD (1998) Root respiration of Douglas fir seedlings: Effects of N concentration. Soil Biol Biochem 30: 331–336

Maijala P, Fagerstedt KF and Raudaskoski M (1991) Detection of extracellular cellulolytic and proteolytic activity in ectomycorrhizal fungi and *Heterobasidion annosum* (Fr.) Bref. New Phytol 117: 643–648

Marschner H (1995) Mineral Nutrition of Higher Plants, 2nd Edition. Academic Press: New York.

Martin F, Ramstedt M and Soderhall K (1987) Carbon and nitrogen metabolism in ectomycorrhizal and ectomycorrhizas. Biochimie 69: 569–581

McCool PM and Menge JA (1984) Influence of ozone on carbon partitioning in tomato: Potential role of carbon flow in regulation of mycorrhizal symbiosis under conditions of stress. New Phytol 94: 241–247

Meier S, Grand LF, Schoeneberger MM, Reinert RA and Bruck RI (1990) Growth, ectomycorrhizae and nonstructural carbohydrates of loblolly pine seedlings exposed to ozone and soil water deficit. Environ Poll 64: 11–27

Menge JA (1984) Inoculum production. In: Powell CL and Bagyaraj DJ (eds) VA Mycorrhiza, pp. 187–203. CRC Press, Inc., Boca Raton

Miller RM, Reinhardt DR and Jastrow JD (1995) External hyphal production of vesicular-arbuscular mycorrhizal fungi in pasture and tallgrass prairie communities. Oecologia 103: 17–23

Modjo HS and Hendrix JW (1986) The mycorrhizal fungus, *Glomus macrocarpum* as a cause of tobacco stunt disease. Phytopath 76: 668–691

Molina R and Chamard J (1983) Use of the ectomycorrhizal fungus *Laccaria laccata* in forestry. II. Effects of fertilizer forms and levels on ectomycorrhizal development and growth of container-grown Douglas-fir and ponderosa pine. Can J For Res 13: 89–95

Nagy S and Nordby HE (1980) Composition of lipids in roots of six citrus cultivars infected with the vesicular-arbuscular mycorrhizal fungus, *Glomus mosseae*. New Phytol 85: 377–384

Newman EI and Reddell P (1987) The distribution of mycorrhizas among families of vascular plants. New Phytol 106: 745–751

Newman EI, Ritz K and Jupp AP (1989) The functioning of roots in the grassland ecosystem. Asp Appl Biol 22: 263–269

Nielson KL, Bouma TJ, Lynch JP and Eissenstat DM (1998) Effect of phosphorus availability and vesicular-arbuscular mycorrhizas on the carbon budget of common bean (*Phaseolus vulgaris*). New Phytol 139: 647–656

Nilsson LO and Wallander H (2003) Production of external mycelium by ectomycorrhizal fungi in a Norway spruce forest was reduced in response to nitrogen fertilizer. New Phytol 158: 409–416

Olsson PA (1999) Signature of fatty acids provide tools for determination of the distribution and interactions of mycorrhizal fungi in soil. FEMS Microbiol Ecol 29: 303–310
Olsson PA and Johansen A (2000) Lipid and fatty acid composition of hyphae and spores of arbuscular mycorrhizal fungi at different growth stages. Mycol Res 104: 429–434
Pacovsky RS and Fuller G (1988) Mineral and lipid composition of *Glycine-Glomus-Bradyrhizobium* symbioses. Physiol Plant 72: 733–746
Palta JA and Nobel PS (1989a) Influences of water status, temperature, and root age on daily patterns of root respiration for two cactus species. Ann Bot 63: 651–662
Palta JA and Nobel PS (1989b) Root respiration for *Agave deserti*: Influence of temperature, water status, and root age on daily patterns. J Exp Bot 40: 181–186
Pang PC and Paul EA (1980) Effects of vesicular-arbuscular mycorrhiza on ^{14}C and ^{15}N distribution in nodulated fababeans. Can J Soil Sci 60: 241–250
Paul EA and Kucey RMN (1981) Carbon flow in plant microbial associations. Science 213: 473–474
Pearson JN and Jakobsen I (1993) Symbiotic exchange of carbon and phosphorus between cucumber and three arbuscular mycorrhizal fungi. New Phytol 124: 481–488
Peng SB, Eissenstat DM, Graham JH, Williams K and Hodge NC (1993) Growth depression in mycorrhizal citrus at high-phosphorus supply. Plant Physiol 101: 1063–1071
Perotto S and Bonfante P (1997) Bacterial association with mycorrhizal fungi: Close and distant friends in the rhizosphere. Trends Microbiol 5: 496–501
Piché Y, Fortin JA and Lafontaine JG (1981) Cytoplasmic phenols and polysaccharides in ectomycorrhizal and nonmycorrhizal short roots of pine. New Phytol 88: 695–703
Plassard C, Barry D, Eltrop L and Mousain D (1994) Nitrate uptake in maritime pine (*Pinus pinaster* Soland in Ait.) and the ectomycorrhizal fungus *Hebeloma cylindrosporum*: Effect of ectomycorrhizal symbiosis. Can J Bot 72: 189–197
Read DJ (1992) The mycorrhizal mycelium. In: MF Allen (ed), Mycorrhizal Functioning. An Integrative Plant-Fungal Process, pp. 102–133. Chapman & Hall, New York
Read DJ and Boyd R (1986) Water relations of mycorrhizal fungi and their host plants. In: Ayres P and Boddy L (eds) Water, Fungi and Plants, pp. 287–303. Cambridge University Press, Cambridge
Reich PB (2002) Root-shoot relations: Optimality in acclimation and adaptation or the 'Emperor's New Clothes?' In: Waisel Y, Eshel A and Kafkafi U (eds) Plant Roots. The Hidden Half. Third Edition, pp 205-220. Marcel Dekker, Inc., New York
Reich PB, Schoettle AW, Stroo HF and Amundson RG (1986) Acid rain and ozone influence mycorrhizal infection in tree seedlings. J Air Pollution Control Association 36: 724–726
Reid CPP, Kidd FA and Ekwebelam SA (1983) Nitrogen nutrition, photosynthesis and carbon allocation to ectomycorrhizal pine. Plant Soil 71: 415–432
Rillig MC, Allen MF, Klironomos JN and Field CB (1998) Arbuscular mycorrhizal percent root infection and infection intensity of *Bromus hordeaceous* grown in elevated atmospheric CO_2. Mycologia 90: 199–205
Robinson D and Fitter AH (1999) The magnitude and control of carbon transfer between plants linked by a common mycorrhizal network. J Exp Bot 50: 9–13
Rouhier H and Read DJ (1998) The role of mycorrhiza in determining the response of *Plantago lanceolata* to CO_2 enrichment. New Phytol 139: 367–373
Rousseau JVD and Reid CPP (1991) Effects of phosphorus fertilization and mycorrhizal development on phosphorus nutrition and carbon balance of loblolly pine. New Phytol 117: 319–326
Rousseau JVD, Sylvia DM and Fox AJ (1994) Contribution of ectomycorrhizas to the potential nutrient absorbing surface of pine. New Phytol 128: 639–644
Rygiewicz PT and Andersen CP (1994) Mycorrhizae alter quality and quantity of carbon allocated below ground. Nature 369: 58–60
Sanders FE, Tinker PB, Black RLB and Palmerley SM (1977) The development of endomycorrhizal root systems. I. Spread of infection and growth promoting effects with four species of vesicular-arbuscular mycorrhizas. New Phytol 78: 257–268
Sanders IR (1996) Plant-fungal interactions in a CO_2-rich world. In: Korner C and Bazzaz FA (eds) Carbon Dioxide, Populations, and Communities, pp. 265–272. Academic Press, New York
Sanders IR, Streitwolf-Engel R, van der Heijden MGA, Boller T and Wiemken A (1998) Increased allocation to external hyphae of arbuscular mycorrhiza fungi under CO_2 enrichment. Oecologia 117: 496–503
Scagel CF and Andersen CP (1997) Seasonal changes in root and soil respiration of ozone-exposed Ponderosa pine (*Pinus ponderosa*) grown in different substrates. New Phytol 136: 627–643
Sieverding E, Toro S and Mosquera O (1989) Biomass production and nutrient concentrations in spores of VA mycorrhizal fungi. Soil Biol Biochem 21: 60–72
Simard SW, Perry DA, Jone MD, Myrold DD, Durall DM and Molina R (1997) Net carbon transfer between ectomycorrhizal tree species in the field. Nature 388: 579–582
Simmons GL and Kelly JM (1989) Influence of O_3, rainfall acidity, and soil Mg status on growth and ectomycorrhizal colonization of loblolly pine roots. Water Air Soil Poll 44: 159–171
Smith SE and Read DJ (1997) Mycorrhizal Symbiosis. Second Edition. Academic Press, New York
Snellgrove RC, Splittstoesser WE, Stribley DP and Tinker PB (1982) The distribution of carbon and the demand of the fungal symbiont in leek plants with vesicular-arbuscular mycorrhizas. New Phytol 92: 75–87
Söderström BE and Read DJ (1987) Respiratory activity of intact and excised ectomycorrhizal mycelial systems growing in unsterilized soil. Soil Biol Biochem 19: 231–236
Son CL and Smith SE (1988) Mycorrhizal growth responses: Interactions between photon irradiance and phosphorus nutrition. New Phytol 108: 305–314
Staddon PL, Fitter AH and Robinson D (1999) Effects of mycorrhizal colonization and elevated atmospheric carbon dioxide on carbon fixation and below-ground carbon partitioning in *Plantago lanceolata*. J Exp Bot 50: 853–860
Stroo HF, Reich PB, Schoettle AW and Amundson RG (1988) Effects of ozone and acid rain on white pine (*Pinus strobus*) seedlings grown in five soils. II. Mycorrhizal infection. Can J Bot 66: 1510–1516
Taylor J and Harrier L (2000) A comparison of nine species of arbuscular mycorrhizal fungi on the development and nutrition of micropropagated *Rubus idaeus* L. cv. Glen Prosen (Red Raspberry). Plant Soil 225: 53–61
Tester M, Smith SE, Smith FA and Walker NA (1986) Effects

of photon irradiance on the growth of shoots and roots, on the rate of initiation of mycorrhizal infection and the growth of infection units in *Trifolium subterraneum* L. New Phytol 103: 375–390

Thomson BD, Robson AD and Abbott LK (1986) Effects of phosphorus on the formation of mycorrhizas by *Gigaspora calospora* and *Glomus fasciculatum* in relation to root carbohydrates. New Phytol 103: 751–765

Tinker PB and Nye PH (2000) Solute Movement in the Rhizosphere. Oxford University Press, Oxford

Tinker PB, Durall DM and Jones MD (1994) Carbon use efficiency in mycorrhizas: Theory and sample calculations. New Phytol 128: 115–122

Tisdall JM and Oades JM (1979) Stabilization of soil aggregates by the root systems of ryegrass. Aust J Soil Res 17: 429–441

Toro M, Azcon R and Barea JM (1997) Improvement of arbuscular mycorrhiza development by inoculation of soil with phosphate bacteria to improve rock phosphate bioavailability (^{32}P) and nutrient cycling. Appl Environ Microbiol 63: 4408–4412

Trappe JM (1987) Phylogenetic and ecological aspects of mycotrophy in the angiosperms from an evolutionary standpoint. In: Safir GR (ed) Ecophysiology of VA Mycorrhizal Plants, pp 5–25. CRC Press, Boca Raton

Van der Heijden MGA, Klironomos JN, Ursic M, Moutoglis P, Streitwolf-Engel R, Boller T, Wiemken A and Sanders IR (1998) Mycorrhizal fungal diversity determines plant biodiversity, ecosystem variability and productivity. Nature 396: 69–72

Veen BW (1981) Relation between root respiration and root activity. Plant Soil 63: 73–76

Vogt KA, Grier CC, Meier CE and Edmonds RL (1982) Mycorrhizal role in net primary production and nutrient cycling in *Abies amabilis* ecosystems in western Washington. Ecology 63: 370–380

Vogt KA, Publicover DA and Vogt DJ (1991) A critique of the role of ectomycorrhizas in forest ecology. Agric, Ecosyst Environ 35: 171–190

Wallander H and Nylund J-E (1991) Effects of excess nitrogen on carbohydrate concentration and mycorrhizal development of *Pinus sylvestris* seedlings. New Phytol 119: 405–411

Wallander H and Nylund J-E (1992) Effects of excess nitrogen and phosphorus starvation on the extramatrical mycelium of ectomycorrhizas of *Pinus sylvestris* L. New Phytol 120: 495–503

Wallander H, Nilsson LO, Hagerberg D and Bååth E (2001) Estimation of the biomass and seasonal growth of external mycelium of ectomycorrhizal fungi in the field. New Phytol 151: 753–760

Watkins NK, Fitter AH, Graves JD and Robinson D (1996) Carbon transfer between C_3 and C_4 plants linked by a common mycorrhizal network, quantified using stable carbon isotopes. Soil Biol Biochem 28: 471–477

Wilcox HE (1996) Mycorrhizae. In: Waisel Y, Eshel A and Kafkafi U (eds) Plant Roots. The Hidden Half. Second Edition, Revised and Expanded, pp 689–721. Marcel Dekker, Inc., New York

Wong KKY, Piché Y, Montpetit D and Kropp BR (1989) Differences in the colonisation of *Pinus banksiana* roots by sib-monokaryotic and dikaryotic strains of ectomycorrhizal *Laccaria bicolor*. Can J Bot 67: 1717–1726

Wong KKY, Piché Y and Fortin JA (1990) Differential development of root colonisation among four closely related genotypes of ectomycorrhizal *Laccaria bicolor*. Mycol Res 94: 876-884

Wright DP, Read DJ and Scholes JD (1998) Mycorrhizal sink strength influences whole plant carbon balance of *Trifolium repens* L. Plant, Cell Environ 21: 881–891

Wright DP, Scholes JD, Read DJ and Rolfe SA (1999) Changes in carbon allocation and expression of carbon transporter genes in *Betula pendula* Roth. colonized by the ectomycorrhizal fungus *Paxillus involutus* (Batsch) Fr. Plant, Cell Environ 23: 39–49

Wu B, Nara K and Hogetsu T (2001) Can ^{14}C-labeled photosynthetic products move between *Pinus densiflora* seedlings linked by ectomycorrhizal mycelia? New Phytol 149: 137–146

Yanai RD, Fahey TJ and Miller SL (1995) Efficiency of nutrient acquisition by fine roots and mycorrhizae. In: Smith WK and Hinckley TM (eds) Resource Physiology of Conifers, pp. 75–103. Academic Press, New York

Zhu H, Guo D and Dancik B (1990) Purification and characterization of an extracellular acid proteinase from the ectomycorrhizal fungus *Hebeloma crustuliniforme*. Appl Environ Microbiol 56: 837–843

Chapter 13

Integrated Effects of Atmospheric CO_2 Concentration on Plant and Ecosystem Respiration

Miquel A. Gonzàlez-Meler* and Lina Taneva
Department of Biological Sciences, University of Illinois at Chicago, 845 West Taylor St, Chicago, IL 60607, U.S.A.

Summary

Atmospheric CO_2 concentrations have been increasing since the industrial revolution due to fossil fuel burning and deforestation. Elevated levels of atmospheric $[CO_2]$ are likely to enhance photosynthesis and plant growth, which, in turn should result in increased specific and whole-plant respiration rates. However, a large body of literature has shown that specific respiration rates of plant tissues can be considerably reduced when plants are exposed to or grown at high $[CO_2]$. Reductions in respiration by $[CO_2]$ have been explained by either direct inhibitory effects of $[CO_2]$ on respiratory processes or by indirect effects associated with changes in the chemical composition of tissues of plants grown at high $[CO_2]$. The observed reductions in plant respiration rates by elevated $[CO_2]$ can represent a large biospheric sink for atmospheric carbon. Although doubling current ambient levels of atmospheric $[CO_2]$ could inhibit some mitochondrial enzymes directly in the short-term, the magnitude of the direct effect of $[CO_2]$ on tissue respiration has now been shown to be largely explained by measurement artifacts, diminishing the impact that direct effects would have on the carbon cycle. A reduction in construction and maintenance costs of tissues of plants grown at high $[CO_2]$ can explain an indirect reduction of respiration. Such indirect effects, however, may be offset by the larger biomass of plants exposed to elevated $[CO_2]$. A lack of clear understanding of the physiological control of plant respiration, of the role(s) of non-phosphorylating pathways, and effects associated with plant size, makes it difficult to predict how respiration and the processes it supports respond to elevated $[CO_2]$. Therefore, the role of plant respiration in augmenting or controlling the sink capacity of terrestrial ecosystems is still uncertain.

*Author for correspondence, email: mmeler@uic.edu

H. Lambers and M. Ribas-Carbo (eds.), Plant Respiration, 225–240.

I. Introduction: Respiration and the Carbon Cycle

Respiration is essential for growth and maintenance of all plant tissues and plays an important role in the carbon balance of individual cells, whole-plants, ecosystems, and the global carbon cycle. Through the processes of respiration, solar energy conserved during photosynthesis and stored as chemical energy in organic molecules is released in a regulated manner for the production of ATP, the universal currency of biological energy transformations, and reducing power (e.g., NADH and NADPH). A quantitatively important by-product of respiration is CO_2, and therefore, plant and ecosystem respiration play a major role in the global carbon cycle.

Terrestrial ecosystems exchange about 120 Gt C per year with the atmosphere, through the processes of photosynthesis and respiration from plants and soils (Schlesinger, 1997). While a large body of literature recognizes the potential consequences on photosynthesis (Drake et al., 1997) considerably less is known about respiratory responses to elevated [CO_2]. Yet, conceptual and theoretical models predict that a small change in global plant respiration could substantially modify the sink capacity of vegetation to fix atmospheric carbon (Drake et al., 1999; Gifford, 2003). It has also been recognized that elevated [CO_2] has two type of effects on plant respiration: i) direct and ii) indirect effects (Gonzàlez-Meler et al., 1996b; Amthor, 1997; Drake et al., 1997). There is, however, little understanding of genetic and biochemical regulators of plant respiratory responses to elevated [CO_2]. This paucity of information is being slowly overcome, but it is still limiting to the development of accurate models to predict long-term effects of elevated [CO_2] on the carbon balance of the terrestrial biosphere.

Roughly, half of the annual photosynthetically fixed CO_2 is released back to the atmosphere by plant respiration (Gifford, 1994; Amthor, 1995). Because terrestrial fluxes of CO_2 between the biosphere and atmosphere far outweigh anthropogenic inputs of CO_2 to the atmosphere, a small change in terrestrial respiration can have a significant impact on the annual increment in atmospheric [CO_2]. A large body of literature has indicated that plant respiration would be reduced in plants grown at high [CO_2]. For example, it is estimated that the observed 15–20% reduction in plant tissue respiration by doubling current atmospheric [CO_2] (Amthor, 1997; Drake et al., 1997; Curtis and Wang, 1998), could increase the sink capacity of global ecosystems by 3.4 Gt of carbon per year (Drake et al., 1999), thus, offsetting an equivalent amount of carbon from anthropogenic CO_2 emissions. Therefore, not only are the gross changes in respiration important for large-scale carbon balance issues, but changes in specific rates of respiration will also have significant impact on basic plant biology such as growth, biomass allocation or nutrient uptake (Amthor, 1991; Wullschleger et al., 1994; Drake et al., 1999). New evidence, however, suggests that the theoretical increase in sink capacity of ecosystems due to reduction in respiration has been overestimated.

II. Effects of CO_2 on Respiration

Amthor (1991) described two different interactions between CO_2 and plant respiration that can be distinguished experimentally: a direct effect, in which the rate of respiration of mitochondria or tissues could be rapidly and reversibly reduced following a rapid increase in CO_2 concentration (Amthor et al., 1992; Amthor, 2000a), and an indirect effect in which high [CO_2] changes the rate of respiration in plant tissues compared to the rate seen for those grown in normal ambient [CO_2], when both are tested at a common background [CO_2] (Azcón-Bieto et al., 1994). Although little effort has been made to distinguish between the direct, reversible effect of elevated [CO_2] and the long-term indirect effect of [CO_2] on respiration, current evidence points to these effects being associated with separate phenomena (Gonzàlez-Meler et al., 1996a; Drake et al., 1999; Jahnke, 2001). Most of the studies on the effects of [CO_2] on respiration rate have focused on the magnitude (from non-significant to 60%) and direction (either stimulation, inhibition or no effect) of these two effects, with little progress in the understanding of the underlying mechanisms. Confounding measurement artifacts are an added complication for establishing the reality and magnitude of direct and indirect effects of [CO_2] on respiration (Gonzàlez-Meler and Siedow, 1999; Jahnke, 2001; Davey et al., 2003).

A. Direct effects

A rapid short-term doubling of current atmospheric CO_2 levels has been reported to inhibit respiration of mitochondria and intact plant tissues by 15–20%, varying from no effect to more than 50% inhibition (Amthor, 1997; Drake et al., 1997; Curtis and Wang 1998). The reported magnitude of the direct effect on non-woody tissues of woody species, but not wood itself (no studies of a direct effect of [CO_2] on wood have been conducted), is similar to that of herbaceous species (Gonzàlez-Meler and Siedow, 1999). Direct effects of [CO_2] on respiration in trees range from about 60% inhibition in *Castanea sativa* shoots (El Kohen et al. 1991) and *Pinus radiata* fine roots (Ryan et al. 1996) to no inhibition in *Pinus ponderosa* seedlings (Griffin et al. 1996a). The magnitude of the direct effect of [CO_2] on intact tissue and whole-organ respiration has now been show to be largely explained by measurement artifacts (Gonzàlez-Meler and Siedow, 1999; Jahnke, 2001), diminishing the impact that direct effects of [CO2] on respiration would have on plant growth and on the carbon cycle.

The direct effect of doubling current levels in [CO_2] inhibits the oxygen uptake of isolated mitochondria and the activity of mitochondrial enzymes (Gonzàlez-Meler et al., 1996b). Dissolved inorganic carbon can also inhibit certain mitochondrial enzymes, although the [CO_2] at which most of these enzymes are inhibited is higher (over 10000 μmol mol^{-1} [CO_2]) than the projected increase in atmospheric [CO_2] (Amthor, 1991; Palet et al., 1992; Gonzàlez-Meler et al., 1996a). Increasing the [CO_2] through the addition of bicarbonate in a reaction medium equivalent to a doubling of the present atmospheric [CO_2], reduces the in vivo activity of cytochrome *c* oxidase and succinate dehydrogenase in mitochondria isolated from *Glycine max L.* cotyledons and roots (Gonzàlez-Meler et al., 1996b) (Fig. 1). In isolated soybean mitochondria, doubling ambient [CO_2] resulted in up to 15% reduction in mitochondrial oxygen uptake (Gonzàlez-Meler et al., 1996b).

Despite the fact that cytochrome oxidase is the primary enzyme of mitochondrial respiration, only up to 50% of total respiratory control (see Kacser and Burns, 1979 for definitions) resides at the level of the mitochondrial electron transport chain (Gonzàlez-Meler and Siedow, 1999; Affourfit et al., 2001). Accordingly, Gonzàlez-Meler and Siedow (1999) argued that direct effects of [CO_2] on respiration exceeding more than 10% were likely due to factors other than inhibition of mitochondrial enzymes. Amthor (1997) pointed out that elevated [CO_2] could increase dark CO_2 fixation catalyzed by phosphoenolpyruvate carboxylase (PEPC), resulting in an apparent reduction of net CO_2 efflux. However, Amthor et al. (2001)

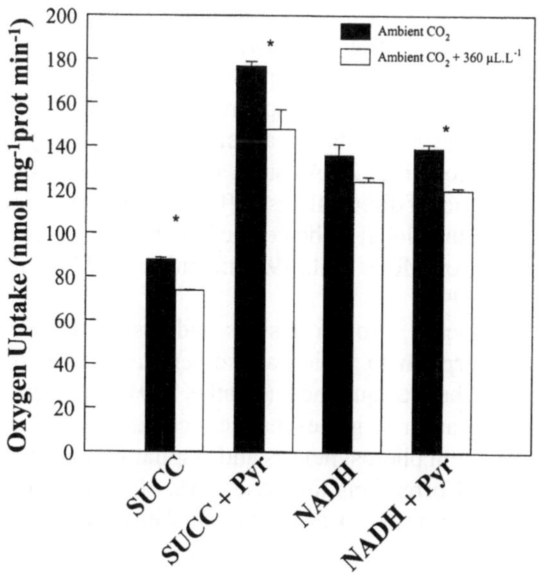

Fig. 1. The direct effect of elevated [CO_2] on soybean cotyledon mitochondrial respiration. Oxidation of either succinate (SUCC), NADH and pyruvate (NADH+PYR), NADH alone (NADH) or succinate and NADH (SUCC+NADH) were measured in the presence of ADP (state 3 conditions). Mitochondria were incubated for 10 min with 0 (open bars) or 0.1 (solid bars) mM Dissolved Inorganic Carbon (DIC) at 25 °C. Values are mean ± SE of 3 to 9 replicates and * indicates a significant difference in the mean ($p<0.05$) using a Student's t test or a Rank Summary test. Modified from Gonzàlez-Meler et al. (1996b).

showed inconsistent small reduction in CO_2 efflux compared to unaltered O_2 uptake rate when CO_2 levels were increased by 300μmol mol^{-1}, suggesting a small effect of rising atmospheric levels of [CO_2] on PEPC activity. Similar results were found for a variety of species, involving 600 measurements, in which CO_2 concentration increases did not alter the CO_2 and O_2 exchanges in the dark (Davey et al., 2003).

As mentioned above, recent evidence indicates that in studying direct effects of CO_2 on respiration, investigators should first be concerned about measurement artifacts. Amthor (1997) and Gonzàlez-Meler and Siedow (1999) showed that gas CO_2 exchange measurement errors could augment and explain the magnitude of the direct effect. Since then, new studies made in this new context have shown that respiration rates are little or not at all inhibited by a doubling of atmospheric [CO_2] (Amthor, 2000; Amthor et al., 2001; Bunce 2001; Tjoelker et al., 2001; Bruhn et al., 2002; Hamilton et al., 2002). Indeed, Jahnke (2001) and Jahnke and Krewitt (2002) confirmed that measurement artifacts due to leakage in CO_2-exchange systems could be as large as the previously reported direct inhibitory effects. They also found the leaks through the intercellular spaces of homobaric leaves will show a significant apparent inhibition of CO_2 efflux that is not due to an inhibition of [CO_2] on respiration (Fig. 2). As such, early conclusions on the impact of direct effects of [CO_2] on plant respiration on the global carbon cycle have been overstated (Gonzàlez-Meler et al., 1996a;Amthor, 1997; Drake et al., 1999).

Gas leaks through gaskets and associated with CO_2 sorption to tubing and other surfaces of the gas-exchange equipment (Jahnke, 2001) represent a complication of gas-exchange techniques than can affect both photosynthesis and respiration measurements. Corrections for these types of leaks can be applied in most cases (Pons and Welschen, 2002). However, leaks occurring through connected leaf air spaces cannot easily be corrected, and measurements in these cases can only be done when the entire leaf is exposed to the same [CO_2] (Jahnke and Krewit, 2001). Applying some corrections for gas exchange leaks, Bunce (2001) reported a significant reduction in respiration after a rapid increase in [CO_2], so not all reports of direct inhibition of respiration by elevated [CO_2] have yet been reconciled with each other.

The observations of Jahnke (2001) and Jahnke and Krewitt (2002) are indications that direct effects of [CO_2] on mitochondrial enzymes may have no consequence on the specific respiratory rate of intact tissues. Gonzàlez-Meler and Siedow (1999) provided two potential mechanisms by which inhibition of enzymes by [CO_2] are not seen at the tissue level: 1) mitochondrial enzymes are 'in excess' of the levels

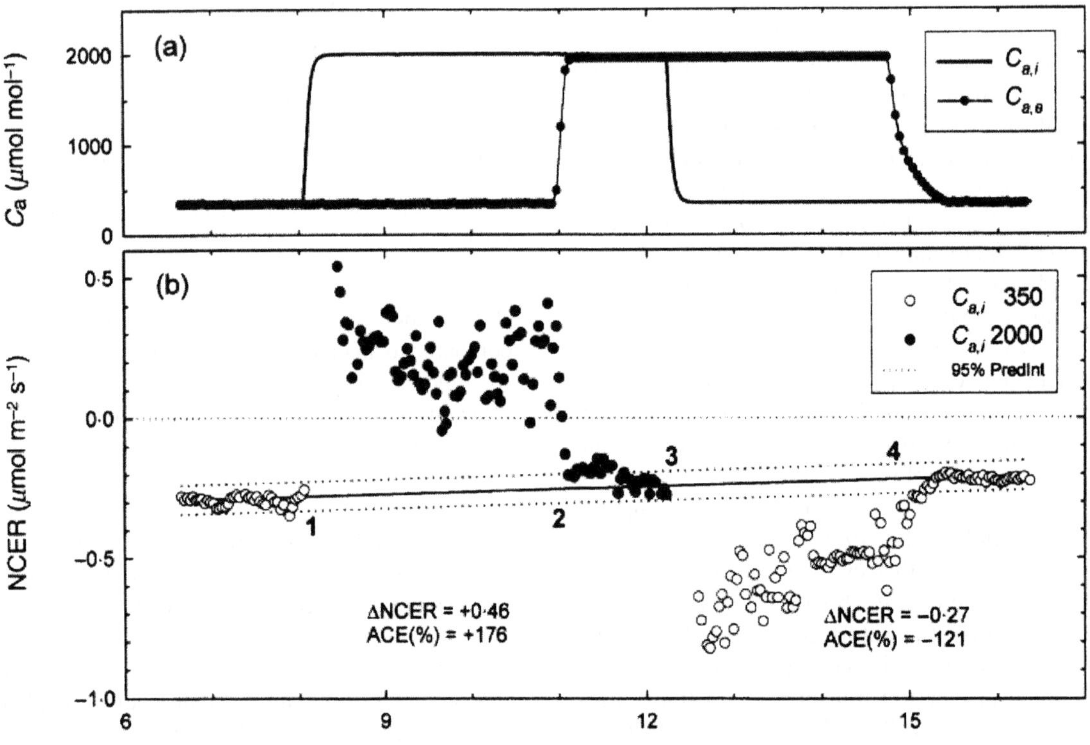

Fig. 2. Measurement artifacts are responsible for apparent direct effects of [CO_2] on tissue respiration. Homobaric leaves of *Nicotiana tabacum* L equilibrate [CO_2] leaf air spaces of the part of the leaf inside the measurement cuvette with that outside. When the CO_2 concentration inside the cuvette ($C_{a,i}$) increases with respect to the outside ($C_{a,e}$), CO_2 moves from the inside (high [CO_2]) to the outside part of the leaf (a), showing an apparent decline in the net carbon exchange rate (NCER) (b). When the whole leaf (inside and outside the measurement cuvette) is exposed to the same [CO_2], no effects of changing [CO_2] are seen on the leaf respiratory rate. $C_{a,i}$ was increased from 350 to 2000 μmol mol^{-1} at time mark 1 and $C_{a,e}$ at time mark 2. $C_{a,i}$ was lowered again to 350 μmol mol^{-1} at time mark 3 and $C_{a,e}$ at time mark 4. Figure 6 from Jahnke and Kewit (2002), published in Plant Cell & Environment and used with the authors' and Blackwell Publishing permission.

required to support normal tissue respiratory activity, and 2) a compensating increase in the activity of the alternative pathway upon inhibition of cytochrome oxidase by $[CO_2]$ will result in unaltered dark respiratory activity. The former mechanism implies that the control coefficient levels of mitochondrial enzymes are very low, and therefore inhibition of enzymatic activity by $[CO_2]$ will have no consequences for the overall tissue respiration rate. Interestingly, recent reports have shown that the number of mitochondria increases in leaves of plants grown at high $[CO_2]$ with no linear changes in the leaf respiratory rate (Griffin et al., 2001; see below). Hence, if the mitochondrial machinery increases in plants grown at high $[CO_2]$ with no changes in respiration rate, then the levels of cytochrome oxidase will presumably be in even more exceeding amounts than in plants grown at ambient $[CO_2]$ for any direct effect of $[CO_2]$ on mitochondrial enzyme activity to affect tissue respiration rate. In this context, it is also important to recognize that the experiments of Jahnke (2001) and Jahnke and Krewitt (2002) were done in tissues exposed to prolonged nights (up to 72 hours), so consequences of inhibition of cytochrome oxidase by $[CO_2]$ on the overall tissue respiration rate will not be expected either, because of an excess cytochrome oxidase enzyme with respect to the respiratory rate.

The latter mechanism, i.e. a compensation by the alternative path, proposed by Gonzàlez-Meler and Siedow (1999) is based upon the competitive nature of the cytochrome and alternative pathways of plant mitochondrial respiration (Chapter 1, Lambers et al.). The activity of the alternative pathway could increase upon a doubling of the $[CO_2]$, masking the direct CO_2 inhibition of the cytochrome pathway (Fig. 3). The oxygen-isotope technique (Chapter 3, Ribas-Carbo et al.) allows for the distinction of direct effects of $[CO_2]$ on the interplay between the cytochrome and the alternative oxidase of plant respiration. Figure 3 shows that mitochondrial electron transport activity is affected when CO_2 (a mild inhibitor) restricts the normal electron flow through one of the pathways (Cyt pathway – v_{cyt}). Under conditions where the alternative pathway activity is low (see Ribas-Carbo et al., 1995 for details), inhibition of the cytochrome pathway by doubling the ambient $[CO_2]$ (18%) is compensated by a similar increase in the activity of the alternative pathway (v_{alt}), resulting in no significant reduction in the overall oxygen uptake of the isolated mitochondria. These results show that increased activity of the alternative pathway upon

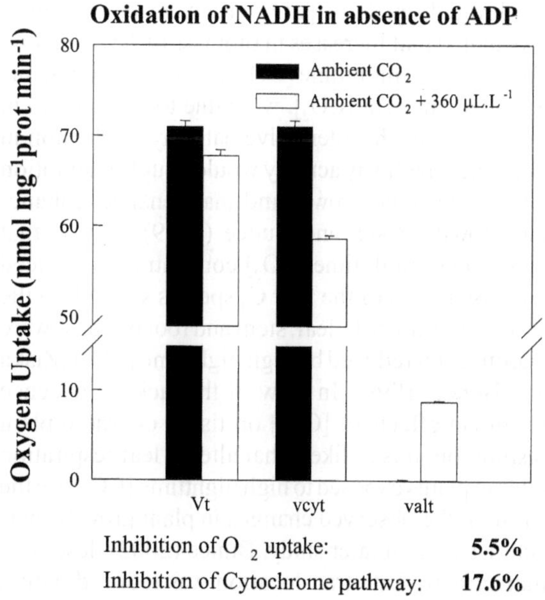

Fig. 3. Effect of doubling the ambient concentration of CO_2 on the oxygen uptake of 5-day old isolated soybean cotyledon mitochondria in state 4 conditions (i.e. ADP-limiting conditions). Experiments were done using the oxygen-isotope technique as described in Ribas-Carbo et al. (1995) and Chapter 3, Ribas-Carbo et al.) applying the treatments as for Figure 1. Results are average of four replicates; bars show standard errors (Gonzàlez-Meler, unpublished observations).

addition of CO_2 can compensate for the direct CO_2 inhibition of the cytochrome oxidase in isolated mitochondria. Although there is no indication that this mechanism is operative in intact tissues of plants exposed to rapid changes in $[CO_2]$ (for reasons exposed above; Gonzàlez-Meler and Siedow, 1999), it has been speculated that increased activity of the non-phosphorylating alternative pathway could be responsible for the altered growth characteristics of plants exposed to elevated $[CO_2]$ only during the nighttime (Reuveni and Gale, 1995, 1997; Griffin et al., 1999, Bunce, 1995, 2001, 2002).

Bunce (1995) observed that the biomass of *Glycine max* L. increased when nighttime CO_2 was elevated. Leaf area ratio increased and photosynthetic rates decreased in plants exposed to high nighttime $[CO_2]$ when compared to plants grown at normal ambient CO_2 all day. It was suggested that the direct inhibitory effect of $[CO_2]$ on leaf respiration caused the alterations in plant development. Also in *Glycine max*, Griffin et al. (1999) found that plants exposed to high nighttime $[CO_2]$ had slower leaf respiration rates and greater biomass than plants grown at normal or elevated levels of [CO_2]. Reuveni et al. (1997) speculated that increases in biomass of *Lemna gibba* grown at high nighttime [CO_2] compared with plants grown at ambient [CO_2], were due to a reduction in the activity of the alternative pathway. Reduction in alternative pathway activity would couple respiration rates with tissue growth and maintenance, enhancing growth. Ziska and Bunce (1999) showed that elevation of nighttime [CO_2] concentration reduced biomass in two of the four C_4 species studied. In the case of *Zea maize* L. leaf, stem and root biomass were significantly reduced by high nighttime [CO_2] (Ziska and Bunce, 1999). In view of the lack of evidence for direct effects of [CO_2] on tissue or whole-plant respiration, it is unlikely that altered leaf respiration rates in plants exposed to high nighttime [CO_2] are the cause of the observed changes in plant growth characteristics. In a later study, Bunce (2002) described that carbohydrate translocation was reduced within two days of exposing plants to elevated nighttime [CO_2] when compared with plants grown at ambient conditions day and night. These results suggest that elevated [CO_2] may have other non-described direct effects of [CO_2] on the plant's physiology that may reduce energy demand for carbohydrate translocation, hence reducing the rate of leaf respiration. If so, these new types of effects cannot be catalogued as direct effects of [CO_2] on respiration (see above), but as indirect effects.

In summary, although rapid changes in [CO_2] can inhibit mitochondrial enzymes directly, previously reported direct effects of [CO_2] on tissue respiration are likely measurement artifacts. Therefore, it would appear that there are no direct effects of [CO_2] on respiration that have any impact on the amount of anthropogenic carbon that vegetation could retain. The fact that effects of [CO_2] on mitochondrial enzymes do not scale to tissue or plant level shows a lack of understanding on how respiration of tissues is regulated. Finally, the role of the alternative pathway in direct respiratory responses to elevated [CO_2], although unresolved, would have little impact, if any, on the general conclusion that direct effects of [CO_2] on plant respiration are not to be considered in plant growth or carbon cycle models.

B. Indirect and Acclimation Effects

The indirect effects represent changes of tissue respiration in response to elevated atmospheric [CO_2], and involve effects on tissue composition, especially carbohydrate accumulation in green tissues and reduced tissue [N] which is a reflection of the reduction of protein content (Drake et al., 1997). Other indirect effects are related to the effects of [CO_2] on growth, response to environmental stress, and other factors that may alter the respiratory demand for energy when compared with plants grown at ambient [CO_2] (Bunce, 1994; Amthor, 2000b). Acclimation of respiration has also been observed in photosynthetic tissues of some plants grown at elevated [CO_2] and represents the downregulation or upregulation of the respiratory machinery (i.e. amounts of cytochrome oxidase, number of mitochondria) irrespective of changes in specific respiration rates (Azcón-Bieto et al., 1994; Griffin et al., 2001). Indirect effects caused by elevated [CO_2] can be measured as reduction in CO_2 emission (or O_2 consumption) from tissues at a common background [CO_2] (Gonzàlez-Meler et al., 1996a).

As atmospheric [CO_2] rises, increased photosynthesis results in higher cellular carbohydrate concentrations (Drake et al., 1997; Curtis and Wang, 1998). Increased carbohydrate concentrations can stimulate the activity of specific dark respiration in the short-term, due to greater availability of respiratory substrates. This happens when photosynthesis is stimulated by increasing light intensity (Azcón-Bieto and Osmond, 1983) or tissues are fed exogenous sucrose (Azcón-Bieto et al., 1983). Increased carbohydrate levels should also increase the respiratory energy demand to support phloem loading and translocation of carbohydrates (Bouma et al., 1995; Amthor, 2000b). Increased carbohydrate content has also been shown to regulate the levels of cytochrome oxidase (Felitti and Gonzalez, 1998). Recent experimental evidence suggests that although carbohydrate levels are increased by elevated [CO_2], specific and whole plant respiration in many plants is reduced (Poorter et al., 1992; Amthor, 1997; Drake et al., 1997, but see below).

It has been generally accepted that leaf respiration is usually reduced as a consequence of plant growth at elevated [CO_2] (Amthor, 1997; Drake et al., 1997; Curtis and Wang, 1998; Norby et al., 1999). Early reports showed that leaf respiration is decreased at high [CO_2] under field or laboratory conditions (El Kohen et al., 1991; Idso and Kimball, 1992; Wullschleger at al., 1992a). Poorter et al. (1992) showed that leaf respiration was reduced, on average, by 14% when expressed on a leaf mass basis, but increased by 16% on a leaf area basis. More recently, Curtis and Wang

(1998) compiled respiratory data for woody plants, and observed that growth at elevated [CO_2] resulted in an 18% inhibition of overall leaf respiration (mass basis). However, many of the data compiled in these two studies compared respiratory rates of plants grown and measured at ambient conditions with plants grown and measured at elevated [CO_2]. These respiratory measurements are also affected by the measurement artifacts described above (Fig. 2). Leaks should not be important when respiratory rates are measured at a common [CO_2] for plants grown at ambient and elevated [CO_2]. An analysis of the literature focused on leaf respiratory responses (on a leaf mass basis) of plants grown at ambient and elevated [CO_2] when respiration rates were only measured at a common [CO_2] suggests that specific leaf respiration rates will be unaltered in plants grown at elevated [CO_2] (Table 1). Therefore, the generally accepted conclusion that respiration of plants grown at elevated [CO_2] will be reduced when compared with plants grown at ambient [CO_2] should be reevaluated. However, there is significant variability in the leaf respiratory response to growth at elevated [CO_2] when compared with plants grown at ambient conditions, ranging from 40% inhibition (Azcón-Bieto et al., 1994) to 50% stimulation (Williams et al., 1992). Considerations on the physiological basis by which the acclimation response of respiration to elevated [CO_2] varies, is considered next.

In boreal species, reduction of respiration of plants grown at high [CO_2] was related to changes in tissue N and carbohydrate concentration (Tjoelker et al., 1999). Tissue N concentration often decreases in plants grown at elevated [CO_2] (Drake et al., 1997; Curtis and Wang, 1998). It is expected that respiration rate would be lower in tissues having lower [N], because the respiratory cost associated with protein turnover and maintenance is a large portion of dark respiration (Bouma et al., 1994; Chapter 10, Bouma). Hence, the metabolic cost (i.e. respiratory energy demand) for construction and maintenance of tissues with high concentrations of protein (high [N]) is greater than the cost for the maintenance of the same tissue with low [N] (assuming no changes in rates of protein turnover between plants grown at ambient and elevated [CO_2]) (Amthor, 1989; Drake et al., 1999). This idea was confirmed for leaves of *Quercus alba* seedlings grown in open-top chambers in the field, where respiration was 21 to 56% lower in elevated [CO_2] than in normal ambient [CO_2] (Wullschleger and Norby, 1992). The growth respiration component of these leaves was reduced by 31% and the maintenance component by 45%. These effects were attributed to the reduced cost of maintaining tissues having lower nitrogen concentrations. Similar results were obtained in leaves and stems of plants of other species (Amthor et al., 1994; Carey et al., 1996; Dvorak and Oplustilova, 1997; Griffin et al., 1996b; Will and Ceulemans, 1997; Wullschleger et al., 1992a,b; 1997), with some exceptions (Wullschleger et al., 1995).

Plants grown at elevated [CO_2] reduce their leaf protein content by 15% on average (Drake et al., 1997). Most of this reduction is attributed to decreases in photosynthetic proteins, and little is known about changes in respiratory proteins of plants grown at elevated [CO_2]. Indirect effects of respiration to elevated [CO_2] on leaves of *Lindera benzoin* were also correlated with a reduction in maximum activity of cytochrome oxidase (Azcón-Bieto et al., 1994). This effect would represent an acclimation response of respiration to elevated [CO_2] analogous to that seen in photosynthesis (Drake et al., 1997). However, reduction of respiratory enzyme activity was not seen in rapidly growing tissues exposed to elevated [CO_2] (Hrubeck et al., 1985; Perez-Trejo, 1981). The general increase in number of mitochondria seen in leaves of adult *Pinus taeda* trees grown at high [CO_2] (Griffin et al., 2001) contrast with the reduction in maximum enzyme activity previously reported for other plants (Azcón-Bieto et al., 1994). No acclimation effect to elevated CO_2 (i.e. reduction in either respiration or

Table 1. Indirect effects of long-term CO_2 enrichment on respiration of leaves on a dry mass basis. Elevated-over-ambient (E/A) refers to the ratio of rate of leaf dark respiration of plants grown in elevated [CO_2] to the rate of plants grown in current ambient CO_2 when measured at a common CO_2 concentration. Under these conditions, effects of gas exchange leaks should be minimal in affecting the comparison of rates of respiration from plants grown at ambient and elevated CO_2.

Reference	E/A	Number of species	Observations
Amthor, 1997	0.96	21	Compiled 26 studies, crop and herbaceous and woody wild species
Davey et al., 2003	1.07	7	Original study, crop and herbaceous and woody wild species
Drake et al., 1997	0.95	17	Compiled 15 studies, crop and herbaceous wild species

cytochrome oxidase) has been observed in leaves of C_4 plants (Azcón-Bieto et al., 1994) or roots of C_3 plants (Gonzàlez-Meler, 1995). With the exception of a few studies (i.e. Azcón-Bieto et al., 1994), there are no reports on the response of membrane-associated mitochondrial enzymes to elevated [CO_2] in leaves or roots of plants. There are, however, studies showing an increased count of mitochondria in leaves of plants grown at elevated [CO_2] (Griffin et al., 2001), which may represent upregulation of mitochondrial enzymes under elevated [CO_2]. The observed increase in mitochondrial number in the study of Griffin et al., (2001), however, had no concomitant increases in leaf respiration. More research is needed in this area.

III. Growth Consequences of the Effects of [CO_2] on Respiration: A Case Study

The fact that respiration does not increase in plants grown at elevated [CO_2] (Table 1) raises a fundamental question: if respiration is important for growth and maintenance of plants, how can unaltered respiration support the increased plant productivity seen when plants are exposed to elevated [CO_2]? In some species, particularly in young, growing tissues, increased dark respiration has been shown to accompany rapid growth in the first developmental stages (see Drake et al., 1999 for references). As tissues reach maturity, respiration slows as relative growth rate (RGR) declines, because of the positive linear relationship between dark respiration per unit of mass and RGR (Hesketh et al., 1971; Amthor, 1989). Elevated [CO_2] enhances growth rates enough to show a significant increase in biomass at the end of the growing season (Kimball et al., 1993; Delucia et al., 1999; Norby et al., 2002; Karnosky et al., 2003). This relationship is also seen in CO_2 studies where faster respiration rates seem to follow a stimulation of photosynthesis by elevated [CO_2] as a result of a transient increase in RGR (see above; Amthor, 2000b; Bunce, 1994). Any factor uncoupling respiration and growth would compromise the supply of energy needed to sustain biosynthesis, and overall growth could be reduced.

The growth component of respiration (Chapter 10, Bouma) could also be reduced in plants grown at elevated [CO_2] as a result of altered tissue chemistry (Griffin et al., 1993). Based on the chemical composition of tissues, Poorter et al. (1997) found that elevated [CO_2] could reduce growth construction costs by 10–20%. Griffin et al. (1993; 1996a) observed reductions in construction costs of *Pinus taeda* seedlings grown at elevated [CO_2]. Hamilton et al. (2001) reported that elevated [CO_2] slightly reduced construction costs of leaves of mature trees (including *P. taeda*) at the top of the canopy, but not at the bottom of the canopy. Such a small reduction could be explained by reductions in tissue [N], as observed in leaves exposed to [CO_2] at the top of the canopy. Changes in construction costs did not result in a decrease in the leaf respiration rates of trees exposed to elevated [CO_2] (Hamilton et al., 2001).

The lack of long-term effects of increased [CO_2] on specific plant respiration rates could also be due to a lower involvement of the alternative pathway (Gonzàlez-Meler and Siedow, 1999; Griffin et al., 1999). Respiration through the alternative pathway bypasses two of the three sites of proton translocation; so the free energy released is lost as heat, and is unavailable for the synthesis of ATP. Respiration associated with this pathway will not support growth and maintenance processes of tissues as efficiently as respiration through the cytochrome path. The activity of the alternative pathway of respiration could decrease upon doubling [CO_2], masking increases in the activity of the cytochrome pathway. If this were the case, unaltered respiration rate in plants grown at high [CO_2] could more efficiently support growth and maintenance processes. On the contrary, excess carbohydrate levels often seen in plants grown at elevated [CO_2] could trigger the activity of the alternative pathway (see chapter 5, Noguchi), although correlations between leaf carbohydrate levels and alternative pathway activity have not been clearly demonstrated (Gonzàlez-Meler et al., 2001). It is important to determine whether inhibitory and/or stimulatory effects of [CO_2] on dark respiration have beneficial (by decreasing carbon losses) or detrimental (by reducing ATP yields per unit of N, see Gonzàlez-Meler et al., 2001) effects on overall plant biomass and allocation to different plant parts.

The oxygen-isotope technique allows for the distinction of [CO_2] effects between the cytochrome and the alternative oxidase of respiration in plants grown at ambient and elevated [CO_2]. Figure 4 illustrates the combined effects of [CO_2] on respiration on a shade-tolerant species *Cornus florida* L. and a shade-intolerant species *Liriodendron tulipifera* L. (Burns and Honkala, 1990). Although [CO_2] did not seem to affect leaf specific respiration rates, oxygen-fractionation data revealed that the growth [CO_2] environment induced important physiological changes at the leaf level. Despite the small effect of elevated [CO_2] on respiratory CO_2 efflux, oxygen-isotope fractionation by respiration increased in plants grown at elevated [CO_2] in both shade-tolerant and shade-intolerant plants. Oxygen-isotope fractionation increased from 21.2 ‰ to 22.8 ‰ in *L. tulipifera*, and from 21.7 ‰ to 23.0 ‰ in *C. florida*. An increase in oxygen-isotope fractionation implies that the activity of the non-phosphorylating alternative pathway of respiration increases. Interestingly, ATP yields of respiration were not reduced in *C. florida* plants grown in elevated [CO_2] when compared with the plants grown at ambient [CO_2] (**Fig. 4**). **How**ever, *L. tulipifera* grown at high [CO_2] reduced ATP yields by 30% when compared with the plants grown at ambient [CO_2] (Fig. 4), as a consequence of a strong inhibition of the cytochrome pathway. If reduction in ATP production is maintained over time in these plants grown at elevated [CO_2], growth may be reduced. Interestingly, annual growth (measured as stem diameter increase) of *C. florida* grown at high [CO_2] increased by 10%, whereas growth of *L. tulipifera* at high [CO_2] was reduced by 35% when compared with control plants (J. Mohan, unpublished). More research is needed to establish the linkages between changes in biomass growth and the altered respiratory metabolism (Fig. 4) in response to [CO_2].

IV. Integrated Effects of Elevated [CO_2] on Respiration at the Ecosystem Level

A. Ecosystem Respiration

Terrestrial ecosystems exchange about 120 Gt C per year with the atmosphere, through the processes of photosynthesis (leading to gross primary production, GPP) and ecosystem respiration (Re) (Schlesinger, 1997). The difference between GPP and Re determines net ecosystem productivity (NEP), the net amount of carbon retained or released by a given ecosystem. An increasing body of evidence derived form direct measurements of net ecosystem exchange (NEE; CO_2 exchange between terrestrial ecosystems and the atmosphere) shows that, in general, the photosynthetic gain of carbon exceeds respiratory losses for a variety of ecosystems (Grace et al., 1995; Katul et al., 1997; 1999; Buchmann and Schultze, 1999; Luo et al., 2000). Currently, the net exchange of C between the terrestrial biosphere and the atmosphere is estimated to result in a global terrestrial sink of about 2 Gt C per year (Gifford, 1994; Schimel, 1995; Steffen et al., 1998). A significant effort has been made to identify the long-term effects of elevated [CO_2] on canopy photosynthesis and ecosystem growth. Unfortunately, the effects of [CO_2] on autotrophic and heterotrophic respiration at the ecosystem level are

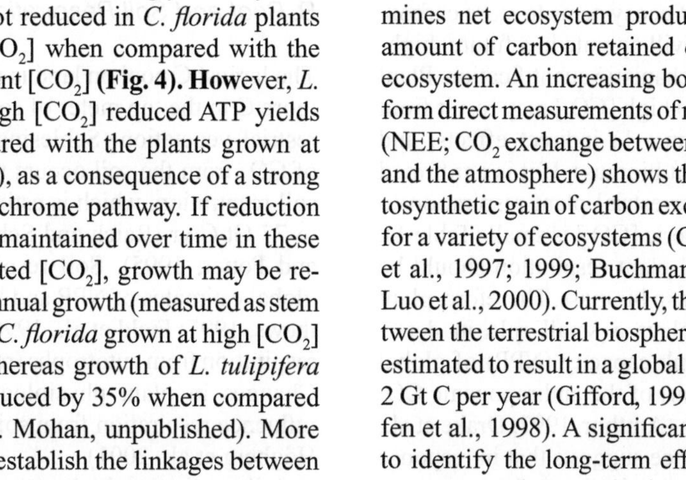

Fig. 4. Respiratory rate of ATP production per unit of tissue N in leaves of a shade-tolerant (*Cornus florida* L.) and a shade-intolerant (*Liquidambar tulipifera* L.) species grown at ambient and at ambient + 200 μmol mol^{-1} [CO_2] in the Duke FACE site. Rates of ATP production were calculated from the activities of the cytochrome (v_{cyt}) and the alternative (v_{alt}) pathways, following Gonzàlez-Meler at al. (2001). Total ATP production values are the addition of the ATP synthesis rates derived from the cytochrome (v_{cyt}) and the alternative (v_{alt}) activity. Rates of total oxygen uptake (in μmol kg^{-1} DM s^{-1}) for plants grown at ambient and elevated [CO_2] were 7.6±1.6 and 8.8±0.5 for *C. florida*, and 10.7±2.1 and 10.2±1.0 for *L. tulipifera*, respectively. Details on plant growing conditions and CO_2 treatment can be found in DeLucia et al. (1999) and Naumberg and Ellsworth (2000). Data are means and SE for three replicates (Gonzàlez-Meler, unpublished).

not known or well understood, despite their potential to control ecosystem carbon budgets (Ryan, 1991; Giardina and Ryan, 2000; Valentini et al., 2000).

Terrestrial plant respiration releases 40 to 60% of the total carbon fixed during photosynthesis (Gifford, 1994; Amthor, 1995) representing about half of the annual input of CO_2 to the atmosphere from terrestrial ecosystems (Schlesinger, 1997). Therefore the magnitude of terrestrial plant respiration and its responses to [CO_2] are important factors governing the intrinsic capacity of ecosystems to store carbon. How plant respiration will operate in a future, high-CO_2 world requires mechanistic quantification. Plant respiration responses to high [CO_2] may result from two distinct mechanisms: 1) indirect effects, and 2) changes in total plant biomass. If it is confirmed that the response of terrestrial plant respiration to an increase in atmospheric [CO_2] is small (Table 1), then changes in global plant respiration should be proportional to changes in biomass. Therefore, plant respiration probably plays a small role in determining the biomass sink capacity (NPP) of global ecosystems to retain part of the anthropogenic CO_2 emissions in vegetation (Amthor, 1997; Drake et al., 1999). However, evidence suggests that in response to elevated [CO_2], plant respiration at the ecosystem level does not necessarily increase with increases in total plant biomass (Drake et al., 1996; Hamilton et al., 2002).

Attempts to scale [CO_2] effects on mitochondrial or tissue respiration to the ecosystem level are problematic, because, unlike photosynthesis, little is known about applicable scaling rules for plant respiration (Gifford, 2003). Attempts to build respiratory carbon budgets at the canopy level require knowledge on maintenance and growth respiration, and tissue respiratory responses to light, temperature and [CO_2] (Amthor 2000b; Gifford 2003). Hamilton et al. (2002) built a model of the carbon balance of a pine-dominated forest exposed to ambient and elevated [CO_2]. Elevated [CO_2] increased forest NPP by 27% without any increase in total plant ecosystem respiration, suggesting that the rates of specific respiration were actually decreased in tissues of plants grown at elevated [CO_2]. Open-top chambers allow for the measurement of canopy respiration at the ecosystem level. Table 2 shows the contribution of CO_2-induced changes in ecosystem respiration to annual NEP of a salt marsh exposed to twice ambient [CO_2]. In this ecosystem, elevated CO_2 consistently reduced night-time ecosystem respiration in C_3 and C_4 community stands (Drake et al., 1996). Inhibition of ecosystem respiration by elevated [CO_2] represented a substantial one fifth to one third of the total extra annual carbon gain (NEP) observed between canopies grown at ambient and elevated [CO_2] (Table 2). Acclimation and other indirect effects of [CO_2] on plant and soil respiration to elevated [CO_2] (Azcón-Bieto et al., 1994; Drake et al., 1996) may explain this consistent reduction in ecosystem respiration.

B. Root and Soil Respiration

The largest C pool on land is in soils, with 2.5 times more C in the top meter of soil than is found in terrestrial vegetation (Schlesinger, 1997). The efflux of CO_2 from the soil occurs through the process of soil respiration, which is estimated to be around 68 to 77 Gt C/yr (Raich and Schlesinger, 1992; Raich and Potter, 1995). Soil respiration is higher than global estimates of NPP and litter production (Matthews, 1997; Field et al., 1998), because it includes respiration from autotrophs (roots) and heterotrophs. Changes in plant physiological activity strongly influence respiration from soils (Schlesinger, 1997; Högberg et al., 2001).

Soil respiration is the result of both autotrophic and heterotrophic below-ground processes, including root respiration and respiration associated with the decomposition of soil organic matter by soil microorganisms. Photosynthetic C uptake is stimulated by elevated [CO_2] (DeLucia et al., 1999; Norby et al., 2002), and enhanced growth at elevated [CO_2] would contribute to increased carbon inputs into the soil through more litter fall and greater root biomass production and turnover (Allen et al., 2000; Matamala and Schlesinger, 2000; King et al., 2001). However, increased C inputs into the soil do not necessarily lead to greater soil C storage (Schlesinger and Lichter, 2001) because elevated [CO_2] has been shown, almost universally, to increase soil respiration rates (Zak et al., 2000; Table 3). Many studies have reported increased root growth under elevated [CO_2] (Johnson et al., 1994; Vose et al., 1995; Hungate et al., 1997; Edwards and Norby, 1999; Matamala and Schlesinger, 2000; Pregitzer et al., 2000; King et al., 2001) which has been correlated with faster soil respiration rates. Greater plant C allocation below ground can result in faster soil respiration rates by 1) increasing the contribution of root respiration to total soil respiration because of greater root biomass relative to ambient [CO_2], or 2) increasing the labile

Table 2. Contribution of CO_2-induced reduction in ecosystem respiration (Re) to total net ecosystem productivity (NEP) in a C_3-dominated salt marsh ecosystem exposed to elevated [CO_2] since 1988. Relative change in CO_2 stimulation was calculated form the annual net ecosystem exchange (NCE) canopy flux at plots exposed to elevated [CO_2] over that of plants growing at ambient [CO_2] (n=5). Modified from Drake et al. (1996) and Gonzàlez-Meler et al. (1995).

Year	NCE	Re	NEP	% NEP gain from change in Re
		% CO_2 stimulation		
1994	31	–57	66	33

Table 3. Percent stimulation of soil respiration in different natural ecosystems by elevated [CO_2] relative to ambient [CO_2] conditions. The table shows the dominant species or ecosystem type in the study, % stimulation of soil respiration (calculated as [(rate at elevated/rate at ambient)*100]), the level of CO_2 enrichment over the ambient CO_2 concentration and the reference.

Species	E/A*100 %	CO_2 treatment	Reference
Acer rubrum	27	+350	Edwards and Norby, 1999
Acer rubrum	15	+350 + 4 °C	Edwards and Norby, 1999
Acer saccharum	5	+350	Edwards and Norby, 1999
Acer saccharum	26	+350 + 4 °C	Edwards and Norby, 1999
Pseudotsuga menziesii	20	+200	Lin et al., 2001
Pseudotsuga menziesii	54	+200 + 4 °C	Lin et al., 2001
Pinus ponderosa	74	+350	Vose et al.,1995
Lindera benzoin	50	+340	Ball et al., 2000
Populus tremuloides	30	+200	Karnosky et al., 2003
P. tremuloides/B. papyrifera	60	+200	Karnosky et al., 2003
P. tremuloides/A. saccharum	10	+200	Karnosky et al., 2003
Pinus taeda			
Dry year	23	+200	Taneva et al., unpublished*
Wet year	12	+200	Taneva et al., unpublished*
Short-grass steppe			
Dry year	85	+360	Pendall et al., 2003
Wet year	25	+360	Pendall et al., 2003
Wetland	15	+340	Ball and Drake, 1995
California grassland	36	+360	Hungate et al., 1997

soil C pool through greater root exudation and fine root turnover, and 3) priming effect. Therefore, long-term soil C sequestration requires that a substantial proportion of the additional C assimilated by plants growing at elevated [CO_2] is allocated to roots and soil C pools that turn over slowly. Little is known about the physiological mechanisms leading to C accumulation in soils.

Although elevated [CO_2] may result in an increased transfer of C to the root and soil pool, other indirect effects associated with elevated [CO_2] or global warming may stimulate root and soil respiration (Schimel et al., 1994; Schlesinger and Andrews, 2000; but see Giardina and Ryan, 2000). For instance, Lin et al. (2001) reported that total soil respiration rates under *Pseudotsuga menziesii* grown in mesocosms were stimulated by 20% by [CO_2] enrichment; the combined effect of elevated [CO_2] and higher temperature, however, increased soil respiration rates by 54% (Table 3). In addition, the authors found that elevated [CO_2] primarily stimulated root respiration and root exudation, whereas elevated temperature had a stronger effect on decomposition of soil organic matter.

In addition to elevated [CO_2] and warming, climate change may include shifts in rainfall patterns.

Soil moisture content affects the response of soil respiration to elevated [CO_2] (Fig. 5). The percent stimulation of soil respiration rates by exposure to elevated [CO_2] in a *Pinus taeda*-dominated forest in North Carolina, USA, was greater during a dry growing season compared with soil respiration rates during a wet growing season (Fig. 5; L.Taneva et al., unpublished). Results shown in Fig. 5 indicate that during the wet year, decomposition of old C (uncoupled from recent plant activity) makes up a larger proportion of soil respiration than that under dry conditions, suggesting that respiration from older soil C pools is more sensitive to soil moisture stress than that from recent C pools, including root and rhizosphere respiration. Similarly, Pendall et al. (2003) reported that, although soil respiration rates were enhanced by elevated [CO_2] in a short-grass steppe ecosystem, the degree of stimulation largely depended on soil moisture content.

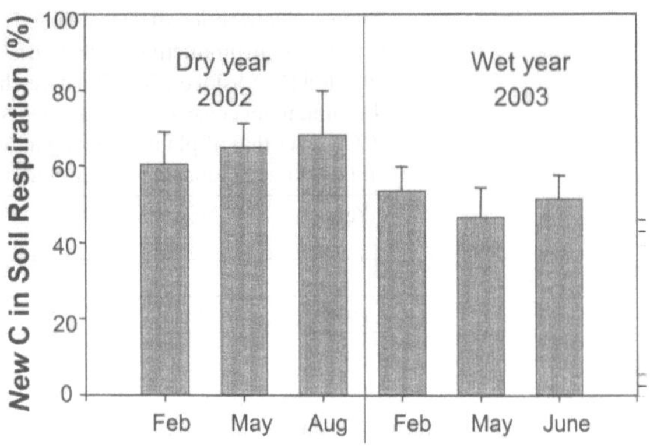

Fig. 5. Respiration of newly fixed C, as a percentage of total soil respiration (R_s), calculated with a $\delta^{13}C$ ecosystem tracer in a loblolly pine-dominated forest in North Carolina exposed to FACE since 1996. Respiration of newly assimilated C is largely due to root and rhizosphere respiration and other very active pools of soil C coupled to plant activity fixed since fumigation was turned on. Respiration of new C represents 64% and 50% of total R_S during the 2002 drought and 2003 wet year, respectively. Bars show means and SE, n=3. (L. Taneva, R. Matamala, J. S. Pippen, W. H. Schlesinger and M. A. Gonzàlez-Meler, unpublished).

V. Conclusions

Contrary to what was previously thought, respiration may not be reduced when plants are grown at elevated [CO_2]. This is because direct effects of [CO_2] on respiratory enzymes are very small. Previous direct effects of [CO_2] on respiration have been confounded with measurement artifacts due to leaks and memory effects in gas-exchange systems, and also due to leaks through leaf air spaces. Such measurement artifacts are also affecting the magnitude of the indirect and acclimation effects of [CO_2] on respiration. A re-analysis of the literature comparing respiration of leaves of plants grown at ambient [CO_2] with leaves of plants grown at elevated [CO_2] when rates are measured at the same [CO_2], indicates that leaf respiration, on average, may not be changed by increasing atmospheric [CO_2]. Increases in growth observed in plants exposed to high [CO_2], appears to be compensated for changes in tissue chemistry that reduce growth and maintenance respiration. In some species an increased activity of the alternative pathway in plants grown at high [CO_2] could counteract a positive plant growth response to elevated [CO_2]. If specific rates of respiration are not affected by growth at elevated [CO_2], respiration from the terrestrial vegetation in a high [CO_2]-world should be proportional to changes in plant mass. However, some studies show that canopy respiration does not follow the increase in biomass (NPP) observed when natural ecosystems are exposed to elevated atmospheric [CO_2]. This can be explained, in part, because root respiration at the ecosystem level seems to increase proportionally to the biomass stimulation by elevated [CO_2]. In addition, increased root exudation in ecosystems exposed to elevated [CO_2] may prompt the oxidation of stored organic carbon in soils, offsetting reductions of plant ecosystem respiration in response to high [CO_2]. The role of plant respiration in augmenting or controlling the sink capacity of terrestrial ecosystems is still uncertain.

Acknowledgments

We thank Jeff Amthor for comments and discussions that greatly improved this manuscript. We also thank Bert Drake, Steve Long and Jim Siedow for past discussions about effects of elevated concentrations of atmospheric CO_2 on respiration of plants and ecosystems.

References

Affourtit C, Krab K and Moore AL (2001) Control of plant mitochondrial respiration. Biochim Biophys Acta 1504, 58–69
Allen AS, Andrews JA, Finnzi AC, Matamala R, Richter DD and

Schlesinger WH, 2000. Effects of free-air CO_2 enrichment on below-ground processes in a loblolly pine forest. Ecol Appl 10: 437–448

Amthor JS (1989) Respiration and Crop Productivity. Springer Verlag, New York

Amthor JS (1991) Respiration in a future, higher CO_2 world. Plant Cell Environ 14: 13–20

Amthor JS (1995) Terrestrial higher-plant response to increasing atmospheric [CO_2] in relation to the global carbon cycle. Global Change Biol 1: 243–274

Amthor JS (1997) Plant respiratory responses to elevated carbon dioxide partial pressure. In: Allen LH, Kirkham MB, Olszyk DM and Whitman CE (eds) Advances in Carbon Dioxide Effects Research, pp 35–77. American Society of Agronomy, Madison

Amthor JS (2000a) Direct effect of elevated CO_2 on nocturnal in situ leaf respiration in nine temperate deciduous trees species is small. Tree Physiol20, 139–144

Amthor JS (2000b) The McCree-de Wit-Penning de Vries-Thornley respiration paradigms: 30 years later. Ann Bot 86: 1–20

Amthor JS, Koch G and Boom AJ (1992) CO_2 inhibits respiration in leaves of *Rumex crispus* L. Plant Physiol 98: 1–4

Amthor JS, Mitchell RJ, Runion GB, Rogers HH, Prior SA and Wood CW (1994) Energy content, construction cost and phytomass accumulation of *Glycine max* (L.) Merr. and *Sorghum bicolor* (L.) Moench grown in elevated CO_2 in the field. New Phytol 128: 443–450

Amthor JS, Koch GW, Willms JR and Layzell DB (2001) Leaf O_2 uptake in the dark is independent of coincident CO_2 partial pressure. J Exper Bot 52: 2235–8

Azcón-Bieto J and Osmond CB (1983) Relationship between photosynthesis and respiration. The effect of carbohydrate status on the rate of CO_2 production by respiration in darkened and illuminated wheat leaves. Plant Physiol 71: 574–581

Azcón-Bieto J, Lambers H and Day DA (1983) Effect of photosynthesis and carbohydrate status on respiratory rates and the involvement of the alternative pathway in leaf respiration. Plant Physiol 72: 598–603

Azcón-Bieto J, Gonzàlez-Meler MA, Doherty W and Drake BG (1994) Acclimation of respiratory O_2 uptake in green tissues of field grown native species after. Plant Physiol 106: 1163–1168

Ball AS, and Drake BG (1998) Stimulation of soil respiration by carbon dioxide enrichment of marsh vegetation. Soil Biol Biochem 30: 1203–1205

Bouma TJ, De Viser R, Janseen JHJA, De Kick MJ, Van Leeuwen PH, and Lambers H (1994) Respiratory energy requirements and rate of protein turnover in vivo determined by the use of an inhibitor of protein synthesis and a probe to assess its effect. Physiol Planta 92: 585–594

Bouma TJ, De Viser R, Van Leeuwen PH, De Kick MJ and Lambers H (1995) The respiratory energy requirements involved in nocturnal carbohydrate export from starch-storing mature source leaves and their contribution to leaf dark respiration. J Exper Bot 46: 1185–1194

Buchmann N and Schulze ED (1999) Net CO_2 and H_2O fluxes of terrestrial ecosystems. Global Biogeochem Cycles 13: 751–760.

Bruhn D, Mikkelsen TN and Atkin OK (2002) Does the direct effect of atmospheric CO_2 concentration on leaf respiration vary with temperature? Responses in two species of *Plantago* that differ in relative growth rate. Physiol Planta, 114: 57–64

Bunce JA (1994) Responses of respiration to increasing atmospheric carbon dioxide concentrations. Physiol Planta 90: 427–430

Bunce JA (1995) Effects of elevated carbon dioxide concentration in the dark on the growth of soybean seedlings. Ann Bot 75: 365–368.

Bunce JA (2001) Effects of prolonged darkness on the sensitivity of leaf respiration to carbon dioxide concentration in C3 and C4 species. Ann Bot 87: 463–468

Bunce JA (2002) Carbon dioxide concentration at night affects translocation from soybean leaves. Ann Bot 90: 399–403

Burns RM and Honkala BH (1990) Silvics of North America: II. Hardwoods. USDA Agriculture Handbook 654. USDA, Washington, D.C.

Carey EV, DeLucia EH and Ball JT (1996) Stem maintenance and construction respiration in *Pinus ponderosa* grown in different concentrations of atmospheric CO_2. Tree Physiol 16: 125–130

Curtis PS and Wang X (1998) A meta-analysis of elevated CO_2 effects on woody plant mass, form, and physiology. Oecologia 113: 299–313

Davey PA, Hunt S, Hymus GJ, DeLucia EH, Drake BG, Karnosky DF and Long SP (2003) Respiratory oxygen uptake is not decreased by an instantaneous elevation of [CO_2], but is increased with long-term growth in the field at elevated [CO_2]. Plant Physiol 134: 520–527

DeLucia EH, Hamilton JG, Naidu SL, Thomas RB, Andrews JA, Finzi A, Lavine M, Matamala R, Mohan JE, Hendrey GR and Schlesinger WH (1999) Net primary production of a forest ecosystem with experimental CO_2 enrichment. Science 284: 1177–1179

Drake BG, Meuhe M, Peresta G, Gonzàlez-Meler MA and Matamala R (1996) Acclimation of photosynthesis, respiration and ecosystem carbon flux of a wetland on Chesapeake Bay, Maryland, to elevated atmospheric CO_2 concentration. Plant Soil 187: 111–118

Drake BG, Gonzàlez-Meler MA and Long SP (1997) More efficient plants: A consequence of rising atmospheric CO_2? Annu Rev Plant Physiol Plant Molec Biol 48: 609–639

Drake BG, Azcón-Bieto J, Berry JA, Bunce J, Dijkstra P, Farrar J, Koch GW, Gifford R, Gonzàlez-Meler MA, Lambers H, Siedow JN, Wullschleger S (1999) Does elevated CO_2 inhibit plant mitochondrial respiration in green plants? Plant Cell Environ 22: 649–657

Dvorak V, Oplustilova M (1997) Respiration of woody tissues of Norway spruce in elevated CO_2 concentrations. In: Mohren GMJ, Kramer K, Sabate S (eds) Impacts of Global Change on Tree Physioland Forests Ecosystems, pp 47–51. Kluver Academic Publishers, Dordrecht

Edwards N, Norby RJ (1999) Belowground respiratory responses of sugar maple and red maple saplings to atmospheric CO_2 enrichment and elevated air temperature. Plant Soil 206: 85–97

El Kohen A, Pontailler Y, Mousseau M (1991) Effect d'un doublement du CO_2 atmosphérique sur la respiration à l'obscurité des parties aeriennes de jeunes chataigniers (*Castanea sativa* Mill.). Comptes Rendus de l'Academie des Sciences 312: 477–481.

Felitti SA and Gonzalez DH (1998) Carbohydrates modulate the expression of the sunflower cytochrome *c* gene at the mRNA level. Planta 206, 410–415

Field CB, Behrenfeld MJ, Randerson JT and Falkowski P (1998)

Primary production of the biosphere: Integrating terrestrial and oceanic components. Science 281: 237–240
Giardina CP and Ryan MG (2000) Evidence that decomposition rates of organic carbon in mineral soil do not vary with temperature. Nature 404: 858–861
Gifford RM (1994) The global carbon cycle: A view point on the missing sink. Aust J Plant Physiol 21: 1–15
Gifford RM (2003) Plant respiration in productivity models: Conceptualization, representation and issues for global terrestrial carbon-cycle research. Func Plant Biol, 30: 171–86
Gonzàlez-Meler MA (1995) Effects of increasing atmospheric concentration of carbon dioxide on plant respiration. Ph.D. Thesis. Universitat Barcelona, Barcelona
Gonzàlez-Meler MA and Siedow JN (1999) Inhibition of respiratory enzymes by elevated CO_2: Does it matter at the intact tissue and whole plant levels? Tree Physiol19: 253–259
Gonzàlez-Meler MA, Drake BG and Azcón-Bieto J (1996a) Rising atmospheric carbon dioxide and plant respiration. In: Breymeyer AI, Hall DO, Melillo JM and Ågren GI (eds) Global Change: Effects on Coniferous Forests and Grasslands, pp 161–181. John Wiley & Sons, New York
Gonzàlez-Meler MA, Ribas-Carbo M, Siedow JN and Drake BG (1996b) Direct inhibition of plant mitochondrial respiration by elevated CO_2. Plant Physiol 112: 1349–1355
Gonzàlez-Meler MA, Giles L, RB Thomas and Siedow JN (2001). Metabolic regulation of leaf respiration and alternative pathway activity in response to phosphate supply. Plant Cell Environ, 24: 205–215
Grace J, Lloyd J, Mcintyre J, Miranda Ac, Meir P, Miranda Hs, Nobre C, Moncrieff J, Massheder J, Malhi Y, Wright I and Gash J (1995) Carbon dioxide uptake by an undisturbed tropical rain forest in Southwest Amazonia, 1992 to 1993. Science 270: 778–780
Griffin KL, Thomas RB and Strain BR (1993) Effects of nitrogen supply and elevated carbon dioxide on construction cost in leaves of *Pinus taeda* (L.) seedlings. Oecologia 95: 575–580
Griffin KL, Ball JT and Strain BR (1996a) Direct and indirect effects of elevated CO_2 on whole-shoot respiration in ponderosa pine seedlings. Tree Physiol16: 33–41
Griffin KL, Winner WE and Strain BR (1996b) Construction cost of loblolly and ponderosa pine leaves grown with varying carbon and nitrogen availability. Plant Cell Environ 19: 729–738
Griffin KL, Sims DA and Seemann JR (1999) Altered night-tome CO_2 concentration affects growth, physiology and biochemistry of soybean. Plant Cell Environ 22: 91–99
Griffin KL, Anderson OR, Gastrich MD, Lewis JD, Lin G, Schuster W, Seemann JR, Tissue DT, Turnbull M and Whitehead D (2001) Plant growth in elevated CO_2 alters mitochondrial number and chloroplast fine structure. Proc Natl Acad Sci USA 98: 2473–2478
Hamilton JG, Thomas RB and Delucia EH (2001) Direct and indirect effects of elevated CO_2 on leaf respiration in a forest ecosystem. Plant Cell Environ 24: 975–982
Hamilton JG, DeLucia EH, George K, Naidu S, Finzi AC and Schlesinger WH (2002) Forest carbon balance under CO_2. Oecologia 131: 250–260
Hellmuth EO (1971) The effect of varying air-CO_2 level, leaf temperature, and illuminance on the CO_2 exchange of the dwarf pea, *Pisum sativum* L. var Meteor. Photosynthetica 5: 190–194
Hesketh JD, Baker DN and Duncan WG (1971) Simulation of growth and yield in cotton: Respiration and the carbon balance. Crop Sci 11: 394–398
Hogberg P, Nordgren A, Buchmann N, Taylor AFS, Ekblad A, Hogberg MN, Nyberg G, Ottosson-Lofvenius M and Read DJ (2001) Large-scale forest girdling shows that current photosynthesis drives soil respiration. Nature 411: 789–792
Hrubec TC, Robinson JM and Donaldson RP (1985) Effects of CO_2 enrichment and carbohydrate content on the dark respiration of soybeans. Plant Physiol 79: 684–689
Hungate BA, Holland EA, Jackson RB, Chapin III FS, Mooney HA and Field CB (1997) The fate of carbon in grasslands under carbon dioxide enrichment. Nature 388: 576–579
Idso SB and Kimball BA (1992) Effects of atmospheric CO_2 enrichment on photosynthesis, respiration and growth of sour orange trees. Plant Physiol 99: 341–343
Jahnke S (2001) Atmospheric CO_2 concentration does not directly affect leaf respiration in bean or poplar. Plant Cell Environ 24: 1139–1151
Jahnke S and Krewitt M (2002) Atmospheric CO_2 concentration may directly affect leaf respiration measurement in tobacco, but not respiration itself. Plant Cell Environ 25: 641–651
Johnson DW, Geisinger D, Walker R, Newman J, Vose JM, Elliott KJ and Ball T (1994) Soil pCO_2, soil respiration, and root activity in CO_2-fumigated and nitrogen-fertilized ponderosa pine. Plant Soil 165: 129–138
Kacser H and Burns JA (1979) Molecular democracy: Who shares control? Biochem Soc Trans 7: 1149–1160
Karnosky DF, Zak DR, Pregitzer KS, Awmack CS, Bockheim JG, Dickson RE, Hendrey GR, Host GE, King JS, Kopper BJ, Kruger EL, Kubiske ME, Lindroth RL, Mattson WJ, Mcdonald EP, Noormets A, Oksanen E, Parsons WFJ, Percy KE, Podila GK, Riemenschneider DE, Sharma P, Thakur R, Sober A, Sober J, Jones WS, Anttonen S, Vapaavuori E, Mankovska B, Heilman W and Isebrands JG (2003) Tropospheric O_3 moderates responses of temperate hardwood forests to elevated CO_2: A synthesis of molecular to ecosystem: Results from the Aspen FACE project. Func Ecol 17: 289–304
Katul GG, Oren R, Ellsworth D, Hsieh CI, Phillips N and Lewin K (1997) A Lagrangian dispersion model for predicting CO_2 sources, sinks, and fluxes in a uniform Loblolly pine (*Pinus taeda* L.) stand. J Geophys Res 102: 9309–9321
Katul GG, Hsieh CI, Bowling D, Clark K, Shurpali N, Turnipseed A, Albertson J, Tu K, Hollinger D, Evans B, Offerle B, Anderson D, Ellsworth D, Vogel C and Oren R (1999) Spatial variability of turbulent fluxes in the roughness sublayer of an even-aged pine forest. Boundary Layer Meteorol 93: 1–28
Kimball BA, Mauney JR, Nakayama FS and Idso SB (1993) Effects of increasing atmospheric CO_2 on vegetation. Vegetatio 104/105: 65–75
King JS, Pregitzer KS, Zak DR, Sober J, Isebrands JG, Dickson RE, Hendrey GR and Karnosky DF (2001) Fine-root biomass and fluxes of soil carbon in young stands of paper birch and trembling aspen as affected by elevated atmospheric CO_2 and tropospheric O_3. Oecologia 128: 237–250
Lin G, Rygiewicz PT, Ehleringer JR, Johnson MG and Tingey DT (2001) Time-dependent responses of soil CO_2 efflux components to elevated atmospheric $[CO_2]$ and temperature in experimental forest mesocosms. Plant Soil 229: 259–270.
Luo Y, Medlyn B, Hui D, Ellsworth D, Reynolds J and Katul G (2001) Gross primary production in Duke Forest: Modeling synthesis of CO_2 experiment and eddy-flux data. Ecol Appls

11, 239–252
Matamala R and Schlesinger WH (2000) Effects of elevated atmospheric CO_2 on fine-root production and activity in an intact temperate forest ecosystem. Global Change Biol, 6: 967–979
Matthews E (1997) Global litter production, pools and turnover times: Estimates from measurement data and regression models. J Geophys Res 102: 18771–18800
Naumburg E and Ellsworth DS (2000) Photosynthetic sunfleck utilization potential of understory saplings growing under elevated CO_2 in FACE. Oecologia 122: 163–174
Norby RJ, Hanson PJ, O'Neill EG, Tschaplinski TJ, Weltzin JF, Hansen RA, Cheng WX, Wullschleger SD, Gunderson CA, Edwards NT and Johnson DW (2002) Net primary productivity of a CO2-enriched deciduous forest and the implications for carbon storage. Ecol Appls 12: 1261–1266
Norby RJ, Wullschleger SD, Gunderson CA, Johnson DW and Ceulemans R (1999) Tree responses to rising CO_2 in field experiments: Implications for the future forest. Plant Cell Environ 22: 683–714
Palet A, Ribas-Carbo M, Argilés JM and Azcón-Bieto J (1991) Short-term effects of carbon dioxide on carnation callus cell respiration. Plant Physiol 96: 467–472
Pendall E, Del Grosso S, King JY, LeCain DR, Milchunas DG, Morgan JA, Mosier AR, Ojima DS, Parton WA, Tans PP and White JWC (2003) Elevated atmospheric CO_2 effects and soil water feedbacks on soil respiration components in a Colorado grassland. Global Biogeochem Cycles 17: 1046
Perez-Trejo MS (1981) Mobilization of respiratory metabolism in potato tubers by carbon dioxide. Plant Physiol 67: 514–517
Pons TL and Welschen RAM (2002) Overestimation of respiration rates in commercially available clamp-on leaf chambers. Complications with measurement of net photosynthesis. Plant Cell Environ 25, 1367
Poorter H, Gifford RM, Kriedemann PE and Wong SC (1992) A quantitative analysis of dark respiration and carbon content as factors in the growth response of plants to elevated CO_2. Aust J Bot 40: 501–513
Poorter H, VanBerkel Y, Baxter R, DenHertog J, Dijkstra P, Gifford RM, Griffin KL, Roumet C, Roy J and Wong SC (1997) The effect of elevated CO_2 on the chemical composition and construction costs of leaves of 27 C-3 species. Plant Cell Environ 20: 472–482
Pregitzer KS, Zak DR, Maziasz J, DeForest J, Curtis PS, and Lussenhop J (2000) Interactive effects of elevated CO_2 and soil N availability on fine roots of *Populus tremuloides*. Ecol Appls 10: 18–33
Raich, JW and Potter CS (1995) Global patterns of carbon dioxide emissions from soils. Global Biogeochem Cycles 9: 23–36
Raich JW and Schlesinger WH (1992) The global carbon dioxide flux in soil respiration and its relationship to vegetation and climate. Tellus 44B: 81–89
Reuveni, J, Gale J and Mayer AM (1993) Reduction of respiration by high ambient CO_2 and the resulting error in measurements of respiration made with O_2 electrodes. Ann Bot 72: 129–131
Reuveni J, Gale J and Mayer AM (1995) High ambient carbon dioxide does not affect respiration by suppressing the alternative cyanide-resistant respiration. Ann Bot 76: 291–295
Reuveni J, Gale J and Zeroni M (1997) Differentiating day from night effects of high ambient [CO_2] on the gas exchange and growth of *Xanthium strumarium* L. exposed to salinity stress. Ann Bot 80: 539–546
Ribas-Carbo M, Berry JA, Yakir D, Giles L, Robinson SA, Lennon AM and Siedow JN (1995) Electron partitioning between the cytochrome and alternative pathways in plant mitochondria. Plant Physiol 109, 829–837
Ryan MG (1991) Effects of climate change on plant respiration. Ecol Appls 1: 157–167
Ryan MG, Hubbard RM, Pongracic S, Raison RJ and McMurtrie RE (1996) Foliage, fine-root, woody-tissue and stand respiration in *Pinus radiata* in relation to nitrogen status. Tree Physiol16: 333–343.
Schimel DS (1995) Terrestrial ecosystems and the carbon cycle. Global Change Biol 1, 77–91
Schimel DS, Braswell BH, Holland EA, McKeown R., Ojima D.S., Painter T.H., Parton WJ and Townsend AR (1994) Climatic, edaphic and biotic controls over storage and turnover of carbon in soils. Global Biogeochem Cycles 8: 279–293
Schlesinger WH (1997) Biogeochemistry: An Analysis of Global Change. Second edition, Academic Press, San Diego
Schlesinger WH and Andrews JA, (2000) Soil respiration and the global carbon cycle. Biogeochemistry 48: 7–20
Schlesinger WH and Lichter J (2000) Limited carbon storage in soil and litter of experimental forest plots under increased atmospheric CO_2. Nature 411: 466–469.
Steffen W, Noble I, Canadell J, Apps M, Schulze ED, Jarvis PG, Baldocchi D, Ciais P, Cramer W, Ehleringer J, Farquhar G, Field CB, Ghazi A, Gifford R, Heimann M, Houghton R, Kabat P, Korner C, Lambin E, Linder S, Mooney HA, Murdiyarso D, Post WM, Prentice IC, Raupach MR, Schimel DS, Shvidenko A and Valentini R (1998) The terrestrial carbon cycle: Implications for the Kyoto protocol. Science 280: 1393–1394
Teskey RO (1995) A field study of the effects of elevated CO_2 on carbon assimilation, stomatal conductance and leaf and branch growth of *Pinus taeda* trees. Plant Cell Environ 18: 565–573
Tjoelker MG, Reich PB and Oleksyn J (1999) Changes in leaf nitrogen and carbohydrates underlie temperature and CO_2 acclimation of dark respiration in five boreal species. Plant Cell Environ 22: 767–778
Tjoelker MG; Oleksyn J, Lee TD and Reich PB (2001) Direct inhibition of leaf dark respiration by elevated CO_2 is minor in 12 grassland species. New Phytologist 150: 419–24
Valentini R, Matteucci G, Dolman AJ, Schulze ED, Rebmann C, Moors EJ, Granier A, Gross P, Jensen NO, Pilegaard K, Lindroth A, Grelle A, Bernhofer C, Grunwald T, Aubinet M, Ceulemans R, Kowalski AS, Vesala T, Rannik U, Berbigier P, Loustau D, Guomundsson J, Thorgeirsson H, Ibrom A, Morgenstern K, Clement R, Moncrieff J, Montagnani L, Minerbi S and Jarvis PG (2000) Respiration as the main determinant of carbon balance in European forests. Nature 404: 861–865.
Vose JM, Elliott KJ, Johnson DW, Walker RF, Johnson MG and Tingey DT (1995) Effects of elevated CO_2 and N fertilization on soil respiration from ponderosa pine (*Pinus ponderosa*) in open-top chambers. Can J For Res 25: 1243–1251.
Will RE and Ceulemans R (1997) Effects of elevated CO_2 concentrations on photosynthesis, respiration and carbohydrate status of coppice *Populus* hybrids. Physiol Planta 100: 933–939
Williams ML, Jones DG, Baxter R and Farrar JF (1992) The effect of enhanced concentrations of atmospheric CO2 on leaf respiration. In: Lambers H, van der Plas LHW (eds), Molecular, Biochimical and Physiological Aspects of Plant Respiration, pp 547–551. SPB Academic Publishing bv, The Hague
Wullschleger SD and Norby RJ (1992) Respiratory cost of leaf

growth and maintenance in white oak saplings exposed to atmospheric CO_2 enrichment. Can J For Res 22: 1717–1721

Wullschleger SD, Norby RJ and Hendrix DL (1992a) Carbon exchange rates, chlorophyll content, and carbohydrate status of two forest tree species exposed to carbon dioxide enrichment. Tree Physiol10: 21–31

Wullschleger SD, Norby RJ and Gunderson CA (1992b) Growth and maintenance respiration in leaves of *Liriodendron tulipifera* L. exposed to long-term carbon dioxide enrichment in the field. New Phytologist 121: 515–523

Wullschleger SD, Ziska LH and Bunce JA (1994) Respiratory response of higher plants to atmospheric CO_2 enrichment. Physiol Planta 90: 221–229.

Wullschleger SD, Norby RJ and Hanson PJ (1995) Growth and maintenance respiration in stems of *Quercus alba* after four years of CO_2 enrichment. Physiol Planta 93, 47–54

Wullschleger SD, Norby RJ, Love JC and Runck C (1997) Energetic costs of tissue construction in yellow-poplar and white oak trees exposed to long-term CO_2 enrichment. Ann Bot 80: 289–297

Zak DR, Pregitzer KS, King JS and Holmes WE (2000) Elevated atmospheric CO_2, fine roots and the response of soil microorganisms: A review and hypothesis. New Phytologist 147: 201–222

Ziska, LH and Bunce JA (1994) Direct and indirect inhibition of single leaf respiration by elevated CO_2 concentrations: Interaction with temperature. Physiol Planta 90: 130–138

Ziska, LH and Bunce JA (1999) Effects of elevated carbon dioxide concentration at night on the growth and gas exchange of selected C_4 species. Aust J Plant Physiol 26: 71–77

Index

A

B

D

E

F

G

M

N

Q

R

S

T

U

Advances in Photosynthesis

Series editor: Govindjee, University of Illinois, Urbana, Illinois, U.S.A.

1. D.A. Bryant (ed.): *The Molecular Biology of Cyanobacteria.* 1994
ISBN Hb: 0-7923-3222-9; Pb: 0-7923-3273-3
2. R.E. Blankenship, M.T. Madigan and C.E. Bauer (eds.): *Anoxygenic Photosynthetic Bacteria.* 1995 ISBN Hb: 0-7923-3681-X; Pb: 0-7923-3682-8
3. J. Amesz and A.J. Hoff (eds.): *Biophysical Techniques in Photosynthesis.* 1996
ISBN 0-7923-3642-9
4. D.R. Ort and C.F. Yocum (eds.): *Oxygenic Photosynthesis: The Light Reactions.* 1996
ISBN Hb: 0-7923-3683-6; Pb: 0-7923-3684-4
5. N.R. Baker (ed.): *Photosynthesis and the Environment.* 1996
ISBN 0-7923-4316-6
6. P.-A. Siegenthaler and N. Murata (eds.): *Lipids in Photosynthesis: Structure, Function and Genetics.* 1998 ISBN 0-7923-5173-8
7. J.-D. Rochaix, M. Goldschmidt-Clermont and S. Merchant (eds.): *The Molecular Biology of Chloroplasts and Mitochondria in Chlamydomonas.* 1998
ISBN 0-7923-5174-6
8. H.A. Frank, A.J. Young, G. Britton and R.J. Cogdell (eds.): *The Photochemistry of Carotenoids.* 1999 ISBN 0-7923-5942-9
9. R.C. Leegood, T.D. Sharkey and S. von Caemmerer (eds.): *Photosynthesis: Physiology and Metabolism.* 2000 ISBN 0-7923-6143-1
10. B. Ke: *Photosynthesis: Photobiochemistry and Photobiophysics.* 2001
ISBN 0-7923-6334-5
11. E.-M. Aro and B. Andersson (eds.): *Regulation of Photosynthesis.* 2001
ISBN 0-7923-6332-9
12. C.H. Foyer and G. Noctor (eds.): *Photosynthetic Nitrogen Assimilation and Associated Carbon and Respiratory Metabolism.* 2002 ISBN 0-7923-6336-1
13. B.R. Green and W.W. Parson (eds.): *Light-Harvesting Antennas in Photosynthesis.* 2003 ISBN 0-7923-6335-3
14. A.W.D. Larkum, S.E. Douglas and J.A. Raven (eds.): *Photosynthesis in Algae.* 2003 ISBN 0-7923-6333-7
15. D. Zannoni (ed.): *Respiration in Archaea and Bacteria.* Diversity of Prokaryotic Electron Transport Carriers. 2004 ISBN 1-4020-2001-5
16. D. Zannoni (ed.): *Respiration in Archaea and Bacteria.* Diversity of Prokaryotic Respiratory Systems. 2004 ISBN 1-4020-2002-3
17. D. Day, A.H. Millar and J. Whelan (eds.): *Plant Mitochondria.* From Genome to Function. 2004 ISBN 1-4020-2399-5
18. *Forthcoming.*

Advances in Photosynthesis

19. G. Papageorgiou and Govindjee (eds.): *Chlorophyll a Fluorescence.* A Signature of Photosynthesis. 2004 ISBN 1-4020-3217-X
20. J.T. Govindjee Beatty, H. Gest and J.F. Allen (eds.): *Discoveries in Photosynthesis.* 2005 ISBN 1-4020-3323-0
21. B. Demmig-Adams, W.W. Adams III and A. Mattoo (eds.): *Photoprotection, Photoinhibition, Gene Regulation and Environment.* 2005 ISBN 1-4020-3564-0

For further information about the series and how to order please visit our Website
http://www.springeronline.com

MIX
Papier aus verantwortungsvollen Quellen
Paper from responsible sources
FSC® C105338
FSC
www.fsc.org

If you have any concerns about our products,
you can contact us on
ProductSafety@springernature.com

In case Publisher is established outside the EU,
the EU authorized representative is:
Springer Nature Customer Service Center GmbH
Europaplatz 3, 69115 Heidelberg, Germany

Printed by Libri Plureos GmbH
in Hamburg, Germany